Finsler Geometry, Relativity and Gauge Theories

Fundamental Theories of Physics

A New International Book Series on the Fundamental Theories of Physics: Their Clarification, Development and Application

Finsler Geometry, Relativity and Gauge Theories

by

G. S. Asanov

Department of Theoretical Physics
Moscow State University

D. Reidel Publishing Company

A MEMBER OF THE KLUWER ACADEMIC PUBLISHERS GROUP

Dordrecht / Boston / Lancaster / Tokyo

Library of Congress Cataloging in Publication Data

Asanov, G. S., 1946–
Finsler geometry, relativity and gauge theories.

(Fundamental theories of physics)
Bibliography: p.
Includes index.
1. Finsler spaces. 2. Relativity (Physics)
3. Gauge fields (Physics) 4. Mathematical physics.
I. Title. II. Series.
QC20.7.D52A83 1985 530.1′1′01516375 85-14304
ISBN 90–277–1960–8

Published by D. Reidel Publishing Company,
P.O. Box 17, 3300 AA Dordrecht, Holland.

Sold and distributed in the U.S.A. and Canada
by Kluwer Academic Publishers,
190 Old Derby Street, Hingham, MA 02043, U.S.A.

In all other countries, sold and distributed
by Kluwer Academic Publishers Group,
P.O. Box 322, 3300 AH Dordrecht, Holland.

Table of contents

Preface

The methods of differential geometry have been so completely merged nowadays with physical concepts that general relativity may well be considered to be a physical theory of the geometrical properties of space-time. The general relativity principles together with the recent development of Finsler geometry as a metric generalization of Riemannian geometry justify the attempt to systematize the basic techniques for extending general relativity on the basis of Finsler geometry. It is this endeavour that forms the subject matter of the present book. Our exposition reveals the remarkable fact that the Finslerian approach is automatically permeated with the idea of the unification of the geometrical space-time picture with gauge field theory – a circumstance that we try our best to elucidate in this book.

The book has been written in such a way that the reader acquainted with the methods of tensor calculus and linear algebra at the graduate level can use it as a manual of Finslerian techniques orientable to applications in several fields. The problems attached to the chapters are also intended to serve this purpose. This notwithstanding, whenever we touch upon the Finslerian refinement or generalization of physical concepts, we assume that the reader is acquainted with these concepts at least at the level of the standard textbooks, to which we refer him or her.

For a first acquaintance with the subject-matter, the reader may skip through the introductory Chapters 1 and 2 and start reading directly that material of the succeding chapters which he/she will deem to be of most interest to him/her, turning to the preceding sections when necessary. Nevertheless, the reader should bear in mind that the book is based on calculations rather than just qualitative reasoning, so that one will not be able fully to understand the underlying idea of the book unless one reads it through, pencil in hand. The author had no intention of forcing the reader under the burden of a heavy volume to wearily agree that to occupy oneself with the field under review is a quite respectable activity. Rather, this book provides a set of necessary analytical means for the investigator who wishes to go beyond the boundaries of the currently ruling opinion that the metric of the real world is a Riemannian metric and to draw a line between what is reality and what is a myth.

We do not expect the reader to have studied the mathematical monograph by H. Rund (1959), in which the principal techniques of Finsler geometry are described in detail, and the information required for our purposes is presented in Section 1 of Chapter 1 (although the reader may well feel advised to read Rund's book). Many results of the monograph by H. Rund and J. H. Beare (1972) are reflected in our exposition, which seems to be expedient because that monograph is comparatively small although its results are of importance to our task. Other books relating to our field have not been available, and we base our analysis on the results described in recent papers.

We do not seek however to review all the suggestions encountered in the literature as to how one should or should not apply Finsler geometry to special or general relativity. The book merely aims at drawing a definite line. The papers which are cited in the Bibliography contain many references to the original publications.

For the convenience of the reader, the basic text of the book is preceded by an introductory Part A in which the purpose and general layout of the book are carefully explained and the principal results are given.

The concluding List of Publications on Finsler Geometry includes the works on the Finsler geometry proper as well as the articles devoted to its generalizations and applications to various physical problems. The author tried and compiled the List to be as exhaustive as possible. The List does not contain the works cited in Bibliography. When looking through the titles of the works in the List, the reader can get an idea of the diversity of the problems studied by nowadays within the scope of the Finsler geometry and find articles on the particular problems which he/she is interested in. Finally, several works cited under the Biographies heading may be of assistance to the reader who may be interested in the history of the evolution of Finsler Geometry.

The logical construction of general relativity synthesized with the necessary mathematical formalism has been described at length in many recent monographs: e.g. Fock (1959); Synge (1960); Bergmann (1962); Adler, Bazin, and Schiffer (1975); Mercier, Treder, and Yourgrau (1979); Treder, Borzeskowski, van der Merwe, and Yourgrau (1980); Sachs (1982). Bearing in mind these references, the author has not provided an introduction to general relativity in the book. Where gauge theories are discussed (Chapter 5), the reader is recommended to acquaint himself with: Drechsler and Mayer (1977); Madore (1981); Konopleva and Popov (1981) and Chaichian and Nelipa (1983).

Although the author has, for several years, striven to understand what untapped resources are possessed by Finsler geometry that can be used for the further development of physical field theory, and had been preparing materials for a special course to be read at the Physics Faculty of Moscow State

University, the author has only now at last felt bold enough to undertake a systematic study along these lines after the kind suggestion by Professor Alwyn van der Merwe, the editor of the present series, to write a book on this subject. This volume is to a considerable degree an exposition and extension of the methods originated by Professor Hanno Rund (Arizona University). It is hard to overestimate the unfailing support of Professor Rund in my research work. Where special Finsler spaces are discussed (Chapter 2), the author much profited from comments of Professor Makoto Matsumoto (Kyoto University) made in our correspondence. Especially, I would like to express my sincere gratitude to my wife, Anna Novikova, whose touching consideration and help made the writing of this book feasible. I also feel indebted to the many colleagues of Moscow State University who have shared my scientific dreams since the time I was a freshman.

Many thanks go to Professor Radu Miron (Universitatea A1. I. Cuza', Iasi) for his support of my work. Special thanks are due to Dr. Peter Holland (Institut Henri Poincaré, Paris) who was so kind as to endeavour to edit the language of my work. I am indebted to Dr. Holland for his efforts, and also for his welcome encouragement which every author needs so badly, since one is deprived of the ability of judging the results of one's own efforts. Last, but not least, I should like to express my deep gratitude to the publishers for their unfailing courtesy and cooperation.

Department of Theoretical Physics
Moscow State University
Moscow, U.S.S.R.
November 1984.

G.S. ASANOV

Part A

Motivation and Outline of the Book

Motivation and Outline of the Book

Finsler geometry is a metric generalization of Riemannian geometry and constitutes a comparatively young branch of classical differential geometry. Although Finsler geometry can be traced back to the work of Finsler (1918), its fundamentals were not completely formulated until the mid-thirties. Later on, the field was rapidly developed by geometricians of many countries. The main results of their studies were presented in the monograph of H. Rund (1959).

The Finslerian generalization arises from the desire to determine the length of vectors by a sufficiently general method which is not restricted by the Riemannian definition of length in terms of the square root of a quadratic form. More precisely, Finsler geometry deals with the case where an N-dimensional differentiable manifold M is endowed with a positive scalar function $F(x, y)$ dependent on the points x^i of M as well as on the vectors y^i tangent to M at these points x^i. The value of $F(x, y)$ is considered to be the length of the vector y^i supported by the point x^i. Apart from certain conditions of smoothness and convexity, the sole restriction imposed on F is that the function must be positively homogeneous of degree one in y^i, i.e. $F(x, ky) = kF(x, y)$, $k > 0$. For example, it seems promising to use the fourth root of a fourth-order form in order to generalize the Riemannian measure of length, an idea set forth in Riemann's fundamental work on the geometry named after him (Riemann, 1892). However, Riemann himself did not attempt to develop this idea further because it was his desire to restrict himself to the logically simplest definition. The homogeneity condition imposed on F has a clear geometrical singificance, namely that if the components of one vector are k times greater than the components of another vector, then the length of the first vector is k times greater than the length of the other. The Finslerian metric tensor is defined in accordance with $g_{ij}(x, y) = \frac{1}{2}\partial^2 F^2(x, y)/\partial y^i \partial y^j$ and, therefore, is dependent on the points x^i as well as on the directions given by the tangent vectors y^i. The reader will find a self-contained introduction to Finsler geometry in Chapters 1 and 2.

As Finsler geometry was gradually turning into a classical branch of differential geometry, works appeared sporadically which, inspired by advances in general relativity, tried to develop a new geometrical picture of the world in terms of Finslerian ideas. Suggestions that Finsler geometry should be applied to general relativity seemed to accumulate. Although several scores of papers containing useful information may be considered relevant to this case, their suggestions overlap but rarely and are often fragmentary. On the other hand, by the early seventies general relativity theory had beed developed to such an extent that it was no longer considered merely as a geometrically formulated theory of the gravitational field. It has become a formalism for constructing the theory of the geometrical properties of space-time and of physical fields in space-time. This development has led to a great number of geometrical and other generalizations of Einstein's initial version of general relativity, each of them being of interest to the theoretical physicist so long as they do not contradict firmly-established experimental facts, and are formulated in a language of rigorous mathematical relations.

Since the concept of metric is one of the geometrical fundamentals, much attention ought to have been attracted to the problem of what are the primary metric properties of real space-time. In reality, however, the Riemannian definition of length dominates the present-day literature devoted to general relativity, which is surely not accidental. The fact is that a light wave front in vacuo can properly be described by the equation $S(\mathrm{d}x) = 0$, where $S(\mathrm{d}x)$ denotes the Riemannian metric of space-time. Because of the high degree of agreement between special relativity theory and observations, the replacement of the Riemannian metric in this equation by some other metric may be hazardous in view of possible discordances with known experimental facts. However, it is by no means the case that the Finslerian generalization is to be formulated in such a straightforward way, that is that the Finslerian metric function of space-time is to be a function which, when equated to zero, yields the equation of a light wave front. Strictly speaking, the light wave equation must be derived from the properly generalized equations of the electromagnetic field, not merely from the form of the metric function, so that the actual problem is how the Finslerian generalization of physical field equations should be formulated.

We shall elaborate two approaches to solving this problem taking advantage of the formal possibilities presented by Finsler geometry. The first approach (Chapter 4) is to resort to the concept of osculation according to which a vector field $y^i(x)$ is to be substituted for the directional variable y^i when a Lagrangian density is constructed. This approach makes use of the so-called invariance identities (Chapter 3). A second appraoch is to use the parametrical representation of the indicatrix of a Finsler space. The parametrical method developed in Chapter 5 clearly suggests that the internal symmetries of physical fields, at least their isotopic invariance, may be described in terms of the invariance properties of the Finslerian indicatrix. Naturally, in all these approaches the usual equation of a light wave front in vacuo is unchanged when all gravitational effects are neglected.

The invariance identities arise as a reflection of the tensorial properties of geometrical objects. Although the method of invariance identities initiated by H. Rund has already ben expounded in monographs (Rund and Beare, 1972; Lovelock and Rund, 1975, Section 8), we begin Section 3.1 of Chapter 3 with a brief exposition ab initio. As a first application of this method we obtain useful identities for the Finslerian metric function and metric tensor. In general, the systematic use of invariance identities may be regarded as a direct method for extracting as much as possible of the information contained in the assumption that a given object has prescribed tensorial properties. Other ways of obtaining such information seem to be rather indirect. For instance, a set of invariance relations for a scalar density L, including the so-called Noether identities often used in the theory of physical fields, can be derived from the invariance of the action integral $\int L \, \mathrm{d}^N x$ under infinitesimal coordinate transformations. Nevertheless, all the invariance relations can be inferred from the invariance identities written directly for L. In particular, the Noether identities are obtainable merely by differentiating the invariance identities (see the final part of Section 3.1). Moreover, reference to the variational properties of an action integral is not possible at all if other geometrical objects, say connection coefficients, are under consideration instead of a scalar density.

It will be shown in Section 3.2 that the invariance identities permit one to obtain the characteristic equations for any set of connection coefficients constructible from the components of a symmetric direction-dependent tensor $g_{ij}(x, y)$ and their first derivatives. In the Finslerian case, the solution of these equations leads directly to the Cartan connection coefficients. Therefore, the use of the invariance identities leads to the conclusion that any connection coefficients constructible from just the Finslerian metric tensor and its first derivatives must be the Cartan connection coefficients, up to any tensor addend dependent on g_{ij} and $\partial g_{ij}/\partial y^k$ only. In this way, the connection coefficients are found to be automatically symmetric and metric, so that neither of these two properties has to be postulated in addition to the form of the functional dependence of the connection coefficients.

Application of the invariance identities is also necessary if one wishes to clarify what tensorial objects can be constructed out of the various derivatives of a given object. In Section 3.3, our interest is in the case when a scalar density L is a given function of a symmetric direction-dependent tensor $g_{ij}(x, y)$ and its derivatives up to the second order, and of a directional variable y^i. The question as to what tensor densities can be constructed from the first-order derivatives of L – we shall call such tensor densities fundamental – was studied systematically in the work by Rund and Beare (1972) whose results are described in Section 3.3. After that, in Section 3.4, we shall confine our attention to just the Finslerian case, namely we shall treat the product $L = JK$ of the complete contraction $K = K^{hk}{}_{hk}$ of the Finslerian curvature tensor $K_l{}^j{}_{hk}(x, y)$ and the Jacobian $J = |\det(g_{ij})|^{1/2}$ as our direction-dependent scalar density. One would expect that simple expressions for the fundamental tensor densities could be obtained in this case, and the calculations in Section 3.4 confirm this: rather simple and, in two cases, vanishing expressions for the fundamental tensor densities have been obtained. Special techniques enable us to avoid extremely laborious calculations.

Even a first acquaintance with Finsler geometry, for example the contents of Section 1.1, encourages one to treat the Finslerian density $L = JK$ as a natural candidate for the role of the Finslerian generalization of its Riemannian prototype used in Einstein's formulation of general relativity as the Lagrangian density of the gravitational field. However, the fact that $L = JK$ depends on the directional variable y^i prevents us from employing such a program immediately. Indeed, the integration of a direction-dependent scalar density over a region of the background differentiable manifold fails to be a coordinate-invariant operation, as can easily be seen (the beginning of Section 4.1 of Chapter 4). A suggestion as to the procedure applicable in this case may be found in Rund and Beare (1972) where the authors generalize Caratheodory's method of equivalent integrals to propose the method of equivalent Lagrangian densities for direction-dependent scalar densities of the type $L = JK$. Although the Rund–Beare equations are elegantly formulated in terms of the fundamental tensor densities associated with L, the application of their treatment to a Finslerian generalization of the gravitational field equations is hampered by the essential circumstance that, as the elaborate analysis in the final part of the work by Rund and Beare has shown, the identities satisfied by the left-hand sides of their equations are far from being similar to the contracted Bianchi identities satisfied by the Einstein tensor of the gravitational field. Therefore, we cannot follow the conventional procedure, used in general relativity, to establish a relationship between the Finslerian equations of the gravitational field and the Finslerian equations of motion of the gravitational field sources. It seems to be a matter of principle, however, that the theory should include such a relationship.

Apparently, this awkward situation cannot be avoided unless the dependence of a Lagrangian density on the directional variable is suppressed. Accordingly, we shall adhere in Chapter 4 to an alternative approach, namely, we shall proceed from the direction-dependent density $L = JK$ to construct a gravitational Lagrangian density L_g by substituting a vector field $y^i(x)$ for the directional variable y^i. In other words, L_g is constructed to be osculating to L along the vector field $y^i(x)$. As a set of gravitational field variables we choose the x-dependent tensors $S^A(x)$ from which the Finslerian metric function can be constructed, while $y^i(x)$ plays the role of an auxiliary quantity (Section 4.1). By definition the function $v(S^A, y)$ which describes the form of the dependence of the Finslerian metric function F on x^i in accordance with $F(x, y) = v(S^A(x), y)$ will enter L_g only as an input function; it is not determined by the gravitational field equations which serve for determining the fields $S^A(x)$ only. Using the invariance identities written in Section 3.1 for the Finslerian metric tensor, we can clarify the explicit structure of the decomposition of the Euler–Lagrange derivative of L_g with respect to S^A in terms of the fundamental tensor densities and covariant derivatives of $y^i(x)$. In this decomposition, the left-hand sides of the Rund–Beare equations appear naturally. As a whole, the approach retains the potentialities of the Finsler geometry involved.

The method based on the concept of osculation may be consistently applied to the formulation of Finslerian generalizations of the equations for any physical field, and application of the method to the electromagnetic field does not lead to any contradictions with the usual equation of the light wave front in vacuo (Section 4.2). The possibility of deriving the Finslerian equations of motion of matter from the gravitational field equations is illustrated in Section 4.3 by using the Noether identities.

General relativity, being a theory developed by means of the techniques of differential geometry, incorporates the principle of covariance which claims that

"in a generally covariant field theory like gravitation theory, a result cannot have physical meaning unless it can be stated in a generally covariant form"

(Pirani, 1962, p. 200). At the same time, the lack of preferred coordinate systems gives rise to many difficulties, for example, when one tries to elucidate the meaning of gravitational energy transfer or of the concept of gravitational radiation. In particular, one frequently uses special coordinate systems or coordinate conditions in order to draw physically significant conlcusions from the covariant gravitational field equations (*op. cit.*).

On the other hand, it is also clear that general covariance prevents one from treating all components of the tensorial field variables as physical observables. Following Bergmann (1962, p. 250), observables are called

"physical quantities that are free from ephemeral aspects of choice of coordinate system and contain information relating exclusively to the physical situation itself."

This suggests the possibility of representing observables geometrically by a set of scalars. Clearly, this possibility will not be realized unless the physical characteristics of space-time enable one to introduce four functionally independent scalar fields which may be used as intrinsic coordinates (see *op. cit.*, Section B).

An ideal approach to this case will be given by those theories in which the needed set of scalars serves as a set of additional field variables. Theories whose Lagrangian densities contain an auxiliary vector field prove to be of such a type. More precisely, suppose we are given a Lagrangian density L constructible from a set of tensor field variables together with a contravariant vector field $y^i(x)$ playing an auxiliary role in L, and let L be a positively homogeneous function of $y^i(x)$ of degree zero (as occurs in the Finslerian case). Then the auxiliary role of $y^i(x)$ can be eliminated. Indeed, considering the Clebsch representation of the field $y^i(x)$ in terms of the scalars which are called Clebsch potentials, we may (Section 4.4) infer from the invariance identities that these scalars may be reinterpreted as additional field variables without affecing the field equations. As a result, the gravitational Lagrangian density constructed to be dependent on an auxiliary vector field will not contain any auxiliary quantities at all. These Clebsch potentials may be used to define a special coordinate system and the concept of world time which the gravitational field equations will describe. Naturally, it is with respect to this coordinate system that the concept of a static gravitational field according to the Finslerian approach will be defined in the logically simplest way, and in such a static case the Finslerian and Riemannian geodesics may be readily compared with one another (Section 4.5).

Since the use of the concept of osculation merely means substituting the vector field $y^i(x)$ for the directional variable y^i, it should be possible to propose, apart from $L = JK$, many other Finslerian densities which will generalize the Einstein gravitational Lagrangian density by involving the Cartan torsion tensor $C_{ijk}(x,y) = \frac{1}{2}\partial g_{ij}(x,y)/\partial y^k$ and the vector field $y^i(x)$. For instance, a density of such a type may be given by the expression $\tilde{L} = J\tilde{K}$, where $\tilde{K}$ is the complete contraction of the Finslerian relative curvature tensor. A variety of possibilities arises for two reasons. First, using the Finslerian torsion tensor C_{ijk} in addition to the metric tensor $g_{ij}(x,y)$ offers the possibility of constructing a wide class of direction-dependent densities. Second, the gravitational field equations constructible in the osculating approach will contain an arbitrary function $v(S^A(x),y)$ which is positively homogeneous of degree one in y^i. Incidentally, Einstein's choice of a Riemannian gravitational Lagrangian density may be regarded as the limiting case of the Finslerian generalization based on the concept of osculation.

An analysis of the possibilities opened up here would have constituted a separate monograph. Thus, the scope of Section 4.6 is restricted to an attempt to demonstrate only some of these possibilities. Accordingly, we shall confine ourselves to the treatment of two Finslerian scalar densities $L = JK$ and $\tilde{L} = J\tilde{K}$. The logically simplest choice for the fields $S^A(x)$ will be made, namely n-tuples $S_i^A(x)$ will be taken as primary x-dependent fields, which means that the Finsler space will be assumed to be of the 1-form type (this type of Finsler space is described in Section 2.4). It is the covariant vectors $S_i^A(x)$ that will be treated as genuine gravitational field variables. We begin Section 4.6 by noting the interesting circumstance that the second-order Lagrangian densities L and $\tilde{L}$ prove to be reducible to the first-order expressions $L_1 = JK_1$ and $\tilde{L}_1 = J\tilde{K}_1$ by separating out ordinary divergences in the initial L and $\tilde{L}$. Since neglecting an ordinary divergence in a Lagrangian density does not affect the associated Euler–Lagrange derivatives, the study of the structure of the gravitational field equations is considerably simplified. Moreover, a phenomenon which has no analogy in the usual Riemannian approach is that the ordinary divergences separated out are of tensorial nature (each being the divergence of a vector density) so that L_1 and $\tilde{L}_1$ are again scalar densities. This fact is advantageous in that it permits us to formulate the conservation laws for the gravitational field in a strictly covariant way (Section 4.7), something which is impossible in the Riemannian approach (see the discussion of the status of conservation laws in Einstein's general relativity in Trautman (1962)). The gravitational field equations obtained are of second order, whereas in general a second-order Lagrangian density will lead to fourth-order equations.

Needless to say, the gravitational field equations obtainable in this way are more complicated than the Einstein equations. This forces us to clarify step by step in Section 4.6 the simplifications arising from particular assumptions concerning the form of the Finslerian metric function, namely $C_k{}^k{}_i = 0$, S3-likeness and the T-condition. When all three of these conditions are satisfied, as is the case for the Berwald–Moór metric function investigated in Section 2.3, we obtain comprehensible explicit expressions for the Euler–Lagrange derivatives of L_1 and $\tilde{L}_1$ with respect to S_i^A. This however by no means exhausts the simplifications which are possible in this case. The fact is that the Berwald–Moór metric function possesses a certain group of scale invariance which enables us to put $S_i^A(x)y^i(x) = 1$ in the field equations without loss of generality. It turns out that the resultant set of equations is sufficiently simple for their structure to be clarified in the static case. On confining ourselves to the case of the density $\tilde{L}_1$, we investigate the gravitational field equations in the spherically symmetric static case and conclude Section 4.6 by finding the post-Newtonian solution.

Another way of approaching the Finslerian generalization of physical field equations is suggested by the concept of the indicatrix which is an important constituent of a Finsler space. By the indicatrices of an N-dimensional Finsler space one means the $(N-1)$-dimensional hypersurfaces which lie in the tangent spaces to the Finsler space and which are defined by the equation $F(x, y) = 1$, where x^i are considered to be fixed and y^i arbitrary (spheres, in the case of a Riemannian space). The indicatrix may be regarded as a tool for defining a one-to-one correspondence between the unit tangent vectors $l^i = y^i/F(x, y)$ and $N-1$ functionally independent scalars $u^a(x, y)$ of zero-degree homogeneity in y^i; these u^a serve as intrinsic coordinates in the indicatrix (see Section 1.2). This parametrical representation $l^i = t^i(x, u)$ of the indicatrix enables one to cast any (x, y)-dependent field of zero-degree homogeneity in y^i into a parametrical form by substituting $t^i(x, u)$ for y^i (Section 5.1). For example, the parametrical representation of the Finslerian metric tensor $g_{ij}(x, y)$ will read $b_{ij}(x, u) = g_{ij}(x, t(x, u))$. The circumstance that u^a are scalars is advantageous in the sense that the dependence of a tensorial object on u^a will not affect the transformation law of its partial derivatives with respect to x^i, in contrast to the case of tensorial objects dependent on x^i and the vector y^i. Therefore, the equations for parametrically represented fields may be formulated by means of conventional Riemannian techniques, u^a being treated merely as parameters. Such equations will be invariant under x-independent transformations $u^a = u^a(\bar{u}^b)$ of the parameters. However, in general the transformations of the coordinates u^a of the points of the indicatrices supported by different points x^i may be admitted to be different, which means that we are to require the invariance of physical field equations under general x-dependent transformations $u^a = u^a(x, \tilde{u})$. The attainment of such an invariance implies the introduction of so-called gauge fields.

It will be recalled in this respect that in recent physical theories fields are characterized not only by their invariance properties under transformations of the points x^i of the background space-time manifold (Lorentz-invariance, space and time parity, etc.) but also by 'internal' symmetries (isotopic invariance, SU(3)-symmetry, charge parity, etc.) which are independent of transformations of x^i. This latter type of symmetry is often associated with the transformations of the points of an 'internal space', the symmetry of which is reflected in the behaviour of physical fields. Comparison of this method with the apparatus of Finsler geometry, which deals with tensorial objects dependent on the pair (x, y) or, as explained above, on the pair (x, u), gives rise to the hope that some of the important classes of internal symmetries of physical fields may be conceived as manifestations of a Finslerian structure inherent in real space-time.

Such expectations are justified by the circumstance that the concept of invariance under transformations of the variables y^i appears naturally in Finsler geometry; this type of invariance can be described in terms of the group G of transformations of y^i leaving the Finslerian metric function $F(x, y)$ invariant, in analogy with the definition of the Lorentz group of symmetries of tangent spaces in the pseudo-Riemannian geometry of space-time. Giving all necessary definitions, we show in Section 1.3 that the metric G-transformations defined as those G-transformations under which the Finslerian metric tensor is subjected to a tensor law transformation have a clear geometrical significance, namely they represent the motions in the indicatrices treated as Riemannian spaces embedded in the tangent spaces to a Finsler space.

These observations lead to the attractive hypothesis that the internal symmetries of physical fields reflect the symmetry properties of the indicatrix of a Finslerian structure assumed to be possessed by real space-time and, accordingly, the indicatrix plays the role of an internal space, the points of the indicatrix the role of internal coordinates, and the group of motions $u^a = u^a(u'^b)$ in the indicatrix the role of a group of internal symmetries. Instead of discussing motions in the indicatrix, one may of course discuss the coordinate transformations $u^a = u^a(x, \tilde{u}^b)$ in the indicatrix, and understand by the internal symmetry group the totality of all coordinate transformations in the indicatrix which leave the indicatrix metric tensor invariant. This second viewpoint is rather convenient for us, and is adopted in Chapter 5.

Following the terminology used in the theory of gauge fields (see, for example, Drechsler and Mayer, 1977), we shall regard the transformations $u^a = u^a(\tilde{u})$ as global transformations, and their x-dependent generalizations $u^a = u^a(x, \tilde{u})$ as local gauge transformations or merely gauge transformations. In Section 5.2 it is shown that if we wish to construct an operator of differentiation with respect to x^i which is covariant under gauge transformations, we have to invoke gauge fields obeying specific transformation laws. The formulae presented in Section 5.2 are quite general and nowhere involve Finslerian tensors. However, if a space is actually assumed to be Finslerian, one may suspect that these gauge fields are not disembodied spirits, and that they are in fact certain concomitants of the constituents of the Finsler space. The purpose of Section 5.3 is to elucidate this possibility in detail, namely we show that the gauge fields are constructible from the projection factors of the indicatrix of a Finsler space in accordance with Equation (5.73). The calculation of the gauge field strength tensors culminates in the remarkable result that these tensors are actually expressible in terms of the Finslerian curvature tensors. The construction of the gauge-covariant differentiation operator can be readily extended to the case of spinor fields if one takes advantage of the well-known formalism of the theory of two-component spinors using the Infeld and van der Waerden connection coefficients (Section 5.4).

In physical theories gauge transformations are usually taken to be linear: $u^a = u^a_b(x)\tilde{u}^b$; an imporant example of such a theory is the Yang–Mills theory which, when restricted to SU(2) gauge transformations, describes the isotopic invariance of strongly interacting fields in terms of gauge fields, or weak interactions in the context of the standard model of unification of electromagnetic and weak interactions (see, e.g., Chaichian and Nelipa, 1983). The ideas outlined above readily account for the geometrical meaning of isotopic symmetry as a manifestation of the Finslerian structure of space-time with a highest-symmetry indicatrix. It will be recalled that isotopic symmetry manifests itself as a symmetry of physical fields under Euclidean rotations in a three-dimensional internal space called the isotopic space. In spite of the fundamental role of isotopic invariance in the theory of physical fields, this invariance, following the initial idea of Heisenberg, has been treated but abstractly as a symmetry relative to Euclidean rotations in an imaginary internal space. The reason for this state of affairs may be found in the circumstance that such a symmetry cannot be explained within the framework of the Riemannian geometrization of space-time that dominates the recent literature, because the tangent spaces to a Riemannian space-time are invariant relative to the Lorentz group and it is unclear why the class of physical interactions considered should be invariant only relative to a subgroup of the Lorentz group. At the same time, the indicatrix of a four-dimensional Finsler space is a three-dimensional space, so that, in the most symmetrical case when the curvature tensor of the indicatrix vanishes, the indicatrix represents a three-dimensional Euclidean space which may be identified with isotopic space. A striking example of a metric function of such a Finsler space is the Berwald–Moór metric function described in detail in Section 2.3.

The case of linear gauge transformations is simple in that it enables one to confine oneself to considering those fields which are tensorial under gauge transformations independent of the points u^a of an internal space. This in turn implies that the gauge fields should also be independent of u^a. In the case when the indicatrix is a Euclidean space, it may be shown (Section 5.5) that the latter condition may be formulated as a certain restriction on a Finsler space, and that the conformally flat 1-form Berwald–Moór Finsler space is an example of a space which is not flat but which admits gauge fields independent of u^a. So, the latter example of a Finsler space reflects both isotopic invariance and some gravitational effects.

On purely formal grounds, we may contend that the important thing added by the Finslerian approach to familiar ideas about isotopic invariance is the notion of embedding isotopic space (read: the indicatrix) into the tangent spaces of the space-time manifold. The embedding is given by the parametrical representation of the indicatrix (Equation (2.64) in the case of the Berwald–Moór metric function). The significance of such a relationship is that it opens up the possibility of unifying the gravitational and Yang–Mills SU(2) gauge fields in terms of space-time geometric objects. Indeed, the projection factors and, hence, gauge fields appear to be constructible from those x-dependent fields $S^A(x)$ which serve as 'building blocks' for the Finslerian metric function and which may naturally be treated as gravitational field variables. Moreover, this dependence has the property that if there exists a coordinate system x^i such that $\partial S^A/\partial x^i = 0$ (i.e. such that the gravitational field is absent), then, with respect to these x^i, the gauge fields vanish identically.

One should, of course, bear in mind that the Finslerian approach to the description of internal symmetries of physical fields is restricted by the circumstance that the indicatrix of a four-dimensional Finsler space is a space of dimension 3. However, as soon as the internal symmetries have been treated as a manifestation of the symmetry of a three-dimensional Riemannian space, we can procced to construct a four-dimensional Finsler space having an indicatrix with the required symmetry properties. For example, it is known that the Lagrangian for describing interactions of soft π-mesons can be obtained by assuming its invariance under the group of motions of a three-dimensional Riemannian isotopic space of constant positive curvature (Meetz, 1969). It appears that the corresponding Finslerian metric function can be deduced explicitly (Section 5.6), so that, by developing a theory of space-time on the basis of such a metric function, it is possible, by a purely deductive method and on the basis of geometrical space-time concepts, to arrive at equations for the description of interactions between such mesons. Again, such considerations assume the use of the concept of osculation, though this time applied to the parameters u^a of the indicatrix (Section 5.7). Generally speaking, the idea of introducing a curvature of isotopic space is suggested by Finsler geometry considerations, so that Finslerian techniques enable one to formulate gauge-invariant equations of physical fields for various internal symmetries in an arbitrary curved isotopic space. This is the point at which the construction of the models of interactions between elementary particles to be based on the ideas of Finsler geometry becomes both possible and necessary. Needless to say, the construction of such models is beyond the scope of this book.

In a book of this kind, one cannot dwell upon the relation of Finsler geometry with classical mechanics. The existence of such a relation is a consequence of the fact that the Finslerian metric function plays the role of the Lagrangian for Finslerian geodesics. Therefore, whenever the Lagrangian L of a dynamical system is homogeneous of degree one in the velocities, the trajectories of the system are the Finslerian geodesics associated with L. Moreover, it appears that the homogeneity condition is not essential to this relationship. Indeed, if a Lagrangian L_1 is nonhomogeneous in the velocities, then one may consider the parameter t used for parametrizing the trajectories as an additional coordinate. This method results in a Lagrangian L of the required homogeneity, the trajectories associated with L_1 then being projections of the Finslerian geodesics associated with L; the remaining component of the equation of these Finslerian geodesics merely states that the Hamiltonian function H_1 related to L_1 is conserved (Section 6.1). Thus, dynamical systems can always be described by homogeneous Lagrangians while remaining within the framework of Finslerian ideas. It is not difficult to extend to the homogeneous case the canonical equations and the Hamilton–Jacobi equation (Section 6.2) usually formulated for nonhomogeneous Lagrangians. The conventional Hamilton–Jacobi theory is formulated in terms of the canonical momenta field subject to the restriction that this field is a gradient of a scalar function, that is that the vorticity tensor of this field vanishes identically. This restriction, however, can be removed which leads to a generalization of the Hamilton–Jacobi theory on the basis of the Clebsch representation of the canonical momenta field (Section 6.3).

Up to this point our considerations have not explicitly assumed any modification of the conventional special theory of relativity. Nevertheless, the possibility of generalizing that theory on the basis of Finsler geometry cannot be discarded. Indeed, the idea suggests itself that, in a Finslerian treatment of kinematics, one should interpret the dependence of the Finslerian metric tensor or tetrads on the tangent vectors as a dependence on the velocities of motion of inertial frames of reference. Accordingly, allowing for the possibility of a deformation of space-time scales under changes of the velocity of motion of an inertial frame of reference, we perform in Chapter 7 a Finslerian refinement of a number of principal kinematic concepts of special relativity theory.

It appears that the Finslerian K-group of linear transformations of tangent spaces which leave the Finslerian metric function invariant plays the role of a proper kinematic transformation group, while the kinematic sense of the Lorentz group is retained only passively, that is, as the symmetry group of an inertial frame of reference with fixed velocity of motion (Section 7.1). This means that an observation of Finslerian effects in kinematic phenomena implies a comparison between the readings of instruments resting on different inertially moving material bodies. Our considerations can naturally be founded on a Finslerian extension of the special principle of relativity, namely, on the postulate that within the scope of kinematics no physical experiment can lead to a preferred Riemannian metric tensor; an absolute sense may be ascribed only to a Finslerian metric tensor (Section 7.2).

Knowledge of the history of the theory of special relativity leads one to the conclusion that this theory has not brought about a radical change in our concepts of space and time in the sense that the older, clear-cut concepts have been replaced by new, as clear, but truer, ideas. On the contrary, the progress achieved by the development of this theory consists, to a large measure, in the refinement of concepts. The theory of special relativity has demonstrated that many basic concepts used before had not been rigorously defined at all and that the resulting confusion of concepts was a reason for many misunderstandings (Whittaker, 1960; Mandelshtam, 1972). It is helpful to bear all this in mind while attempting a new step in the development of ideas concerning the relativity of motion. And one is justified in expecting that a Finslerian generalization will lead to a further refinement of kinematic concepts. Indeed, an immediate analysis reveals (Section 7.3) that the concept of the relative velocity of motion involves in fact two different definitions of relative velocity, which we refer to as the special relative velocity and the proper relative velocity. These two relative velocities have different geometrical and physical meanings, due to the difference between the Lorentz group and the Finslerian K-group of proper kinematic transformations. The first of them is just that relative velocity which is used in the conventional special theory of relativity, whereas the second coincides a priori with the first only if the use of any metric apart from the Riemannian one is prohibited.

On the basis of the concept of the proper relative velocity one can step by step erect the edifice of Finslerian kinematics (Section 7.4), the concept of invariant velocity being an important tool for such a construction. All the features of a Finslerian kinematics which distinguish it from the Lorentzian one may in principle be verified experimentally (Section 7.5). However, the observation of Finslerian corrections to the parameter q (which is the ratio of the relative velocity to the velocity of light) is hampered by the fact that laboratories now used are either at rest on Earth or are moving relative to it with velocities not higher than several kilometers per second, which corresponds to $q \lesssim 10^{-5}$. If such small effects are nevertheless detected experimentally, this would be an unequivocal indication that the Finslerian structure of our real space-time does exist.

The development of kinematic conceptions, no matter how detailed and exact a description of observed phenomena they provide, cannot however give one a feeling of satisfaction until these conceptions are related in a rigorous analytical manner to the equations of motion of matter in a gravitational field. By the nature of things, kinematics may not be postulated independently of these equations. From an analytical viewpoint, an interrelated set of equations describing physical fields and those describing the motion of matter should amount to a set of equations that describe the entire physics of the real world. Kinematics is a totality of consequences following from the equations of motion of matter in the simplest limiting case.

In the present approach, one can readily uncover such a relationship: a Finslerian formulation of the equations of motion of matter dictates a Finslerian kinematics (Section 7.6). Such a relationship clearly originates from the fact that the Finslerian approach is based on a metric tensor for which it is possible to construct a tetrad which is subject to Finslerian parallel transportation along the trajectory of a test body, this transportation being determined by the equations of motion of matter in a gravitational field. It is this tetrad that will geometrically represent a coordinate system of an inertial frame of reference and the behaviour of this tetrad will describe all Finslerian kinematic phenomena.

Only after the kinematics has been perceived as a definite analytical consequence of the equations of motion of matter, can one quite consistently speak about the class of those physical objects in whose kinematics Finslerian properties may manifest themselves. We finish Chapter 7 by arguing that a Michelson-type experiment if conducted in a laboratory moving relative to the Earth may detect the anisotropy of the velocity of light due to the Finslerian distortion of scales.

Finally, in an Appendix A which has the character of a mathematical supplement, we describe the Rund theory, which extends the concept of a direction-dependent connection from the Finslerian case proper, where the connection coefficients are assumed to be symmetric, homogeneous of degree zero with respect to the tangent vectors, and, in Cartan's theory, metric, to the case where all three assumptions are dropped. This theory offers a useful generalized view of the ideas of Finsler geometry. Our exposition is given in terms of the method of differential forms, which long ago became conventional for investigators in general relativity. Rund's generalized direction-dependent connection stands in relation to Cartan's Finslerian connection approximately as the standard concept of affine connection (Eisenhart, 1927) stands in relation to the Riemannian connection. In the general theory of relativity, generalizations involving the affine connection, which in general is nonsymmetric (containing torsion) and nonmetric, have been developed for a long time dating back to Einstein's work (1928). It may be hoped that corresponding direction-dependent approaches generalizing the proper Finslerian extension of general relativity theory will appear in the not too distant future.

In the concluding Appendix B we generalize the Finslerian gauge ideas of Chapter 5 and simultaneously the ideas of the preceding Appendix A. The generalization stems from the observation that in a (x, y)-dependent approach we need the connection coefficients, generally, of four tensorially different types $N_i^p(x, y)$, $L_i{}^j{}_m(x, y)$, $D_p{}^q{}_i(x, y)$, and $Q_p{}^q{}_r(x, y)$ to systematically operate with the derivatives $\partial/\partial x^i$ and $\partial/\partial y^p$. Indeed, given a scalar field $w(x, y)$, we can construct the covariant vector fields by means of the following two tensorially different ways. First, $w_p = \partial w/\partial y^p$. Second, $w_i = \mathrm{D}_i w$, where D_i is a covariant derivative in x^i. The subscript i of w_i may be called *the x-index*, and the subscript p of w_p *the y-index*. To operate with the two types of indices, we need in general two nondegenerate metric tensors $a_{ij}(x,y)$ and $g_{pq}(x,\hat{y})$, so that $w^i = a^{ij}w_j$ and $w^p = g^{pq}w_q$. Putting $\mathrm{D}_i w_j = \mathrm{d}_i w_j - L_j{}^n{}_i w_n$, $\mathrm{D}_i w_p = \mathrm{d}_i w_p - D_p{}^q{}_i w_q$, and $\mathrm{S}_p w_q = \mathrm{d}_p w_q - Q_q{}^r{}_p w_r$, where $\mathrm{d}_p = \partial/\partial y^p$ and $\mathrm{d}_i = \partial/\partial x^i + N_i^p \mathrm{d}_p$, the transformation laws of the gauge fields N_i^p, $L_i{}^j{}_m$, $D_p{}^q{}_i$, and $Q_p{}^q{}_r$ can unambiguously be defined by assuming the operators D_i and S_p to be covariant under both *the general coordinate transformations* $x^i = x^i(\bar{x}^j)$ and *the general gauge transformations* $y^p = y^p(x, y'^q)$. Further evaluating various commutators of D_i and S_p makes it possible to obtain the associated gauge tensors, arriving at the possibility of developing the general theory of gauge fields and (x,y)-dependent generalizations of physical fields carrying any external and internal spins. In the Appendix B we systematically develop such an approach even in a slightly generalized form, namely, we take, instead of y^p, the scalar parameters z^P playing the role of the coordinates of points of an internal space which dimension number M may be equal to any integer.

Clearly, all (x,z)-dependent relations can readily be translated into the language of their counterparts of proper (x,y)-dependent theory by means of the simple formal changes: $z^P \to y^p$, $L_m{}^n{}_i(x,z) \to L_m{}^n{}_i(x,y)$, $D_P{}^Q{}_i(x,z) \to D_p{}^q{}_i(x,y)$, $Q_Q{}^P{}_R(x,z) \to Q_q{}^p{}_r(x,y)$, $\mathrm{d}_p \to \mathrm{d}_p = \partial/\partial y^p$, $\mathrm{d}_i \to \mathrm{d}_i = \partial/\partial x^i + N_i^p(x,y)\mathrm{d}_p$, $\mathrm{D}_i W^P(x,z) \to \mathrm{D}_i W^p(x,y) = \mathrm{d}_i W^p + D_q{}^p{}_i W^q$, $S_R g_{PQ}(x,z) \to \mathrm{S}_r g_{pq}(x,y) = \mathrm{d}_r g_{pq} - Q_p{}^s{}_r g_{sq} - Q_q{}^s{}_r g_{ps}$, $E_P{}^Q{}_{ij}(x^n, z^R) \to E_p{}^q{}_{ij}(x^n, y^r)$, etc., while the $(N+M)$-fold variational principle (B.69) will be replaced by the $2N$-fold variational principle $\delta \int L(x,y) J(x,y) K(x,y)\, \mathrm{d}^N x\, \mathrm{d}^N y = 0$. It should be emphasized that the standard Finsler geometry is a theory which is covariant under only the coordinate transformations $x^i = x^i(\bar{x}^j)$ and not under the gauge transformations $y^p = y^p(x, y'^q)$, for the proper Finslerian relations involve the interlacing of two types of indices, namely, the x-indices $i, j, \ldots$ related to the variable x and the y-indices $p, q, \ldots$ related to the variable y. In particular, the Cartan connection coefficients involve the contractions of the type $\delta^j_p y^p \partial q_{rq}/\partial x^j$, which means in fact that proper Finslerian considerations may refer only to a fixed system $\{y^p\}$.

The identity $\mathrm{D}_i w_P \equiv \partial_i w_P + N_i^Q \mathrm{S}_Q w_P - \mathscr{D}_P{}^Q{}_i w_Q$, where $\mathscr{D}_P{}^Q{}_i = D_P{}^Q{}_i - Q_P{}^Q{}_R N_i^R$, shows that the gauge field N_i^P introduces the interaction of the field w_P carrying the internal spin $s = 1$ with the field $w_{PQ} = S_Q w_P$ of the internal spin $s + 1$ (such a conclusion can be applied to any other internal spin s). Such an interaction, and hence the gauge field N_i^P, has no counterparts in the current Utiyama-type Einstein–Yang-Mills gauge theory (in which the dependence of the fields on the internal z^P is not introduced), while $L_m{}^i{}_n$ and $D_P{}^Q{}_i$ (or the redefined field $\mathscr{D}_P{}^Q{}_i$) represent in fact the gravitational field and the Yang–Mills field, respectively.

Part B

Introduction to Finsler Geometry

Chapter 1

Primary Mathematical Definitions

The basic concepts of Finsler geometry can be easily and lucidly described in the language of classical tensor calculus. Clearly, the more general the geometrical background is taken the thicker will be the forest of tensors which stem from it. In the case of Finsler geometry, however, this forest is not constituted of arbitrarily distributed thickets; on the contrary, all the tensors stem from a single root, namely, from the scalar function $F(x, y)$ treated as a generating metric function. The assumption that the function $F(x, y)$ is positively homogeneous of degree one in y^i is crucial in erecting the edifice of Finsler geometry. The basic concomitants of the Finslerian metric function are described in Section 1.1 of the present chapter. Of these, the metric tensor, the geodesics, the connection coefficients, and, finally, the curvature tensors which we obtain directly by evaluating commutators of various covariant derivatives are the most important. The reader who wishes to revise this material should refer to the monograph by Rund (1959, 1981a).

Several curvature tensors arise in Section 1.1, thereby yielding a nontrivial geometry in not only x^i but also y^i. Therefore, figuratively speaking, Finsler geometry is a geometry with a first floor and, if the proper mathematical language is used, a Finsler space is an example of a fibre space, the fibres being $(N-1)$-dimensional Riemannian spaces called the indicatrices. The geometry of a Finsler space indicatrix will be described in Section 1.2 in some detail; in the first place, the realtionship between the indicatrix curvature tensor and the curvature tensor S_{ijmn} arising in Section 1.1 in terms of the variables x^i and y^i will be demonstrated. Our method will consist in defining the induced Riemannian metric tensor of the indicatrix by means of a certain conformal transformation of the Finslerian metric tensor. The idea of the parametrical representation of the indicatrix employed in this case is well illustrated by the example described in Section 2.3 of the next chapter starting from Equation (2.64). The important applied role of the indicatrix as an internal physical space will be discussed in Chapter 5. In the second part of Section 1.2 we shall derive the interesting particular forms which the curvature tensor S_{ijmn} takes in the cases of dimensions $N = 4$ and $N = 3$, and also in the case when the indicatrix is a space of constant curvature. These particular forms will be referred to in Sections 2.1 and 2.7 of the next chapter as defining interesting special types of Finsler spaces. At the end of Section 1.2 we shall indicate some explicit relationships between the orthonormal n-tuples of the Finslerian metric tensor and the indicatrix metric tensor, which will be of use in Chapter 5.

Finally, in Section 1.3 we set forth the concept of invariance of the Finslerian metric tensor and study its immediate implications. In the case of Riemannian geometry, the tangent spaces are invariant under the group of rotations which may be defined as the totality of all nondegenerate linear transformations of y^i leaving the Riemannian metric function $(a_{ij}(x)y^i y^j)^{1/2}$ invariant. The group is of fundamental significance because it brings out the concept of local invariance inherent in Riemannian geometry. As regards the applications of this concept to general relativity theory, one is quite entitled to treat the current Riemannian gravitational theory as a gauge field theory whose gauge group is given by the rotational group that corresponds to the given physical context, namely, by the Lorentz group (in this respect the useful reference is: Treder (1971)). We shall show how this Riemannian concept can be generalized for Finsler-type metric functions. The corresponding generalization, which we shall call the G-group, will be formed by the transformations of y^i which will be generally nonlinear. We shall derive the action of G-transformations on the metric tensor g_{ij}, on the Cartan torsion tensor C_{ijk}, and on the Hamiltonian function H, thus arriving at the subgroups G_m and G_{mt} under which the objects g_{ij} and C_{ijk} transform as tensors. Proposition 12 of Chapter 2 where an interesting particular case is treated is a good illustration of the definitions introduced in Section 1.3. Concluding the section, we shall prove the assertion which states that the G_m-transformations realize motions in the tangent Minkowskian spaces.

1.1. Concomitants of the Finslerian Metric Function

The subject-matter of Finsler geometry concerns a real[1] N-dimensional differentiable manifold M which is endowed with a non-negative scalar function $F(x, y)$ of two sets of arguments, namely points x^i and contravariant vectors y^i tangent to M at x^i, or, symbolically, $x^i \in M$ and $y^i \in M_x$. Our subsequent considerations will be local in nature so that, remaining within the framework of classical tensor calculus, we shall represent geometrical objects by their components with respect to a local coordinate system $\{x^i\}$ carried by the background manifold M. The small Latin letters $i, j, k, m, n, \ldots$ will be used to denote tensor indices and therefore will range over $1, 2, \ldots, N$. The indices of tensorial objects as well as of the arguments x^i and y^i themselves will be omitted whenever convenient.

It will be sufficient for our purposes to stipulate the smoothness of the function $F(x, y)$ by the following two conditions:

(i) The function $F(x, y)$ is at least of class C^3 with respect to x^i, which makes us assume in turn that the background manifold M itself is at least of class C^3.

(ii) A region M_x^* exists in each tangent space M_x such that, first, M_x^* is conic in the sense that if M_x^* contains some vector y_1^i then M_x^* contains any other vector collinear with y_1^i, and, second, the function $F(x, y)$ is at least of class C^5 with respect to all non-zero vectors $y^i \in M_x^*$. The non-zero vectors y^i from M_x^* will be called *admissible*.

Further, it will be assumed that, for any admissible y^i, $F(x, y) > 0$ and

$$\det(\partial^2 F^2(x, y)/\partial y^i\, \partial y^j) \neq 0. \tag{1.1}$$

Besides this, the function $F(x, y)$ is to be positively homogeneous of degree one with respect to y^i, i.e.,

$$F(x, ky) = kF(x, y) \tag{1.2}$$

for any fixed $k > 0$ and for all $y^i \in M_x^*$. Under these conditions, the triple $(M, M_x^*, F(x, y))$ is called an *N-dimensional Finsler space*, and $F(x, y)$ is called a *Finslerian metric function*. The value of the metric function $F(x, y)$ is treated in Finsler geometry as the length of the tangent vector y^i attached to the point x^i. If a Finsler space allows a coordinate system x^i such that F does not depend on these x^i, the Finsler space and the metric function are called *Minkowskian*.

It will be noted that in mathematical works devoted to Finsler geometry additional conditions are usually imposed on the metric function F which ensure the positive definiteness[2] of the quadratic form $Z^i Z^j \partial^2 F^2\,(x, y)/\partial y^i \partial y^j$ at any point x^i and for any non-zero vector $y^i \in M_x$. However, it is clear already in the Riemannian formulation of general relativity theory that the metric structure of space-time cannot be positively definite, for the space-time metric tensor must be of the indefinite signature $(+ - - -)$. This reason alone makes one expect that it is indefinite metrics that may be of interest in a Finslerian extension of general relativity. Accordingly, we refrain deliberately from imposing the condition of positive definiteness thereby admitting that the Finsler space under study can be indefinite.

As regards the homogeneity condition (1.2) it should be pointed out that the necessity of postulating (1.2) follows from the invariant notions inherent in any centroaffine space, the tangent space M_x being an example of such a space. Indeed, the ratio of the lengths of any two collinear vectors y_1^i and $y_2^i = ky_1^i$ of the centroaffne space may be invariantly defined to be

$$y_1^1/y_2^1 = y_1^2/y_2^2 = \cdots = k$$

which does not involve any metric function. Therefore, (1.2) is nothing but the requirement of consistency of the Finslerian definition of length with the centroaffine definition. The Finslerian metric function is required in order to compare the lengths of non-collinear vectors.

The homogeneity condition (1.2) plays an important constructive role in

Finsler geometry, many Finslerian relations being based on identities ensuing from (1.2). To derive these identities, let us consider any function $Z(x, y)$ which is differentiable and positively homogeneous of degree r with respect to y^i; that is, $Z(x, ky) = k^r Z(x, y)$ for any $k > 0$, where the degree r may be any real number. On differentiating the latter equality with respect to k and putting $k = 1$, we find that

$$Z(x, ky) = k^r Z(x, y) \quad \textit{implies} \quad y^i \partial Z(x, y)/\partial y^i = rZ(x, y). \tag{1.3}$$

The assertion (1.3) is known in the literature as *the Euler theorem* on homogeneous functions.

The application of (1.3) to F^2 yields

$$F^2(x, y) = g_{ij}(x, y) y^i y^j \tag{1.4}$$

where

$$g_{ij}(x, y) = \tfrac{1}{2}\partial^2 F^2(x, y)/\partial y^i \partial y^j \tag{1.5}$$

is called *the Finslerian metric tensor*. Introducing the covariant tangent vector

$$y_i = g_{ij}(x, y) y^j \tag{1.6}$$

we obtain from the definition (1.5) and the Euler theorem (1.3) with $r = 1$ that

$$y_i = F(x, y) \partial F(x, y)/\partial y^i. \tag{1.7}$$

Then *the unit tangent vector*

$$l_i = y_i / F(x, y) \tag{1.8}$$

will merely be

$$l_i = \partial F(x, y)/\partial y^i. \tag{1.9}$$

The definitions (1.5) and (1.6) imply in view of (1.7) that the following useful identity holds:

$$g_{ij}(x, y) = \partial y_i / \partial y^j, \tag{1.10}$$

whereas

$$\partial l_i / \partial y^j = h_{ij}(x, y)/F(x, y), \tag{1.11}$$

where

$$h_{ij}(x, y) = g_{ij}(x, y) - l_i(x, y) l_j(x, y) \tag{1.12}$$

is the so-called *angular metric tensor*. The latter tensor has the properties

$$y^i h_{ij} = 0, \qquad g^{ij} h_{ij} = N - 1, \tag{1.13}$$

as a consequence of (1.6) and $l^i l_i = 1$. The formulae (1.4)–(1.13) represent an immediate generalization of well-known Riemannian prototypes, this generalization being the result of but a single homogeneity condition (1.2).

A Finsler geometry will reduce to a Riemannian geometry in the case when the metric tensor $g_{ij}(x,y)$ is assumed to be independent of y^n, which reads $C_{ijk} = 0$ in terms of the notation

$$C_{ijk}(x,y) = \tfrac{1}{2}\partial g_{ij}(x,y)/\partial y^k. \tag{1.14}$$

This latter entity is a tensor because differentiation of a tensor with respect to a vector argument gives again a tensor (cf. Equation (3.13)), in contrast to differentiation with respect to the coordinates x^i. From (1.5) and (1.14)

$$C_{ijk} = \tfrac{1}{4}\partial^3 F^2/\partial y^i \partial y^j \partial y^k, \tag{1.15}$$

which says that this tensor is symmetric in all its subscripts. $C_{ijk}(x,y)$ is called *the Cartan torsion tensor*. The Euler theorem (1.3) applied to the Finslerian metric tensor (1.5) indicates, since $r = 0$, that the Cartan torsion tensor satisfies the following important identity

$$y^i C_{ijk}(x,y) = 0. \tag{1.16}$$

Similarly, the tensor

$$C_{ijkm}(x,y) = \partial C_{ijk}(x,y)/\partial y^m \tag{1.17}$$

will also be symmetric in all its subscripts and will satisfy the identity

$$y^i C_{ijkm}(x,y) = -C_{jkm}(x,y) \tag{1.18}$$

as a result of (1.3) with $r = -1$. Needless to say, all Finslerian relations generalize their Riemannian analogs as a result of the presence of the Cartan torsion tensor C_{ijk} and the derivatives thereof.

The requirement (1.1) makes it possible to introduce the contravariant reciprocal tensor $g^{ij}(x,y)$ of the covariant metric tensor (1.5) in the usual way:

$$g^{ki}(x,y)g_{ij}(x,y) = \delta^k_j. \tag{1.19}$$

The designation δ^k_j will be used for the Kronecker symbol. The differentiation of (1.19) with respect to y^n yields

$$g^{ki}\,\partial g_{ij}/\partial y^n + g_{ij}\,\partial g^{ki}/\partial y^n = 0$$

or, remembering the definition (1.14),

$$\tfrac{1}{2}\partial g^{ki}(x,y)/\partial y^n = -C_n{}^{ki}(x,y). \tag{1.20}$$

Here and in what follows indices are raised and lowerd by means of the Finslerian metric tensor, for example

$$C_m{}^{ni} = C_{mkj}g^{nk}g^{ij}, \qquad C^i_{jkm} = C_{njkm}g^{in}, \qquad l^i = g^{ij}l_j.$$

The geodesics of a Finsler space can be defined in a way essentially similar to that used in Riemannian geometry. To this end, we consider the functional

$$I(C) = \int_{C^{P_1}}^{P_2} F(x, \mathrm{d}x), \tag{1.21}$$

where the integration is carried out along a curve C joining two fixed points P_1 and P_2 of the manifold M. The stationary curves of the variational problem $\delta I(C) = 0$ are called *Finslerian geodesics.* Although in positive-definite Finsler spaces the geodesics afford the minimum of the functional (1.21), in the case of indefinite Finsler spaces we may speak only of stationarity.

The Finslerian metric function provides one with a natural parametrization of curves by means of the parameter s of the Finslerian arc-length defined by

$$\mathrm{d}s = F(x, \mathrm{d}x). \tag{1.22}$$

It follows immediately from the homogeneity of $F(x, y)$ that the integral considered is parameter-independent, that is,

$$\int F(x, \mathrm{d}x) = \int F(x, \mathrm{d}x/\mathrm{d}s)\mathrm{d}s = \int F(x, \mathrm{d}x/\mathrm{d}t)\,\mathrm{d}t$$

for any parameter $t = t(s)$ subject to the condition $\mathrm{d}t/\mathrm{d}s \neq 0$. Therefore, for any such parameter t, our variational problem takes the form

$$\delta \int_{C^{P_1}}^{P_2} F(x, \dot{x})\,\mathrm{d}t = 0 \tag{1.23}$$

where $\dot{x}^i = \mathrm{d}x^i/\mathrm{d}t$. It is well known from the calculus of variations that the condition for a curve $x^i(t)$ to be the stationary curve of the variational problem (1.23) reads

$$\mathrm{d}(\partial F(x, \dot{x})/\partial \dot{x}^i)/\mathrm{d}t - \partial F(x, \dot{x})/\partial x^i = 0. \tag{1.24}$$

In other words, Finslerian geodesics satisfy the Euler-Lagrange equations (1.24) associated with the Lagrangian $F(x, \dot{x})$.

The equation (1.24) is called *the equation of Finslerian geodesics.* This equation can be readily rewritten as

$$\mathrm{d}^2x^i/\mathrm{d}t^2 + \gamma_{m\ n}^{\ i}(x, \dot{x})\dot{x}^m\dot{x}^n - \dot{x}^i\mathrm{d}(\ln F(x, \dot{x}))/\mathrm{d}t = 0, \tag{1.25}$$

where the object

$$\gamma_{m\ n}^{\ i}(x, y) = \tfrac{1}{2}g^{ik}(x, y)(\partial g_{mk}(x, y)/\partial x^n + \partial g_{nk}(x, y)/\partial x^m - \\ - \partial g_{mn}(x, y)/\partial x^k) \tag{1.26}$$

is called *the Finslerian Christoffel symbols.* In order to verify the representation (1.25) we substitute the equality (1.4) (with $y^i = \dot{x}^i$) in (1.24) and get

$$\mathrm{d}(g_{ij}(x, \dot{x})\dot{x}^j/F(x, \dot{x}))/\mathrm{d}t - \tfrac{1}{2}F^{-1}(x, \dot{x})\dot{x}^m\dot{x}^n\,\partial g_{mn}(x, \dot{x})/\partial x^i = 0$$

or, multiplying the latter by $F(x, \dot{x})$,

$$g_{ij}(x, \dot{x})(\mathrm{d}^2x^j/\mathrm{d}t^2 - \dot{x}^j\,\mathrm{d}(\ln F(x, \dot{x}))/\mathrm{d}t) + \\ + \dot{x}^j\dot{x}^n\,\partial g_{ij}(x, \dot{x})/\partial x^n - \tfrac{1}{2}\dot{x}^m\dot{x}^n\,\partial g_{mn}(x, \dot{x})/\partial x^i = 0, \tag{1.27}$$

where we have made use of the circumstance that

$$\mathrm{d}g_{ij}(x,\dot{x})/\mathrm{d}t = \dot{x}^n\,\partial g_{ij}(x,\dot{x})/\partial x^n + 2C_{ijn}(x,\dot{x})\,\mathrm{d}\dot{x}^n/\mathrm{d}t$$

and, therefore,

$$\dot{x}^j\mathrm{d}g_{ij}(x,\dot{x})/\mathrm{d}t = \dot{x}^j\dot{x}^n\,\partial g_{ij}(x,\dot{x})/\partial x^n$$

because of (1.16). So, raising the index i in (1.27) and noting that

$$\dot{x}^m\dot{x}^n\gamma_{\min}(x,\dot{x}) = \dot{x}^m\dot{x}^n(\partial g_{im}(x,\dot{x})/\partial x^n - \tfrac{1}{2}\partial g_{mn}(x,\dot{x})/\partial x^i)$$

as a consequence of the definition (1.26), we find that relation (1.25) is indeed valid.

In the case when the parameter t is chosen to be the Finslerian arc-length s defined by (1.22) we find, using the notation

$$x'^i = \mathrm{d}x^i/\mathrm{d}s, \tag{1.28}$$

that the vector x'^i has unit Finslerian length

$$F(x,x') = 1 \tag{1.29}$$

So that equation (1.25) for the Finslerian geodesics takes the simpler form

$$\mathrm{d}^2x^i/\mathrm{d}s^2 + \gamma_m{}^i{}_n(x,x')x'^m x'^n = 0. \tag{1.30}$$

Despite the fact that the Finslerian Christoffel symbols are defined by the same rule (1.26) as in the Riemannian case, their transformation law under coordinate transformations $x^i = x^i(\bar{x}^j)$ differs essentially from the transformation law of the Riemannian Christoffel symbols. This fact may be inferred from the dependence of the Finslerian Christoffel symbols $\gamma^i_{mn}(x,y)$ not only on the points x^i but also on the tangent vectors y^i. However, the covariant partial derivative of any tensor which depends on x^i alone, for example of the tensor field of the form $X_i{}^k(x)$, can easily be constructed by comparing the transformation laws of $\partial X_i{}^k/\partial x^j$ and $\gamma_m{}^i{}_n(x,y)$. This procedure, described in detail in Rund (1959, Chapter 2), leads to the following definition of the partial covariant derivative:

$$\begin{aligned} X_i{}^k(x)_{;j} = \partial X_i{}^k(x)/\partial x^j + \Gamma_n{}^k{}_j(x,y)X_i{}^n(x) - \\ - \Gamma_i{}^n{}_j(x,y)X_n{}^k(x), \end{aligned} \tag{1.31}$$

where

$$\Gamma_i{}^k{}_j(x,y) = \gamma_i{}^k{}_j - C_i{}^k{}_n G_j{}^n - C_j{}^k{}_n G_i{}^n + C_{ijn}G^{kn}, \tag{1.32}$$

$$G_j{}^n(x,y) = -C_j{}^n{}_m 2G^m + y^m\gamma_m{}^n{}_j = y^m\Gamma_j{}^n{}_m \equiv \partial G^n/\partial y^j, \tag{1.33}$$

$$2G^m(x,y) = y^n y^l \gamma_n{}^m{}_l = y^n G_n{}^m = y^n y^l \Gamma_n{}^m{}_l. \tag{1.34}$$

The coefficients (1.32) are called *the Cartan connection coefficients*. They are obviously symmetric in their subscripts:

$$\Gamma_i{}^k{}_j = \Gamma_j{}^k{}_i. \tag{1.35}$$

Incidentally, in the mathematical literature, the Cartan connection coefficients are often denoted by the starred symbol $\Gamma^{*k}_{i\ j}$; we omit the asterisk in order to simplify the notation. We shall present a self-consistent construction of the Cartan connection coefficients by the method of invariance identities in Section 3.2.

Following the terminology used in Rund (1959, Chapter 2), we shall call the covariant derivative of the form (1.31) *the δ-derivative*. It will be noted that the δ-derivative of an x-dependent tensor results in a tensor depending on both x^i and y^i, which is a distinctive feature of Finsler geometry. The δ-process may be extended to the differentiation of tensors depending not only on the point x^i but also on a contravariant vector field $q^i(x)$. The partial δ-derivative of such a tensor, say of the form $X_i^{\ k}(x, q(x))$, with respect to x^j in the direction y^i is defined by the formula

$$\begin{aligned} X_{i\ ;j}^{\ k} = \partial X_i^{\ k}(x, q)/\partial x^j + q^n_{\ ,j}\, \partial X_i^{\ k}(x, q)/\partial q^n + \\ + \Gamma_n^{\ k}{}_j(x, y) X_i^{\ n}(x, q) - \Gamma_i^{\ n}{}_j(x, y) X_n^{\ k}(x, q), \end{aligned} \tag{1.36}$$

where $q^n_{,j} = \partial q^n(x)/\partial x^j$. In particular, if a tensor $X_i^{\ k}(x, q(x))$ is differentiated in the direction of the field $q^i(x)$, we get

$$\begin{aligned} X_{i\ ;j}^{\ k}(x, q) = \partial X_i^{\ k}(x, q)/\partial x^j + q^n_{,j}\, \partial X_i^{\ k}(x, q)/\partial q^n + \\ + \Gamma_n^{\ k}{}_j(x, q) X_i^{\ n}(x, q) - \Gamma_i^{\ n}{}_j(x, q) X_n^{\ k}(x, q). \end{aligned} \tag{1.37}$$

The peculiarity of Finslerian covariant differentiation is that the δ-derivative of the Finsler metric tensor does not in general vanish, this being indicated by the equality

$$g_{ij;\,k}(x, q(x)) = 2C_{ijm}(x, q(x)) q^m_{;k} \tag{1.38}$$

which is obtainable directly by applying definition (1.37) to $g_{ij}(x, q)$. Yet, owing to the identity (1.16), the equality (1.38) implies the relation

$$g_{ij;\,k}(x, q) q^i = 0 \tag{1.39}$$

which shows that (1.38) does not prevent one from taking the Finslerian metric tensor outside the δ-derivative of a vector field $q^i(x)$, i.e.

$$q_{n;\,i}(x, q) = q^m_{;i}(x, q) g_{mn}(x, q). \tag{1.40}$$

When considering tensors which are functions of the independent variables x^i and y^i, an important part is played in Finsler geometry by the so-called *Cartan covariant derivative* defined as

$$\begin{aligned} X_i^{\ k}(x, y)_{|j} = \partial X_i^{\ k}(x, y)/\partial x^j - G_j^{\ n}(x, y) \partial X_i^{\ k}(x, y)/\partial y^n + \\ + \Gamma_n^{\ k}{}_j(x, y) X_i^{\ n}(x, y) - \Gamma_i^{\ n}{}_j(x. y) X_n^{\ k}(x, y) \end{aligned} \tag{1.41}$$

for any (x, y)-dependent tensor $X_i^k(x, y)$. In particular,

$$X^k_{\ |j} = \partial X^k/\partial x^j - G_j^{\ n} \partial X^k/\partial y^n + \Gamma_n^{\ k}{}_j X^n, \tag{1.42}$$

$$X_{i|j} = \partial X_i/\partial x^j - G_j{}^n \partial X_i/\partial y^n - \Gamma_i{}^n{}_j X_n \tag{1.43}$$

for contravariant and covariant vector fields, respectively, and

$$Q_{|j} = \partial Q/\partial x^j - G_j^n \partial Q/\partial y^n \tag{1.44}$$

for a scalar field $Q(x,y)$. Comparison of (1.41) with (1.31) indicates that the Cartan covariant differentiation and the δ-differentiation give the same result in the particular case when the differentiated tensor depends on x^i alone.

Since the covariant differentiations (1.36) and (1.41)–(1.44) are linear operations the conventional laws of covariant differentiation are valid; for example, the derivative of a sum is the sum of the derivatives and the product rule of ordinary differentiation holds. Moreover, as in Riemannian geometry, the Cartan covariant derivative is *metric*, i.e.

$$g_{ik|j} = 0, \tag{1.45}$$

which is obtained by substituting (1.14) and (1.32) in the definition

$$g_{ik|j} \stackrel{def}{=} \partial g_{ik}/\partial x^j - G_j{}^n \partial g_{ik}/\partial y^n - \Gamma_{ikj} - \Gamma_{kij}. \tag{1.46}$$

In addition,

$$l^k{}_{|j} = y^k{}_{|j} = F_{|j} = 0. \tag{1.47}$$

Indeed, since $\partial y^k/\partial x^j \equiv 0$ and $\partial y^k/\partial y^n = \delta^k_n$, definition (1.42) yields $y^k{}_{|j} = -G_j{}^k + \Gamma_n{}^k{}_j y^n$ which is zero in view of (1.33). The last identity in (1.47) follows from definition (1.44) together with relations (1.9) and $y_n G_j{}^n = y^m y^n \gamma_{mnj}$ (see (1.33)). The identities (1.45) and (1.47) show that the metric tensor g_{ij}, the metric function F, and the tangent vectors y^i or l^i may be taken outside the Cartan covariant derivative; in particular

$$X^k(x,y)_{|j} = X_i(x,y)_{|j} g^{ik}(x,y), \qquad (y^i X_i)_{|j} = X_{i|j} y^i.$$

The Cartan covariant derivative of the torsion tensor C_{ijk} does not in general vanish. Also, definition (1.32) entails the relation

$$\partial \Gamma_i{}^h{}_j/\partial y^r = C_j{}^h{}_{r|i} + C_i{}^h{}_{r|j} - g^{hk} C_{ijr|k} - \\ - (C_j{}^h{}_l C_i{}^l{}_{r|k} + C_i{}^h{}_l C_j{}^l{}_{r|k} - C_i{}^l{}_j C_l{}^h{}_{r|k}) y^k \tag{1.48}$$

which explicitly reflects the fact that the partial derivative of the connection coefficients with respect to the tangent vectors is a tensor.

The Berwald connection coefficients are defined as follows

$$G_k{}^i{}_j = \partial G_k{}^i/\partial y^j \tag{1.49}$$

where $G_k{}^i$ are given by (1.33). They are symmetric in their subscripts since $G_k{}^i = \partial G^i/\partial y^k$ (see (1.33)). It can be readily deduced from (1.32) and (1.33) together with the definition (1.49) that

$$G_i{}^k{}_j = \Gamma_i{}^k{}_j + C_i{}^k{}_{j|r} y^r \tag{1.50}$$

which indicates explicitly that $G_i{}^k{}_j$, being the difference between a set of connection coefficients and a tensor, represent connection coefficients. *The Berwald covariant derivative* is defined in a way similar to the Cartan covariant derivative (1.41) and is usually denoted by a parenthesis, namely

$$X_i{}^k{}_{(j)} = \partial X_i{}^k/\partial x^j - G_j{}^n \partial X_i{}^k/\partial y^n + G_n{}^k{}_j X_i{}^n - G_i{}^n{}_j X_n{}^k. \tag{1.51}$$

It follows from (1.45), (1.47) and (1.50) that

$$F_{(k)} = l^i{}_{(k)} = l_{i(k)} = 0, \qquad g_{ij(k)} = -2C_{ijk|r}y^r. \tag{1.52}$$

The relation (1.50) shows that Cartan's and Berwald's connection coefficients are equal if and only if $C_{ikj|r}y^r = 0$, in which case the Finsler space is often called a *Landsberg space.*

It may happen that the Cartan covariant derivative of the torsion tensor C_{ikj} vanishes. A Finsler space possessing the property $C_{ikj|h} = 0$ is called an *affinely-connected* space or *Berwald space.* In such a case we get from (1.48) that the connection coefficients are independent of y^n:

$$G_i{}^k{}_j = \Gamma_i{}^k{}_j = \Gamma_i{}^k{}_j(x) \quad \textit{if} \quad C_{ikj|h} = 0. \tag{1.53}$$

Conversely[3),]

$$\partial\Gamma_i{}^k{}_j/\partial y^n = 0 \quad \textit{implies} \quad C_{ikj|h} = 0.$$

Knowledge of the partial covariant derivative may be used to construct the covariant derivative of an x-dependent tensor, say $X^k(x)$, along any vector field $v^i(x)$ in a natural way:

$$X^k{}_{;j}(x, v(x))v^j(x) \equiv (\partial X^k(x)/\partial x^j + \Gamma_n{}^k{}_j(x, v(x)))v^j(x). \tag{1.54}$$

Further more, the covariant derivative $\delta X^k/\delta t$ of $X^k(x)$ along a curve C with some parameter t may be defined as follows:

$$\delta X^k/\delta t = \mathrm{d}X^k/\mathrm{d}t + \Gamma_n{}^k{}_j(x, \dot{x})X^n\dot{x}^j \equiv \mathrm{d}X^k/\mathrm{d}t + G_n{}^k(x, \dot{x})X^n. \tag{1.55}$$

In the second step here we have used the identity (1.33); $\dot{x}^j = \mathrm{d}x^j/\mathrm{d}t$. The definition (1.55) is obtainable from (1.54) by a formal replacement of $v^j(x)$ by $\dot{x}^j(t)$.

If the curve C is parametrized by the Finslerian arc-length s and the unit vector (1.28) tangent to C is transported along C, the covariant δ-derivative (1.55) takes the form

$$\delta x'^k/\delta s = \mathrm{d}x'^k/\mathrm{d}s + G_n{}^k(x, x')x'^n$$

or, in view of (1.34),

$$\delta x'^k/\delta s = \mathrm{d}x'^k/\mathrm{d}s + x'^m x'^n \gamma_m{}^k{}_n(x, x'). \tag{1.56}$$

The curve C is called *autoparallel* if $\delta x'^k/\delta s = 0$. A mere glance at the equation for Finslerian geodesics (1.30) is sufficient to see that the geodesics of a Finsler space are autoparallel curves.

A significant feature of Riemannian geometry is that the Riemannian metric tensor endows the tangent spaces M_x to the background manifold with the structure of Euclidean space. This concept will now be generalized to Finsler geometry by means of the following definition: the pair $(M_x^*, F(x, y))$, where x^i is fixed and y^i is arbitrary, is called *the tangent Minkowskian space* at this x^i and will be denoted by $M(x)$. According to this definition the dependence of the Finslerian metric tensor $g_{ij}(x, y)$ on y^i at a fixed x^i may be treated as a dependence on the coordinates y^i adopted in $M(x)$. Therefore, the Finslerian metric tensor $g_{ij}(x, y)$ at a point x^i will play the role of *the metric tensor of the tangent Minkowskian space* $M(x)$ and the square of the length of an infinitesimal displacement $\mathrm{d}y^i$ in $M(x)$ will be given by

$$\mathrm{d}s^2_{\mathrm{Mink}} = g_{ij}(x, y)\,\mathrm{d}y^i\,\mathrm{d}y^j, \tag{1.57}$$

that is, by the usual Riemannian rule as applied to $M(x)$.

The Christoffel symbols of the tangent Minkowskian space

$$q_i{}^k{}_j = \tfrac{1}{2}g^{kn}(\partial g_{in}/\partial y^j + \partial g_{jn}/\partial y^i - \partial g_{ij}/\partial y^n) \tag{1.58}$$

reduce to the Cartan torsion tensor, namely

$$q_i{}^k{}_j(x, y) = C_i{}^k{}_j(x, y), \tag{1.59}$$

as a consequence of the symmetry of the tensor $C_i{}^k{}_j$ (given by equation (1.14)) in its subscripts. The associated geodesic equations of the tangent Minkowskian space at a point x^i read

$$\mathrm{d}^2y^i/\mathrm{d}s^2 + C_m{}^i{}_n(x, y)(\mathrm{d}y^m/\mathrm{d}s)(\mathrm{d}y^n/\mathrm{d}s) = 0, \tag{1.59$'$}$$

where $\mathrm{d}s = F(x, \mathrm{d}y)$. Because of the identity (1.16) the equations (1.59′) admit the straight line solution $y^i(x, s) = a^i(x)s$, with $a^i(x)$ an arbitrary vector field, and $y_i\,\mathrm{d}^2y^i/\mathrm{d}s^2 = 0$ is an integral of the equations. If we consider *the curvature tensor of the tangent Minkowskian space*

$$F^{-2}S_i{}^j{}_{mn} \overset{\mathrm{def}}{=} \partial q_i{}^j{}_m/\partial y^n - \partial q_i{}^j{}_n/\partial y^m + q_k{}^j{}_n q_i{}^k{}_m - q_k{}^j{}_m q_i{}^k{}_n \tag{1.60}$$

then (1.59), together with the identity

$$\partial C_i{}^j{}_m/\partial y^n = C_i{}^j{}_{mn} - 2C_k{}^j{}_n C_i{}^k{}_m \tag{1.61}$$

which follows (1.17) and (1.20), implies that

$$S_i{}^j{}_{mn} = (C_k{}^j{}_m C_i{}^k{}_n - C_k{}^j{}_n C_i{}^k{}_m)F^2 \tag{1.62}$$

where the tensor $C_i{}^j{}_{mn}$ disappears because of the symmetry property $C_i{}^j{}_{mn} = C_i{}^j{}_{nm}$.

As in Euclidean space proper, any (sufficiently smooth) non-degenerate non-linear transformation of the form $y^i = y^i(z^P)$ at some fixed point x^i may be interpreted as the transformation in $M(x)$ from rectilinear coordinates y^i to a system of curvilinear coordinates z^P. Nothing prevents one from defining the transform of the tensor g_{ij} under such transformations by means of the usual tensor law, namely

$$g_{PQ}(x,z) = g_{ij}(x,y)(\partial y^i/\partial z^P)(\partial y^j/\partial z^Q) \tag{1.63}$$

at the point x^i, which entails in turn that the quantity (1.60) will transform like a tensor and that the transform of the object (1.58) will be identical with that of the Riemannian Christoffel symbols. As regards relations (1.59) and (1.62), they will not be valid with respect to curvilinear coordinates z^P because C_{ikj} defined by (1.14) must transform like a partial derivative of the metric tensor.

Once the above prescription is adopted, the tangent Minkowskian space can be regarded as a Riemannian space (and may be called the tangent Riemannian space), though of a particular type for there always exists in $M(x)$ a coordinate system y^i such that relations (1.5) and (1.2) hold with respect to this y^i. Hence, covariant differentiation in $M(x)$ will be governed by the Riemannian laws. In particular, the partial covariant derivative of a tensor of the form $X_i{}^k(x,y)$ with respect to y^j will be given in $M(x)$ by

$$D_j^{\text{Mink}}\, X_i^{\,k} = \partial X_i{}^k/\partial y^j + C_j{}^k{}_n X_i{}^n - C_j{}^n{}_i X_n{}^k \tag{1.64}$$

which takes into account the equality (1.59). Since equation (1.60) represents the Riemannian curvature tensor associated with the metric tensor of $M(x)$, the Riemannian commutation relation

$$(D_h^{\text{Mink}}\, D_k^{\text{Mink}} - D_k^{\text{Mink}}\, D_h^{\text{Mink}})X^i = -F^{-2}S_j{}^i{}_{kh}X^j \tag{1.65}$$

must be valid. If one wishes relation (1.65) may be regarded as the definition of the tensor $S_j{}^i{}_{kh}$.

It will be noted that the Minkowskian derivative (1.64) lowers the degree of homogeneity of objects with respect to y^i. If we wish to retain the degree of homogeneity, we may adhere to the notation $|_j$ used in Rund (1959, pp. 99–100) which is related to D_j^{Mink} by

$$|_j = FD_j^{\text{Mink}}. \tag{1.66}$$

Since the Riemannian covariant derivative of a Riemannian metric tensor vanishes identically, the definition (1.66) entails

$$g_{mn}|_j = 0 \tag{1.67}$$

similarly to (1.45). However, in contrast to (1.47),

$$\begin{aligned} &F|_j = y_j, \qquad y^i|_j = F\delta^i_j, \qquad y_i|_j = Fg_{ij}, \\ &h_{mn|j} = -h_{mj}l_n - h_{nj}l_m, \qquad h^{ij}|_j = -(N-1)l^i. \end{aligned} \tag{1.68}$$

In particular calculations it may be useful to represent the Cartan covariant derivative in terms of $|_n$ instead of the partial derivative $\partial/\partial y^n$, for example

$$X^i{}_{|h} = \partial X^i/\partial x^h - F^{-1}G_h{}^m X^i|_m + G_k^{*i}{}_h X^k, \tag{1.69}$$

$$X_{i|h} = \partial X_i/\partial x^h - F^{-1}G_h{}^m X_{i|m} - G_i^{*k}{}_h X_k, \tag{1.70}$$

$$C_{ikj|h} = \partial C_{ikj}/\partial x^h - F^{-1}G_h{}^n C_{ikj}|_n - G_i^{*m}{}_h C_{mkj} - \\ - G_k^{*m}{}_h C_{imj} - G_j^{*m}{}_h C_{ikm}, \tag{1.71}$$

where

$$G_i^{*m}{}_h = \Gamma_i{}^m{}_n + C_i{}^m{}_l G_h{}^l. \tag{1.72}$$

A straightforward way of deriving expressions for curvature tensors is to evaluate the commutators of covariant derivatives. In contrast to Riemannian geometry, Finslerian techniques suggest several definitions for covariant derivatives, thereby entailing that several tensors play the role of Finslerian curvature tensors. One of them, namely the tensor $S_i{}^j{}_{mn}$ describing the curvature of the tangent Minkowskian space, has appeared in the commutator (1.65). The commutator of the derivative (1.66) with the Cartan covariant derivative given by (1.41) can readily be represented in the form

$$X_i|_{h|k} - X_{i|k}|_h = -P_{jikh}X^j + X_i|_j C_h{}^j{}_{k|r}y^r + X_{i|j}C_h{}^j{}_k F, \tag{1.73}$$

where the curvature tensor P_{jikh} is given by

$$F^{-1}P_{jikh} = C_{kih|j} - C_{jkh|i} - C_{kim}C_j{}^m{}_{h|r}y^r + C_j{}^m{}_k C_{mih|r}y^r. \tag{1.74}$$

This tensor is skew-symmetric with respect to the first pair of indices:

$$P_{jikh} = -P_{ijkh}, \tag{1.75}$$

whereas $P_{jikh} \neq -P_{jihk}$ in general. Contracting (1.74) with y^j shows that

$$P_{jikh} = 0 \quad \textit{implies} \quad C_{kih|r}y^r = 0.$$

The curvature tensors S_{ijmn} and P_{ijmn} vanish in the Riemannian special case $C_i{}^k{}_j = 0$. In order to obtain Finslerian generalizations of the Riemann curvature tensor, one may consider the commutator of the δ-derivatives or of the Cartan covariant derivatives. Straightforward calculations based on the definition (1.36) of the δ-derivative yield

$$X^i{}_{;h;k} - X^i{}_{;k;h} = \tilde{K}_j{}^i{}_{hk}(x,q)X^j, \tag{1.76}$$

where

$$\tilde{K}_j{}^i{}_{hk} = (\partial\Gamma_j{}^i{}_h/\partial x^k + q^1{}_{,k}\partial\Gamma_j{}^i{}_h/\partial y^l) - \\ - (\partial\Gamma_j{}^i{}_k/\partial x^h + q^l{}_{,h}\partial\Gamma_j{}^i{}_k/\partial y^l) + \Gamma_m{}^i{}_k\Gamma_j{}^m{}_h - \Gamma_m{}^i{}_h\Gamma_j{}^m{}_k \tag{1.77}$$

and $\Gamma_j{}^i{}_h = \Gamma_j{}^i{}_h(x,y)$. The tensor (1.77) is called *the relative curvature tensor* in view of the fact that it depends on an arbitrary vector field $q^l(x)$ together with its derivatives $q^l{}_{,k} = \partial q^l/\partial x^k$. However, $q^l{}_{,k}$ has the same transformation law as $G_k{}^l(x,q)$ because the δ-derivative $q^l_{;k} = q^l{}_{,k} + G_k{}^l(x,q)$ (see Equations (1.31) and (1.33)) is a tensor. This circumstance permits one to conclude from (1.77) that the object

$$K_j{}^i{}_{hk}(x,y) = (\partial\Gamma_j{}^i{}_h/\partial x^k - G_k{}^l\partial\Gamma_j{}^i{}_h/\partial y^l) - \\ - (\partial\Gamma_j{}^i{}_k/\partial x^h - G_h{}^l\partial\Gamma_j{}^i{}_k/\partial y^l) + \Gamma_m{}^i{}_k\Gamma_j{}^m{}_h - \Gamma_m{}^i{}_h\Gamma_j{}^m{}_k \tag{1.78}$$

is a tensor. This tensor, defined as a function of two independent variables x^i and y^i, is called *the K-tensor of curvature.* Further, utilising the Cartan covariant derivative (1.41),

$$X^i{}_{|h|k} - X^i{}_{|k|h} = R_j{}^i{}_{hk} X^j - K_r{}^j{}_{hk} l^r X^i|_j, \tag{1.79}$$

where

$$R_j{}^i{}_{hk} = K_j{}^i{}_{hk} + C_j{}^i{}_m K_r{}^m{}_{hk} y^r. \tag{1.80}$$

The tensors S_{ijmn}, P_{ijmn}, and R_{ijmn} are called *the curvature tensors of Cartan.* The equality (1.80) shows that R_{ijmn} is expressible in terms of the simpler tensor K_{ijmn}. It is also of interest to note that, owing to the skew-symmetry of $K_r{}^j{}_{hk}$ and the symmetry of $C^{hk}{}_m$ in the indices h, k, Equation (1.80) implies that

$$R = K, \tag{1.81}$$

where $R = R^{hk}{}_{hk}$ and $K = K^{hk}{}_{hk}$.

It is apparent from the above definitions that all the curvature tensors, except for P_{ijmn}, are skew-symmetric in the last two indices. Moreover, we have

$$R_{ijhk} = -R_{jihk}, \qquad S_{ijhk} = -S_{jihk}, \tag{1.82}$$

together with (1.75). Equations (1.80) and (1.82) show that

$$K_{jihk} = -K_{ijhk} - 2C_{ijl} K_r{}^l{}_{hk} y^r. \tag{1.83}$$

On the other hand the curvature tensor K_{ijhk}, in contrast to R_{ijhk}, satisfies the cyclic identity

$$K_{jihk} + K_{kijh} + K_{hikj} = 0 \tag{1.84}$$

which is of the same form as in Riemannian geometry. In general, $R_{ijhk} \neq R_{hkij}$. Nevertheless, the equality

$$(R_{ijhk} - R_{hkij}) y^i y^h = 0$$

is valid.

Finally, commutating the Berwald covariant derivative (1.51) leads to *the Berwald curvature tensor*

$$\begin{aligned} H_h{}^i{}_{jk} = {} & \partial G_h{}^i{}_j / \partial x^k - \partial G_h{}^i{}_k / \partial x^j + G_h{}^r{}_j G_r{}^i{}_k - \\ & - G_h{}^r{}_k G_r{}^i{}_j + G_j{}^r \partial G_r{}^i{}_h / \partial y^k - G_k{}^r \partial G_r{}^i{}_h / \partial y^j \equiv \\ & \equiv K_h{}^i{}_{jk} + y^r \partial K_r{}^i{}_{jk} / \partial y^h \end{aligned} \tag{1.85}$$

(cf. (1.80)) which also satisfies a cyclic identity of the form (1.84).

Up to now the vector y^i has been treated as a variable independent of x^i. If, however, we have a vector field $y^i(x)$ the Finslerian metric tensor gives rise to *the osculating Riemannian metric tensor*

$$a_{ij}(x) = g_{ij}(x, y(x)). \tag{1.86}$$

The Riemannian Christoffel symbols

$$\left\{ {i \atop k\ j} \right\} = \tfrac{1}{2} a^{ih}(\partial a_{jh}/\partial x^k + \partial a_{kh}/\partial x^j - \partial a_{kj}/\partial x^h) \tag{1.87}$$

will be related to the Finslerian Christoffel symbols (1.26) as follows

$$\begin{aligned} \left\{ {i \atop k\ j} \right\} &= \gamma_k{}^i{}_j(x, y(x)) + y^l{}_{,k} C_j{}^i{}_l(x, y(x)) + \\ &\quad + y^l{}_{,j} C_k{}^i{}_l(x, y(x)) - g^{ih}(x, y(x)) y^l{}_{,h} C_{kjl}(x, y(x)), \end{aligned} \tag{1.88}$$

where $y^l{}_{,k} = \partial y^l(x)/\partial x^k$. The subsequent substitution of (1.88) in the Cartan connection coefficients (1.32) when $y^i = y^i(x)$ gives

$$\begin{aligned} \Gamma_{ikj}(x, y(x)) &= \left\{ ikj \right\} - C_{jkh}(x, y(x)) y^h_{;i} - \\ &\quad - C_{ikh}(x, y(x)) y^h_{;j} + C_{ijh}(x, y(x)) y^h_{;k} \end{aligned} \tag{1.89}$$

which permits one to compare the δ-derivative (1.36) with the Riemannian covariant derivative D_i associated with the osculating Riemannian metric tensor (1.86), for example

$$y^h_{;i}(x, y(x)) = \mathrm{D}_i y^h(x) - C_i{}^h{}_m(x, y(x)) y^n(x) \mathrm{D}_n y^m(x). \tag{1.90}$$

1.2. The Indicatrix

In each tangent space M_x the Finslerian metric function $F(x, y)$ defines a $(N-1)$-dimensional hypersurface

$$F(x, y) = 1, \tag{1.91}$$

where x^i are assumed to be fixed any y^i are arbitrary, called *the indicatrix*. We see from this definition that the indicatrix is formed by the end points of the unit vectors $l^i(x, y) \stackrel{\text{def}}{=} y^i/F(x, y)$ supported by the point x^i. Owing to the homogeneity condition (1.2), the indicatrix at a point x^i unambiguously defines the Finslerian metric function $F(x, y)$ at this x^i. In the Riemannian case, the definition (1.91) evidently represents a sphere.

Let u^a, $a = 1, 2, \ldots, N-1$, be a coordinate system on the indicatrix supported by some point x^i. Since the Finslerian metric function is assumed to be single-valued, there should exist a one-to-one correspondence between u^a and the unit vectors l^i. Therefore, u^a, like l^i itself, may be treated as zero-degree homogeneous functions of y^i.

This remark suggests the idea that a parametrical representation of the indicatrix may be obtained through a set of $N-1$ scalars

$$u^a = u^a(x, y), \qquad a = 1, 2, \ldots, N-1 \tag{1.92}$$

where the following conditions are assumed:

(i) The positive zero-degree homogeneity of u^a with respect to y^i, i.e.

$$u^a(x, ky) = u^a(x, y), \qquad k > 0. \tag{1.93}$$

(ii) The matrix having entries $\partial u^a/\partial y^i$ is of the highest rank:

$$\operatorname{rank}(\partial u^a/\partial y^i) = N - 1. \tag{1.94}$$

As regards the smoothness of u^a, it is appropriate for our purposes to stipulate that the functions (1.92) obey the same differentiability conditions as those imposed on the Finslerian metric function in preceding Section 1.1, namely we assume

(iii) Smoothness of at least classes C^3 and C^5 with respect to x^i and y^i respectively.

If we add the function $\ln F(x, y)$ to the set of our $N - 1$ functions u^a in order to get a complete set of N functions $z^P(x, y)$, $P = 0, 1, \ldots, N - 1$, so that

$$z^a(x, y) = u^a(x, y), \qquad z^0(x, y) = \ln F(x, y), \tag{1.95}$$

then the matrix having entries $\partial z^P/\partial y^i$ will be of the highest rank N:

$$\operatorname{rank}(\partial z^P/\partial y^i) = N. \tag{1.96}$$

Indeed, let us assume the opposite, i.e. that $\operatorname{rank}(\partial z^P/\partial y^i) < N$ for some x^i and y^i. This assumption means that there exists a set of nonvanishing numbers A_P such that $A_P \partial z^P/\partial y^i = 0$ for our x^i and y^i. Contraction of the latter equality with l^i gives $A_0 = 0$ as a consequence of $l^i \partial z^0/\partial y^i = l^i F^{-1} \partial F/\partial y^i = 1/F$ together with the identity $y^i \partial u^a/\partial y^i = 0$ which follows from the homogeneity (1.93) of u^a. Therefore $A_a \partial u^a/\partial y^i = 0$ which contradicts the condition (1.94). Thus (1.96) is valid.

By virtue of (1.95), the functions z^P satisfy condition (iii). Therefore the inverse function theorem may be employed to conclude from (1.96) that the functions $z^P(x, y)$ are invertible with respect to y^i for any x^i, that is that there exist functions $w^i(x, z)$ of at least class C^3 with respect to x^i and of at least class C^5 with respect to y^i such that

$$y^i = w^i(x, z) \tag{1.97}$$

and

$$z^P(x, w(x, z)) = z^P. \tag{1.98}$$

As a result of our choice $z^0 = \ln F$ relation (1.97) gives the unit y^i in the case when the functions w^i in (1.97) are evaluated for $z^0 = 0$. Therefore, denoting

$$t^i(x, u) = w^i(x^n, 0, u^a), \tag{1.99}$$

we get the relation

$$l^i = t^i(x, u) \tag{1.100}$$

which represents the indicatrix parametrically. Thus we have the following:

THEOREM. Any set of $N - 1$ scalars of the form $u^a(x, y)$ satisfying the conditions (i)–(iii) may be treated as the representation of a coordinate system of the indicatrix. With the help of these scalars the indicatrix may be represented in the parametrical form (1.100) where the functions $t^i(x, u)$ are at least of classes C^3 and C^5 with respect to x^i and u^a respectively.

The quantities

$$t^i_a = \partial t^i/\partial u^a, \qquad u^a_i = \partial u^a/\partial y^i \tag{1.101}$$

are called *the projection factors*[4]. They obey the following series of important identities

$$t^i_a l_i = 0, \qquad u^a_i y^i = 0, \qquad t^i_a u^b_i = \delta^b_a/F, \qquad t^j_a u^a_i = h_i{}^j/F \tag{1.102}$$

the first and the third of which are obtained by differentiating the identities $F(x, t(x, u)) = 1$ and $u^a(x, t(x, u)) = u^b$ with respect to u^a, while the second identity follows from the homogeneity condition (1.93). As regards the fourth identity in (1.102), we can verify it as follows:

$$t^j_a u^a_i = (\partial t^j/\partial u^a)(\partial u^a/\partial y^i) = \partial l^j/\partial y^i = h_i{}^j/F,$$

where in the last step we have used (1.11). Also, the representation (1.100) entails, in view of the equalities $l^i = y^i/F$ and $z^0 = \ln F$, that

$$w^i = t^i \exp z^0 \tag{1.103}$$

from which we find

$$\partial w^i/\partial z^0 = w^i, \qquad \partial w^i/\partial z^a = t^i_a \exp z^0. \tag{1.104}$$

Since the rank of the matrix $(\partial z^P/\partial y^i)$ equals N (Equation (1.96)), nothing prevents us from treating the relation

$$z^P = z^P(x, y) \tag{1.105}$$

as a coordinate transformation made in the tangent space M_x at point x^i, namely as the transformation in M_x from the rectilinear coordinate system y^i to the curvilinear coordinate system z^P. The behaviour of the Finslerian metric tensor under such a transformation may naturally be defined by means of the tensor law, namely

$$g_{PQ}(x, y) \stackrel{\text{def}}{=} g_{ij}(x, t(x, z))\frac{\partial w^i}{\partial z^P}\frac{\partial w^j}{\partial z^Q}, \tag{1.106}$$

where the functions w^i are defined by (1.97). In writing (1.106) we have made use of the circumstance that the metric tensor g_{ij} is homogeneous of degree zero with respect to y^i so that $g_{ij}(x, y)$ is equal to $g_{ij}(x, l)$ or, in view of (1.100), to $g_{ij}(x, t)$. It is obvious from the definition (1.106) and equations (1.102)–(1.104) that

$$g_{00} = e^{2z^0}, \qquad g_{0a} = 0. \tag{1.107}$$

The remaining components of tensor (1.106) will be of the form

$$g_{ab} = e^{2z^0} g_{ij} t^i_a t^j_b \tag{1.108}$$

as a consequence of (1.104).

Formulae (1.107) and (1.108) indicate that the tensor g_{PQ} is conformal to the simpler tensor g^*_{PQ}, that is

$$g_{PQ} = e^{2z^0} g^*_{PQ}, \tag{1.109}$$

where

$$g^*_{ab} = g_{ij}(x, t(x, u)) t^i_a(x, u) t^j_b(x, u) \tag{1.110}$$

and

$$g^*_{00} = 1, \qquad g^*_{0a} = 0 \tag{1.111}$$

which in turn implies that

$$g^{*00} = 1, \qquad g^{*0a} = 0. \tag{1.112}$$

We observe from (1.110) and (1.111) that the tensor g^*_{PQ} is independent of z^0 and therefore that the tensor g_{PQ} depends on z^0 only through the conformal factor $\exp 2z^0$ entering (1.109).

The object g^*_{ab} defined by (1.110) is called *the induced Riemannian metric tensor of the indicatrix*. This tensor is intimately related to the angular metric tensor h_{ij} (Equation (1.12)). In fact, the first identity in (1.102) permits us to rewrite (1.110) as

$$g^*_{ab} = h_{ij} t^i_a t^j_b. \tag{1.113}$$

Conversely, upon contracting (1.110) with $u^a_m u^b_n$ and taking into account the fourth identity in (1.102), we get

$$g^*_{ab} u^a_m u^b_n = g_{ij} h^i_m h^j_n / F^2 = h_m{}^i h_{in} / F^2$$

or, by virtue of (1.13),

$$g^*_{ab} u^a_m u^b_n = h_{mn} / F^2. \tag{1.114}$$

Equations (1.113) and (1.114) imply in turn that

$$F g^*_{ab} u^a_m = g_{mn} t^n_b. \tag{1.115}$$

Let us denote by $R_{PQRS}$and R^*_{PQRS} the Riemannian curvature tensors constructed from the tensors, $g_{PQ}(x, z)$ and $g^*_{PQ}(x, z)$ respectively, by differentiating them with respect to z^P, the variable x^i being treated as a parameter. Taking equalities (1.111) together with the fact that the tensor (1.110) is independent of z^0 evidently implies that the only non-vanishing components of the tensor R^*_{PQRS} are

$$R^*_{abcd} = R^*_{abcd}(x, u) \tag{1.116}$$

which means that R^*_{abcd} equals the Riemannian curvature tensor constructed from $g^*_{ab}(x, u)$ by differentiating it with respect to u^a. Accordingly, quantity (1.116) *plays the role of the Riemannian curvature tensor of the indicatrix*.

The tensor R_{PQRS} may be readily expressed in terms of R^*_{abcd} by means of the well-known relation between the Riemannian curvature tensors of conformal Riemannian spaces[5] which in our notation reads

$$e^{2U}R^{PQ}{}_{RS} = R^{*PQ}{}_{RS} + Y^Q_R\delta^P_S - Y^P_R\delta^Q_S - Y^Q_S\delta^P_R + Y^P_S\delta^Q_R, \tag{1.117}$$

where

$$\begin{aligned} Y^Q_R = \partial U^Q/\partial z^R + \tfrac{1}{2}g^{*QS}(\partial g^*_{RS}/\partial z^T + \partial g^*_{TS}/\partial z^R - \\ - \partial g^*_{RT}/\partial z^S)U^T - U^Q U_R + \tfrac{1}{2}\delta^Q_R U_T U^T \end{aligned} \tag{1.118}$$

and

$$U_R = \partial U/\partial z^R, \qquad U^R = g^{*RP}U_P, \qquad U = z^0.$$

The last equality implies that $U_R = \delta^0_R$ so that, as a consequence of (1.111)–(1.112) and the fact that g^*_{ab} is independent of z^0, the first two terms on the right-hand side of (1.118) vanish leaving us with just

$$Y^Q_R = -\delta^Q_0\delta^0_R + \tfrac{1}{2}\delta^Q_R.$$

This permits us to conclude from (1.117) without too much labour that

$$e^{-2z^0}R_{abcd} = R^*_{abcd} - g^*_{ac}g^*_{bd} + g^*_{ad}g^*_{bc} \tag{1.119}$$

while the other components of the tensor R_{PQRS} vanish as a result of the disappearance of all the components of R^*_{PQRS} other than R^*_{abcd}.

If our definition of the tensor R_{PQRS} is compared with the expression (1.60) for the tensor $F^{-2}S_{ijmn}$, we observe that these two entities are the Riemannian curvature tensors respectively constructed from $g_{PQ}(x,z)$ and $g_{ij}(x,y)$ by differentiation with respect to z^P and y^i. Since g_{PQ} is related to g_{ij} through the tensor transformation (1.106) these curvature tensors must also be related to each other in accordance with the tensor law, namely,

$$F^{-2}S_{ijmn} = R_{PQRS}z^P_i z^Q_j z^R_m z^S_n, \tag{1.120}$$

where $z_i{}^P = \partial z^P/\partial y^i$. In other words, $F^{-2}S_{ijmn}$ and R_{PQRS} are representations of the same curvature tensor of the tangent Minkowskian space, with respect to two distinct coordinate sets y^i and z^P.

Since R_{abcd} are the only non-vanishing components of R_{PQRS} we may rewrite (1.120) as

$$F^{-2}S_{ijmn} = R_{abcd}u^u_i u^b_j u^c_m u^d_n, \tag{1.121}$$

where we have substituted $z^a = u^a$ in accordance with definition (1.95). If the tensor R_{abcd} is replaced in (1.121) by the curvature tensor R^*_{abcd} of the indicatrix using Equation (1.119) together with the equality $F^{-2} = e^{-2z^0}$ (see (1.95)) we obtain the following result

$$F^{-4}S_{ijmn} = (R^*_{abcd} - g^*_{ac}g^*_{bd} + g^*_{ad}g^*_{bc})u^a_i u^b_j u^c_m u^d_n \tag{1.122}$$

showing the geometrical significance of the curvature tensor S_{ijmn}. If we take into account (1.114) and use the notation

$$R^*_{ijmn} = R^*_{abcd} u^a_i u^b_j u^c_m u^d_n F^4 \tag{1.123}$$

we can rewrite (1.122) in the following elegant form

$$S_{ijmn} = R^*_{ijmn} - h_{im} h_{jn} + h_{in} h_{jm}. \tag{1.124}$$

This last relation entails that for any fixed point x^i and any two vectors Y^a and X^a tangent to the indicatrix at a given point u^a, the curvature

$$R(u, Y, X) = R^*_{abcd} Y^a Y^c X^b X^d / (g^*_{bd} g^*_{ac} - g^*_{bc} g^*_{ad}) Y^a Y^c X^b X^d \tag{1.125}$$

of the indicatrix at u^a with respect to the orientation defined by the two vectors Y^a and X^a is related to the scalar

$$S^*(y, Y, X) = S_{jihk} Y^j Y^h X^i X^k / (g_{jh} g_{ik} - g_{ji} g_{hk}) X^j Y^h X^i X^k \tag{1.126}$$

in accordance with

$$R = 1 + S^*. \tag{1.127}$$

The relations (1.122)–(1.124) indicate that the tensor S_{ijmn} inherits all the distinguishing features of the Riemannian curvature tensor R^*_{abcd} of the indicatrix. An interesting implication arises from this in the case when the Finsler space is four-dimensional so that the indicatrix is a three-dimensional space. Indeed, it is well known that the Riemannian curvature tensor of three-dimensional space has, in terms of our notation, the following form

$$\begin{aligned} R^*_{abcd} = & -\tfrac{1}{2} R^* (g^*_{ac} g^*_{bd} - g^*_{ad} g^*_{bc}) + g^*_{ac} R^*_{bd} - g^*_{bc} R^*_{ad} + g^*_{bd} R^*_{ac} - \\ & - g^*_{ad} R^*_{bc} \end{aligned} \tag{1.128}$$

where $R^*_{ac} = R^{*b}_{a\ \ cb}$ and $R^* = R^{*a}_a$. Equation (1.128) reads simply that the Weyl conformal tensor of three-dimensional Riemannian space vanishes identically[6]. The substitution of (1.128) in (1.122) yields the following result: *the tensor* S_{ijmn} *of a four-dimensional Finsler space has the form*

$$S_{ijmn} = h_{im} M_{jn} - h_{jm} M_{in} + h_{jn} M_{im} - h_{in} M_{jm}. \tag{1.129}$$

Here, $M_{in} = R^*_{in} - \frac{1}{4}(R^* + 2) h_{in}$ where $R^*_{in} = R^{*j}_{i\ \ nj}$.

Similarly, the curvature tensor of a two-dimensional Riemannian space takes the form[7]

$$R^*_{abcd} = \tfrac{1}{2} R^* (g^*_{ac} g^*_{bd} - g^*_{ad} g^*_{bc}). \tag{1.130}$$

Substituting (1.130) in (1.122) permits one to conclude that *the tensor* S_{ijmn} *of a three-dimensional Finsler space is of the form*

$$S_{ijmn} = \tfrac{1}{2} S (h_{im} h_{jn} - h_{in} h_{jm}) \tag{1.131}$$

where

$$S = R^* - 2 \tag{1.132}$$

and the notation $S = S^{ij}{}_{ij}$ has been used. A similar result may be obtained in the case when the indicatrix is a space of constant curvature, because the curvature tensor of a Riemannian space of constant curvature is known to be of the form (1.130) with constant R^{*8}. Therefore, *the indicatrix of a Finsler space with the dimension* $N \geq 3$ *is a space of constant curvature if and only if the tensor* $S_{ijmn}(x, y)$ *is of the form* (1.131) *with the scalar* S *independent of* y^i.

It is noted for future use that the projection factors (1.101) may be used to relate the orthonormal n-tuples of the Finslerian metric tensor $g_{ij}(x, y)$ to the orthonormal n-tuples of the induced metric tensor $g^*_{ab}(x, u)$ of the indicatrix. To this end, let us denote by $g^{*(a)}_{\ b}$ some orthonormal covariant n-tuple of the tensor g^*_{ab}, so that the relation

$$g^*_{ab} = \sum_{c=1}^{N-1} q_c g^{*(c)}_{\ a} g^{*(c)}_{\ b} \tag{1.133}$$

holds, where the set of constants q_c equaling 1 or -1 reflects the signature of the tensor g^*_{ab}. Then it is clear from (1.106), (1.109) and (1.111) that the set of $N - 1$ covariant vectors

$$g^{(a)}_{\ i} = F u^b_i g^{*(a)}_{\ b} \tag{1.134}$$

will be orthonormal with respect to the Finslerian metric tensor. Therefore, the following set of N vectors

$$g^P_i = (l_i, g^{(a)}_{\ i}) \tag{1.135}$$

will represent the orthonormal covariant n-tuple of the Finslerian metric tensor, i.e.

$$g_{ij} = \sum_{P=0}^{N-1} q_P g^P_i g^P_j, \tag{1.136}$$

where $q_0 = 1$.

Similarly, if we denote by $g^{*\ a}_{(c)}$ the contravariant reciprocal to $g^{*(c)}_{\ a}$, so that

$$g^{*ab} = \sum_{c=1}^{N-1} q^c g^{*\ a}_{(c)} g^{*\ b}_{(c)} \tag{1.137}$$

where $q^c = q_c$, and put

$$g^{\ i}_{(a)} = t^i_b g^{*\ b}_{(a)}, \tag{1.138}$$

we find that

$$g^i_P = (l^i, g^{\ i}_{(a)}) \tag{1.139}$$

represents the orthonormal contravariant n-tuple of the Finslerian metric tensor, i.e.

$$g^{ij} = \sum_{P=0}^{N-1} q^P g^i_P g^j_P, \tag{1.140}$$

where $q^0 = 1$. The reciprocity relations

$$g^{*(c)}_{\ a} g^{*\ a}_{\ (b)} = \delta^c_b, \qquad g^{*(c)}_{\ a} g^{*\ b}_{\ (c)} = \delta^b_a, \qquad g^i_P g^Q_i = \delta^Q_P,$$
$$g^i_P g^P_j = \delta^i_j, \qquad g^{\ i}_{(a)} g^{(a)}_{\ j} = \delta^i_j - l_j l^i \tag{1.140'}$$

hold.

1.3. The Group of Invariance of the Finslerian Metric Function

Let us fix for the moment a point x^i and consider some transformation $y'^i = y'^i(y^j)$ of the tangent vectors attached to x^i. Then the requirement that the Finslerian metric function remains invariant under the transformation will read $F(y'(y)) = F(y)$. However, as soon as the limitation that the point x^i is fixed is suppressed, the last equality, as well as the transformation of the form $y'^i(y^j)$, both become meaningless and the correct relations will be

$$y'^i = Z^i(x, y), \tag{1.141}$$

$$F(x, Z(x, y)) = F(x, y) \tag{1.142}$$

for any admissible vectors y^i and y'^i.

To avoid misunderstanding, it should be emphasized that $F(x, Z)$ and $F(x, y)$ in (1.142) is the same function, though of different sets of arguments. In other words, the formal replacement $Z^i \to y^i$ results in $F(x, Z) \to F(x, y)$. Since the Finslerian metric function was defined in Section 1.1 to be a function of at least class C^3 with respect to x^i and of at least class C^5 with respect to y^i, we should impose the same differentiability requirement on the functions $Z^i(x, y)$ in (1.141). Furthermore, we tacitly assume that the range of definition of $Z^i(x, y)$ is contained in the range of definition of $F(x, y)$.

On the assumption that the transformations (1.141) are invertible, i.e. that there exist functions $Z^{*i}(x, y')$ of at least class C^5 with respect to y^i such that the identity

$$y'^i = Z^i(x, Z^*(x, y')) \tag{1.143}$$

holds for any admissible x^i and y'^i, the set of all transformations (1.141) satisfying the condition (1.142) will evidently constitute a group.

In order that we be sure that the transformations (1.141) maintain the homogeneity (1.2) of the Finslerian metric function, it is necessary to explicitly constrain (1.141) by the homogeneity condition, i.e. to require that

$$Z^i(x, ky) = kZ^i(x, y), \qquad k > 0, \tag{1.144}$$

and the same for the inverse functions Z^{*i}.

Accordingly, we introduce the following:

DEFINITION. The group of all invertible transformations (1.141) of at least class C^3 with respect to x^i and of at least class C^5 with respect to y^i which satisfy the homogeneity condition (1.144) and leave the Finslerian metric function invariant in the sense of Equation (1.142) is called *the group of invariance of the Finslerian metric function*, or for brevity *the* G-*group*. The transformations from the G-*group* will be called G-*transformations.*

The group of those G-transformations which are *linear* in y^n will be referred to as *the* K-*group* of invariance of the metric function. According to this definition, the transformations

$$y'^i = K^i_j(x)y^j \tag{1.145}$$

belong to the K-group if and only if

$$F(x, K^i_j(x)y^j) = F(x, y^i) \tag{1.146}$$

from which it directly follows that

$$y'_i = K^{*j}_i(x)\, y_j \tag{1.147}$$

and

$$g^{ij}(x, y') = g^{mn}(x, y)K^i_m(x)K^j_n(x), \tag{1.148}$$

$$g_{ij}(x, y') = g_{mn}(x, y)K^{*m}_i(x)K^{*n}_j(x). \tag{1.149}$$

Such transformations will be called K-*transformations*. The starred coefficients represent the K-transformations inverse to (1.145) so that

$$K^j_i(x)K^{*n}_j(x) = \delta^n_i. \tag{1.150}$$

Generally speaking, the G-transformations need not be linear. The effect of an arbitrary G-transformation on Finslerian tensors may be found by considering implications of the restriction (1.142). First of all, differentiation of the relation $\frac{1}{2}F^2(x, y') = \frac{1}{2}F^2(x, y)$ with respect to y'^i yields

$$y'_i = y_n y^n_i, \tag{1.151}$$

where $y'_i = \frac{1}{2}\partial F^2(x, y')/\partial y'^i$. The following notation will be used below

$$y^n_i = \partial Z^{*n}/\partial y'^i, \qquad y^n_{ij} = \partial y^n_i/\partial y'^j, \qquad y^n_{ijk} = \partial y^n_{ij}/\partial y'^k. \tag{1.152}$$

These are evidently tensors and, by the homogeneity condition (1.144), satisfy the identities

$$y'^i y^n_i = y^n, \qquad y'^i y^n_{ij} = 0, \quad y'^i y^n_{ijk} = -y^n_{jk}. \tag{1.153}$$

The differentiation of (1.143) with respect to y'^j yields

$$y^n_j \partial y'^i/\partial y^n = \delta^i_j. \tag{1.154}$$

Equation (1.151) says that the covariant tangent vectors y_n transform as vectors under G-transformations. Such a conclusion, however, cannot be extended to the Finslerian metric tensor because the differentiation of (1.151) with respect to y'^j shows that

$$g_{ij}(x, y') = g_{mn}(x, y)y_i^m y_j^n + y_n y_{ij}^n. \tag{1.155}$$

We observe from Equation (1.155) that the metric tensor obeys the tensor transformation law

$$g_{ij}(x, y') = g_{mn}(x, y)y_i^m y_j^n \tag{1.156}$$

under some G-transformation only if

$$y_n y_{ij}^n = 0. \tag{1.157}$$

We shall call such G-transformations *metric*. The condition (1.157) selects from the G-group *the metric subgroup* G_m containing all metric G-transformations. Evidently, any G-transformation linear in y^j in each tangent space (for example, the Euclidean rotations in Riemannian geometry) will be metric in view of $y_{ij}^n = 0$. An interesting example of non-linear metric G-transformations will be given in Section 2.3 where the Berwald-Moór metric function will be considered.

Further, differentiating equation (1.155) with respect to y'^k yields

$$\begin{aligned} C_{ijk}(x, y') = {} & C_{mnl}(x, y)y_i^m y_j^n y_k^l + \tfrac{1}{2}g_{mn}(x, y)(y_i^m y_{jk}^n + \\ & + y_j^m y_{ik}^n + y_k^m y_{ij}^n) + \tfrac{1}{2}y_n y_{ijk}^n, \end{aligned} \tag{1.158}$$

which reduces in the case of metric G-transformations to

$$C_{ijk}(x, y') = C_{mnl}(x, y)y_i^m y_j^n y_k^l + \tfrac{1}{2}g_{mn}(x, y)(y_i^m y_{jk}^n + y_j^m y_{ki}^n). \tag{1.159}$$

The condition

$$g_{mn}(x, y)(y_i^m y_{jk}^n + y_j^m y_{ik}^n) = 0 \tag{1.160}$$

is necessary and sufficient for the Cartan torsion tensor C_{mnl} to have tensor character under metric G-transformations. The group of all metric G-transformations satisfying Equation (1.160) will be denoted by G_{mt}.

Another series of identities can be obtained by appropriate differentiations with respect to x^i. For instance, differentiating (1.142) at a fixed value of y'^i yields

$$\partial F(x, y')/\partial x^i = \partial F(x, y)/\partial x^i + l_m \partial Z^{*m}(x, y')/\partial x^i.$$

On substituting in the latter equation the relations

$$\partial F(x, y)/\partial x^i = l_m G_i{}^m(x, y)$$

(see the definition (1.44) and the third equality in (1.47)) and

$$\partial F(x, y')/\partial x^i = l'_m G_i{}^m(x, y') = l_m y_j^m G_i{}^j(x, y')$$

(Equations (1.142) and (1.151) imply $l'_i = l_m y_i^m$), we get the following result:

$$l_m(\partial Z^{*m}(x, y')/\partial x^i + G_i{}^m(x, y) - y_j^m G_i{}^j(x, y')) = 0. \tag{1.161}$$

It will be noted that G-transformations leave the Hamiltonian function $H(x, y_i) \stackrel{\text{def}}{=} F(x, y^n(x, y_i))$ (Equation (6.18)) invariant. Indeed, if $y'^i = Z^i(x, y^n)$, then

$$H(x, y'_i) = F(x, y'^i) = F(x, y^i) = H(x, y_i),$$

so that the invariance property

$$H(x, y'_i) = H(x, y_i) \tag{1.162}$$

is certainly valid. Similarly to Equations (1.151) and (1.155) we find, after differentiating the square of relation (1.162) with respect to y'_i and then y'_j, that

$$y'^i = y^n \partial y_n / \partial y'_i, \tag{1.163}$$

$$g^{ij}(x, y') = g^{mn}(x, y) \frac{\partial y_m}{\partial y'_i} \frac{\partial y_n}{\partial y'_j} + y^n \frac{\partial^2 y_n}{\partial y'_i \partial y'_j}, \tag{1.164}$$

where we have substituted $y^n = H \partial H / \partial y_n$ (Equation (6.19)) and $g^{mn} = \frac{1}{2} \partial^2 H^2 / \partial y_m \partial y_n$ (Equation (6.20)). By (1.151),

$$\partial y'_i / \partial y_m = \partial y^m / \partial y'^i + y_n \partial^2 y^n / \partial y_m \partial y'^i, \tag{1.165}$$

whence the following assertion is valid; *if a G-transformation is metric, then*

$$\partial y'_i / \partial y_m = \partial y^m / \partial y'^i, \qquad \partial y_m / \partial y'_i = \partial y'^i / \partial y^m. \tag{1.166}$$

Indeed, the Equation (1.157) characteristic of the metric case results in the vanishing of the second term in the right-hand side of (1.165), which leads to the first equality in (1.166). The second equality in (1.166) follows from the first one by virtue of (1.154).

Moreover, on substituting (1.163) in the right-hand side of the second equality in (1.166), we may conclude that *the characteristic equation* (1.157) *of the metric case is equivalent to*

$$y^n \partial^2 y_n / \partial y'_i \partial y'_j = 0.$$

This implies that Equation (1.164) reduces in the metric case to

$$g^{ij}(x, y') = g^{mn}(x, y) \frac{\partial y_m}{\partial y'_i} \frac{\partial y_n}{\partial y'_j}$$

which is similar to (1.156).

In accordance with the definition (1.91) of the indicatrix, the G-transformations defined by Equations (1.141) and (1.142) may be understood as representations of mappings of the indicatrix in itself. In terms of the parameters u^a entering the parametrical representation (1.100) of the indicatrix, any such mapping may be represented by $N - 1$ functions

$$u^a = u^a(x, u') \tag{1.167}$$

which may be arbitrary except for the conditions that they are of at least classes C^3 and C^5 with respect to x^i and u'^a respectively and that the matrix having entries $\partial u^a/\partial u'^b$ is of the highest rank $N-1$. The last condition ensures the existence of the inverse of (1,167):

$$u'^a = u'^a(x, u) \tag{1.168}$$

which is of the same smoothness as the functions (1.167). Therefore knowledge of the explicit form of at least one parametrical representation (1.100) of the indicatrix permits one to write the following explicit form of *any* G-transformation:

$$y'^i = F(x, y)t^i(x, u'^a(x, u(x, y))). \tag{1.169}$$

The metric G-group proves to be intimately related to the group of motions (isometries) of the tangent Minkowskian spaces $M(x)$. Indeed, the equality (1.156) means that the metric G-transformation induces motions[9] in $M(x)$. Conversely, any transformation of the form (1.141) represents a G-transformation under the conditions that the transformation (1.141) is at least of classes C^3 and C^5 with respect to x^i and y^i respectively, is homogeneous (Equation (1.144)), and induces motions in $M(x)$, i.e. satisfies Equation (1.156). To prove this assertion, that is that Equation (1.142) holds under these conditions, it is sufficient to note that the contraction of Equation (1.156) with $y'^i y'^j$ gives $F^2(x, y) = F^2(x, y')$ because of the relation (1.4) and the first identity in (1.153) resulting from the homogeneity condition (1.144).

Following the introduction to Finsler geometry given in this chapter we shall proceed in the next chapter to describe a number of interesting special types of Finsler spaces.

Problems

1.1. A tensor of the form $g_{ij}(x, y)$ will not necessarily be a Finslerian metric tensor if it satisfies only the following four conditions: the tensor is symmetric in its subscripts, positively homogeneous of degree zero in y^i, non-degenerate (i.e. $\det(g_{ij}) \neq 0$), and $y^i y^j g_{ij}(x, y) > 0$ for all non-zero vectors y^i. What fifth condition is required?

1.2. Given the field of unit vectors $l^i(x)$ tangent to a congruence of Finslerian geodesics, and writing $l_k(x) = g_{ki}(x, l(x))l^i(x)$, show that

$$l_{k;\,j}(x, l)l^k = 0.$$

Verify that this relation together with the equation

$$l^k_{;j}(x, l)l^j = 0$$

(the congruence consists of geodesics) imply that

$$l^j(x)(\partial l_k(x)/\partial x^j - \partial l_j(x)/\partial x^k) = 0,$$

i.e. that a congruence of Finslerian geodesics can be described by an equation of just the same form as in the case of Riemannian geodesics.

1.3. A Riemannian space is said to be symmetric if the covariant derivative of the Riemannian curvature tensor vanishes. When applied to the indicatrix this reads $D^*_e R^*_{abcd} = 0$. Following up the considerations of Section 1.2, show that the indicatrix is a symmetric space if and only if

$$S_{ijmn}|_k = -l_i S_{kjmn} - l_j S_{ikmn} - l_m S_{ijkn} - l_n S_{ijmk}.$$

1.4. Using Equations (1.107)–(1.110), find explicitly the relationship between the Killing equations written in the indicatrix and in the tangent Minkowskian space.

1.5. Prove that the tensor S_{ijmn} vanishes identically in two-dimensional Finsler spaces.

1.6. Prove the assertion: if tangent Minkowskian spaces are of constant curvature, than the tensor S_{ijmn} vanishes identically.

1.7. Consider the case when the indicatrix is a conformally flat space.

1.8. In general relativity theory an analysis is often based on the concept of an extended frame of reference. Geometrically such a frame is represented by a vector field whose covariant derivative is decomposed in terms of acceleration, rotation, shear, and expansion (see e.g. Equation (6.13) in Kramer *et al.* (1980)). Perform the corresponding decomposition in the Finslerian case by using the δ-derivative.

1.9. Given a field $X_i(x, y)$ which is tensorial under the metric G-transformations, i.e. $X_i(x, y') = X_j(x, y) y^j_i$, show that the Minkowskian covariant derivative (the definition (1.64)) of X_i is a tensor under such transformations, i.e. $X_{ij}(x, y') = X_{mn}(x, y) y^m_i y^n_j$, where $X_{mn} = D^{\text{Mink}}_n X_m$.

1.10. Formulate the Finslerian scalar, spinor and electromagnetic field equations in the tangent Minkowskian space.

1.11. Considering the metric functions $F_1 = [(y^1)^f + (y^2)^f + (y^3)^f + (y^4)^f]^{1/f}$ and $F_2 = [(y^1)^f - (y^2)^f - (y^3)^f - (y^4)^f]^{1/f}$ (cf. the definition (2.9)), indicate a parametrical representation of the indicatrix such that the metric tensor g^*_{ab} of the indicatrix becomes diagonal.

1.12. Indicate the metric G-transformations for the metric functions F_1 and F_2 mentioned in the preceding Problem 1.11.

Notes

[1] The apparatus of Finsler geometry is generalized to complex manifolds in Rund (1967).

[2] The theory of indefinite Finsler spaces has been developed by Busemann (1967), Beem (1970, 1973, 1976), Beem and Kishta (1974).

[3] Berwald (1946); Vagner (1949, p. 123).
[4] With the notation used in Section 8 of Chapter V of Rund (1959) we should have written b^i_α and B^α_i instead of our t^i_a and u^a_i.
[5] See e.g. Kramer *et al.* (1980, p. 55).
[6] See e.g. Eisenhart (1950).
[7] *Op. cit.*
[8] *Op. cit.*
[9] For the general definition of motion in a Riemannian space see, for example, Eisenhart (1950) or Kramer *et al.* (1980).

Chapter 2

Special Finsler Spaces

The definition of the Finslerian metric function given in Chapter 1 is of course too general for specific applications. Thus, for example, Riemannian geometry results from a specific choice of the form of the Finslerian metric function. It would be natural to isolate, for the most careful study, those subtypes of Finsler spaces which are nontrivial, yet sufficiently simple for the purposes of particular applications. Such an isolation can be achieved by various methods: by postulating a particular form for a typical Finslerian tensor, a symmetry property of the indicatrix, or the nature of the dependence of the Finslerian metric function on x^i or y^i, etc., not mentioning the choice of any particular metric function. Many special cases have been studied in the literature. In the present chapter, we shall describe a set of special Finsler spaces suitable for the applications studied in this volume, it being necessary for this purpose to develop a rather detailed level of exposition. Most of the formulae that are derived are used in succeeding chapters, while the remainder serve to provide the reader with a helpful guide. All the geometrical definitions and results which will be used in Part C are contained in Sections 2.1–2.4 of the present chapter, the essence of these sections consisting in 19 Propositions proved in all rigour. Three additional Sections 2.5–2.7 present a brief description of some other speical types of Finsler spaces which are attractive in view of their applicability to physical theories.

2.1. S3-like Finsler Spaces

The circumstance that the tensor S_{ijmn} of any three-dimensional Finsler space has form (1.131) naturally suggests defining a special type of Finsler space according to the following:[1]

DEFINITION. A Finsler space of dimension $N \geqslant 4$ is called *S3-like* if there exists a scalar S such that the tensor S_{ijmn} is of the form

$$S_{ijmn} = \frac{S}{(N-1)(N-2)}(h_{im}h_{jn} - h_{in}h_{jm}). \tag{2.1}$$

The contraction of (2.1) yields, after using Equations (1.12) and (1.13),

$$S_{im} = Sh_{im}/(N-1), \qquad S^i_i = S \tag{2.2}$$

where $S_{im} \overset{\text{def}}{=} S_i{}^j{}_{mj}$. Since $F^{-2}S_{ijmn}$ given by (1.60) is the Riemannian curvature tensor of the tangent Minkowskian space, the contracted Bianchi identities applied to the tensor $S'^{ij} = F^{-2}(S^{ij} - \frac{1}{2}Sg^{ij})$ will read simply $D^{\text{Mink}}_j S'^{ij} = 0$ or, in terms of the derivative $|_j = FD^{\text{Mink}}_j$ (Equation (1.66)),

$$(F^{-2}(S^{ij} - \tfrac{1}{2}Sg^{ij}))|_j = 0$$

which, after substituting (2.2) and then using the fourth identity (1.68) together with $F^{-2}|_j = -2l_jF^{-2}$, takes the following simple form

$$F^{-1}\left(\frac{1}{N-1}h^{ij} - \tfrac{1}{2}g^{ij}\right)\partial S/\partial y^j = 0.$$

Since the scalar S in (2.1), as well as the tensor S_{ijmn} itself, is homogeneous of degree zero with respect to y^i, the contraction $l^j\partial S/\partial y^j$ will vanish and the above equality implies $\partial S/\partial y^j = 0$. Thus the scalar S entering (2.1) is independent of y^i:

$$S = S(x). \tag{2.3}$$

The condition of $S3$-likeness has a clear geometrical significance. The necessary and sufficient condition for a Finsler space to be $S3$-like is that the indicatrix should be a space of constant curvature, as is clear from the result stated below equation (1.132). A comparison of (2.1) with (1.126) shows that the scalar S^* determining the curvature R (Equation (1.125)) of the indicatrix in accordance with (1.127) is equal to

$$S^* = S/(N-1)(N-2). \tag{2.4}$$

It will also be recalled that any Riemannian space of constant curvature is conformally flat and symmetric[2] from which it directly follows that all indicatrices associated with an $S3$-like Finsler space are conformally flat and symmetric and, therefore, that the derivative $S_{ijmn}|_k$ is given by the relation indicated in Problem 1.3.

Additional information concerning $S3$-like Finsler spaces suggests the following propostion[3].

PROPOSITION 1. *If the tangent Minkowskian spaces of a Finsler space of dimension $N \geqslant 4$ are conformally flat and the conformal multiplier $p(x, y)$ is of the form $p = z(F(x, y), x)$ (z being a scalar function), then*

$$p(x, y) = b(x)(F(x, y))^{2a(x)} \tag{2.5}$$

where $a(x)$ and $b(x)$ are scalar functions and the Finsler space is $S3$-like with

$$S^* = a(a+2). \tag{2.6}$$

Proof. The representation $p = z(F, x)$ of the conformal multiplier p makes it possible to express the condition that the Weyl conformal tensor[4] vanishes in the following form

$$F^{-2}S_{mnij} + ((q')^2 + 2q'F^{-1})(h_{mj}h_{ni} - h_{mi}h_{nj}) + \\ + (q'' + q'F^{-1})(h_{mj}l_n l_i + h_{ni}l_m l_j - h_{mi}l_n l_j - h_{nj}l_m l_i) = 0 \tag{2.7}$$

where $2q = \ln p$, $q' = \partial q/\partial F$, $q'' = \partial q'/\partial F$. Since the tensor S_{mnij} must satisfy the identity $S_{mnij}l^m = 0$ (see Equations (1.62) and (1.16)) and, at the same time, $l^m h_{mj} = 0$ in accordance with (1.13), Equation (2.7) entails $q'' + q'F^{-1} = 0$, which may be readily integrated to give (2.5). Under these conditions a comparision of (2.7) with (2.1) shows that the Finsler space is indeed $S3$-like and that the scalar (2.4) equals

$$((q')^2 + 2q'F^{-1})F^2. \tag{2.8}$$

Because of (2.5), the expression (2.8) turns out to be (2.6). The proof is complete.

At present, a fairly broad class of metric functions of $S3$-like Finsler spaces is known. Let $S^A_m(x)$ be a field of covariant n-tuples. Denote by $f(x)$ and $r^A(x)$ scalar functions such that $f \neq 0, 1, 2$ and $r^A \neq 0$ at any point x^i. Then the following proposition is valid[5].

PROPOSITION 2. *Finsler spaces of dimension $N \geqslant 4$ with metric functions of the form*

$$F(x, y) = \left[\sum_{A=1}^{N} r^A(x)(S^A_m(x)y^m)^{f(x)} \right]^{1/f(x)} \tag{2.9}$$

or

$$F(x, y) = \prod_{A=1}^{N} (S^A_m(x)y^m)^{r^A(x)}, \qquad \sum_{A=1}^{N} r^A(x) = 1 \tag{2.10}$$

are $S3$-like.

Proof. Let us fix some point x^i and consider the n-tuple $S^A_m(x)$ as a basis in the tangent space M_x. Then we obtain for (2.9) and (2.10) simultaneously:

$$l_A = \partial F/\partial y^A = r^A(l^A)^{f-1} = g_{AB}l^B, \tag{2.11}$$

$$g_{AB} = \tfrac{1}{2}\partial^2 F^2/\partial y^A \partial y^B = (2-f)l_A l_B + (f-1)\delta_{AB}l_A(l^A)^{-1}, \tag{2.12}$$

$$g^{AB} = \frac{f-2}{f-1}l^A l^B + \frac{1}{f-1}\delta^{AB}l^A(l_A)^{-1}, \tag{2.13}$$

$$C_{ABD} = \tfrac{1}{2}\partial g_{AB}/\partial y^D = \frac{2-f}{2F}(g_{AB}l_D + g_{BD}l_A + g_{DA}l_B + \\ + (f-4)l_A l_B l_D - (f-1)\delta_{AB}\delta_{BD}\delta_{DA}l_A(l^A)^{-2}). \tag{2.14}$$

The case $f = 0$ corresponds to the metric function (2.10). By substituting (2.11)–(2.14) in the general expression (1.62) for the tensor S_{ijmn}, we find after straightforward calculations that the equality (2.1) holds with

$$S^* = (f - 2)^2/4(f - 1). \tag{2.15}$$

The proof is complete.

As a consequence of Equations (2.15) and (1.127), the curvature R of the indicatrix will be

$$R = f^2/4(f - 1). \tag{2.16}$$

Since $f = 0$ corresponds to the case (2.10), we infer from (2.16) the following result.

PROPOSITION 3. Metric function of the form (2.10) represent non-trivial examples of Finsler spaces with indicatrices of vanishing curvature.

The expression (2.15) obeys the inequalities $S \geqslant 0$ if $f > 1$ and $S < -1$ if $f < 1$. Since $a(a + 2) \geqslant -1$, we can conclude after equating (2.6) to (2.15) that the function (2.5) with

$$a = -1 \pm (1 + (f - 2)^2/2(f - 1))^{1/2}$$

is the conformal multiplier for the metric function (2.9) when $f > 1$ and for the metric function (2.10) when $f = 0$. If a $S3$-like Finsler space has a metric function (2.9) with $f < 1$, then the conformal multiplier cannot be of the form (2.6). Examples of $S3$-like Finsler spaces with $-1 < S < 0$ are not yet known; they should have a conformal multiplier of the form (2.6) with $-1 < a < 1$.

2.2. Spaces with Quadratic Dependence of the Finslerian Metric Tensor on the Unit Tangent Vectors

A Finslerian metric tensor g^{ij} is said to be *quadratic in the unit tangent vectors* $l^n \overset{\text{def}}{=} y^n/F$ if g^{ij} can be represented as[6]

$$g^{ij} = Q^{ij}_{mn}(x) l^m l^n, \tag{2.17}$$

where $Q^{ij}_{mn}(x)$ is a tensor symmetric in the index pair (i,j) as well as in (m,n). Since $l^i = g^{ij} l_j$, we get

$$l^i = Q^{ij}_{mn} l^m l^n l_j. \tag{2.18}$$

It can readily be proved that (2.17) implies a quadratic dependence of g_{mn} on l_i. that is

$$g_{mn} = Q^{ij}_{mn}(x) l_i l_j. \tag{2.19}$$

In fact, differentiating (2.17) with respect to y^k yields

$$-2C_k{}^{ij} = 2Q^{ij}_{mn} l^m h_k{}^n/F. \tag{2.20}$$

On contracting Equation (2.20) with l_i, we obtain $l_j Q^{ij}_{mn} l^m h_k{}^n = 0$ or, substituting Equations (2.17) and (2.18),

$$l^i l_k = Q^{ij}_{mk} l^m l_j. \tag{2.21}$$

From the contraction of (2.21) with l_i it then follows that, similarly to (2.18),

$$l_k = Q^{ij}_{mk} l^m l_i l_j. \tag{2.22}$$

Further, if Equation (2.22) is differentiated with respect to y^n, we find

$$h_{kn} = 2Q^{ij}_{mk} l^m l_i h_{jn} + Q^{ij}_{mk} h_n{}^m l_i l_j. \tag{2.23}$$

On substituting (2.17) in (2.23) and using (2.17), (2.21) and (2.22), we indeed get (2.19).

The converse statement, that (2.19) implies (2.17), can be proved in the same way. Therefore, we get the following

PROPOSITION 4. The conditions (2.17) and (2.19) are equivalent.

Let us introduce the tensor

$$B^{ij}_{mnk} \stackrel{\text{def}}{=} \partial^3 (F^2 g^{ij})/\partial y^m \partial y^n \partial y^k$$

which is of interest to us here since (2.17) is equivalent to $B^{ij}_{mnk} = 0$. The substitution of the obvious equality

$$\begin{aligned}\partial^2 (F^2 g^{ij})/\partial y^m \partial y^n = 2g_{mn} g^{ij} - 4y_m C_n{}^{ij} - 4y_n C_m{}^{ij} - \\ -2F^2 \partial C_n{}^{ij}/\partial y^m\end{aligned} \tag{2.24}$$

in the identity

$$y_i B^{ij}_{mnk} = \partial[y_i \partial^2 (F^2 g^{ij})/\partial y^m \partial y^n]/\partial y^k - g_{ik} \partial^2 (F^2 g^{ij})/\partial y^m \partial y^n$$

gives after some calculations the following result:

$$l_i B^{ij}_{mnk} = 4T^j{}_{mnk}, \tag{2.25}$$

where

$$\begin{aligned}T_{mnij} = (\partial C_{nij}/\partial y^m - C_m{}^k{}_n C_{kij} - C_m{}^k{}_i C_{nkj} - C_m{}^k{}_j C_{nik})F + \\ + l_m C_{nij} + l_j C_{mni} + l_i C_{jmn} + l_n C_{ijm}\end{aligned} \tag{2.26}$$

is called the *T-tensor*.

This last tensor is obviously symmetric in all its subscripts and satisfies the identity $y^m T_{mnij} = 0$ Recalling the definition (1.66) of the covariant derivative in the tangent Minkowskian space, we may rewrite (2.26) as

$$T_{mnij} = C_{nij}|_m + l_m C_{nij} + l_j C_{mni} + l_i C_{jmn} + l_n C_{ijm}. \tag{2.27}$$

A Finsler space is said to satisfy the *T-condition* if the *T*-tensor vanishes[7]. In such a case, we shall also say that the metric function satisfies the T-condition. The identity (2.25) shows that the following proposition is valid.

PROPOSITION 5. If a Finslerian metric tensor is quadratic in the unit tangent vectors, then the T-condition is satisfied.

It is still unclear whether the converse statement is true. Nevertheless, a partial converse is possible. Indeed, the tensor B^{ij}_{mnk} can be expressed explicitly as

$$\begin{aligned} B^{ij}_{mnk} = & -2F\,\partial T^{ij}{}_{mn}/\partial y^k + 2(l^i T^j_{mnk} + l^j T^i_{mnk} - l_m T^{ij}{}_{nk} - \\ & - l_n T^{ij}{}_{km} - l_k T^{ij}{}_{mn}) + 2F(C_m{}^{ip} T^j{}_{nkp} + C_m{}^{jp} T^i{}_{nkp} + \\ & + C_n{}^{ip} T^j{}_{kmp} + C_n{}^{jp} T^i{}_{kmp} - C_{mn}{}^p T^{ij}{}_{kp} - C^{ijp} T_{mnkp}) + \\ & + 2(-C_{km}{}^p R^{*ij}_{p\ n} - C_{kn}{}^p R^{*ij}_{p\ m} + C_k{}^{ip} R^{*}_{pnm}{}^i + \\ & + C_m{}^{ip} R^{*}_{pkn}{}^j + C_n{}^{jp} R^{*}_{pkm}{}^i), \end{aligned} \tag{2.28}$$

where R^*_{ijmn} is the tensor that describes the curvature of the indicatrix in accordance with (1.124). The verification of the identity (2.28) involves a straightforward but tedious procedure and is therefore omitted here. From (2.28) we obtain the following result.

PROPOSITION 6. If the tensors T_{mnij} and R^*_{ijmn} are both vanishing, then the Finslerian metric tensor is quadratic in the unit tangent vectors.

Since vanishing R^*_{ijmn} means that the indicatrix is a space of vanishing curvature, we get

PROPOSITION 7. A quadratic dependence of the metric tensor on the unit tangent vectors is tantamount to the T-condition, provided the indicatrix is a space of vanishing curvature.

The last proposition may be applied to the metric function (2.10) in view of Proposition 3. Moreover, since the metric tensor (2.13) with $f = 0$ is apparently quadratic in l^A because of the equality (2.11), we get

PROPOSITION 8. Any metric function of the form (2.10) satisfies the T-condition.

A fairly broad class of metric functions satisfying the quadratic relation (2.17) will be given in Problem 2.3.

2.3. Properties of the Berwald-Moór Metric Function

Let us take some contravariant n-tuple $S^m_A(x)$ and any tangent vector y^m supported by a fixed point x^i and construct an N-dimensional parallelotope whose ribs are parallel to $S^m_A(x)$, and whose diagonal coincides with y^m in the tangent space M_x. After dividing its volume by the volume of the basic parallelotope spanned by the vectors $S^m_A(x)$, we obtain the following expression

$$\prod_{A=1}^{N} (S^A_m(x)y^m), \tag{2.29}$$

where the covariant n-tuple S^A_m is reciprocal to S^m_A. This ratio has a constant sign in each of the 2^N sectors formed by the N vectors $S^m_A(x)$ in M_x. Let us confine ourselves to considering one of them, say the sector $S^+_x \subset M_x$ formed by all the vectors $y^m \in M_x$ for which $y^m S^A_m(x) > 0$ for any value of the index A. The N-degree homogeneity of the expression (2.29) with respect to y^m suggests that the Finslerian metric function should be defined by extracting the N-th root of (2.29):

$$F(x,y) = \left[\prod_{A=1}^{N} (S^A_m(x)y^m)\right]^{1/N}. \tag{2.30}$$

Obviously, a Finsler space with such a metric function is based on the notion of volume, the lengths of vectors in this space being measured by N-dimensional volumes.

If $S^m_A(x)$ is chosen to be a basis in M_x, the metric function (2.30) becomes Minkowskian:

$$F(y^A) = \left[\prod_{A=1}^{N} y^A\right]^{1/N} \tag{2.31}$$

which is called *the Berwald-Moór metric function*[8]. The Finslerian function (2.30) will be called *the 1-form Berwald-Moór metric function* in agreement with the terminology adopted in the next section. A Finsler space with such a metric function will be called, for brevity, a *1-form Berwald-Moór Finsler space.*

In the following only the case $N \geqslant 3$ will be of interest to us because the metric function (2.30) is Riemannian for $N = 2$. The definition (1.5) of the Finslerian metric tensor yields for the case (2.30):

$$g_{ij} = (2y_i y_j - N \sum_{A=1}^{N} y_A y_A S^A_i S^A_j)/F^2 \tag{2.32}$$

where $y_A = S^i_A(x)y_i$. The Finslerian correspondence (1.7) between y^i and $y_i = g_{ij}y^j$ becomes

$$y_A = F^2/Ny^A \tag{2.33}$$

where $y^A = S^A_i(x)y^i$. The reader can readily verify that the corresponding Hamiltonian function $H(x, y_m) = F(x, y^n(x, y_m))$ (Equation (6.18)) and the reciprocal metric tensor $g^{ij} = \frac{1}{2}\partial^2 H^2/\partial y_i\, \partial y_j$ (Equation (6.20)) are given by

$$H(x,y) = N\left[\prod_{A=1}^{N} (S^m_A(x)y_m)\right]^{1/N}, \tag{2.34}$$

$$g^{ij} = (2y^i y^j - N \sum_{A=1}^{N} y^A y^A S^i_A S^j_A)/F^2 \tag{2.35}$$

respectively.

The metric function (2.30) belongs to the class (2.10) with $r^A = 1/N$. Therefore, since $f = 0$ in (2.15) corresponds to the metric (2.10), we obtain $S^* = -1$ in the present case which reduces the tensor (2.1) to just

$$S_{ijmn} = h_{in}h_{jm} - h_{im}h_{jn}. \tag{2.36}$$

According to Proposition 3, the indicatrix of the Finsler space under consideration will be a space of vanishing curvature. Furthermore, the metric tensor (2.35) is quadratic in the unit tangent vectors (cf. the definition (2.17)) so that according to Proposition 8 the T-tensor (2.26) must vanish, i.e.

$$\begin{aligned} \partial C_{nij}/\partial y^m = C_{m\ n}^{\ k}C_{kij} + C_{m\ i}^{\ k}C_{nkj} + C_{m\ j}^{\ k}C_{nik} - \\ - (l_m C_{nij} + l_j C_{mni} + l_i C_{jmn} + l_n C_{ijm})/F. \end{aligned} \tag{2.37}$$

Moreover, the metric function under study has so far been the only example of a Finslerian metric function with the following property.

PROPOSITION 9. The determinant of the Finslerian metric tensor associated with the Berwald-Moór metric function is independent of y^i.

In fact, considering (2.32), the reader may readily verify that

$$\det(g_{ij}) = (-1)^{N+1} N^{-N} \det{}^2 (S^A_m)$$

from which Proposition 9 follows. This result may be reformulated as[9]

$$C_{k\ m}^{\ k} = 0 \tag{2.38}$$

because the definition (1.5) entails the equality

$$C_{k\ m}^{\ k} = \tfrac{1}{2}\partial \ln|\det(g_{ij})|/\partial y^m \tag{2.39}$$

for any Finslerian metric tensor g_{ij}.

The above observations permit one to obtain simple expressions for various contractions involving the Cartan torsion tensor. First of all,

$$S_{in} = C^{tj}_{\ \ i}C_{tjn}F^2 = -(N-2)h_{in}, \tag{2.40}$$

as is obvious from (2.2), (1.62), and (2.38). Next,

$$C_{irs}C^r_{\ jt}C_k^{\ st} = -(N-3)C_{ijk}/F^2. \tag{2.41}$$

Indeed by (2.38) the left-hand side of Equation (2.41) may be written as $C_{irs}(C^r_{\ jt}C_k^{\ st} - C_{j\ k}^{\ r}C_t^{\ st})$ which, after substituting (2.36), reduces to the right-hand side of (2.41). Similarly, the identity

$$\begin{aligned} C_{j\ k}^{\ n}C_t^{\ ij}C_{r\ p}^{\ k}C_q^{\ tr} = C_{j\ p}^{\ n}C_q^{\ ij}F^{-2} - (N-3)C^{rni}C_{rpq}F^{-2} + \\ + (N-2)h^{in}g_{pq}F^{-4} \end{aligned} \tag{2.42}$$

is obtained via the identity

$$C_{r\ p}^{\ k}C_q^{\ tr} = (C_{r\ p}^{\ k}C_q^{\ tr} - C_{r\ p}^{\ t}C_q^{\ kr}) + C_{r\ p}^{\ t}C_q^{\ kr}$$

together with (2.36), (2.41) and (2.40). Also, Equations (2.37), (2.41) and (2.40) yield

$$C_m{}^k{}_p \partial C^{mij}/\partial y^k = (3N-8)C_p{}^{ij}F^{-2} + (N-2)(l^i h_p{}^j + l^j h_p{}^i)F^{-3} \tag{2.43}$$

and furthermore

$$C^{lik}C_{ikjn} = -(3N-8)C_j{}^l{}_n F^{-2} + (N-2)(l_j h_n{}^l + l_n h_j{}^l)F^{-3} \tag{2.44}$$

where $C_{ikjn} = \partial C_{ikj}/\partial y^n$.

Our next observation is:

PROPOSITION 10. The Finslerian metric tensor associated with the Berwald-Moór metric function has the signature $(+ - - \cdots)$.

In proving this proposition it is useful to remember that, owing to the quadratic dependence of the metric tensor (2.35) on l^m, an orthonormal n-tuple $h_P^m(x, y)$ associated with the metric may be constructed in a form which is linear in l^m. This form of h_P^m is given by

$$h_0^m = l^m, \tag{2.45}$$

$$h_a^m = \left[\frac{N}{(N-a+1)(N-a)}\right]^{1/2} \sum_{A=a}^{N-1} (L_{A+1}^m - L_a^m), \qquad a = 1, 2 \ldots, N-1 \tag{2.46}$$

where $L_A^m = l^i S_i^A \cdot S_A^m$ (the point indicates that the summation convention does not apply here to the index A). Indeed, it is easily shown that upon the substitution of (2.45) and (2.46) in

$$g^{ij}(x, y) = q^{PQ} h_P^i(x, y) h_Q^j(x, y), \tag{2.47}$$

where the indices P and Q run from 0 to $N-1$ and the only nonvanishing members among q^{PQ} are

$$q^{00} = -q^{11} = -q^{22} = \cdots = 1, \tag{2.48}$$

the right-hand side of (2.47) is identical with (2.35), and therefore Proposition 10 is valid.

The representation

$$h_m^0 = l_m, \tag{2.49}$$

$$h_m^a = \left[\frac{N}{(N-a+1)(N-a)}\right]^{1/2} \sum_{A=a}^{N-1} (L_m^{A+1} - L_m^a) \qquad a=1, 2, \ldots, N-1, \tag{2.50}$$

which is evidently similar to (2.45)–(2.46), will be valid for the covariant orthonormal n-tuples $h_m^P(x, y)$ reciprocal to $h_P^m(x, y)$. In (2.50) $L_m^A = l_i S_A^i \cdot S_m^A$ (no summation over A). The reciprocity condition

$$h_P^m(x, y) h_m^Q(x, y) = \delta_P^Q \tag{2.51}$$

can be readily verified directly with the help of the above expressions.

The metric tensor and the orthonormal n-tuples under consideration satisfy a series of interesting identities involving their derivatives in y^i, namely

$$\partial g^{ij}/\partial y^i = 0, \tag{2.52}$$

$$\partial h^i_P/\partial y^i = (N-1)\delta^0_P/F, \tag{2.53}$$

$$\partial h^P_i/\partial y^j - \partial h^P_j/\partial y^i = (l_j h^P_i - l_i h^P_j)/F. \tag{2.54}$$

Also

$$\partial((g^{ij} + 2l^i l^j)/F^2)/\partial y^i = 0 \qquad \textit{in the case } N = 4. \tag{2.55}$$

Moreover, if we introduce the following two differential operators of the second and N-th orders, respectively:

$$\hat{g} = F^2(x,y) g^{ij}(x,y)\, \partial^2/\partial y^i \partial y^j, \tag{2.56}$$

$$\hat{S} = (NF)^N \prod_{A=1}^{N} (S^i_A(x)\partial/\partial y^i), \tag{2.57}$$

then, in addition to the identities

$$\hat{g} l^m = (1-N) l^m, \qquad \hat{g} l_m = (1-N) l_m + 2FC_k{}^k{}_m \tag{2.58}$$

valid in any Finsler space, another set of identities

$$\hat{S} l^m = (-1)^{N-1}(N-1) l^m, \qquad \hat{S} l_m = (1-N) l_m \tag{2.59}$$

will hold in the Finsler space under study. The simplest way of verifying (2.59) is to differentiate the relation $l^m = S^m_A y^A/F$ represented as

$$l^m = S^m_A y^A \prod_{B=1}^{N} (y^B)^{-1/N}.$$

Since the n-tuples h^m_P and h^P_m given by Equations (2.45)–(2.46) and (2.49)–(2.50) are linear functions of l^i and l_i, respectively, the identities (2.58) and (2.59) will entail

$$\hat{g} h^P_m = \hat{S} h^P_m = (1-N) h^P_m,\ \hat{g} h^m_P = (-1)^N \hat{S} h^m_P = (1-N) h^m_P. \tag{2.60}$$

These identities (see also Problem 2.6) clearly give a number of eigenfunctions of the operators (2.56) and (2.57). Finally, the above formulae can be used to demonstrate the validity of

$$\hat{g} g^{ij} = (8-4N) h^{ij} = -2F^2\, \partial C^{mij}/\partial y^m,$$

$$\hat{g} C_{ijk} = (24-10N) C_{ijk} + 4(2N-1)(l_i h_{jk} + l_j h_{ik} + l_k h_{ij})/F + $$
$$+ 12(N-2) l_i l_j l_k/F.$$

Attention is drawn to the interesting fact that the 1-form Berwald-Moór metric function (2.30) is invariant under the following group of *special scale transformations of* n-tuples:

$$S^A_m(x) \to \tilde{S}^A_m(x) = Q^A(x) S^A_m(x), \tag{2.61}$$

where $Q^A(x)$ are assumed to be any scalars satisfying the equality

$$\prod_{A=1}^{N} Q^A(x) = 1. \tag{2.62}$$

This observation is valid also for the generalization (2.10) of the Berwald-Moór metric function if the condition (2.62) is replaced by

$$\prod_{A=1}^{N} (Q^A(x))^{rA(x)} = 1.$$

Apparently, any geometrical object constructible on the basis of the considered metric functions will be invariant under the transformations of such a group.

The following result is valid for the Berwald-Moór metric.

PROPOSITION 11. Let $y^i(x)$ be any contravariant vector field. Then the special scale transformation (2.61) of n-tuples $S^A_m(x)$ can be chosen in such a way that the scalars $y^A(x) = S^A_i(x) y^i(x)$ become equal to

$$y^A(x) = Q(x) \tag{2.63}$$

for any value of the index A, where

$$Q(x) = \left[\prod_{A=1}^{N} y^A(x) \right]^{1/N}$$

is the value of the Berwald-Moór metric function.

Indeed, Equation (2.63) is obtained immediately by putting $Q^A(x) = Q(x)/y^A(x)$ in (2.61).

For the metric function under study the parametrical representation $l^A = t^A(x, u)$ of the indicatrix (Equation (1.100)) can be chosen as follows

$$l^A = \exp(C^A_a u^a) \tag{2.64}$$

where C^A_a denotes a set of real constants which for the present can be entirely arbitrary except for the requirement that

$$\sum_{A=1}^{N} C^A_a = 0 \tag{2.65}$$

for any value of the index $a = 1, 2, \ldots, N - 1$. In fact, as follows from the form of the metric function (2.30), we have

$$F(\exp(C^A_a u^a)) = \exp(N^{-1} u^a \sum_{A=1}^{N} C^A_a)$$

which, as a consequence of (2.65), is actually unity. This circumstance makes it legitimate to treat the variables u^a in (2.64) as a system of coordinates on the indicatrix.

In order to invert (2.64), it is appropriate to extend the set of constants C_a^A by adding the components

$$C_0^A = 1. \tag{2.66}$$

The square matrix having entries C_P^A thus obtained will be assumed to be non-singular, so that the inverse matrix $||C_A^P||$ can be defined conventionally:

$$C_A^P C_Q^A = \delta_Q^P. \tag{2.67}$$

The indices P and Q range from 0 to $N - 1$.

It can be shown that the constants C_A^0 are the same for any value of the index A. In fact, the minors of the elements C_1^0 and, say, C_2^0 in the matrix $||C_A^P||$ are the same because of the equality (2.65) which allows the replacement of the first column in the minor of C_1^0 with the negative of the first column in the minor of C_2^0, the difference between the minors being a determinant containing linearly dependent columns which is vanishing. This last observation, together with the fact that the equality (2.67) with $P = Q = 0$ reads

$$\sum_{A=1}^{N} C_A^0 = 1$$

(cf. Equation (2.66)) shows that

$$C_A^0 = 1/N. \tag{2.68}$$

Moreover, putting $P = 1$ and $Q = 0$ in (2.67) and using (2.66) we get

$$\sum_{A=1}^{N} C_A^a = 0 \tag{2.69}$$

similarly to (2.65). It is convenient to restrict the remaining arbitrariness in the set of C_a^A by stipulating that

$$\sum_{A=1}^{N} C_P^A C_Q^A = N\delta_{PQ}. \tag{2.70}$$

By virtue of Equations (2.65) and (2.66) the condition (2.70) is satisfied identically if P or Q equals zero. Therefore, Equation (2.70) imposes $(N - 1)^2$ conditions on $N(N - 1)$ constants. Since the difference $N(N - 1) - (N - 1)^2 = N - 1$ is equal to the number of the conditions (2.65), the $N(N - 1)$ constants C_a^A are constrained by exactly $N(N - 1)$ conditions. Hence, any arbitrariness in C_a^A may consist only in the possibility of such transformations which leave the conditions (2.65) and (2.70) invariant. The condition (2.65) is clearly invariant with respect to any three-dimensional transformation. As regards the condition (2.70) we can see that since it is fulfilled identically with P or Q equal to zero, this condition is, due to the specific form of its right-hand side, invariant only with respect to three-dimensional rotations. Thus we may conclude that the above conditions define the quantities C_a^A up to arbitrary three-dimensional rotations. The set of equations for C_a^A can be readily solved explicitly for any dimension number N. In the particular case $N = 4$, which is of interest to physical applications, the solution can be expressed in the following form

$$||C_P^A|| = \left\| \begin{matrix} 1 & 1 & 1 & 1 \\ -\sqrt{3} & 1/\sqrt{3} & 1/\sqrt{3} & 1/\sqrt{3} \\ 0 & \sqrt{8/3} & -\sqrt{2/3} & -\sqrt{2/3} \\ 0 & 0 & -\sqrt{2} & \sqrt{2} \end{matrix} \right\| \tag{2.71}$$

where the index P indicates the row number and the index A the column number. It can easily be verified that the matrix inverse to (2.71) is

$$||C_A^P|| = \tfrac{1}{N}\, ||C_P^A||^T, \tag{2.71'}$$

i.e. that $C_A^P C_Q^A = \delta_Q^P$ if $C_A^P = C_Q^B \delta_{BA} \delta^{PQ}/N$ with δ denoting the Kronecker symbol.

The set of constants C_A^a permits us to obtain the inverse of (2.64) in the form

$$u^a = C_B^a \ln l^B, \tag{2.72}$$

where summation over B is assumed. Therefore, following (1.169), the general form of the G-transformations for the case of the Berwald-Moór metric function may be written explicitly as

$$y'^A = F(y^A) \exp(C_b^A u'^b (C_B^a \ln l^B)). \tag{2.73}$$

These G-transformations are generated by $N - 1$ functions $u'^b(u^a)$ of class C^5 subject to the sole condition that the inverse functions $u^a\ (u'^b)$ exist. The inverse of (2.73) reads

$$y^A = F(y'^A) \exp(C_a^A u^a (C_B^b \ln l'^B)). \tag{2.74}$$

The particular case $u'^b(u^a) = u^b$ corresponds to the identical transformation $y'^A = y^A$

Let us examine the restrictions imposed on the functions $u^a(u'^b)$ by the requirement that the G-transformation be metric, i.e. that

$$y_A y_B{}^A{}_C = 0 \tag{2.75}$$

(Equation (1.157)). The notation

$$u_b^a = \partial u^a/\partial u'^b, \qquad u_{bc}^a = \partial u_b^a/\partial u'^c, \qquad u^A = C_a^A u^a, \qquad u_B^A = u_b^a C_a^A C_B^b,$$

etc. will be used in the remaining part of the present section. The requirement (2.65) implies

$$\sum_{A=1}^{N} u_B^A = 0, \qquad \sum_{A=1}^{N} u_B{}^A{}_C = 0. \tag{2.76}$$

By (2.72),

$$\partial u'^b / \partial y'^B = \sum_{A=1}^{N} C_A^b \frac{1}{l'^A} \partial l'^A / \partial y'^B$$

or

$$\partial u'^b / \partial y'^B = N F^{-1} C_B^b l'_B, \tag{2.77}$$

as a consequence of (2.69) and (2.33). Therefore

$$\partial u^b / \partial y'^B = N F^{-1} u_B^a l'_B, \qquad \partial u_B^A / \partial y'^C = N F^{-1} u_B{}^A{}_C l'_C. \tag{2.78}$$

Considering (2.74) we have

$$y^A = F \exp u^A \tag{2.79}$$

which, together with the first equality in (2.78), enables one to conclude that the derivative $y_B^A = \partial y^A / \partial y'^B$ is

$$y_B^A = l^A (l'_B + N u_B^A l'_B).$$

Further, differentiation of the last equality with respect to y'^C yields after using (2.79) and the second equality (2.78) the following expression for the second derivative $y_{BC}^A = \partial y_B^A / \partial y'^C$:

$$\begin{aligned} y_B{}^A{}_C = F^{-1} l^A ((1 + N u_B^A)(g'_{BC} - l'_B l'_C) + u_C^A l'_B l'_C + \\ + N^2 l'_B l'_C (u_B^A u_C^A + u_B{}^A{}_C)). \end{aligned} \tag{2.80}$$

Since $y_A = F^2 / N y^A$ (Equation (2.33)) and the identities (2.76) hold, it follows after contracting (2.80) with y_A that condition (2.75) is reducible to

$$g'_{BC} - l'_B l'_C + N l'_B l'_C \sum_{A=1}^{N} u_B^A u_C^A = 0. \tag{2.81}$$

Finally, taking into account that

$$g_{BC} = l_B l_C (2 - N \delta_{BC})$$

(see Equation (2.32)) which entails that

$$l'^B l'^C g'_{BC} = 2N^{-2} - N^{-1} \delta_{BC}$$

for any value of the indices B and C, we find, upon multiplying Equation (2.81) by $l'^B l'^C$ without summation over B or C, that Equation (2.81) is identical with

$$\sum_{A=1}^{N} u_B^A u_C^A = \delta_{BC} - \frac{1}{N}. \tag{2.82}$$

The second derivatives $u_B{}^A{}_C$ have been eliminated because of (2.76).

The condition (2.82) may be simplified by contracting it with $C^B_b C^A_c$ and taking into account (2.65) and (2.70). This procedure gives

$$u^a_b u^d_c \sum_{A=1}^{N} C^A_a C^A_d = N\delta_{bc}.$$

After using (2.70) again, we finally obtain the result:

$$\sum_{a=1}^{N-1} u^a_b u^a_c = \delta_{bc}. \tag{2.83}$$

Thus we get:

PROPOSITION 12. In the case of the Berwald-Moór metric function the metric G-transformations are generated by those mappings $u^a(u'^b)$ of the indicatrix on itself under which the Euclidean metric

$$(\mathrm{d}l)^2 = \sum_{a=1}^{N-1} (\mathrm{d}u^a)^2$$

given on the indicatrix remains invariant.

In a similar fashion, we can elucidate the meaning of the condition (1.160) under which a metric G-transformation belongs to the group G_{mt}. A calculation similar to that leading from (2.77) to (2.83) shows that a metric G-transformation belongs to the group G_{mt} if and only if

$$\sum_{A=1}^{N} u^A_B u^A_C u^A_D = \delta_{BC}\delta_{CD},$$

which proves to amount to the condition

$$C_{abc} u^a_d u^b_e u^c_f = C_{def}$$

where

$$C_{abc} = \frac{1}{N} \sum_{A=1}^{N} C^A_a C^A_b C^A_c. \tag{2.84}$$

Thus G_{mt}-transformations are those metric G-transformations which keep invariant the cubic form

$$(\mathrm{d}l)^3 = C_{abc}\,\mathrm{d}u^a\,\mathrm{d}u^b\,\mathrm{d}u^c$$

given on the indicatrix.

Proposition 12 arouses the suspicion that the induced metric tensor g^*_{ab} of the indicatrix is Euclidean in the case considered. This may be verified as follows. By (2.64) the projection factors $t^A_a = \partial l^A/\partial u^a$ (Equation (1.101)) will be

$$t^A_a = C^A_a l^A. \tag{2.85}$$

Substituting (2.85) and (2.32) in the definition (1.110) of g^*_{ab} and then using the identity $l_A t^A_a = 0$ (the first identity in (1.102)) together with (2.77), we find that

$$g^*_{ab} = -\delta_{ab}. \tag{2.86}$$

Thus we have proved the following:

PROPOSITION 13. For the case of the Berwald-Moór metric function the induced Riemannian metric tensor of the indicatrix is equal to the negative of the Euclidean metric tensor.

Similarly, it can easily be verified that the projection of the Cartan torsion tensor on the indicatrix is given by

$$C_{ABC}t^A_a t^B_b t^C_c = C_{abc}$$

where C_{ABC} denotes the Cartan torsion tensor of the space considered and C_{abc} is the set of constants (2.84).

2.4. 1-form Finsler Spaces

Any Minkowskian metric function $F_{\text{Mink}}(y^A)$ may be used to construct the following 1-*form Finslerian metric function* with the help of a set of n-tuples $S^A_m(x)$:

$$F(x^i, y^m) = F_{\text{Mink}}(S^A_m(x^i)y^m) \tag{2.87}$$

which defines *the* 1-*form Finsler space*[10].

According to this definition the tangent Minkoskian spaces to a given 1-form Finsler space are identical at any point x^i. Although (2.87) gives a rather broad class of Finslerian metric functions, far from every Finsler space may be treated locally as a space of the 1-form type. For example, among $S3$-like metric functions (2.9) and (2.10) the case (2.9) does not belong to the 1-form type unless f = const, whereas (2.10) represents 1-form metric functions if and only if r^A = const for any value of the index A.

In what follows quantities of the type V_A and V_m will be assumed to be interrelated by $V_A = S^m_A V_m$, $V_m = S^A_m V_A$. In particular,

$$M^A{}_{BC} = S^m_B S^n_C M^A_{mn}, \qquad g^{AB} = S^A_m S^B_n g^{mn}, \qquad g_{ij} = S^A_i S^B_j g_{AB}, \text{ etc.}$$

Writing y^A means the scalar function of x^m and y^n:

$$y^A = S^A_n(x)y^n. \tag{2.88}$$

The partial derivative $\partial/\partial x^i$ will be denoted simply as ∂_i.

A set of n-tuples $S^A_i(x)$ gives rise to the connection coefficients

$$L_j{}^k{}_i \overset{\text{def}}{=} S^k_A\,\partial_i S^A_j = -S^A_j\,\partial_i S^k_A \tag{2.89}$$

which are non-symmetric in their subscripts j, i. Such connection coefficients have been used in general relativity theory[11] since the well-known work of Einstein (1928) in which they were called *the connection coefficients of absolute parallelism* because the covariant derivatives of S^A_i and S^i_A with respect to these

$L_j{}^k{}_i$ vanish identically:

$$\partial_i S_j^A - L_j{}^k{}_i S_k^A = 0, \qquad \partial_i S_A^j + L_k{}^j{}_i S_A^k = 0. \tag{2.90}$$

This implies in turn that the curvature tensor constructed out of the connection coefficients (2.89) vanishes:

$$\partial_k L_l{}^j{}_h - \partial_h L_l{}^j{}_k + L_m{}^j{}_k L_l{}^m{}_h - L_m{}^j{}_h L_l{}^m{}_k = 0. \tag{2.91}$$

The definition (2.87) implies that the Finslerian metric tensor of a 1-form Finsler space can be written as

$$g_{ij}(x, y^n) = S_i^A(x) S_j^B(x) g_{AB}(y^D(x, y^n)), \tag{2.92}$$

where

$$g_{AB}(y^D) = \tfrac{1}{2}\partial^2 F^2_{\text{Mink}}(y^D)/\partial y^A\, \partial y^B \tag{2.93}$$

is the Minkowskian metric tensor associated with F_{Mink}. Since (2.93) depends on x^k only through y^A as given by (2.88) we get

$$\partial_k g_{AB} = 2C_{ABD}\, \partial_k y^D, \qquad \partial_k g^{AB} = -2C^{AB}{}_D\, \partial_k y^D \tag{2.94}$$

where Equations (1.15) and (1.20) have been used. Therefore the partial derivatives of g_{ij} with respect to x^m can be expressed in terms of the derivatives of g_{ij} with respect to the tangent vectors y^h, namely

$$\partial_m g_{ij} = g_{AB}\, \partial_m(S_i^A S_j^B) + 2y^k L_k{}^h{}_m C_{ijh}. \tag{2.95}$$

If we consider the Jacobian

$$J(x, y) \overset{\text{def}}{=} |\det(g_{ij}(x, y))|^{1/2} \tag{2.96}$$

we find from (2.94), together with the obvious equality

$$\partial_m \ln J = \tfrac{1}{2} g^{ij}\, \partial_m g_{ij},$$

that

$$\partial_m \ln J = \tfrac{1}{2} g_{AB} g^{ij}\, \partial_m(S_i^A S_j^B) + y^k L_k{}^h{}_m C_i{}^i{}_h.$$

Here, the first term on the right-hand side is

$$\tfrac{1}{2} g_{AB} g^{ij}(S_i^A\, \partial_m S_j^B + S_j^B\, \partial_m S_i^A) = S_B^j\, \partial_m S_j^B = L_j{}^j{}_m$$

so that

$$\partial_m \ln J = L_j{}^j{}_m + y^k L_k{}^j{}_m C_i{}^i{}_j. \tag{2.97}$$

As a consequence of the definition (2.89)

$$L_j{}^j{}_m = \partial_m \ln \det(S_i^A). \tag{2.98}$$

Further, Equation (2.92) implies that the Jacobian (2.96) may be represented as

$$J = |\det(g_{AB})|^{1/2} \det(S_i^A). \tag{2.99}$$

It follows directly from (2.99) that

$$\partial J/\partial S_i^A = (S_A^i + C_A{}^B{}_B y^i)J \tag{2.100}$$

and

$$\partial J/\partial y^i = JC_i{}^B{}_B \tag{2.101}$$

where we have used (2.39) and (2.88) together with the obvious identity

$$\partial \det(S_n^B)/\partial S_i^A = S_A^i \det(S_n^B). \tag{2.102}$$

As with g_{AB}, the object $C_{ABC} = \frac{1}{2}\partial g_{AB}/\partial y^C$ depends on x^m only through y^D. Therefore, similarly to (2.95),

$$\partial_m C_{ijk} = C_{ABC}\, \partial_m(S_i^A S_j^B S_k^C) + y^l L_l{}^h{}_m C_{ijkh} \tag{2.103}$$

where $C_{ijkh} = \partial C_{ijk}/\partial y^h$. Relations of the type (2.95) and (2.103) make it possible to examine the explicit form of the dependence of the Finslerian objects on $S_i^A(x)$. First of all, we observe from (2.95) that the Christoffel symbols (1.26) of any 1-form Finsler space may be written as

$$\gamma_i{}^k{}_j = L_j{}^k{}_i + Z_{ij}{}^k + (C_i{}^k{}_h L_l{}^h{}_j + C_j{}^k{}_h L_l{}^h{}_i - C_{ijh} L_l{}^{hk})y^l \tag{2.104}$$

where

$$Z_{ijh} = \tfrac{1}{2}(M_{ijh} + M_{jih} - M_{hij}) \tag{2.105}$$

and

$$M^A{}_{jh}(x) = \partial_j S_h^A - \partial_h S_j^A \equiv -M^A{}_{hj} \tag{2.106}$$

are tensors. The notation $M^m{}_{jh} = S_A^m M^A{}_{jh}$, $M_{ijh} = g_{im} M^m{}_{jh}$, etc. has been used. The tensor (2.105) can be written as

$$Z_{ijh}(x, y) = g_{AB}(x, y) Z_{ijh}^{AB}(x) \tag{2.107}$$

where

$$Z_{ijh}^{AB}(x) = \tfrac{1}{2}(S_i^A M^B{}_{jh} + S_j^A M^B{}_{ih} - S_h^A M^B{}_{ij}). \tag{2.108}$$

The following handy identities are valid:

$$Z_{ijh} = -Z_{ihj}, \qquad Z_{ij}{}^i = M^i{}_{ji}(x), \tag{2.109}$$

$$Z_{ij}{}^j = Z_{imn} y^m y^n = 0, \qquad Z_{mni} y^m y^n = M_{mni} y^m y^n, \tag{2.110}$$

$$Z_{ijh} + Z_{jih} = M_{ijh} + M_{jih} = 2Z_{ijh} + M_{hij}, \tag{2.111}$$

$$Z_{ijh} - Z_{jih} = -M_{hij}, \tag{2.112}$$

$$4y_m y^n Z^{kmt} Z_{knt} = y_m y^n (2M_{knt} M^{kmt} + 2M_{knt} M^{tmk} + M_{nkt} M^{mkt}). \tag{2.113}$$

Moreover, (2.104) and $y^j C_{ijh} = 0$ imply

$$\gamma_i{}^k{}_n y^n = (L_n{}^k{}_i + Z_{in}{}^k + C_{ih}{}^k L_j{}^h{}_n y^j) y^n \tag{2.114}$$

and

$$\gamma_{m\ n}^{\ k}y^m y^n = (L_{m\ n}^{\ k} + Z_{mn}^{\ \ k})y^m y^n. \tag{2.115}$$

It follows directly from Equations (2.104), (2.114), and (2.115) that the Cartan connection coefficients (1.32) have the following form in the 1-form case:

$$\Gamma_{i\ j}^{\ k} = L_{i\ j}^{\ k} + V_{i\ j}^{\ k} \tag{2.116}$$

where

$$V_{i\ j}^{\ k} = M^k_{\ ij} + Z_{ij}^{\ \ k} - C_{i\ m}^{\ k}w_j^{\ m} - C_{j\ m}^{\ k}w_i^{\ m} + C_{ijm}w^{km} \tag{2.117}$$

denotes a tensor nonsymmetric in its subcripts, and

$$w_i^{\ k} = Z_{ip}^{\ \ k}y^p - C_{i\ m}^{\ k}Z_{pq}^{\ \ m}y^p y^q \equiv y^n V_{n\ i}^{\ k}. \tag{2.118}$$

The contraction $G_i^{\ k} = y^j \Gamma_{i\ j}^{\ k}$ is

$$G_i^{\ k} = (L_{i\ p}^{\ k} + M^k_{\ ip} + Z_{ip}^{\ \ k} - C_{i\ m}^{\ k}Z_{pq}^{\ \ m}y^q)y^p \equiv w_i^{\ k} + L_{p\ i}^{\ k}y^p \tag{2.119}$$

which entails that

$$G_i^{\ i} = (L_{i\ p}^{\ i} - C_{i\ m}^{\ i}Z_{pq}^{\ \ m}y^q)y^p \tag{2.120}$$

where Equation (2.109) is taken into account. Differentiating (2.119) with respect to y^j we find that the Berwald connection coefficients (1.49) take the following form

$$G_{i\ j}^{\ k} = L_{j\ i}^{\ k} + Z_{ij}^{\ \ k} + U_{i\ j}^{\ k} \tag{2.121}$$

with

$$\begin{aligned} U_{i\ j}^{\ k} = &-C_{i\ m}^{\ k}(Z_{jp}^{\ \ m} + Z_{pj}^{\ \ m})y^p - C_{j\ m}^{\ k}(Z_{ip}^{\ \ m} + Z_{pi}^{\ \ m})y^p + \\ &+ C_{ijm}(Z_{hp}^{\ \ m} - Z_{ph}^{\ \ m})g^{kh}y^p - y^p y^q Z_{pqh}\partial C_i^{\ kh}/\partial y^j. \end{aligned} \tag{2.122}$$

The representations (2.116) and (2.121) show that the Cartan and Berwald connection coefficients of a 1-form Finsler space are equal to the sum of the connection coefficients (2.89) of absolute parallelism and tensor terms containing the Cartan torsion tensor C_{ijm}. In contrast to the Berwald coefficients (2.121), the Cartan coefficients (2.116) do not involve the derivatives of C_{ijm} but, instead, contain products of the tensors C_{ijm}. In these two coefficients the term which does not involve C_{ijm} is the same:

$$E_{i\ j}^{\ k} \stackrel{\text{def}}{=} L_{j\ i}^{\ k} + Z_{ij}^{\ \ k} \equiv E_{j\ i}^{\ k} \tag{2.123}$$

and itself represents connection coefficients since $L_{j\ i}^{\ k}$ are connection coefficients and $Z_{ij}^{\ \ k}$ is a tensor. They may be treated as a generalization of the Riemannian Christoffel symbols to the case of the 1-form Finsler space. Although the coefficients (2.123) are not metric, that is the covariant derivative of the Finslerian metric tensor relative to the connection coefficients $E_{i\ j}^{\ k}$ does not vanish, the covariant derivative of the metric function is zero:

$$\partial_i F - y^j E_i{}^k{}_j \partial F/\partial y^k = 0, \tag{2.124}$$

similarly to the case of the Berwald connection coefficients (see (1.52)). Indeed, differentiating the 1-form metric function given by (2.87) with respect to x^i directly yields

$$\partial_i F = y^j L_j{}^k{}_i \partial F/\partial y^k$$

whence Equation (2.124) follows in view of the relations (1.9) and $E_{ikj}y^k y^j = L_{ikj}y^k y^j$ (cf. (2.123) and (2.110)).

From (2.116) and (2.119) we obtain

$$G^{*k}_i{}_j \overset{\text{def}}{=} \Gamma_i{}^k{}_j + C_i{}^k{}_n G_j{}^n = L_i{}^k{}_j + C_i{}^k{}_n L_p{}^n{}_j y^p + B_i{}^k{}_j \tag{2.125}$$

where

$$B_i{}^k{}_j = Z_{ji}{}^k - C_j{}^k{}_m Z_{ip}{}^m y^p + C_j{}^m{}_i Z^k{}_{pm} y^p + S_{tj}{}^k{}_i Z_{pq}{}^t l^p l^q \tag{2.126}$$

($M^k{}_{ij} + Z_{ij}{}^k = Z_{ji}{}^k$ in accordance with Equation (2.112)). The complexity of the last term in Equation (2.126) is diminished in the $S3$-like case. The relation (2.125) may be used to calculate the Cartan covariant derivatives when they are taken in the form (1.69)–(1.71). For example, Equations (1.71) and (2.125) together with (2.103) yield

$$\begin{aligned} C_{ikj|h} = &- T_{ikjm} w_h{}^m F^{-1} - (B_i{}^m{}_h - l_i w_h{}^m F^{-1}) C_{mkj} - \\ &- (B_k{}^m{}_h - l_k w_h{}^m F^{-1}) C_{imj} - (B_j{}^m{}_h - l_j w_h{}^m F^{-1}) C_{ikm} \end{aligned} \tag{2.127}$$

where we have used the definition (2.27) of the tensor T_{ikjm}. Contracting Equation (2.127) with y^h leads to

$$\begin{aligned} C_{ikj|h} y^h = &- T_{ikjm} Z_{pq}{}^m y^p y^q F^{-1} - C_{mkj}(Z_{pi}{}^m - l_i Z_{pq}{}^m l^q) y^p - \\ &- C_{imj}(Z_{pk}{}^m - l_k Z_{pq}{}^m l^q) y^p - C_{ikm}(Z_{pj}{}^m - l_j Z_{pq}{}^m l^q) y^p. \end{aligned} \tag{2.128}$$

From Equation (2.128) we may infer the following result concerning the possibility of a 1-form Finsler space being the Landsberg space characterized by the vanishing of $C_{ikj|h}y^h$.

PROPOSITION 14. *Let the following three conditions hold:*

$$\begin{aligned} &T_{ikjm} = 0, \qquad C_{kjm} C_i{}^k{}_n - C_{kjn} C_i{}^k{}_m = (h_{in} h_{jm} - h_{im} h_{jn}) F^{-2}, \\ &C_i{}^m{}_m = 0 \end{aligned} \tag{2.129}$$

as in the case of the 1*-form Berwald-Moór metric function. Then the tensor* $C_{ikj|h}y^h$ *vanishes if and only if*

$$l^p (Z_{pim} - l_i l^q Z_{pqm} + l_m l^q Z_{pqi}) = 0. \tag{2.130}$$

Proof. The sufficiency is evident from the representation (2.128). To prove the necessity, we contract (2.128) with C^{ijn} which yields

$$C^{ijn}C_{ikj|h}y^h = -2C^{ijn}C^m_{kj}Z_{pim}y^p - C^{ijn}C_{imj}(Z_{pk}{}^m - l_kZ_{pq}{}^m l^q)y^p. \tag{2.131}$$

By virtue of the skew-symmetry of the tensor Z_{pim} in the last two indices (see the definition (2.105)), we may represent the first term on the right-hand side of (2.131) as follows

$$-(C^{ijn}C^m{}_{kj} - C^{mjn}C^i{}_{kj})Z_{pim}y^p$$

and then simplify it by using the second identity given in (2.129). The second term on the right-hand side of (2.131) is evaluated by substituting $C^{ijn}C_{imj} = -(N-2)h_m{}^n/F^2$ (ensuing from the last two identities in (2.129)). In so doing, we get

$$C^{ijn}C_{ikj|h}y^h = NF^{-2}(Z_{pk}{}^n + Z_{pqk}l^q l^n - Z_{pq}{}^n l^q l_k)$$

which completes the proof.

An n-tuple $S^m_A(x)$ gives rise to *a preferred vector field*

$$S^m(x) = \sum_{A=1}^{N} S^m_A(x) \tag{2.132}$$

which is symmetric with respect to the vectors constituting the n-tuple:

$$S^m S^A_m = 1 \tag{2.133}$$

for any value of the index A. This vector field can be used to define the osculating Riemannian metric tensor

$$a_{ij}(x) = g_{ij}(x, S(x)) \tag{2.134}$$

in accordance with the definition (1.86). By (2.88) and (2.133),

$$y^A(x, S(x)) = 1. \tag{2.135}$$

The last equality, together with the fact that g_{AB} and C_{ABD} are functions of y^A alone, imply the following

PROPOSITION 15. The quantities $a_{AB} = g_{AB}(x, S(x))$ and $C_{ABD}(x, S(x))$ are constants.

It can be easily verified that the connection coefficients (2.123) exhibit the remarkable property

$$E_i{}^k{}_j(x, S(x)) = \left\{{}_i{}^k{}_j\right\} \tag{2.136}$$

where the right-hand side is the Riemannian Christoffel symbols (1.88) associated with the osculating tensor (2.134). Further, owing to Equations (2.136) and (2.123), we have

$$\mathrm{D}_j S^i_A = (\partial_j S^i_A + L_k{}^i{}_j S^k_A) + Z_{jA}{}^i(x, S(x)),$$

where D_j denotes the Riemannian covariant derivative taken with respect to the Christoffel symbols $\{{}^{k}_{ij}\}$. The definition (2.89) of L^{i}_{kj} immediately implies that the sum in parenthesis in this expression vanishes. Therefore,

$$S^{A}_{k}\,\mathrm{D}_{j}S^{i}_{A} = Z_{jk}{}^{i}(x, S(x)). \tag{2.137}$$

As a consequence of Proposition 14, a representation of the form

$$a_{AB} = a_{PQ}R^{P}_{A}R^{Q}_{B} \tag{2.138}$$

must exist, (R^{B}_{A}) being a nonsingular matrix of real constants and the factors a_{PQ} being zero except for a_{PP} equalling $+1$ or -1 depending on the signature of the metric tensor; $P, Q = 0, 1, \ldots, N-1$. It follows from Equations (2.134) and (2.128) that

$$h^{P}_{i}(x) \overset{\text{def}}{=} R^{P}_{A}S^{A}_{i}(x) \tag{2.139}$$

plays the role of the covariant orthonormal n-tuple for the osculating metric tensor (2.134), i.e.

$$a_{ij}(x) = a_{PQ}h^{P}_{i}(x)h^{Q}_{j}(x). \tag{2.140}$$

As a consequence of the definition (2.139)

$$S^{A}_{k}\,\mathrm{D}_{j}S^{i}_{A} = h^{P}_{k}\,\mathrm{D}_{j}h^{i}_{P} \tag{2.141}$$

where the n-tuple h^{i}_{P} is reciprocal to h^{P}_{i}. The right-hand side of Equation (2.141), and hence of (2.137), is nothing but the Ricci rotation coefficients of the osculating Riemannian space defined by the tensor (2.134).

In view of (2.98) and (2.115) the contraction (2.120) may be expressed in the form

$$G_{m}{}^{m} = y^{m}\,\partial_{m}\ln|\det(S^{A}_{i})| - C_{k}{}^{k}{}_{m}(\gamma_{p}{}^{m}{}_{q} - L_{p}{}^{m}{}_{q})y^{p}y^{q} \tag{2.142}$$

in which simplifactions will occur if the coordinates x^{i} are chosen such that

$$\det(S^{A}_{i}) = \text{const.} \tag{2.143}$$

Such a choice is always possible locally because of the fact that $\det(S^{A}_{i})$ is a scalar density of weight $+1$. Indeed, let $W(x)$ be any scalar density of weight $+1$, which means that the transform of W under arbitrary coordinate transformations $x^{i} = x^{i}(\bar{x}^{j})$ reads $\bar{W}(\bar{x}) = J(x)W(x)$ where $J(x) = \det(\partial x^{i}/\partial\bar{x}^{j})$. Choosing the coordinates $\bar{x}^{j}$ to be

$$\bar{x}^{1} = C^{-1}\textstyle\int W(x)\,\mathrm{d}x^{1}, \qquad \bar{x}^{2} = x^{2}, \ldots, \bar{x}^{N} = x^{N}$$

with C being a constant, we get $J(x) = C/W$ and hence $\bar{W}(\bar{x}) = C$. Thus we have established[12]

PROPOSITION 16. For any given n-tuple $S^{A}_{i}(x)$ the system of local coordinates x^{i} can be chosen such that the equality (2.143) will hold with respect to these x^{i}.

The equality (2.143) eliminates the first term in the right-hand side of

Equation (2.142). If also $C_k{}^k{}_m = 0$, as is so for the Berwald-Moór metric function (Equation (2.38)), then (2.143) results in $G_m{}^m = 0$. Also, it is worth noting that the condition $C_k{}^k{}_m = 0$ entails the following interesting identities for the curvature tensors in the 1-form case:

$$K_i{}^i{}_{jk} = R_i{}^i{}_{jk} = H_i{}^i{}_{jk} = 0, \qquad K_{mn} = K_{nm}, \qquad H_{mn} = H_{nm},$$

where $K_{mn} = K_m{}^i{}_{ni}$ and $H_{mn} = H_m{}^i{}_{ni}$. These identities can be readily verified by using the coordinates x^i obeying the property (2.143) and noting that $C_k{}^k{}_m = 0$ together with (2.143) imply $G_m{}^m{}_n = \Gamma_m{}^m{}_n = 0$.

In order to have a sufficiently simple subtype of a 1-form Finsler space we introduce the following:

DEFINITION. A 1-form Finsler space is called *conformally flat* if the space admits a coordinate system x^i such that, relative to these x^i, the n-tuples $S_i^A(x)$ are of the form

$$S_i^A(x) = M(x)\delta_i^A \tag{2.144}$$

where δ_i^A is the Kronecker symbol.

It should be clear that relative to an arbitrary coordinate system x^i the n-tuples (2.144) will take a form proportional to gradients:

$$S_i^A(x) = M(x)\,\partial f^A(x)/\partial x^i$$

where $f^A(x)$ are N independent scalar functions; $M(x)$ transforms like a scalar. This last representation may also be regarded as the definition of a conformally flat 1-form space.

To simplify calculations in the remainder of the present Section we shall assume that the coordinates x^i are chosen such that Equation (2.144) holds. In this case we have the remarkable relation

$$\partial_i S_j^A = K_i S_j^A \tag{2.145}$$

where $K_i = \partial \ln M/\partial x^i$, which leads to many simplifications. Indeed, the tensors (2.106) take the form

$$M_{jh}^A = K_j S_h^A - K_h S_j^A$$

so that

$$M^i_{jh} = K_j\delta^i_h - K_h\delta^i_j, \qquad M_{ijh} = K_j g_{ih} - K_h g_{ij}$$

which reduces the expression (2.105) of the tensor Z_{ijh} to the simple form

$$Z_{ijh} = K_j g_{ih} - K_h g_{ij}.$$

The last equality in turn results in

$$l^i Z_{ijh} = K_j l_h - K_h l_j, \qquad l^j Z_{ijh} = l^j K_j g_{ih} - K_h l_i,$$

$$l^i l^j Z_{ijh} = l^j K_j l_h - K_h,$$

which shows that

$$l^p Z_{pim} = l_i Z_{pqm} l^p l^q - l_m Z_{pqi} l^p l^q \tag{2.146}$$

(cf. Equation (2.130)). Relation (2.146) reduces (2.128) to

$$C_{ikj|h} y^h = F K_p T^p_{ikj}. \tag{2.147}$$

Further, Equation (2.145) entails

$$\partial_i g_{mn} = 2K_i g_{mn} + (\partial g_{mn}/\partial y^A)\partial_i y^A = 2K_i g_{mn} + 2C_{mnA} y^k \partial_i S_k{}^A =$$
$$= 2K_i g_{mn} + 2K_i C_{mnA} y^A = 2K_i g_{mn}.$$

So,

$$\partial_i g_{mn} = 2K_i g_{mn}. \tag{2.148}$$

Also,

$$\partial_i g^{mn} = -2K_i g^{mn} \tag{2.149}$$

because of

$$\partial_i S^m_A = -S^n_A S^m_B \partial_i S^B_n = -K_i S^m_A. \tag{2.150}$$

Similarly,

$$\partial_i C_{mnk} = 2K_i C_{mnk}, \qquad \partial_i C^n_{mk} = 0, \qquad \partial_i C^{mn}_k = -2K_i C^{mn}_k, \tag{2.151}$$
$$\partial_i l_k = K_i l_k, \qquad \partial_i l^A = \partial_i l_A = 0.$$

From Equations (2.148) and (2.149), the Christoffel symbols (1.26) will be equal to just

$$\gamma^i_{mn} = K_m \delta^i_n + K_n \delta^i_m - g_{mn} g^{ip} K_p, \tag{2.152}$$

from which we get

$$G^i \stackrel{\text{def}}{=} \tfrac{1}{2} y^m y^n \gamma_m{}^i{}_n = y^i y^p K_p - \tfrac{1}{2} F^2 g^{ip} K_p. \tag{2.153}$$

The differentiation of (2.153) yields simple expressions for the Berwald connection coefficients $G_k{}^i{}_h = \partial G_k{}^i/\partial y^h$ where $G_k{}^i = \partial G^i/\partial y^k$. Namely,

$$G_k{}^i = \delta^i_k y^p K_p + y^i K_k - y_k g^{ip} K_p + F^2 C_k{}^{ip} K_p, \tag{2.154}$$

$$G_k{}^i{}_h = \delta^i_k K_h + \delta^i_h K_k - Q^{ip}_{kh} K_p, \tag{2.155}$$

where

$$Q^{ip}_{kh} = -F^2 \partial C_k{}^{ip}/\partial y^h - 2y_k C_h{}^{ip} - 2y_h C_k{}^{ip} + g_{kh} g^{ip} \equiv$$
$$\equiv \tfrac{1}{2}\partial^2(F^2 g^{ip})/\partial y^k \partial y^h$$

(cf. (2.24)) which implies

$$g^{ip}(x, y) = Q^{ip}_{kh}(x, y) l^k l^h \tag{2.156}$$

because of the homogeneity of g^{ip} with respect to y^n.

The relation (2.155) shows that *a space is affinely-connected* (i.e. that the connection coefficients $G_{k\ h}^{\ i}$ are independent of y^n (see (1.53)) if and only if $\partial Q_{kh}^{ip}/\partial y^n = 0$. The latter condition means, by virtue of Equation (2.156), that the Finslerian metric tensor must be quadratic in the unit tangent vectors (in the sense of Section 2.2). Thus we get the following result.

PROPOSITION 17. A conformally flat 1-form Finsler space is affinely-connected if and only if the Finslerian metric tensor is quadratic in the unit tangent vectors.

According to Proposition 5, this possibility is realized only if the T-condition is satisfied which is so, for example, in the 1-form Berwald-Moór Finsler space. Hence we obtain:

PROPOSITION 18. The conformally flat 1-form Berwald-Moór Finsler space is affinely-connected.

Equations (2.154)–(2.155) yield

$$\begin{aligned} G_{k\ h}^{\ i} + C_{k\ q}^{\ i} G_h^{\ q} = \delta_k^i K_h + \delta_h^i K_h - g_{kh} g^{ip} K_p + \\ + y_k C_h^{\ ip} K_p - y^i C_{k\ h}^{\ p} K_p + K_p (F T_{kh}^{\ \ ip} + S^i_{\ kh}{}^p). \end{aligned} \tag{2.157}$$

The complexity of the last term involving the T-tensor (2.27) and the S-tensor (1.62) will be lessened if the T-condition (Section 2.2) and the $S3$-like condition (Section 2.1) are satisfied. In particular, in the case of the Berwald-Moór metric function, we have $T_{kh}^{\ \ ip} = 0$ and Equation (2.36) which lead to:

PROPOSITION 19. Cartan's and Berwald's connection coefficients for the conformally flat 1-form Berwald-Moór Finsler space are equal, dependent on x^i alone, and given by

$$\begin{aligned} \Gamma_{k\ h}^{\ i}(x) = G_{k\ h}^{\ i}(x) = & -C_{k\ q}^{\ i} G_h^{\ q} + \delta_k^i K_h + \\ & + l_k K_p (\delta_h^i l^p - l_h g^{pi} + F C_h^{\ ip}) + l^i K_p (\delta_k^p l_h - \\ & - g_{kh} l^p - F C_{k\ h}^{\ p}) \end{aligned} \tag{2.158}$$

where $G_h^{\ q}$ are given by Equation (2.154).

From Equations (2.146) and (2.128) it may be concluded that a conformally flat 1-form Finsler space has the property $C_{ikj|h} y^h = 0$ (specific to a Landsberg space) if and only if the space satisfies the T-condition. In particular, because of Proposition 8 of the present chapter, any Finsler space with a conformally flat 1-form metric function of the generalized Berwald-Moór type (2.10) possesses the property $C_{ikj|h} y^h = 0$ and, therefore, the property $\Gamma_{i\ j}^{\ k} = G_{i\ j}^{\ k}$ (see Equation (1.50)).

In view of Equations (2.153), (1.50), and (2.146) the Cartan connection coefficients of any conformally flat 1-form Finsler space are of the form

$$\begin{aligned} \Gamma_{k\ h}^{\ i} = -C_{k\ q}^{\ i} G_h^{\ q} + \delta_k^i K_h + \delta_h^i K_k - g_{kh} g^{ip} K_p + \\ + y_k C_h^{\ ip} K_p - y^i C_{kh}^{\ p} K_p + K_p S_{\ kh}^{i\ \ p} \end{aligned} \tag{2.159}$$

and are simpler than $G_{kh}^{\ i}$ in that (2.159) does not involve the T-tensor which enters the last term in (2.157). Substituting (2.154) in (2.159) we obtain

$$\Gamma_{k\ h}^{\ i}(x,y) = K_p(x)E_{kh}^{ip}(y) \tag{2.160}$$

where

$$\begin{aligned} E_{kh}^{ip} = \delta_k^i\delta_h^p + \delta_h^i\delta_k^p - g_{kh}g^{ip} + y_kC_h^{\ ip} + y_hC_k^{\ ip} - \\ - y^iC_{k\ h}^{\ p} - y^pC_{k\ h}^{\ i} - F^2C_{km}^{\ \ i}C_h^{mp} + S_{\ kh}^{i\ \ p}; \end{aligned} \tag{2.161}$$

the fact that these E_{kh}^{ip} are independent of x^n:

$$\partial E_{kh}^{ip}/\partial x^n = 0$$

is evident from the relations (2.148)–(2.51). Equation (2.160) implies that the Cartan curvature tensor can be represented in the form

$$R_{j\ kh}^{\ i}(x,y) = K_{pq}(x)R_{(1)}{}^{pq}{}_{j\ kh}^{\ i}(y) + K_p(x)R_{(2)}{}^{pq}{}_{j\ kh}^{\ i}(y). \tag{2.162}$$

In this relation $K_{pq} = \partial K_p/\partial x^q$ and

$$R_{(1)}{}^{pq}{}_{j\ kh}^{\ i} = \tfrac{1}{2}(\delta_k^pN_{jh}^{iq} + \delta_k^qN_{jh}^{ip} - \delta_h^qN_{jk}^{iq} - \delta_h^qN_{jk}^{ip})$$

with

$$N_{kh}^{ip}(y) = \delta_k^i\delta_h^p + \delta_h^i\delta_k^p - g_{kh}g^{ip} + y_kC_h^{\ ip} - y^iC_{k\ h}^{\ p} + S_{\ kh}^{i\ \ p}$$

(see the definitions (1.80) and (1.78) and Equation (1.59)), while the tensor $R_{(2)}$ is of a more complicated structure

It is also of interest to note that in the particular case when K_p are independent of x^i, that is when (2.144) takes the form

$$S_i^A(x) = \delta_i^A \exp K_nx^n, \ K_n \text{ being constants,} \tag{2.163}$$

the above formulae (2.160)–(2.162) imply that the Cartan connection coefficients $\Gamma_{k\ h}^{\ i}$, as well as the Berwald connection coefficients $G_{k\ h}^{\ i}$ (Equation (2.155)), and therefore the curvature tensors $R_{j\ kh}^{\ i}$, $K_{j\ kh}^{\ i}$, $H_{j\ kh}^{\ i}$, are all independent of x^i:

$$\begin{aligned} \partial\Gamma_{k\ h}^{\ i}/\partial x^n = \partial G_{k\ h}^{\ i}/\partial x^n = \partial K_{j\ kh}^{\ i}/\partial x^n = \partial R_{j\ kh}^{\ i}/\partial x^n = \\ = \partial H_{j\ kh}^{\ i}/\partial x^n = 0 \quad \text{if} \quad \partial K_n/\partial x^i = 0. \end{aligned} \tag{2.164}$$

2.5. The Randers Metric Function

Denoting

$$a(x,y) = (a_{mn}(x)y^my^n)^{1/2}, \qquad b(x,y) = b_i(x)y^i \tag{2.165}$$

where $a_{mn}(x)$ is a Riemannian metric tensor and $b_i(x)$ is a covariant vector field, we can construct the Finslerian metric function

$$F(x,y) = a(x,y) + b(x,y) \tag{2.166}$$

which is called *the Randers metric function*[13]. A Finsler space with such a metric function is called *a Randers space*. The associated equations of geodesics will be written in terms of the Lagrangian function

$$F(x,\dot{x}) = a(x,\dot{x}) + b(x,\dot{x}) \tag{2.167}$$

where $\dot{x}^i = \mathrm{d}x^i/\mathrm{d}q$ and q is a parameter along a curve. The calculation of the Finslerian metric tensor related to the metric function (2.166) gives the following result

$$g_{ij} = F(a_{ij} - u_i u_j)/a + p_i p_j, \tag{2.168}$$

$$g^{ij} = (a^{ij} - p^i b^j - p^j b^i + (b_n b^n + b_n u^n) p^i p^j) a/F \tag{2.169}$$

where we use the notation

$$b^i = a^{ij} b_j, \qquad u_i = \partial a/\partial y^i = a_{ij} y^j/a,$$

$$p_i = \partial F/\partial y^i = u_i + b_i, \qquad p^i = a^{ij} p_j.$$

Differentiating Equation (2.168) with respect to y^m shows that the Cartan torsion tensor (1.14) has the following form

$$C_{ijm} = \frac{1}{N+1}(h_{ij} C_m + h_{jm} C_i + h_{mi} C_j) \tag{2.170}$$

where h_{ij} is the angular metric tensor (1.12),

$$C_i \overset{\text{def}}{=} C_n{}^n{}_i = \frac{N+1}{2F}(b_i - u_i b/F) \tag{2.171}$$

and N is the dimension number of the space.

If we restrict consideration of the definition (2.167) to the four-dimensional case $N = 4$, identifying $a_{mn}(x)$ with the (pseudo-) Riemannian metric tensor of space-time and putting

$$b_i(x) = eA_i(x)/m_0 c^2$$

where e and m_0 are the electric charge and the rest mass of a test particle respectively, c is the light velocity, and $A_i(x)$ denotes the electromagnetic vector potential, then the function (2.167) takes the form

$$F(x,\dot{x}) = (a_{mn}(x)\dot{x}^m \dot{x}^n)^{1/2} + \frac{e}{m_0 c^2} A_i(x)\dot{x}^i.$$

This last expression is, up to a constant factor, *the Lagrangian function of a test electric charce*[14] in the electromagnetic and gravitational fields described by the vector potential $A_i(x)$ and the Riemannian metric tensor $a_{mn}(x)$.

The reader can readily verify that, taking the Riemannian arc-length to be the parameter q along a curve, so that

$$\mathrm{d}q = a(x, \mathrm{d}x),$$

the equations (1.24) of the geodesics of the present space may be written in the form

$$\mathrm{d}\dot{x}^m/\mathrm{d}q + 2T^m(x, \dot{x}) = 0$$

with

$$2T^m(x, \dot{x}) = \{{}^{\,m}_{ij}\}(x)\dot{x}^i\dot{x}^j + a(x, \dot{x})\dot{x}^j F_j^m(x)$$

where $F_j^m = a^{mn}F_{jn}$ and

$$F_{jn}(x) \doteq \partial A_n/\partial x^j - \partial A_j/\partial x^n$$

which in a physical context plays the role of the electromagnetic field strength tensor; $\{{}_i{}^m{}_j\}$ denote the Riemannian Christoffel symbols associated with $a_{mn}(x)$.

From T^m we can construct connection coefficients $T_i{}^m{}_j$ in the same way as the Berwald connection coefficients $G_i{}^m{}_j$ are constructed from G^m (see Equations (1.33) and (1.49)); namely we write

$$T_i^m \overset{\text{def}}{=} \partial T^m/\partial \dot{x}^i = \{{}_i{}^m{}_j\}\dot{x}^j + u_i T^m/a^2 + \tfrac{1}{2}aF_i{}^m,$$

$$T_i{}^m{}_j \overset{\text{def}}{=} \partial T_i{}^m/\partial \dot{x}^j = \{{}_i{}^m{}_j\} + \tfrac{1}{2}(a_{ij}\dot{x}^k F_k^m + u_i F_j{}^m + u_j F_i{}^m)/a - $$
$$- \tfrac{1}{2}u_i u_j \dot{x}^k F_k{}^m/a^3,$$

where $a = a(x, \dot{x})$ and $u_i = u_i(x, \dot{x}) \equiv a_{ij}\dot{x}^j/a(x, \dot{x})$. In terms of this notation the geodesics of the Randers space may be represented by equations of the form

$$\mathrm{d}\dot{x}^m/\mathrm{d}q + T_{ij}^m(x, \dot{x})\dot{x}^i\dot{x}^j = 0.$$

With these connection coefficients we can associate the curvature tensor $\bar{H}_h{}^i{}_{jk}(x, \dot{x})$, by substituting in (1.85) the T-objects instead of the G-objects, the result being

$$\bar{H}_h{}^i{}_{jk}(x, \dot{x}) = M_h{}^i{}_{jk}(x) + \mathscr{A}_{[jk]}[(u_h \nabla_k F_j{}^i + \dot{x}^m a_{hj} \nabla_k F_m{}^i +$$
$$+ u_j \nabla_k F_h{}^i)a^{-1} - \dot{x}^m u_h u_j a^{-3} \nabla_k F_m{}^i]$$

with

$$M_h{}^i{}_{jk}(x) = R'_h{}^i{}_{jk} + \tfrac{1}{2}\mathscr{A}_{[jk]}(F_h{}^i F_{jk} + a_{hk}F_j{}^m F_m{}^i - F_{hj}F_k{}^i).$$

Here, R'^i_{hjk} and ∇_k are respectively the Riemannian curvature tensor and covariant derivative associated with $a_{ij}(x)$ and $\mathscr{A}_{[jk]}$ stands for the skew-symmetrization operator with factor $\frac{1}{2}$, for example $\mathscr{A}_{[jk]}Z_{mjk} = (Z_{mjk} - Z_{mkj})/2$.

The interesting fact is that the tensor $H_h{}^i{}_{jk}$ has a part $M_h{}^i{}_{jk}$ which does not depend on the velocities $\dot{x}^n$. Contraction yields

$$M_h{}^i{}_{ji}a^{hj} = R' + \text{const}F_{mn}F^{mn}$$

($F^{mn} = a^{mi}F_i{}^n$), which is exactly the Lagrangian of the classical gravitational and electromagnetic fields.

2.6. The Kropina Metric Function

Using the scalars (2.165) we can construct a Finslerian metric function of the form

$$F(x, y) = a^2(x, y)/b(x, y) \tag{2.172}$$

which is called *the Kropina metric function*[15]. The associated Finslerian metric tensor is found to be

$$g_{ij} = a^2(2a_{ij} - l_i b_j - l_j b_i)/b^2 + l_i l_j,$$

$$g^{ij} = [(ra^{ij} + 2(l^i b^j + l^j b^i) - b^i b^j + 2(ra^2/b^2 - 2)l^i l^j]b^2/2ra^2$$

where

$$r = a^{ij}b_i b_j, \qquad b^i = a^{ij}b_j, \qquad l_i = \partial F/\partial y^i, \qquad l^i = g^{ij}l_j.$$

The tensor C_{ijm} is given by the expression (2.170) with

$$C_i = (N + 1)(l_i - a^2 b_i/b^2)/2F. \tag{2.173}$$

Such a metric function is of physical interest in the sense that it describes the general dynamical system represented by the following Lagrangian function[16]:

$$L(x^a, t, \dot{x}^a) = \tfrac{1}{2}g_{ab}(x^c, t)\dot{x}^a\dot{x}^b + K_b(x^a, t)\dot{x}^b - U(x^a, t)$$

where the indices a, b, c range over $1, 2, \ldots, N - 1$; the parameter t represents the time, so that $\dot{x}^a = \mathrm{d}x^a/\mathrm{d}t$; the object g_{ab} has the physical meaning of the mass tensor; K_b is the so-called attenuation coefficient; and U is the potential energy of the system. Let us treat t as an additional coordinate $x^N = t$ and choose any admissible parameter q such that x^a, x^N could be considered as functions of q along any trajectory of the considered dynamical system. Then we shall have $\dot{x}^a = x'^a/x'^N$ where the prime denotes that differentiation with respect to q. Under these conditions, specifying that the indices i, j, n range over $1, 2, \ldots, N$, and putting

$$g_{ab} = 2a_{ab}(x^n)/b_N(x^n), \qquad K_b = 2a_{bN}(x^n)/b_N(x^j),$$

$$U = -a_{NN}(x^n)/b_N(x^j),$$

where b_N is an arbitrary nonvanishing function and $a_{bN} = a_{Nb}$, we get

$$F(x^n, x'^n) \overset{\text{def}}{=} Lx'^N = (a_{mn}(x^i)x'^m x'^n)/b_N(x^j)x'^N \tag{2.173'}$$

which indeed agrees with the definition (2.172)[17].

2.7. C-reducible Finsler Spaces

An interesting special type of Finsler space arises when one assumes a particular form for the Cartan torsion tensor in accordance with the following definition[18].

DEFINITION. A Finsler space of dimension N is called *C-reducible* if the following three conditions are satisfied:

(i) The Finsler space is not Riemannian.
(ii) The dimension number N is higher than 2.
(iii) The Cartan torsion tensor C_{ijm} can be written in the form

$$C_{ijm} = (N+1)^{-1}(h_{ij}C_m + h_{jm}C_i + h_{mi}C_j) \tag{2.174}$$

where $C_i = C_k{}^k{}_i$.
It will be noted that postulating the form

$$C_{ijm} = h_{ij}M_m + h_{jm}M_i + h_{mi}M_j$$

would not result in a more general type of Finsler space. Indeed, the identities $y^i h_{ij} = 0$ and $y^i C_{ijm} = 0$ (see Equations (1.13) and (1.16)) require that the condition $y^i M_i = 0$ be imposed on M_i. Therefore, contracting C_{ijm} with g^{jm} will give merely

$$M_i = C_i/(N+1); \tag{2.175}$$

hence, we again get the representation (2.174). The introduction of item (ii) in the definition is explained by the well-known fact that the tensor C_{ijm} of any two-dimensional Finsler space has form[19] $FC_{ijk} = Jm_i m_j m_k$ where J is some scalar and m_i is a covariant vector.

C-reducible Finsler spaces exhibit many interesting properties. Thus, substituting (2.174) in the general expression (1.62) of the curvature tensor S_{ijmn} yields

$$F^{-2}S_{ijmn} = h_{in}M_{jm} - h_{im}M_{jn} + h_{jm}M_{in} - h_{jn}M_{im} \tag{2.176}$$

where

$$M_{ij} = \tfrac{1}{2}M_n M^n h_{ij} + M_i M_j \tag{2.177}$$

and M_i is defined by Equation (2.175). As follows from Equations (2.176) and (1.129), the tensor S_{ijmn} of a C-reducible Finsler space of any dimension $N \geqslant 5$ has the same form as that which the tensor S_{ijmn} has in any 4-dimensional Finsler space. On the other hand, a C-reducible Finsler space cannot be $S3$-like. Indeed, contracting Equation (2.176) yields

$$S_{im} = -(N-3)F^2 M_{im} - F^2 M_n{}^n h_{im} \tag{2.178}$$

as a consequence of $l^i M_{ij} = 0$ together with the equalities $h^{ij} = g^{ij} - l^i l^j$ and $h_i^i = N - 1$ (see Equations (1.12) and (1.13)). If we assume that the Finsler space under consideration is $S3$-like, we shall have $S_{im} = Sh_{im}/(N-1)$ (Equation (2.2)) and equating this to (2.178) gives an equation of the form

$$Bh_{im} + M_i M_m = 0, \tag{2.179}$$

where we have used Equation (2.177) and the inequality $N > 3$ entering the definition of $S3$-like Finsler spaces (see Section 2.1). Since the rank of the angular metric tensor h_{im} defined by Equation (1.12) is evidently equal to $N-1$ the factor B in (2.179) must vanish, in which case (2.179) implies $M_i = 0$ or, in view of (2.175), $C_i = 0$. If $C_i = 0$ the tensor C_{ijm} of the postualted form (2.174) vanishes and, therefore, the space is Riemannian. Thus we have established that the expressions (2.178) and (2.2) with $N > 3$ exclude each other. As a particular consequence of this observation we get:

PROPOSITION 20. There does not exist a Finsler space which is simultaneously $S3$-like and C-reducible.

Further, the condition (2.174) reduces the T-tensor (2.27) to the following elegant form

$$T_{mnij} = FM(h_{ni}h_{jm} + h_{nj}h_{mi} + h_{nm}h_{ij}) \tag{2.180}$$

where M is a scalar defined by the relation

$$M_i|_j + (M_i l_j + M_j l_i)/F = Mh_{ij}.$$

Also, it can be easily verified that Equation (2.174) entails

$$C_{nij}|_m y^m = h_{ni}P_j + h_{ij}P_n + h_{jn}P_i \tag{2.181}$$

where $P_i = M_i|_j y^j$. This last relation could be used to write down the explicit form of the second curvature tensor (1.74) of Cartan.

We see from Equations (2.170), (2.171), and (2.173) that a space having a Randers or Kropina metric function in C-reducible. The inverse statement is also valid:

PROPOSITION 21. The Finslerian metric function of any C-reducible Finsler space is of either the Randers form or the Kropina form.

This recent result[20] reduces the study of C-reducible Finsler spaces to the investigation of two comparatively simple metric functions.

The results of this chapter will often be referred to in the subsequent parts of the book.

Problems

2.1. Verify that every tangent Minkowskian space of an N-dimentional $S3$-like Finsler space can be locally embedded in an $(N+1)$-dimensional Euclidean space.

2.2. Prove that none of the metric functions (2.9) satisfies the T-condition. Calculate the explicit form of the T-tensor for the case of the metric functions (2.9).

2.3. Examine the signatures of the Finslerian metric tensors associated with the metric functions (2.9). Verify, in particular, that the following assertion is valid in the four-dimensional case $N = 4$: with $r^A = (1, 1, 1, 1)$,

$$\operatorname{sign}(g_{ij}) = \begin{cases} (+ \ + \ + \ +) \text{ iff } > 1 \\ (+ \ - \ - \ -) \text{ iff } < 1 \end{cases}.$$

with $r^A = (1, -1, -1, -1)$,

$$\operatorname{sign}(g_{ij}) = \begin{cases} (+ \ + \ + \ +) \text{ iff } < 1 \\ (+ \ - \ - \ -) \text{ iff } > 1 \end{cases}.$$

2.4. Consider N independent linear 1-forms Y^A whose values at x^n with respect to the tangent vector y^i are given by $Y^A(x^n, y^i) = S_i^A(x^n)y^i$ where S_i^A is a set of covariant n-tuples. Let $f^\beta(x, Y^{A_{\beta-1}+1}, \ldots, Y^{A_\beta})$ be a collection of M arbitrary scalar functions which are positively homogeneous of degree one in their arguments Y^A; the index β ranges over $1, \ldots, M$ and $0 = A_0 < A_1 < \cdots < A_{M-1} < A_M = N$. Let M be a positive integer not greater than N. The designation I_β will be adopted for the interval $(A_{\beta-1} + 1, \ldots, A_\beta)$. Assuming that $g^\beta_{AB} = \frac{1}{2}\partial^2(f^\beta)^2/\partial Y^A \partial Y^B$ with $A, B \in I_\beta$ are each quadratic in $l^\beta_A = \partial f^\beta/\partial Y^A$, prove that the Finslerian metric tensor g_{ij} associated with the following composite Finslerian metric function

$$F(x^n, y^i) = \prod_{\beta=1}^{M} [f^\beta(x^n, Y^{A_{\beta-1}+1}(x^n, y^i), \ldots, Y^{A_\beta}(x^n, y^i))]^{e_\beta(x)}$$

is quadratic in the unit tangent vectors l_n, where $e_\beta(x)$ are arbitrary nonvanishing scalars subject to the condition $\Sigma^M_{\beta=1} e_\beta = 1$.

2.5. Extend the relations (2.21)–(2.23) to the case when the components g^{ij} of the Finslerian metric tensor are positively homogeneous of an arbitrary degree r in l^n.

2.6. Not every tensor $Q^{ij}_{mn}(x)$ symmetric in its subscripts and superscripts may serve in Equation (2.19) to represent the Finslerian metric tensor. Find out the corresponding necessary and sufficient condition.

2.7. Given two tangent vectors y^i_1 and y^i_2 and a nonvansihing scalar $q(x)$, prove that the equality

$$g_{im}(x, y_1)y^m_2 = q(x)g_{im}(x, y_2)y^m_2$$

implies that $q = 1$ and $y^i_1/F(x, y_1) = y^i_2/F(x, y_2)$ in the case of the 1-form Berwald-Moór metric function.

2.8. Verify that

$$\hat{g}F^q = (q(q-1) + q(N-1))F^q$$

in any Finsler space, and that

$$\hat{S}F^q = q^N F^q, \qquad \hat{S}(l_m F^q) = (q+1)^{N-1}(q+1-N)l_m F^q,$$

$$\hat{S}(l^m F^q) = (q-1)^{N-1}(q-1+N)l^m F^q$$

in the Finsler space with the 1-form Berwald-Moór metric function. Here, the notation q is used for the power exponent which takes on any real values. The operators $\hat{g}$ and $\hat{S}$ are given by (2.56) and (2.57).

2.9. Consulting the description of the projective and conformal properties of Finsler spaces in the monograph by Rund (1959), find out what simplifications occur in these properties for 1-form Finsler space under the assumption $C_k{}^k{}_m = 0$. Consider the projective properties of the Randers space.

2.10. Calculate the explicit form of the K-curvature tensor (1.78) and the Cartan curvature tensor (1.80) of the conformally flat 1-form Berwald-Moór Finsler space.

2.11. In Riemannian geometry the vanishing of the Riemann curvature tensor implies that the space is Euclidean. In Finsler geometry one cannot, however, state that the vanishing of the K-tensor, R-tensor, or H-tensor of curvature, or even their simultaneous vanishing, will imply that the space is Minkowskian, i.e. that there exists a coordinate system x^i such that the Finslerian metric tensor is independent of x^i. What additional condition is required? May it be that $K_j{}^i{}_{kh} = R_j{}^i{}_{kh} = H_j{}^i{}_{kh} = C_{ijk|h} = 0$ and that at the same time the Finslerian metric tensor does depend on x^i relative to any coordinate system?

2.12. Find the scalar M entering the T-tensor (2.170) for the cases of the Randers and Kropina metric functions.

2.13. Prove for the case of the Randers metric function that either of the conditions $S_{ijmn} = 0$ or $T_{ijmn} = 0$ implies that $b_i = 0$, i.e. that the space is Riemannian.

2.14. Any fixed field of n-tuples $h_i^P(x,y)$ endows a space with the structure of absolute parallelism in the same way as the purely Riemannian case[11]. That is, a vector field $X^i(x,y)$ may be called absolutely parallel if the projections $X^i h_i^P$ are constants. Construct the covariant derivatives given rise to by such a parallelism and show that all three associated curvature tensors vanish.

2.15. Verify the assertion: given any metric function of the form $F(x,y) = (a_{ijk}(x)y^i y^j y^k)^{1/3}$, the tensor (1.17) satisfies the relation $FC_{ijkm} + l_i C_{jkm} + l_j C_{ikm} + l_k C_{ijm} + l_m C_{ijk} + \frac{1}{2}F^{-1}(h_{ij}h_{km} + h_{ik}h_{jm} + h_{im}h_{jk}) = 0$.

Notes

[1]This definition was introduced by Matsumoto (1971, p. 203); the property (2.3) has been given therein. Our definition of the scalar S differs from that used in the cited paper which in fact is identical with our S^* given by Equation (2.4). This change in notation has been made in order to obtain the equality $S_i^i = S$ in Equation (2.2).
[2]See, for example, Eisenhart (1950, Chapter 2).
[3]Asanov, 1979a, p. 332.
[4]See, for example, Eisenhart (1950, Equation (28.12)).
[5]The first example of an $S3$-like Finsler space was given in Asanov (1976, p. 295) where it was shown by direct calculation that the metric function (2.10) with $r^A = 1/N$, that is the 1-form Berwald-Moór metric function, satisfies the condition of $S3$-likeness; this calculation has been repeated in Matsumoto and Shimada (1978b). Proposition 2 was proved in Asanov (1979a, p. 331); another proof of this for the metric function (2.9) with $r^A = 1$ was given by Okubo (1979). The Minkowskian version of the metric function (2.10) was considered by Busemann (1967, pp. 41–44) in developing this theory of timelike Finslerian metrics.
[6]This definition was introduced and investigated in the paper by Asanov and Kirnasov (1982) which we follow in the present section.
[7]The T-tensor first appeared in papers of Matsumoto (1972b) and Kawaguchi (1972). The metric function (2.10) with $r^A = 1/N$ was the first example for which the T-condition was shown to hold (Matsumoto and Shimada, 1978b).
[8]The metric function (2.32) was first given by Berwald (1939) for Cartan spaces and, after that, by Moór (1954, p. 187) for Finsler spaces, precisely as an example of a space possessing the property $C_k{}^k{}_m = 0$.
[9]As is clear from Proposition 10, the metric function considered is not positive-definite, so that Deicke's theorem (Deicke, 1953), which claims that $C_k{}^k{}_i = 0$ entails that the space is Riemannian provided the metric function is positively definite and convex, cannot be applied to our case.
[10]This terminology was introduced in the paper of Matsumoto and Shimada (1978a). The present section follows the work by Asanov (1979c, 1983b).
[11]See Treder (1971, Chapter 4). The reader may find a mathematical description of absolute parallelism in the monograph by Slebodziński (1970, Chapter X, Section 129 and the historical note in Section 133). The suggestion to use Finslerian tetrads to define the connection coefficients of absolute parallelism in Finslerian space-time is made in a recent paper by Holland (1982).
[12]We have borrowed this kind of reasoning from Eisenhart (1927, p. 104).
[13]Randers (1941) proposed the metric function (2.166) in view of its application in general relativity. This asymmetrical (i.e. non-symmetrical relative to the replacement $y^m \to -y^m$) metric function seemed to him to reflect the unidirection of time in the real world. Randers deduced the equations of geodesics which proved to be of a form which, in the four-dimensional case, is identical with the equations of motion of a test electric charge. It is just this circumstance that has led many authors to become interested in this metric function: Stephenson (1953, 1957), Stephenson and Kilmister (1953), Ingarden (1957, 1976), Eliopoulos (1965), McKiernan (1966), Asanov (1977b), Holland (1982), a.o. Randers himself tried to reduce the study of space-time with his metric function to a sort of 5-dimensional Kaluza-Klein geometry where Riemannian methods plus a special method of tensor projection are used. No reference to Finsler geometry was made in this paper.
[14]See, for example, Landau and Lifshitz (1962, Section 16).
[15]Kropina (1961). In the two-dimensional case, this metric function was considered earlier by Berwald (1941).
[16]Such a dynamical interpretation of the Kropina metric function was proposed by Shibata (1978a, p. 117). This interpretation follows, in fact, the general line of applying geometrical methods to descriptions of dynamical systems (see Synge, 1936).
[17]The procedure used here is nothing but an application of the method of additional coordinates (see Section 6.1) according to which, given a Lagrangian function inhomogeneous with respect to

velocities, one may treat the parameter of time as an additional coordinate in order to obtain a Lagrangian function obeying the property of homogeneity of degree one in the velocities.
[18]Matsumoto (1971, p. 203). Formulae (2.176)–(2.181) were first given in Matsumoto (1974); Proposition 20 is known from Matsumoto (1972a).
[19]Equation (6.6.5) of Rund (1959).
[20]Matsumoto and Hōjō (1978). The idea that governed these authors in proving this conclusive theorem on C-reducible Finsler spaces was to use the characteristic tensor equation (2.174) of C-reducibility in order to construct a complete set of partial differential equations for the metric function.

Part C
Basic Equations

Chapter 3

Implications of the Invariance Identities

3.1. Invariance Identities

Differential geometry deals with those geometrical objects which obey definite transformation laws under admissible local coordinate transformations

$$x^i = x^i(\bar{x}^j). \tag{3.1}$$

The transformation laws are expressible in terms of the derivatives

$$B^i_j = \partial x^i/\partial \bar{x}^j, \qquad B_{j\,n}^{\,i} = \partial B^i_j/\partial \bar{x}^n, \qquad b^i_j = \partial \bar{x}^i/\partial x^j, \ldots. \tag{3.2}$$

The evident identity

$$B^i_j b^j_n = \delta^i_n \tag{3.3}$$

implies immediately that

$$\partial b^i_j/\partial B^m_n = -b^i_m b^n_j. \tag{3.4}$$

It is frequently assumed in geometrical considerations that a given geometrical object can be expressed functionally in terms of a set of other geometrical objects. In contrast to functional analysis where the form of the dependence of a function on other functions may be freely postulated, in considering the dependence of a geometrical object on others care should be taken to avoid contradictions with the transformation laws. On the other hand, one is entitled to expect that the differential consequences of the prescribed transformation laws should yield valuable information about the structure of the functional dependence of geometrical objects.

Such information is contained in the so-called *invariance identities*.[1] Let A_a and B_b be two sets of geometrical objects, the first of which is assumed to be functionally expressible through the second, i.e. $A_a = A_a(B_b)$. If the objects A_a and B_b and the functions $A_a(B_b)$ are sufficiently smooth, we may differentiate the transform $\bar{A}_a = \bar{A}_a(\bar{B}_b)$ with respect to the derivatives (3.2) at any fixed point x^i to obtain the invariance identities satisfied by the derivatives $\partial A_a/\partial B_b$. In turn, after differentiating the equality $\bar{A}_a = \bar{A}_a(\bar{B}_b)$ with respect to B_b and taking into account that $\bar{B}_b$ is expressible in terms of B_b according to the transformation law of B_b, we get an invariance identity of another type giving the transformation law of the derivatives $\partial A_a/\partial B_b$ under the coordinate transformations (3.1).

As an example which is of interest to us here, let us consider the case when a Finslerian metric function $F(x, y)$ is constructed by means of a set of x-dependent tensors $S^{Aj\cdots}_{i\ldots}(x)$, $A = 1, 2, \ldots$, of arbitrary types (p, q), so that

$$F(x, y) = v(S^{Aj\cdots}_{i\ldots}(x), y) \tag{3.5}$$

where $v(S^A, y)$ is a scalar function of degree one positive homogeneity in y^n:

$$v(S^A, ky) = kv(S^A, y), \qquad k > 0. \tag{3.6}$$

The type (p, q) may be different for different values of A.

By hypothesis, v obeys the following transformation law under (3.1):

$$\bar{v}(\bar{S}^A, y) = v(S^A, y) \tag{3.7}$$

where

$$\bar{y}^i = b^i_j y^j \tag{3.8}$$

because y^i form the components of contravariant vector, and

$$\bar{S}^{Aj_1\cdots j_p}_{i_1\cdots i_q} = S^{Ak_1\ldots k_p}_{m_1\ldots m_q} B^{m_1}_{i_1} \ldots B^{m_q}_{i_q} b^{j_1}_{k_1} \ldots b^{j_p}_{k_p}. \tag{3.9}$$

After differentiating Equation (3.7) with respect to B^n_m at any fixed point x^i, using the relations (3.3) and (3.5)–(3.9), and putting $\bar{x}^i$ equal to the initial x^i (which enatils writing B^n_m and b^j_i as δ^n_m and δ^j_i, respectively), we immediately find the following *invariance identity for the Finslerian metric function*:

$$T_i^j = y_i y^j / F \tag{3.10}$$

where y_i is related to y^j by the Finslerian rule (1.7) and

$$\begin{aligned} T_i^j = {} & S^{An\ldots}_{im\ldots} F^{jm\ldots}_{An\ldots} + S^{An\ldots}_{mik\ldots} F^{mjk\ldots}_{An\ldots} + \cdots - \\ & - S^{Ajm\ldots}_{n\ldots} F^{n\ldots}_{Aim\ldots} - S^{Amjk\ldots}_{n\ldots} F^{n\ldots}_{Amik\ldots} - \cdots . \end{aligned} \tag{3.11}$$

The notation

$$F^{j\cdots}_{An\ldots} = \partial v / \partial S^{An\cdots}_{j\ldots} \tag{3.12}$$

is used here. If we differentiate Equation (3.7) with respect to S^A and take into account (3.9), we get another invariance identity

$$B^{m_1\cdots m_q}_{i_1\ldots i_q} b^{j_1\cdots j_p}_{k_1\ldots k_p} \partial \bar{v} / \partial \bar{S}^{Aj_1\cdots j_p}_{i_1\ldots i_q} = \partial v / \partial S^{Ak_1\cdots k_p}_{m_1\ldots m_q} \tag{3.13}$$

which indicates merely that the derivative $\partial v / \partial S^A$ is a tensor. It is noted that the last observation is an example of the general rule: the derivative of a tensor with respect to another tensor is again a tensor.

By virtue of the definition (1.5), the Finslerian metric tensor is constructed using the same set of x-dependent variables as that used for the function $F(x, y)$, namely

$$g_{mn}(x, y) = v_{mn}(S^{Aj\cdots}_{i\dots}(x), y) \tag{3.14}$$

where

$$v_{mn} = \tfrac{1}{2}\partial^2 v^2/\partial y^m\, \partial y^n.$$

The same procedure which led from (3.7) to (3.11) yields the following *invariance identity for the Finslerian metric tensor*:

$$S^{Ak\cdots}_{i\dots}\,\partial v_{mn}(S^B, y)/\partial S^{Ak\cdots}_{j\dots} + \cdots - S^{Aj\cdots}_{k\dots}\,\partial v_{mn}(S^B, y)/\partial S^{Ai\cdots}_{k\dots} - \\ - \cdots = \delta^j_m g_{in}(x, y) + \delta^j_n g_{im}(x, y) + 2y^j C_{mni}(x, y). \tag{3.15}$$

Bearing in mind the definition (1.5), we may also obtain the identity (3.15) by direct differentiation, with respect to y^m and y^n, of the invariance identity (3.10) multiplied by $F(x, y)$.

Another example is provided by the scalar density L of the form

$$L = L(a_{mn}, w^A, w^A_i, w^A_{ij}). \tag{3.16}$$

Here, $a_{mn} = a_{mn}(x)$ is a Riemannian metric tensor, $w^A = w^A(x)$ are scalars, $w^A_i = \partial w^A/\partial x^i$, $w^A_{ij} = \partial w^A_i/\partial x^j$, so that

$$\bar{a}_{mn} = a_{ij} B^i_m B^j_n, \qquad \bar{w}^A = w^A, \quad \bar{w}^A_i \equiv \partial w^A/\partial \bar{x}^i = w^A_m B^m_i, \\ \bar{w}^A_{ij} \equiv \partial \bar{w}^A{}_i/\partial \bar{x}^j = w^A_{mn} B^m_i B^n_j + w^A_m B_i{}^m{}_j. \tag{3.17}$$

Therefore, the transformation law of L:

$$\bar{L}(\bar{a}_{mn}, \bar{w}^A, \bar{w}^A_i, \bar{w}^A_{ij}) = \det(B^i_j)\, L(a_{mn}, w^A, w^A_i, w^A_{ij}) \tag{3.18}$$

will imply two invariance identities:

$$- L\delta^j_i + \frac{\partial L}{\partial w^A_j} w^A_i + 2\frac{\partial L}{\partial w^A_{jn}} w^A_{in} = -2\frac{\partial L}{\partial a_{jn}} a_{in}, \tag{3.19}$$

$$\frac{\partial L}{\partial w^A_{jn}} w^A_i = 0 \tag{3.20}$$

obtained by differentiating (3.18) with respect to B^i_j and $B_j{}^i{}_n$ in turn, taking the derivatives of the barred arguments in (3.17), and then putting $\bar{x}^i = x^i$. In writing Equation (3.19), we have used evident equality

$$\partial \det(B^n_m)/\partial B^i_j = b^i_j \det(B^n_m) \tag{3.21}$$

(see Equation (3.3)).

Results proved in general relativity and gauge field theories are often demonstrated by using the so-called *Noether identities* (or Bianchi-type identities) satisfied by the Euler-Lagrange derivatives of a scalar density L chosen as a Lagrangian density. If L is of the form (3.16) such identities will read

$$\mathrm{d}_j\left(E_A w^A - 2\frac{\partial L}{\partial a_{jn}} a_{in} \right) = E_A w^A_i - \frac{\partial L}{\partial a_{mn}} \mathrm{d}_i a_{mn}. \tag{3.22}$$

Here,

$$E_A = -d^2_{mn}(\partial L/\partial w^A_{mn}) + d_m(\partial L/\partial w^A_m) - \partial L/\partial w^A \tag{3.23}$$

is *the Euler-Lagrange derivative* of L with respect to w^A and $d_m = \partial/\partial x^m$. The Noether identities are usually derived by considering the implications of the invariance of the action integral under infinitesimal coordinate displacements[2]. However, such an indirect approach is not necessary for the Noether identities are directly derivable from the invariance identities by taking the divergence (with respect to x^j) of the invariance identity of the first type (e.g. (3.19)) and then using other invariance identities. In particular, the application of such a procedure to Equations (3.19) and (3.20) yields the identity (3.22).

We leave it to the reader to write down the three invariance identities (obtainable by differentiating the transformation law $\bar{L} = \det(B^n_m)L$ with respect to B^i_j, $B^{\ i}_{j\ n}$ and $B^{\ i}_{j\ nm}$) for the scalar density of the following general form

$$L = L(y(x), d_m y(x), S^A(x), d_m S^A(x), d^2_{mn} S^A(x)) \tag{3.24}$$

(where $y^i(x)$ is a vector field and $S^A(x)$ is a set of tensors) and to verify that, after taking the divergence of the first invariance identity thus obtained and using the two other identities, we get

$$d_j(N_i^{\ j} - E_i y^j) = E_{Aj\dots}^{\ n\dots} d_i S^{Aj\dots}_{\ \ n\dots} + E_j d_i y^j \tag{3.25}$$

which is the Noether identity for the density (3.24). Here,

$$E_{Aj\dots}^{\ i\dots} = -d^2_{mn}\frac{\partial L}{\partial(d^2_{mn}S^{Aj\dots}_{\ \ i\dots})} + d_m\frac{\partial L}{\partial(d_m S^{Aj\dots}_{\ \ i\dots})} - \frac{\partial L}{\partial S^{Aj\dots}_{\ \ i\dots}}, \tag{3.26}$$

$$E_i = d_m\frac{\partial L}{\partial(d_m y^i)} - \frac{\partial L}{\partial y^i} \tag{3.27}$$

are the Euler-Lagrange derivatives of the density (3.24) with respect to $S^{Aj\dots}_{\ \ i\dots}$ and $y^i(x)$, while

$$N_i^{\ j} = S^{An\dots}_{\ \ i\dots} E_{An\dots}^{\ j\dots} + \cdots - S^{Aj\dots}_{\ \ n\dots} E_{Ai\dots}^{\ n\dots} - \cdots \tag{3.28}$$

(cf. (3.11)). The reader may also readily verify that, for any scalar density (3.47) constructed from the components of a Riemannian metric tensor and its derivatives up to the second order, the identities (3.25) state that the Riemannian covariant divergence of the Euler-Lagrange derivative (3.53) vanishes identically (Equation 3.54)).

It should be emphasized that the Noether identities, as well as the primary invariance identities, are valid independently of the field equations associated with the Lagrangian density. In the literature devoted to general relativity such identities are frequently called *the strong conservation laws*, in contrast to *the weak conservation laws* which hold only in virtue of the field equations[3]. Since the Noether identities explicitly contain the Euler-Lagrange derivatives they reduce to the weak conservation laws when the Euler-Lagrange derivatives with respect to the field variables are put equal to zero.

For example, in the Lagrangian density of the form (3.24) let the tensors S^A be treated as the field variables while $y^i(x)$ play an auxiliary role. Then it follows from the Noether identities (3.25) that the weak conservation law

$$-\mathrm{d}_j(E_i y^j) = E_j \mathrm{d}_i y^j \tag{3.29}$$

is valid whenever the field equations $E_{Aj\ldots}^{\;i\ldots} = 0$ are satisfied. Taking into account (3.27), we can rewrite (3.29) as follows

$$\mathrm{d}_j H_i^{\;j} = -(\partial L/\partial y^j)\,\mathrm{d}_i y^j - (\partial L/\partial \mathrm{d}_k y^j)\,\mathrm{d}_{ik}^2 y^j \tag{3.30}$$

where

$$H_i^{\;j} = -E_i y^j - (\partial L/\partial \mathrm{d}_j y^k)\,\mathrm{d}_i y^k \tag{3.31}$$

is called *the Hamiltonian complex.*

Similarly, considering the Lagrangian density (3.16) and treating the scalars w^A as the field variables and the tensor a_{ij} as the auxiliary quantity, we get from the Noether identity (3.22) the following weak conservation law

$$\mathrm{d}_j H_i^{\;j} = -(\partial L/\partial a_{mn})\mathrm{d}_i a_{mn} \tag{3.32}$$

with the Hamiltonian complex

$$H_i^{\;j} = -2(\partial L/\partial a_{jn})a_{in}. \tag{3.33}$$

These examples illustrate the general rule that the ordinary divergence of the Hamiltonian complex is equal to the negative of the total derivative of the Lagrangian density with respect to x^i at constant field variables. If this weak conservation law is derived from the Noether identities, the Hamiltonian complex is constructed from the derivatives of the Lagrangian density with respect to auxiliary quantities.

It will be noted that in the calculus of variations the Hamiltonian complex is defined in another way, namely in such a way that the definition of the complex $H_i'^{j}$ involves the derivatives of the Lagrangian density with respect to the field variables; for example, in the case of the Lagrangian density (3.16) with auxiliary a_{mn}, the definition reads[4]

$$H_i'^{j} = -\delta_i^j L + w_i^A(\partial L/\partial w_j^A) - w_i^A\,\mathrm{d}_n(\partial L/\partial w_{jn}^A) + w_{in}^A\,\partial L/\partial w_{jn}^A. \tag{3.34}$$

The relationship between $H_i'^{j}$ given by the definition (3.34) and $H_i^{\;j}$ obtained from the Noether identities can be found by using the invariance identities. In the case (3.34), the invariance identity (3.20) implies immediately that the right-hand side of (3.34) is identical with the left-hand side of the invariance identity (3.19), hence $H_i'^{j}$ is identical with $H_i^{\;j}$. In more complicated situations, the difference between $H_i'^{j}$ and $H_i^{\;j}$ may be of the form $\mathrm{d}_n H_i^{\;jn}$ where the object $H_i^{\;jn}$ is skew-symmetric in the indices j, n (see, for example, Equations (4.208) and (4.210)).

The general theory of invariance identities outlined in this section may be applied in various specialized ways. The aim of the succeeding section is two-fold. On the one hand, the material presented will illustrate with a nontrivial example how the invariance identity technique works. On the other hand, the subject of our examination will be what may be called a dissection of the concept of direction-dependent connection coefficients, an important task in the context of the problems studied in this book.

3.2. Construction of the Connection Coefficients with the Help of the Invariance Identities

Let us assume for the sake of generality that some symmetric non-degenerate tensor $g_{mn}(x, y)$ of class C^5 in the local positional coordinates x^i and of class C^4 in the directional coordinates y^i plays the role of the primary geometrical quantity, suppressing temporarily the assumption that g_{mn} is a Finslerian metric tensor in the sense of Section 1.1. The requirement that g_{mn} is a tensor reads

$$\bar{g}_{hk}(\bar{x}, \bar{y}) = g_{ij}(x, y) B^i_h B^j_k. \tag{3.35}$$

The transformation law of the partial derivatives $g_{ij,m} = \partial g_{ij}/\partial x^m$ can be found directly by differentiating (3.35) with respect to $\bar{x}^l$ for constant y^l, which gives

$$\begin{aligned}\bar{g}_{hk,1} = g_{ij,p} B^p_l B^i_h B^j_k + g_{ij}(B_h{}^i{}_l B^j_k + B^i_h B_k{}^j{}_l) + \\ + 2C_{ijq} y^m B_m{}^q{}_1 B^i_h B^j_k\end{aligned} \tag{3.36}$$

where

$$C_{ijq} = \tfrac{1}{2}\partial g_{ij}/\partial y^q \tag{3.37}$$

is a tensor symmetric in i,j but, generally speaking, not symmetric in all its subscripts. The notation (3.2) will be adopted.

It will be remembered that, according to the general definition of the connection coefficients, the objects $\Gamma_l{}^i{}_m(x, y)$ are called *connection coefficients* if their transformation law under the coordinate transformations (3.1) is of the form

$$B_l{}^t{}_m = B^t_i \bar{\Gamma}_l{}^i{}_m - \Gamma_q{}^t{}_p B^q_l B^p_m. \tag{3.38}$$

The simplest limitation which can be imposed on the functional dependence of $\Gamma_l{}^i{}_m$ on the tensor g_{mn} consists in the following:

$$\Gamma_l{}^i{}_m(x, y) = T_l{}^i{}_m(g_{rs}(x, y), g_{rs,k}(x, y), C_{rsk}(x, y)) \tag{3.39}$$

where $T_l{}^i{}_m$ is a sufficiently smooth function of the indicated arguments. It is obvious from the definition (3.38) that the dependence of $T_l{}^i{}_m$ on $g_{rs,k}$ cannot be avoided. Since the transform (3.36) of $g_{rs,k}$ involves the tensor C_{rsk}, the assumption that $T_l{}^i{}_m$ is independent of C_{rsk} would be too restrictive as well.

Differentiating the transformation law (3.38) with respect to $B_i{}^n{}_j$ yields the following invariance identity

$$\tfrac{1}{2}\delta^t_n(\delta^i_l \delta^j_m + \delta^i_m \delta^j_l) = B^t_q(\partial \bar{T}_l{}^q{}_m/\partial \bar{g}_{hk,p})(\partial \bar{g}_{hk,p}/\partial B_i{}^n{}_j)$$

which, after taking the derivative $\partial\bar{g}_{hk,p}/\partial B_{i\ j}^{\ n}$ of Equation (3.36) and then putting $\bar{x}^i = x^i$, becomes

$$\tfrac{1}{2}\delta^t_n(\delta^i_l\delta^j_m + \delta^i_m\delta^j_l) = (\partial T_{l\ m}^{\ t}/\partial g_{hk,p})((\delta^i_p\delta^j_h + \delta^i_h\delta^j_p)g_{nk} + \\ + (y^i\delta^j_p + y^j\delta^i_p)C_{hkn}),$$

or equivalently,

$$\frac{1}{2}\left(\frac{\partial}{\partial g_{ik,j}} + \frac{\partial}{\partial g_{jk,i}}\right)T_{l\ m}^{\ t} = \tfrac{1}{4}g^{tk}(\delta^i_l\delta^j_m + \delta^i_m\delta^j_l) - \\ - \tfrac{1}{2}C_{rs}^{\ \ k}\left(y^i\frac{\partial T_{l\ m}^{\ t}}{\partial g_{rs,j}} + y^j\frac{\partial T_{l\ m}^{\ t}}{\partial g_{rs,i}}\right). \tag{3.40}$$

On using the evident identity

$$\frac{\partial}{\partial g_{ik,j}} + \frac{\partial}{\partial g_{jk,i}} + \frac{\partial}{\partial g_{ij,k}} + \frac{\partial}{\partial g_{kj,i}} - \frac{\partial}{\partial g_{ki,j}} - \frac{\partial}{\partial g_{ji,k}} = 2\frac{\partial}{\partial g_{jk,i}},$$

we find from Equation (3.40) that

$$\frac{\partial T_{l\ m}^{\ t}}{\partial g_{jk,i}} = \tfrac{1}{4}(g^{tk}(\delta^i_l\delta^j_m + \delta^i_m\delta^j_l) + g^{tj}(\delta^i_l\delta^k_m + \delta^i_m\delta^k_l) - \\ - g^{ti}(\delta^k_l\delta^j_m + \delta^k_m\delta^j_l)) - \tfrac{1}{2}C_{rs}^{\ \ k}\left(y^i\frac{\partial T_{l\ m}^{\ t}}{\partial g_{rs,j}} + y^j\frac{\partial T_{l\ m}^{\ t}}{\partial g_{rs,i}}\right) - \\ - \tfrac{1}{2}C_{rs}^{\ \ j}\left(y^i\frac{\partial T_{l\ m}^{\ t}}{\partial g_{rs,k}} + y^k\frac{\partial T_{l\ m}^{\ t}}{\partial g_{rs,i}}\right) + \\ + \tfrac{1}{2}C_{rs}^{\ \ i}\left(y^k\frac{\partial T_{l\ m}^{\ t}}{\partial g_{rs,j}} + y^j\frac{\partial T_{l\ m}^{\ t}}{\partial g_{rs,k}}\right). \tag{3.41}$$

These are the characteristic equations to be satisfied by the connection coefficients having the form (3.39). Equations (3.41) have a tensor form, for the invariance identities obtainable by differentiating Equation (3.38) with respect to $g_{ij,k}$ indicate that the derivative $\partial T_{l\ m}^{\ t}/\partial g_{ij,k}$ is a tensor.

The number of terms which do not contain derivatives of $T_{l\ m}^{\ t}$ can be enlarged on the right-hand side of (3.41) by substituting (3.41) in the last three terms in (3.41), which gives

$$\frac{\partial T_{l\ m}^{\ t}}{\partial g_{jk,i}} = P^{tjki}_{lm} + P^{tkji}_{lm} - P^{tjik}_{lm} \tag{3.42}$$

where

$$P^{tjki}_{lm} = \tfrac{1}{4}g^{tk}(\delta^i_l\delta^j_m + \delta^i_m\delta^j_l) + y^iy^jC_{rs}^{\ \ k}C_{pq}^{\ \ s}\frac{\partial T_{l\ m}^{\ t}}{\partial g_{pq,r}} - \\ - \tfrac{1}{4}y^iP^{tkj}_{lm} - \tfrac{1}{4}y^jP^{tki}_{lm}$$

and

$$P^{tkj}_{lm} = C_l^{\ tk}\delta^j_m + C_m^{\ tk}\delta^i_l - C_{lm}^{\ \ k}g^{tj} -$$
$$- y^r C_{rs}^{\ \ k}\left(C_{pq}^{\ \ s}\frac{\partial T^{\ t}_{l\ m}}{\partial g_{pq,j}} + C_{pq}^{\ \ j}\frac{\partial T^{\ t}_{l\ m}}{\partial g_{pq,s}}\right).$$

This procedure may be continued to give a chain of an infinite number of equations for $\partial T^{\ t}_{l\ m}/\partial g_{jk,i}$. Since the condition which allows breaking off such a chain is unknown in the general case, simplifying assumptions about the tensor C_{ijk} are required. A close inspection of the structure of Equations (3.41) and (3.42) suggests the following assumption

$$y^r C_{rsk} = 0 \tag{3.43}$$

in order to interrupt the chain. Indeed, Equations (3.42) and (3.43) imply directly that

$$C_{rs}^{\ \ k} C_{pq}^{\ \ s}\frac{\partial T^{\ t}_{l\ m}}{\partial g_{pq,r}} = \tfrac{1}{2}(C_l^{\ ts}\delta^r_m + C_m^{\ ts}\delta^r_l - C_{lm}^{\ \ s}g^{tr})C_{rs}^{\ \ k},$$

so that Equation (3.42) is reduced to the following form:

$$\frac{\partial T^{\ t}_{l\ m}}{\partial g_{jk,i}} = Q^{tjki}_{lm} + Q^{tkji}_{lm} - Q^{tjik}_{lm} \tag{3.44}$$

with

$$\begin{aligned} Q^{tjki}_{lm} = {} & \tfrac{1}{4}g^{tk}(\delta^i_l\delta^j_m + \delta^i_m\delta^j_l) - \tfrac{1}{4}y^i(C_l^{\ tk}\delta^j_m + C_m^{\ tk}\delta^j_l - C_{lm}^{\ \ k}g^{tj}) - \\ & - \tfrac{1}{4}y^j(C_l^{\ tk}\delta^i_m + C_m^{\ tk}\delta^i_l - C_{lm}^{\ \ k}g^{ti}) + \\ & + \tfrac{1}{2}y^i y^j(C_l^{\ ts}C_{ms}^{\ \ \ k} + C_m^{\ ts}C_{ls}^{\ \ k} - C_{lms}C^{tsk}). \end{aligned} \tag{3.45}$$

It is remarkable that the right-hand side of Equation (3.44) depends on g_{rs} and C_{rsk} only, and does not involve the derivatives $g_{rs,k}$. Therefore, *in the Finslerian case any connection coefficients of the form* (3.39) *depend on* $g_{rs,k}$ *linearly*, i.e.

$$T^{\ i}_{l\ m} = g_{rs,k}\,\partial T^{\ i}_{l\ m}/\partial g_{rs,k} + T^{*\ i}_{l\ m}(g_{rs}, C_{rsk}). \tag{3.46}$$

Equations (3.44)–(3.46) can effectively be used to find the explicit form of the connection coefficients (3.39). The substitution of (3.44) and (3.45) in (3.46) shows that *the first term in Equation* (3.46) *represents the Cartan connection coefficients* (1.32), which can be readily verified in a direct manner. As regards the second term in (3.46), $T^{*\ i}_{l\ m}$ is the difference between two connection coefficients and so must be a tensor (cf. the transformation law (3.38)).

Thus we obtain the following result.

PROPOSITION 1. In the Finslerian case, any connection coefficients of the form (3.39) are the Cartan connection coefficients, up to an arbitrary tensor term dependent on g_{rs} and C_{rsk} only.

Incidentally, no symmetry assumption has been imposed on the connection coefficients in the course of our consideration. Thus, as soon as the form (3.39) of the functional dependence of the connection coefficients on the Finslerian metric tensor has been postulated, the invariance identities rigorously require that, up to an arbitrary tensor term depending on g_{rs} and C_{rsk} only, the connection coefficients are simultaneously symmetric and metric because the Cartan connection coefficients have both these properties. In essence, the symmetry of the thus obtained connection coefficients in their subscripts is an immediate consequence of the symmetry in l, m of the $B_{l\,m}^{t}$ entering the transformation law (3.38).

It will be noted that the condition (3.43) does not allow one to deal with the tensors $g_{ij}(x, y)$ of types more general than Finslerian (cf. the way of reasoning in solving Problem 1.1), so that it is to the Finslerian case that the expressions (3.44)–(3.45) must be addressed. The significance of the Finslerian specification of $g_{mn}(x, y)$ can be thought of as arising from the necessity of imposing the condition (3.43) on C_{rsk} in order to solve Equations (3.42) for the connection coefficients.

3.3. Fundamental Tensor Densities Associated with Direction-Dependent Scalar Densities

In studying the concomitants of some geometrical object the question may arise as to what tensorial objects can be constructed from its derivatives. The direct method for solving a problem of this type consists in a detailed examination of the structure of the invariance identities for the derivatives of the object. Suppose, in particular, that we are given a scalar density (of weight +1) of the form

$$L = L(a_{ij}, a_{ij,m}, a_{ij,mn}) \tag{3.47}$$

where $a_{ij}(x)$ denotes a Riemannian metric tensor. The notation $a_{ij,m} = \partial a_{ij}/\partial x^m$ and $a_{ij,mn} = \partial a_{ij,m}/\partial x^n$ is used. The expression (3.47) comprises numerous gravitational Lagrangian densities studied in general relativity theory. Considering the invariance identities for the derivatives

$$L^{ij} = \partial L/\partial a_{ij}, \qquad L^{ij,k} = \partial L/\partial a_{ij,k}, \qquad L^{ij,mn} = \partial L/\partial a_{ij,mn}$$

we can obtain the following results[5]:

(i) $L^{ij,mn}$ is a tensor density with the following symmetry properties: in (i, j), in (m, n), in the pair (ij, mn), and also

$$L^{ij,mn} + L^{in,jm} + L^{im,nj} = 0. \tag{3.48}$$

(ii) $L^{ij,k}$ is not of tensorial nature. The identity

$$\Pi^{ij,k} \overset{\text{def}}{=} L^{ij,k} + 2\left\{ {i \atop m\;l} \right\} L^{mj,kl} + 2\left\{ {j \atop m\;l} \right\} L^{mi,kl} + \left\{ {k \atop m\;l} \right\} L^{ij,ml} = 0, \tag{3.49}$$

where $\{^{\ i}_{m\ l}\}$ denote the Riemannian Christoffel symbols, holds.

(iii) The quantities

$$L^{ij} = \Pi^{ij} + \tfrac{1}{3}(a^{ih}L^{mn,jq} + a^{jh}L^{mn,iq})R_{nhmq} \tag{3.50}$$

and

$$\Pi^{ij} = \tfrac{1}{2}a^{ij}L + \tfrac{2}{3}L^{mn,jq}R_{n\ mq}^{\ i} \tag{3.51}$$

where $R_{n\ mq}^{\ i}$ denotes the Riemannian curvature tensor, are tensor densities, and the symmetry relation

$$\Pi^{ij} = \Pi^{ji} \tag{3.52}$$

holds.

(iv) The Euler-Langrange derivative E^{ij} can be expressed as follows:

$$E^{ij} \overset{\text{def}}{=} -\mathrm{d}^2_{mn}L^{ij,mn} + \mathrm{d}_m\mathrm{L}^{ij,m} - \mathrm{L}^{ij} = -\Pi^{ij} - \mathrm{D}_m\mathrm{D}_nL^{ij,mn} \tag{3.53}$$

where $\mathrm{d}_m = \partial/\partial x^m$ and D_m stands for the Riemannian covariant derivative.

(v) The covariant divergence of E^{ij} vanishes identically:

$$\mathrm{D}_jE^{ij} = 0. \tag{3.54}$$

These results can be generalized to the case when a density L is considered to be constructible from a direction-dependent symmetric tensor $g_{ij}(x, y)$ and its derivatives. As in the preceding section, we shall not suppose below that g_{ij} is the Finslerian metric tensor.

By differentiating Equation (3.36) with respect to $\bar{x}^m$ for a fixed $\bar{y}^l$ we obtain the following transformation law of the second derivatives $g_{ij,pq} = \partial g_{ij,p}/\partial x^q$:

$$\begin{aligned}\bar{g}_{hk,lm} = {} & g_{ij}(B_{h\ lm}^{\ i}B_k^j + B_h^iB_{k\ lm}^{\ j} + B_{h\ l}^{\ i}B_{k\ m}^{\ j} + B_{h\ m}^{\ i}B_{k\ l}^{\ j}) + \\ & + g_{ij,p}(B_{l\ m}^{\ p}B_h^iB_k^j + B_l^pB_{h\ m}^{\ i}B_k^j + B_l^pB_h^iB_{k\ m}^{\ j}) + \\ & + g_{ij,p}B_m^p(B_{h\ l}^{\ i}B_k^j + B_h^iB_{k\ l}^{\ j}) + \\ & + g_{ij,pq}B_l^pB_m^qB_h^iB_k^j + 2C_{ijq,p}\bar{y}^rB_{r\ m}^{\ q}B_l^pB_h^iB_k^j + \\ & + 2C_{ijq}\bar{y}^rB_{r\ m}^{\ q}(B_{h\ l}^{\ i}B_k^j + B_h^iB_{k\ l}^{\ j}) + \\ & + 2C_{ijq,p}\bar{y}^rB_m^pB_{r\ l}^{\ q}B_h^iB_k^j + 2C_{ijqt}\bar{y}^r\bar{y}^sB_{r\ m}^{\ t}B_{s\ l}^{\ q}B_h^iB_k^j + \\ & + 2C_{ijq}B_{r\ lm}^{\ q}\bar{y}^rB_h^iB_k^j + 2C_{ijq}\bar{y}^rB_{r\ l}^{\ q}(B_{h\ m}^{\ i}B_k^j + B_h^iB_{k\ m}^{\ j})\end{aligned} \tag{3.55}$$

where $B_{h\ lm}^{\ i} = \partial B_{h\ l}^{\ i}/\partial\bar{x}^m$, $C_{ijq,p} = \partial C_{ijq}/\partial x^p$ and

$$C_{ijqt} = \partial C_{ijq}/\partial y^t. \tag{3.56}$$

The objects (3.37) and (3.56), being the derivatives of tensors with respect to contravariant vectors, will naturally form the components of covariant tensors, which means that

$$\bar{C}_{hkl} = C_{ijp}B_h^iB_k^jB_l^p, \tag{3.57}$$

$$\bar{C}_{hklm} = C_{ijpq} B^i_h B^j_k B^p_l B^q_m. \tag{3.58}$$

The relation (3.57) implies that

$$\begin{aligned}\bar{C}_{hkl,m} &= C_{ijp,q} B^i_h B^j_k B^p_l B^q_m + C_{ijpq} B^i_h B^j_k B^p_l B_r{}^q{}_m \bar{y}^r + \\ &\quad + C_{ijp}(B_h{}^i{}_m B^j_k B^p_l + B^i_h B_k{}^j{}_m B^p_l + B^i_h B^j_k B_l{}^p{}_m).\end{aligned} \tag{3.59}$$

Suppose we have a scalar density L whose explicit functional dependence is as follows[6]:

$$L = L(y, g_{hk}, g_{hk,l}, g_{hk,lm}, C_{hkl}, C_{hkl,m}, C_{hklm}). \tag{3.60}$$

Considering the implications of the transformation law of L:

$$\bar{L} = BL \tag{3.61}$$

where

$$B = \det(B^j_h), \tag{3.62}$$

and exploiting the transformation laws of the arguments of L, one can deduce the explicit form of the fundamental tensor densities linear in the first derivatives

$$L^{ij} = \frac{\partial L}{\partial g_{ij}}, \qquad L^{ij,m} = \frac{\partial L}{\partial g_{ij,m}}, \qquad L^{ij,mn} = \frac{\partial L}{\partial g_{ij,mn}}, \tag{3.63}$$

$$\Omega^{ijk} = \frac{\partial L}{\partial C_{ijk}}, \qquad \Omega^{ijk,l} = \frac{\partial L}{\partial C_{ijk,l}}, \qquad \Omega^{ijkl} = \frac{\partial L}{\partial C_{ijkl}}. \tag{3.64}$$

These last objects possess the obvious symmetry properties

$$L^{ij,mn} = L^{ji,mn} = L^{ij,nm}, \qquad L^{ij} = L^{ji}, \qquad L^{ij,m} = L^{ji,m},$$

$$\Omega^{ijk} = \Omega^{jik} = \Omega^{ikj}, \text{ etc.},$$

which are to be taken into account in the subsequent calculations.

First of all, it is clear from Equation (3.55) that *the derivatives $L^{hk,lm}$ form the components of a tensor density of contravariant valency* 4. In fact, the only argument in $\bar{L}$ on the left-hand side of (3.61) which involves the second-order derivatives $g_{ij,hk}$ is $\bar{g}_{hk,lm}$, so that we find, after differentiating Equation (3.61) with respect to $g_{rs,tu}$, that

$$\bar{L}^{hk,lm} \partial\bar{g}_{hk,lm}/\partial g_{rs,tu} = BL^{rs,tu}. \tag{3.65}$$

But taking the symmetries of $\bar{g}_{hk,lm}$ into account, we have from (3.55):

$$\partial\bar{g}_{hk,lm}/\partial g_{rs,tu} = \tfrac{1}{4}(B^r_h B^s_k + B^s_h B^r_k)(B^t_l B^u_m + B^u_l B^t_m),$$

so that Equation (3.65) shows that the tensorial transformation law

$$\bar{L}^{hk,lm} B^r_h B^s_k B^t_l B^u_m = BL^{rs,tu} \tag{3.66}$$

is valid indeed.

However, $L^{ij,mn}$ is the only tensor density amongst the derivatives (3.63)–(3.64), because all the arguments of L other than $g_{hk,lm}$ appear in more than one argument of $\bar{L}$. For example, let us examine the transformation law of $\Omega^{ijk,l}$. Differentiating Equation (3.61) with respect to $C_{rst,u}$ yields

$$B\Omega^{rst,u} = \bar{L}^{hk,lm}\frac{\partial \bar{g}_{hk,lm}}{\partial C_{rst,u}} + \bar{\Omega}^{hkl,m}\frac{\partial \bar{C}_{hkl,m}}{\partial C_{rst,u}} \tag{3.67}$$

where we have taken into account that $\bar{g}_{hk,lm}$ and $\bar{C}_{hkl,m}$ are the only arguments of $\bar{L}$ which depend on $C_{rst,u}$ (see Equations (3.35), (3.36), (3.55) and (3.57)–(3.59)). Substituting for the derivatives in (3.67) from (3.55) and (3.58) we get the following result

$$\begin{aligned} B\Omega^{rst,u} = \bar{\Omega}^{hkl,m} B^r_h B^s_k B^t_l B^u_m + \tfrac{4}{3}\bar{L}^{hk,lm} B^u_l \bar{y}^p (B^r_h B^s_k B_p{}^t{}_m + \\ + B^t_h B^r_k B_p{}^s{}_m + B^s_h B^t_k B_p{}^r{}_m), \end{aligned} \tag{3.68}$$

from which it is evident that $\Omega^{hkl,m}$ cannot form the components of a tensor density.

To be able to construct a tensor density with the help of the relation (3.68), we have to introduce a set of connection coefficients. Let $\Gamma_l{}^i{}_m(x, y)$ denote an arbitrary set of symmetric connection coefficients of class C^2 in both x^i and y^i, so that $\Gamma_l{}^i{}_m = \Gamma_m{}^i{}_l$ and $\Gamma_l{}^i{}_m$ satisfies the transformation law (3.38) governing the connection coefficients. No relationship between $\Gamma_l{}^i{}_m$ and g_{ij} will be assumed in the present section until otherwise stated. The contraction of Equation (3.38) with $\bar{y}^l$ yields

$$\bar{y}^l B_l{}^t{}_m = B^t_i \Gamma_0{}^i{}_m - \bar{\Gamma}_0{}^t{}_p B^p_m \tag{3.69}$$

where we have used the notation

$$\Gamma_0{}^t{}_p = y^l \Gamma_l{}^t{}_p. \tag{3.70}$$

It can be readily verified with the help of Equations (3.68), (3.69) and (3.66) that *the objects*

$$\Psi^{hkl,m} = \Omega^{hkl,m} + \tfrac{4}{3}(\Gamma_0{}^h{}_j L^{kl,mj} + \Gamma_0{}^k{}_j L^{lh,mj} + \Gamma_0{}^l{}_j L^{hk,mj}) \tag{3.71}$$

form the components of a tensor density of contravariant valency 4; that is, the objects (3.71) obey the transformation law of the form

$$\bar{\Psi}^{hkl,m} B^r_h B^s_k B^t_l B^u_m = B\Psi^{rst,u}. \tag{3.72}$$

This procedure may be continued. The results obtained may be summarized as follows. *The fundamental tensor densities associated with a scalar density of the form* (3.60) *are given by*

$$\Psi^{ijk,l} = \Omega^{ijk,l} + 4\mathscr{A}^{(ijk)}(\Gamma_0{}^i{}_n L^{jk,ln}), \tag{3.73}$$

$$\Psi^{ijkl} = \Omega^{ijkl} + \mathscr{A}^{(ijkl)}(\Gamma_0{}^i{}_n \Psi^{jkl,n} - 2\Gamma_0{}^i{}_m \Gamma_0{}^j{}_n L^{kl,mn}), \tag{3.74}$$

$$\Psi^{ijk} = \Omega^{ijk} + \mathscr{A}^{(ijk)}(2\Gamma_{0\ m}^{\ i}\Pi^{jk,m} + 3\Gamma_{m\ n}^{\ i}\Psi^{jkm,n} + \\ + 2y^p M_{p\ mn}^{\ i} L^{jk,mn} - 8\Gamma_{0\ m}^{\ i}\Gamma_{s\ n}^{\ j} L^{ks,mn}), \tag{3.75}$$

$$\Pi^{ij,m} = L^{ij,m} + \Gamma_{r\ s}^{\ m} L^{ij,rs} + 2\Gamma_{r\ s}^{\ i} L^{rj,ms} + 2\Gamma_{r\ s}^{\ j} L^{ri,ms}, \tag{3.76}$$

$$\Pi^{ij} = L^{ij} + 2\mathscr{A}^{(ij)}(M_{r\ ls}^{\ i} L^{jr,ls} + \Gamma_{r\ s}^{\ i}\Pi^{jr,s} - \Gamma_{r\ s}^{\ i}\Gamma_{m\ n}^{\ j} L^{mr,ns}), \tag{3.77}$$

$$E^{ij} = d'_m(L^{ij,m} - d'_n L^{ij,mn}) - L^{ij} = \\ = -L^{ij,mn}{}_{|m|n} + \Pi^{ij,m}{}_{|m} - \Pi^{ij}, \tag{3.78}$$

plus $L^{ij,mn}$. Here, $|m$ denotes the covariant derivative with respect to $\Gamma_{m\ n}^{\ i}$ written in accordance with the same rule (1.41) as holds in the Finslerian case. The object

$$M_{j\ lm}^{\ i} = \partial\Gamma_{j\ l}^{\ i}/\partial x^m - \Gamma_{0\ m}^{\ h}\partial\Gamma_{j\ l}^{\ i}/\partial y^h - \Gamma_{l\ h}^{\ i}\Gamma_{j\ m}^{\ h} - \Gamma_{j\ h}^{\ i}\Gamma_{l\ m}^{\ h} \tag{3.79}$$

is not a tensor although the skew-symmetric part $M_{j\ lm}^{\ i} - M_{j\ ml}^{\ i}$ is. The symbol $\mathscr{A}^{(\cdots)}$ stands for the symmetrization operator with factor $1/n!$, i.e.

$$\mathscr{A}^{(i_1\cdots i_n)} X^{i_1\cdots i_n\cdots} = \textit{summation of } n! \textit{ terms } X^{\cdots} \textit{ with} \\ \textit{the permuted indices divided by } n!; \tag{3.79'}$$

for example,

$$\mathscr{A}^{(ij)} X^{ijmn} = \tfrac{1}{2}(X^{ijmn} + X^{jimn}).$$

The symbol d'_m stands for the operator whose action on any direction-dependent object $w(x, y)$ is defined as follows[7]:

$$\mathrm{d}'_m w(x, y) = \partial w/\partial x^m - \Gamma_{0\ m}^{\ k}\partial w/\partial y^k. \tag{3.80}$$

Moreover, if a new tensor density is formed according to

$$\Pi^{ij,mn} = L^{ij,mn} + y^j C_{p\ q}^{\ i} L^{pq,mn}, \tag{3.81}$$

then the cyclic identity

$$\Pi^{ij,mn} + \Pi^{im,nj} + \Pi^{in,jm} = 0 \tag{3.82}$$

(cf. Equation (3.48)) will be satisfied.

In the case when g_{rs} in the scalar density (3.60) is taken to be a Finslerian metric tensor and $\Gamma_{i\ j}^{\ k}$ is the Finslerian Cartan connection coefficients, the tensor density (3.76) reduces to

$$\Pi^{ij,h} = \tfrac{1}{2}\Phi^{lm}\{y^h(y^i C_{l\ m}^{\ j} + y^j C_{l\ m}^{\ i}) - y^i y^j C_{l\ m}^{\ h}\} + \\ + \tfrac{1}{4}\{y^i(\Phi^{hj} - \Phi^{jh}) + y^j(\Phi^{hi} - \Phi^{ih}) - y^h(\Phi^{ij} + \Phi^{ji})\} + \\ + \tfrac{3}{4}(\Phi^{hji} + \Phi^{hij}) \tag{3.83}$$

where

$$\Phi^{ijl} = C_{h\ k}^{\ i}\Psi^{hkj,l} - C_{h\ k}^{\ l}(\Psi^{hkj,i} + \Psi^{hki,j}), \tag{3.84}$$

$$\Phi^{ij} = 3C_{h\ k}^{\ i}\Phi^{jhk} + 4C_{h\ k|m}^{\ i}L^{hk,mj} + C_{h\ km}^{\ i}\Psi^{hkm,j}. \tag{3.85}$$

These formulae show that $\Pi^{ij,h}$ is a linear homogeneous combination of the tensor densities $L^{hk,ml}$ and $\Psi^{hkm,l}$. Since, at the same time, $\Pi^{ij,m}$ given by Equation (3.76) is linear in $L^{ij,h}$ and $L^{hk,mj}$, while $\Psi^{hkm,l}$ is in turn linear in $\Omega^{hkm,l}$ and $L^{hk,ml}$ (see Equation (3.73)), we can infer from Equations (3.83)–(3.85) the following result.

PROPOSITION 2. For any scalar density L of the type (3.60) in which g_{rs} is a Finslerian metric tensor the derivatives $L^{hk,l} = \partial L/\partial g_{hk,l}$ are homogeneous linear combinations of the derivatives $\Omega^{hkl,m} = \partial L/\partial C_{hkl,m}$ and $L^{hk,lm} = \partial L/\partial g_{hk,lm}$.

This proposition allows one to conclude immediately that the following assertion is valid (cf. Problem 3.6).

PROPOSITION 3. There exists no nonvanishing scalar density depending on just the Finslerian tensors g_{hk}, C_{hkl}, C_{hklm} and the first derivatives $g_{hk,l}$.

Clearly, Proposition 3 will remain valid for scalar densities depending solely on the Finslerian tensors g_{hk} and C_{hkl}.

It is noted for the sake of completeness that, in addition to Proposition 2, the following result is valid.

PROPOSITION 4. For a scalar density L of the type (3.60), in which g_{hk} is assumed to be a Finslerian metric tensor, the derivatives $L^{hk} = \partial L/\partial g_{hk}$ are given by the sum of a linear homogeneous combination of the derivatives of L with respect to (y^k, $g_{hk,lm}$, C_{hkl}, $C_{hkl,m}$, C_{hklm}) and $\frac{1}{2}g^{hk}L$.

For the proof of Proposition 4, as well as for more details on direction-dependent scalar densities, the reader is referred to Rund and Beare (1972).

Thus we have found all the basic tensor densities associated with scalar densities of the general functional type (3.60), which generalizes its Riemannian prototype (3.47). With a view to formulating the gravitational field equations, however, it is natural to take an appropriate particular case of the density (3.60), namely the density (3.86), by analogy with the manner in which a representative of Lagrangian densities of the form (3.47) is selected when the Riemannian formulation is applied to the gravitational field equations.

3.4. Choice of the Finslerian Scalar Density $L = JK$

According to the remarkable equality (1.81), the complete contraction $R = R^{hk}{}_{hk}$ of the Cartan curvature tensor $R_j{}^i{}_{hk}$ proves to be identical to the complete contraction $K = K^{hk}{}_{hk}$ of the Finslerian K-tensor of curvature $K_j{}^i{}_{hk}$, the latter being of an essentially simpler structure than $R_j{}^i{}_{hk}$. Therefore, it seems certain that we should regard the scalar density

$$L = JK \tag{3.86}$$

where $J = |\det(g_{ij}(x, y))|^{1/2}$ and g_{ij} is a Finslerian metric tensor, as the natural Finslerian generalization of its Riemannian prototype[8].

The relations given in the previous section provide all that we need to calculate the explicit form of the fundamental tensor densities associated with (3.86). However, despite the circumstance that the scalar density (3.86) is of the form (3.60) treated in the previous section, it would be erroneous to believe that all expressions for the fundamental tensor densities indicated in the previous section hold in the Finslerian case (3.86) without modifications. Indeed, the results of the previous section have been obtained in the case when the arguments of the scalar density (3.60) are entirely independent, that is, when the object L forms a scalar density without any assumptions being made about the functional inter-dependence of its arguments. The expression (3.86), however, does not have such a property; the object (3.86) represents a scalar density under the assumption that the tensor g_{ij} is a Finslerian metric tensor, a circumstance which implies the validity of the identities

$$y^i C_{ijk} = 0, \qquad y^i C_{ijk,m} = 0, \qquad y^i C_{ijkm} = -C_{jkm} \tag{3.87}$$

(see Equations (1.16) and (1.18)), which are just functional relationships between the arguments y^i, C_{ijk}, $C_{ijk,m}$ and C_{ijkm}. The object (3.86) obeys the scalar density transformation law only under the assumption of the functional relationships (3.87).

In view of Equation (3.87), the Ω-type derivatives (3.64) of the Finslerian density (3.86) fail to obey the same transformation law as in the case when L of the form (3.60) is a density without any additional assumptions. This fact, in turn, will prevent us from obtaining the tensor densities by calculating $\Psi^{ijk,m}$, Ψ^{ijkm}, and Ψ^{ijk} on the basis of (3.86). The corresponding tensorial redefinitions $\Psi^{*ijk,m}$, Ψ^{*ijkm}, and Ψ^{*ijk} will be given below. As to the quantities $L^{ij,mn}$, $\Pi^{ij,m}$, Π^{ij}, and E^{ij}, their tensorial character is retained because their definitions do not involve the Ω-type derivatives (3.64).

The effect of the additional condition on the transformation law of the derivatives of L can most clearly be seen if we take the derivative $\Omega^{ijk,m}$ as an example. Let this derivative be calculated using the Finslerian scalar density (3.86). Then the relation (3.67) turns out to be invalid, the correct transformation law of $\Omega^{ijk,m}$ following from the general.

PROPOSITION 5. *Let L of the form* (3.60) *be a scalar density with the assumption* (3.87) *and the arguments of L obey the same transformation law as in the Finslerian case Denote by* $L^{*ij,mn}$ *that part of* $L^{ij,mn}$ *which remains after deleting all the terms proportional to* y^i *or* y^j, *i.e. after omitting the terms of the form* $y^i Z^{jmn} + y^j Z^{imn}$ *with any* Z^{imn}. *Then the transformation law of the derivatives* $\Omega^{rst,u} = \partial L/\partial C_{rst,u}$ *will be of the form*

$$B\Omega^{rst,u} = \bar{L}^{*hk,lm}\frac{\partial \bar{g}_{hk,lm}}{\partial C_{rst,u}} + \bar{\Omega}^{hkl,m}\frac{\partial \bar{C}_{hkl,m}}{\partial C_{rst,u}}. \tag{3.88}$$

Proof. Following the method described in the preceding section, we shall obtain the transformation law of $\Omega^{rst,u}$ by treating $\bar{L}$ as a function of its barred arguments, replacing the barred arguments by their transforms and, after that, differentiating the identity (3.61) with respect to $C_{rst,u}$. Since, by hypothesis, L is a scalar density under the additional constraint (3.87), we have to set the contractions $y^i C_{ijk,m}$ contained in $\bar{L}$ equal to zero *before* differentiating $\bar{L}$ with respect to $C_{rst,u}$.

Among the arguments of $\bar{L}$, the $C_{ijm,n}$ and $g_{ij,mn}$ are the only ones where the transformation law involves $C_{ijm,n}$, namely

$$\bar{C}_{hkl,m} = C_{ijp,q} B^i_h B^j_k B^p_l B^q_m + \dots, \tag{3.89}$$

$$\bar{g}_{hk,lm} = 2C_{ijq,p} B^i_h B^j_k (B_t{}^q{}_m B^p_l + B_t{}^q{}_l B^p_m)\bar{y}^t + \dots \tag{3.90}$$

(cf. Equations (3.59) and (3.55)) where the dots denote terms not involving $C_{ijm,n}$. The $C_{ijm,n}$ extering $\bar{L}$ through the relation (3.89) is reflected in the second terms on the right-hand sides of (3.67) and (3.88). Consider now how $C_{ijm,n}$ enters into $\bar{L}$ through the relation (3.90). To this end, we shall separately examine two cases: (A) the $g_{ij,mn}$ enters into $\bar{L}$ in the contractions $y^i g_{ij,mn}$ or $y^j g_{ij,mn}$; (B) the $g_{ij,mn}$ enters into $\bar{L}$ in any other way. The case (B) contributes to $L^{*ij,mn}$, but not to $L^{**ij,mn} \overset{\text{def}}{=} L^{ij,mn} - L^{*ij,mn}$, and the case (A) to $L^{**ij,mn}$, and not to $L^{*ij,mn}$. The way by which $C_{ijm,n}$ becomes an argument of $\bar{L}$ through case (B) of the dependence of $\bar{L}$ on $\bar{g}_{ij,mn}$ is reflected by the first term on the right-hand side of Equation (3.88). As regards the remaining possibility, namely the $C_{ijm,n}$ entering $\bar{L}$ through case (A) of the dependence of $\bar{L}$ on $\bar{g}_{ij,mn}$, it is suppressed by the identity $y^i C_{ijm,n} = 0$ because, by Equation (3.90),

$$\bar{y}^h \bar{g}_{hk,lm} = 2y^i C_{ijq,p} B^j_k (B_t{}^q{}_m B^p_l + B_t{}^q{}_l B^p_m)\bar{y}^t + \dots = 0 + \dots.$$

Thus, under the above assumptions, it is $\bar{L}^{*hk,lm}$ (not $\bar{L}^{hk,lm}$) that should be taken as the coefficients of $\partial\bar{g}_{hk,lm}/\partial C_{rst,u}$ on the right-hand side of Equation (3.88). The proof is complete.

It can be easily verified in a direct fashion with the help of the explicit form (3.96) of the derivative $\Omega^{ijk,m}$ of the Finslerian scalar density (3.86) that $\Omega^{ijk,m}$ indeed satisfies the transformation law (3.88). The transformation laws of the derivatives Ω^{ijkl} and Ω^{ijk} subject to the conditions (3.87) can be elucidated in a similar fashion.

In the course of the subsequent calculations of the fundamental tensor densities associated with the scalar density (3.86) it will be beneficial to use the following commutation formulae which are obvious from the structure of the Cartan connection coefficients (1.32):

$$\frac{\partial(\partial\Gamma_{psq}/\partial x^r)}{\partial g_{ij,mn}} = \mathscr{A}^{(mn)}\left[\delta^n_r \frac{\partial\Gamma_{psq}}{\partial g_{ij,m}}\right], \tag{3.91}$$

$$\frac{\partial(\partial\Gamma_{psq}/\partial y^t)}{\partial C_{ijkl}} = \mathscr{A}^{(ijkl)}\left[\delta^l_t \frac{\partial\Gamma_{psq}}{\partial C_{ijk}}\right], \tag{3.92}$$

$$\frac{\partial(\partial\Gamma_{psq}/\partial x^r)}{\partial C_{ijk,m}} = \delta_r^m \frac{\partial\Gamma_{psq}}{\partial C_{ijk}}, \tag{3.93}$$

$$\frac{\partial(\partial\gamma_{psq}/\partial y^t)}{\partial C_{ijk,m}} = 2\mathscr{A}^{(ijk)}\left[\delta_t^k \frac{\partial\gamma_{psq}}{\partial g_{ij,m}}\right]. \tag{3.94}$$

By virtue of Equation (3.91) and of the structure of the K-tensor (1.78), we have from (3.86) that

$$L^{ij,mn} = J\mathscr{A}^{(mn)}((g^{pq}g^{ns} - g^{nq}g^{ps})\partial\Gamma_{psq}/\partial g_{ij,m}). \tag{3.95}$$

Similarly, by (3.92),

$$\Omega^{ijkl} = -J\mathscr{A}^{(ijkl)}(G_m{}^l(g^{pq}g^{ms} - g^{mq}g^{ps})\partial\Gamma_{psq}/\partial C_{ijk}). \tag{3.96}$$

Further, Equations (1.33) and (3.93)–(3.94) imply that

$$\begin{aligned}\Omega^{ijk,m} = {} & J(g^{pq}g^{ms} - g^{mq}g^{ps})\partial\Gamma_{psq}/\partial C_{ijk} + \\ & + 2J\mathscr{A}^{(ijk)}((g^{nq}g^{ps} - g^{pq}g^{ns})G_n{}^k\partial\gamma_{psq}/\partial g_{ij,m} + \\ & + y^m G_n{}^k(C^{nij} - C^i g^{nj}))\end{aligned} \tag{3.97}$$

where $C^i = C^{im}{}_m$.

The derivatives appearing in (3.95) can easily be calculated to get the following result:

$$L^{ij,mn} = L^{*ij,mn} + L^{**ij,mn}, \tag{3.98}$$

$$\begin{aligned}J^{-1}L^{*ij,mn} = {} & \tfrac{1}{2}(g^{mj}g^{ni} + g^{nj}g^{mi}) - g^{ij}g^{mn} - \\ & - \mathscr{A}^{(mn)}\mathscr{A}^{(ij)}(y^m(C^{nij} - C^i g^{nj})),\end{aligned} \tag{3.99}$$

$$J^{-1}L^{**ij,mn} = -y^i y^j C^{mn} + \tfrac{1}{2}(y^i T^{jmn} + y^j T^{imn}) \tag{3.100}$$

where we have used the following notation:

$$T^{imn} = \mathscr{A}^{(mn)}(2y^m C^{ni} + C^i g^{mn} - C^m g^{ni}) \tag{3.101}$$

and

$$C^{mn} = C^{mpq}C_{pq}{}^n - C_p C^{pmn} \equiv S^{mn}/F^2. \tag{3.102}$$

From Equations (3.98)–(3.100) and (3.87),

$$C_{ijk}L^{**ij,mn} = 0, \qquad C_{ijkl}L^{**ij,mn} = -C_{jkl}T^{jmn}, \tag{3.103}$$

$$J^{-1}C_p{}^i{}_q L^{pq,mn} = C^{imn} - g^{mn}C^i - \tfrac{1}{2}y^m C^{in} - \tfrac{1}{2}y^n C^{im}. \tag{3.104}$$

The equality (3.104) permits one to obtain the explicit form of the tensor density (3.81).

Considering the relations (3.95)–(3.100), we find, after a comparatively short calculation, that the objects $\Psi^{ijk,n}$ and Ψ^{ijkl} given by the formulae (3.73) and (3.74) are equal to

$$\Psi^{ijk,n} = 4\mathscr{A}^{(ijk)}(L^{**ij,mn}G_m{}^k), \tag{3.105}$$

$$\Psi^{ijkl} = 2\mathscr{A}^{(ijkl)}(L^{**ij,mn}G_m{}^kG_n{}^l). \tag{3.106}$$

These objects are not of tensorial nature, the corresponding tensorial parts being

$$\Psi^{*ijk,n} = 0, \qquad \Psi^{*ijkl} = 0. \tag{3.107}$$

As regards the remaining fundamental tensor densities $\Pi^{ij,m}$ and Π^{ij} given by (3.76) and (3.77), and the tensorial part Ψ^{*ijk} of the object Ψ^{ijk} given by (3.75), their calculation for the Finslerian scalar density (3.86) would have required an enormous number of elementary operations. However, since we know that the desired object is a tensor density, such laborious calculations may be avoided if the following calculational technique is used.

(1) Omit from the expression for the desired tensor density divided by the Jacobian J all the terms which are of the form of the contraction of the Cartan connection coefficients with geometrical objects, that is, for brevity, the terms proportional to the Cartan connection coefficients.

(2) In the remaining terms, replace the partial derivatives with respect to x^k by the Cartan covariant derivatives $|k$. The skew-symmetric combinations of the partial derivatives of the connection coefficients with respect to x^k are replaced by the K-tensor (1.78) taken with the appropriate sign.

Such a method generalizes, to the Finslerian case, a device which is frequently used in Riemannian geometry and according to which the coordinates are chosen to be normal with the pole at some point[9]. The validity of the method follows from the circumstance that a sum of terms each of which is proportional to the connection coefficients cannot be a nonvanishing tensor. A rigorous proof of this kind of assertion consists in writing the invariance identities for such an expression and considering their implications. The following example clarifies the picture.

PROPOSITION 6. *Let an object of the form*

$$\Gamma_m{}^h{}_kA_h{}^{mk} + \Gamma_m{}^h{}_k\Gamma_i{}^l{}_jD_{hl}{}^{mkij}, \tag{3.108}$$

where A and D denote tensors and $\Gamma_m{}^h{}_k$ are connection coefficients, be a tensor. Then $A_h{}^{mk}$ and $D_{hl}{}^{mkij}$ should vanish.

Proof. Consider the transform of (3.108) under the coordinate transformations $x^i(\bar{x}^j)$ and take into account the transformation law (3.38) of the connection coefficients. Differentiating the transform of (3.108) with respect to $B_m{}^h{}_k$ and then with respect to $B_i{}^l{}_j$, we immediately find after returning to the initial coordinates x^i that $D_{hl}{}^{mkij} = 0$. Therefore, the entity (3.108) reduces to the first term. Again, differentiating the transform of the first term in (3.108) with respect to $B_m{}^h{}_k$ and then putting $\bar{x}^i = x^i$, we get $A_h{}^{mk} = 0$. The proof is complete.

Let us apply this technique to the calculation of the tensor density $\Pi^{ij,m}$. According to item (1) of the above method, all terms in (3.76) except for the first should be omitted. After that, the derivative $L^{ij,m}$ is to be calculated using our density (3.86) excluding, again by virtue of item (1), the terms proportional to the connection coefficients. The expression (3.86) implies that

$$J^{-1}L^{ij,m} = (g^{pq}g^{rs} - g^{rq}g^{ps})\frac{\partial(\partial_r\Gamma_{psq})}{\partial g_{ij,m}} + \\ + g^{pq}\left[-\frac{\partial G_r^{\ t}}{\partial g_{ij,m}}\frac{\partial \Gamma_{p\ q}^{\ r}}{\partial y^t} + \frac{\partial G_p^{\ t}}{\partial g_{ij,m}}\frac{\partial \Gamma_{q\ s}^{\ s}}{\partial y^t}\right] + \text{t.p.c.c.} \tag{3.109}$$

where $\partial_r = \partial/\partial x^r$ and the abbrevation

t.p.c.c. = terms proportional to the connection coefficients

has been used. It is immediately seen from (1.32) that, up to terms proportional to the connection coefficients, the first term on the right-hand side of (3.109) is equal to

$$2g^{pq}g^{rs}\left[-C_{pst,r}\frac{\partial G_q^{\ t}}{\partial g_{ij,m}} + C_{pqt,r}\frac{\partial G_s^{\ t}}{\partial g_{ij,m}} + \\ + 2(C_{ps}^{\ \ n}C_{nqt,r} - C_{pq}^{\ \ n}C_{nst,r})\frac{\partial G^t}{\partial g_{ij,m}}\right]. \tag{3.110}$$

Replacing the partial derivative ∂_r in (3.110) by the Cartan covariant derivative $|r$, we get

$$(g^{pq}g^{rs} - g^{rq}g^{ps})\frac{\partial(\partial_r\Gamma_{psq})}{\partial g_{ij,m}} = -2C^{qr}_{\ \ t|r}\frac{\partial G_q^{\ t}}{\partial g_{ij,m}} + \\ + 2g^{rs}C_{t|r}\frac{\partial G_s^{\ t}}{\partial g_{ij,m}} + \\ + 4g^{rs}(C^{nq}_{\ \ s}C_{nqt|r} - C^nC_{nst|r})\frac{\partial G^t}{\partial g_{ij,m}} + \\ + \text{t.p.c.c.} \tag{3.111}$$

Further, from the expression (1.33) we have

$$\frac{\partial G_q^{\ t}}{\partial g_{ij,m}} = \tfrac{1}{2}\mathscr{A}^{(ij)}[y^m(-2y^iC_q^{\ tj} + g^{ti}\delta_q^j) + \\ + y^i(C_q^{\ tm}y^j + g^{tj}\delta_q^m - g^{tm}\delta_q^j)]. \tag{3.112}$$

Since (3.112) is a tensor and the derivative $\partial\Gamma_{p\ q}^{\ r}/\partial y^t$ is also a tensor (Equation (1.48)), the quantity $J^{-1}\Pi^{ij,m}$, being a tensorial part of $J^{-1}L^{ij,m}$, equals

$$J^{-1}\Pi^{ij,m} = \left[-g^{pq}\left(\frac{\partial\Gamma_{p\ q}^{\ r}}{\partial y^t} + 2C^r_{\ pt|q}\right) + g^{rq}\left(\frac{\partial\Gamma_{q\ s}^{\ s}}{\partial y^t} + \right.\right.$$
$$\left.\left. + 2C_{t|q}\right)\right]\frac{\partial G_r^{\ t}}{\partial g_{ij,m}} + 4(C^{nqr}C_{nqt|r} - C_nC_t^{\ nr}{}_{|r})\frac{\partial G^t}{\partial g_{ij,m}}. \tag{3.113}$$

Considering Equation (3.112) and replacing $\partial\Gamma_{p\ q}^{\ r}/\partial y^t$ by the expression (1.48), we find the decomposition

$$\Pi^{ij,m} = \Pi^{*ij,m} + \Pi^{**ij,m} \tag{3.114}$$

with

$$\Pi^{*ij,m} = Jy^m\mathscr{A}^{(ij)}(-2C^{ijp}{}_{|p} + 2C^{i|j} + C^{ips}C^j_{\ ps|0} - C_sC^{ijs}{}_{|0}) \tag{3.115}$$

where

$$C^j_{\ ps|0} = C^j_{\ ps|k}y^k$$

and

$$J^{-1}\Pi^{**ij,m} = y^iQ^jy^m + Q^iy^jy^m - y^iy^jQ^m + y^iQ^{jm} + y^jQ^{im}. \tag{3.116}$$

Here,

$$Q^m = C^{nqr}C^m_{\ nq|r} - C_nC^{mnr}{}_{|r} - C^{mtr}(-2C^p_{\ tr|p} + 2C_{t|r} +$$
$$+ C_{rlq}C^{lq}{}_{t|0} - C^lC_{lrt|0}), \tag{3.117}$$

$$Q^{jm} = C^{j|m} - C^{m|j} + \tfrac{1}{2}C^j_{\ lq|0}C^{mlq} - \tfrac{1}{2}C^m_{\ lq|0}C^{jlq}. \tag{3.118}$$

The identities

$$C_{ijk}\Pi^{**ij,m} = 0, \qquad Q^{jm} = -Q^{mj}$$

are valid.

It will be noted that the explicit form of $\Pi^{ij,m}$ can be obtained also through the representation (3.83). Indeed, our Equations (3.99) and (3.105) together with (3.84) and (3.85) imply that

$$C_{h\ k}^{\ i}\Psi^{hkj,l} = \tfrac{4}{3}y^jT^{hlm}C_{h\ k}^{\ i}G_m^{\ k}, \tag{3.119}$$

$$3C_{h\ k}^{\ i}\Phi^{jhk} = -3C_{h\ k}^{\ i}C_{r\ s}^{\ k}\Psi^{rsj,h} = -4y^jC_{h\ k}^{\ i}C_{r\ s}^{\ k}T^{smh}G_m^{\ r}, \tag{3.120}$$

$$C_{h\ km}^{\ i}\Psi^{hkm,j} = -4C_{k\ m}^{\ i}T^{kjr}G_r^{\ m}. \tag{3.121}$$

Substituting the relations (3.119)–(3.121) in (3.83) and cancelling similar terms, we get merely

$$\Pi^{ij,h} = \tfrac{1}{2}\Phi'^{lm}\{y^h(y^iC_{l\ m}^{\ j} + y^jC_{l\ m}^{\ i}) - y^iy^jC_{l\ m}^{\ h}\} +$$
$$+ \tfrac{1}{4}\{y^i(\Phi'^{hj} - \Phi'^{jh}) + y^j(\Phi'^{hi} - \Phi'^{ih}) - y^h(\Phi'^{ij} + \Phi'^{ji})\} \tag{3.122}$$

where

$$\Phi'^{ij} = 4C_{h\ k|m}^{\ i} L^{hk,mj} \equiv 4C_{h\ k|m}^{\ i} L^{*hk,mj}.$$

The expression (3.122) is but another form of Equation (3.114), as can be readily verified directly.

The tensor density Π^{ij} given by (3.77) may be found in a similar manner. Indeed, from Equations (3.86) and (1.32) we get:

$$\begin{aligned}\frac{\partial K}{\partial g_{ij}} = \left[\frac{\partial}{\partial g_{ij}}(g^{pq}g^{rs} - g^{rq}g^{ps})\right]\partial_r \Gamma_{psq} - \\ -2C_t^{\ pr}\frac{\partial(\partial_r G_p^{\ t})}{\partial g_{ij}} + 2C_t g^{pr}\frac{\partial(\partial_r G_p^{\ t})}{\partial g_{ij}} + \text{t.p.c.c.}\end{aligned} \tag{3.123}$$

Performing the differentiations with respect to g_{ij}, we obtain from (3.123) that

$$\begin{aligned}\frac{\partial K}{\partial g_{ij}} = -\mathscr{A}^{(ij)}\{g^{pi}g^{qj}[\partial_r \Gamma_{p\ q}^{\ r} - \partial_p \Gamma_{q\ r}^{\ r}] + \\ + g^{lh}g^{pi}[\partial_p \Gamma_{l\ h}^{\ j} - \partial_l \Gamma_{p\ h}^{\ j}] - \\ -2(C^{qri} - C^i g^{qr})\,\partial_r G_q^{\ j} + 4C^{ri}\,\partial_r G^j\} + \text{t.p.c.c.}\end{aligned} \tag{3.124}$$

By adding appropriate terms proportional to the connection coefficients, the first brackets in (3.124) may be made equal to $K_{p\ ql}^{\ r}$, and the second to $K_{l\ hp}^{\ j}$. Thus,

$$\begin{aligned}J^{-1}L^{ij} = \tfrac{1}{2}g^{ij}K - \tfrac{1}{2}(K^{ij} + K^{ji} + K'^{ij} + K'^{ji}) + \\ +2\mathscr{A}^{(ij)}[(C^{qri} - C^i g^{qr})\,\partial_r G_{q}^{\ j} - 2C^{ri}\,\partial_r G^j] + \text{t.p.c.c.}\end{aligned} \tag{3.125}$$

where

$$K_{ij} = K_{i\ jr}^{\ r}, \qquad K'_{ij} = K^h_{\ ihj}. \tag{3.126}$$

The first term on the right-hand side of (3.125) has emerged as a result of differentiating the Jacobian J with respect to g_{ij}.

Terms which are not proportional to the connection coefficients also appear in the addends in (3.77) involving $M_{r\ ls}^{\ i}$. Using the expression (3.79) for $M_{r\ ls}^{\ i}$, we obtain from the definition of Π^{ij} that

$$\Pi^{ij} = L^{ij} + 2\mathscr{A}^{(ij)}[L^{ir,ls}\,\partial_s \Gamma_{r\ l}^{\ j}] + \text{t.p.c.c.} \tag{3.127}$$

The cyclic identity (3.82) can be very conveniently used here. Indeed, it follows directly from Equation (3.82) that

$$3\Pi^{ir,ls} = (\Pi^{ir,ls} - \Pi^{il,rs}) + (\Pi^{ir,ls} - \Pi^{is,lr}) \tag{3.128}$$

so that, owing to the symmetry of the connection coefficients $\Gamma_{r\ l}^{\ j}$ in r, l,

$$\Pi^{ir,ls}\,\partial_s \Gamma_{r\ l}^{\ j} = \tfrac{1}{3}(\Pi^{ir,ls} - \Pi^{is,lr})\,\partial_s \Gamma_{r\ l}^{\ j}$$

or

$$\Pi^{ir,ls}\,\partial_s\Gamma_{r\,l}^{\,j} = \tfrac{1}{3}\Pi^{ir,ls}(\partial_s\Gamma_{r\,l}^{\,j} - \partial_r\Gamma_{s\,l}^{\,j}). \tag{3.129}$$

The expression in parentheses on the right-hand side of (3.129) equals $K_{l\,rs}^{\,j}$ up to terms proportional to the connection coefficients. Thus, when going over from $L^{ir,ls}$ to $\Pi^{ir,ls}$ in (3.127) in accordance with (3.81), we shall get

$$\begin{aligned}\Pi^{ij} = {} & L^{ij} + 2\mathscr{A}^{(ij)}(\tfrac{1}{3}\Pi^{ir,ls}K_{l\,rs}^{\,j}) - \\ & -2\mathscr{A}^{(ij)}[-C_{m\,n}^{\,i}L^{*mn,ls}\,\partial_s G_l^{\,j}] + \text{t.p.c.c.}\end{aligned} \tag{3.130}$$

where we have written $L^{*mn,ls}$ instead of $L^{mn,ls}$, which is legitimate because of the first of the identities (3.103). After substituting (3.99) in (3.130), it can easily be verified that the sum of the terms entering the brackets in (3.125) and (3.130) are equal, on cancelling similar terms, to

$$\tfrac{1}{2}C^{im}y^n y^k(\partial_n\Gamma_{k\,m}^{\,j} - \partial_m\Gamma_{n\,k}^{\,j}).$$

This is expressible through the K-tensor; a mere glance at the expression of the K-tensor (1.78) is sufficient to see that

$$y^p y^q K_{p\,qr}^{\,k} = y^p y^q(\partial_r\Gamma_{p\,q}^{\,k} - \partial_p\Gamma_{r\,q}^{\,k}) + \text{t.p.c.c.} \tag{3.131}$$

So, the eventual result is

$$\begin{aligned}\Pi^{ij} = {} & -\tfrac{1}{2}J(K^{ij} + K^{ji} + K'^{ij} + K'^{ji}) + \tfrac{1}{2}Jg^{ij}K + \\ & + \tfrac{1}{3}\Pi^{ir,ls}K_{l\,rs}^{\,j} + \tfrac{1}{3}\Pi^{jr,ls}K_{l\,rs}^{\,i} + \tfrac{1}{2}J(C^{im}K_{p\,mq}^{\,j} + \\ & + C^{jm}K_{p\,mq}^{\,i})y^p y^q.\end{aligned} \tag{3.132}$$

This formula represents the Finslerian generalization of the similar Riemannian relation found in Lovelock and Rund (1975, p. 311, formula (4.40)).

We leave it to the reader to verify that for the case of the Finslerian scalar density (3.86) the tensorial part Ψ^{*ijk} of the object Ψ^{ijk} given by Equation (3.75) is

$$\begin{aligned}\Psi^{*ijk} = {} & \Omega^{ijk} + \mathscr{A}^{(ijk)}(2G_m^{\,i}\Pi^{*jk,m} + 2y^p M_{p\,nm}^{\,i}L^{*jk,nm} - \\ & -8G_n^{\,i}\Gamma_{s\,m}^{\,j}L^{*ks,mn} - 2JT^{imn}G_m^{\,j}G_n^{\,k}).\end{aligned} \tag{3.133}$$

To calculate this tensor density with the help of our method, we write

$$\Psi^{*ijk} = \Omega^{ijk} + \mathscr{A}^{(ijk)}(2y^p M_{p\,nm}^{\,i}L^{*jk,nm}) + \text{t.p.c.c.} \tag{3.134}$$

From Equations (1.32)–(1.34) we have

$$\begin{aligned}J^{-1}\Omega^{ijk} = {} & (g^{pq}g^{rs} - g^{rq}g^{ps})\,\partial(\partial_r\Gamma_{psq})/\partial C_{ijk} + \text{t.p.c.c.} = \\ & = 2\mathscr{A}^{(ijk)}(-g^{iq}g^{jr}\,\partial_r G_q^{\,k} + g^{ij}g^{rs}\,\partial_r G_s^{\,k} + 2(C^{ijr} - C^i g^{jr})\,\partial_r G^k) + \\ & + \text{t.p.c.c.}\end{aligned} \tag{3.135}$$

The second term on the right-hand side of (3.134) can be written as

$$2y^p M_p{}^{in}{}_m L^{*jk,nm} = 2\partial_m G_n{}^i L^{*jk,nm} + \text{t.p.c.c.}, \tag{3.136}$$

according to Equation (3.79).

Substituting (3.135) and (3.136) in (3.134) and taking account of the relation (3.99), we find after cancelling similar terms:

$$J^{-1}\Psi^{*ijk} = \mathscr{A}^{(ijk)}((C^{ijr} - C^i g^{jr})(\partial_r \Gamma_p{}^k{}_q - \partial_p \Gamma_r{}^k{}_q) y^p y^q) + \\ + \text{t.p.c.c.} \tag{3.137}$$

The second parenthetic expression in (3.137) can be expressed with the help of the equality (3.131) in terms of the K-tensor. Thus we conclude from (3.137) that

$$J^{-1}\Psi^{*ijk} = \mathscr{A}^{(ijk)}((C^{ijr} - C^i g^{jr}) K_p{}^k{}_{qr} y^p y^q) + \text{t.p.c.c.} \tag{3.138}$$

Since Ψ^{*ijk} is a tensor density, we obtain from (3.138) the eventual result:

$$\Psi^{*ijk} = J\mathscr{A}^{(ijk)}((C^{ijr} - C^i g^{jr}) K_p{}^k{}_{qr} y^p y^q). \tag{3.139}$$

Many significant simplifications in the structure of the fundamental tensor densities result if the Finslerian metric function satisfies the identities

$$C_i = 0, \qquad C^{mn} = -(N-2)F^{-2}h^{mn} \tag{3.140}$$

(such is the case for the 1-form Berwald-Moór metric function; see Equations (2.38) and (2.40) and the definition (3.102) of C^{mn}). Namely, in view of the identities (1.47) and the second equality in (3.140), the Cartan covariant derivative of C^{mn} will vanish identically:

$$C^{mn}|_k = 0. \tag{3.141}$$

From (3.140) and (3.102) we have $C^{mn} = C^{mpq} C^n{}_{pq}$, so that Equation (3.141) may be rewritten as

$$C^{mpq}|_k C^n{}_{pq} + C^{mpq} C^n{}_{pq|k} = 0. \tag{3.142}$$

The identities $C_i = 0$, (3.141) and (1.47) make it possible to conclude from (3.97)–(3.102) that

$$L^{**ij,mn}{}_{|k} = 0 \tag{3.143}$$

and

$$L^{ij,mn}{}_{|m|n} = L^{*ij,mn}{}_{|m|n} = -\tfrac{1}{2}J(C^{ijn}{}_{|m|n} + C^{ijn}{}_{|n|m}) y^m. \tag{3.144}$$

Further, the substitution of $C_i = 0$ and (3.142) in (3.115) gives just

$$\Pi^{*ij,m} = -2Jy^m C^{ijp}{}_{|p}, \tag{3.145}$$

$$Q^m = C^{mtr} C^p{}_{tr|p}, \qquad Q^{jm} = \tfrac{1}{2}(C^j{}_{lq|t} C^{mlq} - C^m{}_{lq|t} C^{jlq}) y^t. \tag{3.146}$$

It follows from Equations (3.145), (3.114) and (3.116) that

$$J^{-1}\Pi^{ij,m}{}_{|m} = -2C^{ijp}{}_{|p|t} y^t + l^i Z^j + l^j Z^i - l^i l^j Z \tag{3.147}$$

where

$$Z^i = F(Q^i{}_{|t} y^t + Q^{im}{}_{|m}), \qquad Z = F^2 Q^m{}_{|m}. \tag{3.148}$$

The explicit form of the tensor density E^{ij} can be obtained by substituting Equations (3.144), (3.147), and (3.132) in the definition (3.78) of E^{ij}.

To summarize: given a Finslerian scalar density (3.86), the associated tensor densities have been found in an explicit form which is not difficult to calculate, as indicated in Proposition 6. Two tensor densities vanish identically according to Equation (3.107). Numerous simplifications arise from particular specifications of the Finslerian metric function, as is illustrated by Equations (3.140)–(3.148).

At this stage we have finished our preparatory work and we proceed to elaborate a formulation of gravitational and other field equations, against the background of Finsler geometry.

Problems

3.1. Prove that it is impossible to construct a tensor quantity, other than one of the form $S^{i \ldots m} = \text{const } y^i \ldots y^m$, which depends only on the contravariant vector y^i. Thus there does not exist a scalar $S(y)$ other than a constant, or a nonvanishing tensor of the form $T_i{}^j(x)$ other than const δ_i^j.

3.2. Prove rigorously that the concept of a 1-form Finsler space introduced in Section 2.4 may be redefined as follows: a Finsler space is of the 1-form type if the x-dependent fields $S^A(x)$ entering the representation $F(x, y) = v(S^A(x), y)$ (Equation (3.5)) of the Finslerian metric function are given by a set of n-tuples.

3.3. If a Riemannian metric tensor $a_{ij}(x)$ plays the role of the x-dependent tensors $S^A(x)$ in the representation (3.5) of the Finslerian metric function, then the metric function is of Riemannian type. Prove this assertion.

3.4. Verify that, for any scalar density constructible from the components of a Riemannian metric tensor and their derivatives up to the second order, the Noether identities (3.25) read (3.54).

3.5. Compare the construction of the Cartan connection coefficients based on the invariance identities (Section 3.2) with that given by Rund and Cartan (Section 2.4 and 3.1 in Rund, 1959).

3.6. There exists no nonvanishing tensor depending just on the Finslerian tensors g_{hk} and C_{hkl} and the first-order derivatives $g_{hk,l}$. Prove this assertion. Compare it with Proposition 2 in Section 3.3.

3.7. Using Equation (3.104), verify directly that the cyclic identity (3.82) for the tensor density (3.81) is valid indeed.

3.8. Show that the identity

$$\Pi^{hk,lm} K_{hksm} = 0$$

holds in the Finslerian case.

3.9. Suppose we are given connection coefficients $\Gamma_{k\ n}^{\ m}(x, y)$ constructed from y^j and a mixed tensor $S_i^j(x)$ together with its first derivatives $S_{i,p}^j = \partial S_i^j / \partial x^p$; that is, there exist connection coefficients $T_{k\ n}^{\ m}(y^j, S_i^j, S_{i,p}^j)$ such that

$$\Gamma_{k\ n}^{\ m}(x, y) = T_{k\ n}^{\ m}(y, S_i^j(x), S_{i,p}^j(x)).$$

Prove that the object

$$T_{knj}^{mi} = \left(\frac{\partial}{\partial S_i^j} + T_{p\ q}^{\ i} \frac{\partial}{\partial S_{q,p}^j} - T_{p\ j}^{\ q} \frac{\partial}{\partial S_{i,p}^q} \right) T_{k\ n}^{\ m}$$

is a tensor.

3.10. Verify that the object Ψ^{*ijk} given by Equation (3.133) is a tensor density.

3.11. Solve the problems included at the end of Chápter 8 of Lovelock and Rund (1975).

Notes

[1] The invariance identities have been proposed by H. Rund, and used in the monographs by Rund and Beare (1972) and Lovelock and Rund (1975, Chapter 8).
[2] See, for example, Konopleva and Popov (1981).
[3] See Trautman (1962).
[4] Lovelock and Rund (1975, formula (6.7.13)).
[5] *Op. cit.*, Chapter 8.
[6] In the remainder of this section we describe the results obtained by Rund and Beare (1972). We retain the notation used by these authors, except for writing the Latin L instead of the Greek Λ. To avoid confusion, it is noted that Rund-Beare's analog (2.19) of our formula (3.66) contains an obvious misprint, namely the density B is omitted on the right-hand side.
[7] In Rund and Beare (1972) such an operator was denoted merely by d_m. However, we reserve the symbol d_m for the total derivative exemplified by (4.8).
[8] This section and Chapter 4 expound the results obtained by Asanov (1982, 1983a).
[9] Cf. the way of reasoning used by Lovelock and Rund (1975, p. 311) in the Riemannian case.

Chapter 4

Finslerian Approach Based on the Concept of Osculation

4.1. Formulation of Gravitational Field Equations in Terms of the Fundamental Tensor Densities

In theoretical physics the equations of physical fields are formulated in terms of Lagrangian scalar densities which are assumed to be functions of the field variables and, perhaps, auxiliary fields, both depending on x^i alone. In attempting to exploit the potentialities of Finsler geometry with the aim of generalizing the equations of physical fields, one is faced with the new circumstance that the Finslerian generalizations of Lagrangian densities will involve a dependence on the tanget vector y^i treated as a directional variable essentially independent of the positional argument x^i.

Therefore, there arises the difficulty that the principle of action cannot be formulated in the usual way unless the dependence of the Lagrangian density on the directional variable y^i is suppressed. Indeed, the integration of a direction-dependent scalar density over a finite simply-connected N-dimensional region G of the N-dimensional background manifold is not a coordinate-invariant procedure. Let, for example, $w^A(x, y)$ be a set of arbitrary tensor fields. Then whichever scalar density $L = L(y, w^A(x, y))$ we choose, the integral $I(y) = \int_G L(y, w^A(x, y))\, \mathrm{d}^N x$ fails to be a geometrical scalar (as it would be in the case of a scalar density involving no directional variables). This assertion follows simply from the fact that there does not exist a non-constant scalar function constructible solely from a contravariant vector, because such a possibility would contradict the invariance identities (see Problem 3.1). Thus, it seems to be impossible to retain the usual formalism of the principle of action by treating y^i simply as parameters (which would be possible in the case when the parameters are scalars, as occurs in Chapter 5)[1].

It should be clear from the above that a strictly covariant formulation of the field equations on the basis of a direction-dependent scalar density requires some ingenious methods. One would be right to expect that such a formulation could be achieved in terms of the fundamental tensor densities associated with the scalar density. An interesting attempt in this direction was made by Rund and Beare (1972) who, proceeding from the method of equivalent integrals for the multiple. problem in the calculus of variations, proposed the method of equivalent Lagrangian densities for direction-dependent scalar densities L of the type (3.60). This yields the following elegant system of vacuum field equations:

$$E^{ij} = 0, \tag{4.1}$$

$$\Psi^{ijk,m}{}_{|m} - \Psi^{ijk} = 0, \tag{4.2}$$

$$\Psi^{ijkl} = 0 \tag{4.3}$$

called the principal field equations, associated principal field equations, and the third set of field equations, respectively. Here, the tensor $g_{ij}(x, y)$ entering L is assumed to be Finslerian, and the connection coefficients are chosen to be Cartan's. However, if L forms a scalar density subject to the Finslerian identities (3.87), we should replace the Ψ-objects (Equations (3.73)–(3.75)) in (4.2) and (4.3) by their tensorial redifinitions $\Psi^{*ijk,m}$, Ψ^{*ijkl}, and Ψ^{*ijk} (see Section 3.4) in order to convert Equations (4.2) and (4.3) to tensorial form. In particular, for the Finslerian scalar density (3.86) we have the identities (3.107), according to which we get

$$\Psi^{*ijk} = 0 \tag{4.4}$$

instead of the equation (4.2), while the third set (4.3) of equations will be reduced to an identity.

In view of Equation (4.1) we could attempt to treat the tensor density E^{ij} as a Finslerian generalization of the Einstein tensor density of the gravitational field. Such a treatment, however, is hampered by an essential circumstance. Despite the striking formal resemblance of the definition (3.78) of E^{ij} to its Riemannian prototype (3.53) it was found by Rund and Beare (*op. cit.*) by an elaborate analysis that the identities satisfied by the Finslerian E^{ij} are far from similar to the Bianchi-type identities $\mathrm{D}_j E^{ij} = 0$ satisfied by Einstein's E^{ij}. Had E^{ij} obeyed Finslerian identities of the type $E^{ij}{}_{|j} = 0$ it would have been possible to try to use them in establishing a relationship between the Finslerian equations of the gravitational field and the Finslerian equations of motion of the gravitational field sources, as in the procedure used in general relativity. The fact that the identities satisfied by the Finslerian E^{ij} prove to be very complicated makes such a relationship obscure[2]. Nevertheless, there exists an alternative approach which we outline below.

Let us continue to treat $g_{ij}(x, y)$ as a tensor (which, as in Section 3.3, is not necessarily a Finslerian metric tensor) constructible from a set of x-dependent tensors $S^A(x)$ in accordance with a relation of the form (3.14). Then, after substituting (3.14) in (3.60) the scalar density L takes the form

$$L = L(y, S^A(x), \mathrm{d}_m S^A(x), \mathrm{d}^2_{mn} S^A(x)) \tag{4.5}$$

($\mathrm{d}_m = \partial/\partial x^m$) which does depend on the directional variable y^i. However, we can introduce *the osculating scalar density* L_g by substituting a vector field

$y^i(x)$ for the variable y^i in Equation (4.5):

$$L_g = L(y(x), S^A(x), \mathrm{d}_m S^A(x), \mathrm{d}^2_{mn} S^A(x)) \tag{4.6}$$

which involves x-dependent objects only. It is L_g that will be treated in the present chapter as the Finslerian generalization of the Lagrangian density of the gravitational field. The tensors $S^A(x)$ will be considered to be field variables, while $y^i(x)$ will play the role of an auxiliary vector field. The homogeneity fo L_g in $y^i(x)$ will be assumed, namely

$$\begin{aligned} L_g(t(x)y(x), S^A(x), \mathrm{d}_m S^A(x), \mathrm{d}^2_{mn} S^A(x)) = \\ = L_g(y(x), S^A(x), \mathrm{d}_m S^A(x), \mathrm{d}^2_{mn} S^A(x)), \end{aligned} \tag{4.7}$$

$t(x)$ being an arbitrary positive scalar.

The symbol d_m will be used for the operator of *the total derivative* with respect to x^m; that is, for any object $w(x, y)$, the action of this operator will read

$$\mathrm{d}_m w = \mathrm{d}_m(w(x, y(x))) \equiv \partial w/\partial x^m + (\partial w/\partial y^k)\partial y^k(x)/\partial x^m. \tag{4.8}$$

A comparison of this operator with the definition (3.80) of the operator d'_m leads to the useful relation

$$\mathrm{d}_m w = \mathrm{d}'_m w + y^k_m \, \partial w/\partial y^k \tag{4.9}$$

where $\mathrm{d}'_m w$ and $\partial w/\partial y^k$ are evaluated at $y^i = y^i(x)$; y^k_m is the covariant derivative

$$y^k_m = \partial y^k(x)/\partial x^m + \Gamma_m{}^k{}_n(x, y(x))y^n(x) \equiv \partial y^k(x)/\partial x^m + G_m{}^k(x, y(x)). \tag{4.10}$$

The second covariant derivative of $y^k(x)$ is

$$y^{\;k}_{mn} = \mathrm{d}_n y^k_m - \Gamma_n{}^i{}_m(x, y(x))y^k_i + \Gamma_n{}^k{}_i(x, y(x))y^i_m. \tag{4.11}$$

We are now in a position to clarify the tensorial structure of the Euler-Lagrange derivative $E_A{}^{i\ldots}_{j\ldots}$ given by the definition (3.26) which will be applied henceforth to the Lagrangian scalar density L_g of the form (4.6). To do this, we shall use direct calculations to find the explicit form of $E_A{}^{i\ldots}_{j\ldots}$ in terms of our notation (3.63) and (3.64). This leads to

$$E_A{}^{q\ldots}_{p\ldots} = \frac{\partial v_{ij}}{\partial S^{A p\ldots}_{\;\;q\ldots}} M^{ij} + \frac{\partial v_{ijk}}{\partial S^{A p\ldots}_{\;\;q\ldots}} M^{ijk} - \frac{\partial v_{ijkl}}{\partial S^{A p\ldots}_{\;\;q\ldots}} M^{ijkl} \tag{4.12}$$

where v_{ij} is defined by the relation of the form (3.14); $v_{ijk} = \frac{1}{2}\partial v_{ij}/\partial y^k$ and $v_{ijkl} = \partial v_{ijk}/\partial y^l$;

$$M^{ij} = -\mathrm{d}^2_{mn} L^{ij,mn} + \mathrm{d}_m L^{ij,m} - L^{ij}, \tag{4.13}$$

$$\begin{aligned} M^{ijk} = \mathrm{d}_m \Omega^{ijk,m} - \Omega^{ijk} + \\ + 2\mathscr{A}^{(ijk)}((L^{ij,m} - 2\mathrm{d}_n L^{ij,mn})y^k_{,m} - L^{ij,mn}\,\mathrm{d}^2_{mn} y^k), \end{aligned} \tag{4.14}$$

$$M^{ijkl} = \Omega^{ijkl} - \mathscr{A}^{(ijkl)}(\Omega^{ijk,m} y^l_{,m} - 2L^{ij,mn} y^k_{,m} y^l_{,n}), \tag{4.15}$$

and where $\mathscr{A}^{(\ldots)}$ is the symmetrization operator including the factor $n!$ (Equation (3.79′)).

Going over at this point from $y^k_{,m} = \mathrm{d}_m y^k$ and $\mathrm{d}^2_{mn} y^k$ to the covariant derivatives (4.10) and (4.11), and from the total derivative d_m to the operator d'_m in accordance with (4.9), and then taking into account the expressions (3.73)–(3.78) for the fundamental tensor densities, we obtain after some simple straightforward calculations that the following tensorial representations are valid:

$$\begin{aligned} M^{ij} = {} & E^{ij} - y^k_{mn}\,\partial L^{ij,mn}/\partial y^k - y^k_m y^l_n\,\partial^2 L^{ij,mn}/\partial y^k\,\partial y^l + \\ & + y^k_l(-(\partial L^{ij,ml}/\partial y^k)_{|m} - \partial(L^{ij,ml}{}_{|m})/\partial y^k + \\ & + \partial\Pi^{ij,l}/\partial y^k - 2\mathscr{A}^{(ij)}(L^{sj,ml}\,\partial\Gamma_s{}^i{}_m/\partial y^k)), \end{aligned} \tag{4.16}$$

$$\begin{aligned} M^{ijk} = {} & \Psi^{ijk,m}{}_{|m} - \Psi^{ijk} + \mathscr{A}^{(ijk)}(-2y^k_{mn}L^{ij,mn} - \\ & - 4y^k_m y^l_n\,\partial L^{ij,mn}/\partial y^l - 4y^k_l L^{ij,nl}{}_{|n} + y^r_l(\partial\Psi^{ijk,l}/\partial v^r - \\ & - 2L^{ij,ml}C_m{}^k{}_{r|n}y^n) + 2y^k_l\Pi^{ij,l}), \end{aligned} \tag{4.17}$$

$$M^{ijkl} = \Psi^{ijkl} + \mathscr{A}^{(ijkl)}(2L^{ij,mn}y^k_m y^l_n - y^l_m\Psi^{ijk,m}). \tag{4.18}$$

Each term in Equations (4.16)–(4.18) represents a tensor density. Hence, the objects M^{ij}, M^{ijk}, and M^{ijkl} are tensor densities.

To calculate the tensor density $N_i{}^j$ entering the Noether identities (3.25), we can substitute the decomposition (4.12) in the definition (3.28) of $N_i{}^j$ and take into account the invariance identities (3.15) for v_{ij} together with two subsequent invariance identities for v_{ijk} and v_{ijkl} (of which the first identity is obtainable by differentiating Equation (3.15) with respect y^k and the second by differentiating with respect to y^k and y^l), which yields

$$\begin{aligned} N_i{}^j = {} & (2\delta^j_m g_{ni} + 2y^j C_{mni})M^{mn} + (3\delta^j_m C_{nik} + y^j C_{mnik})M^{mnk} - \\ & - (4\delta^j_m C_{nikl} + y^j C_{mnikl})M^{mnkl} \end{aligned} \tag{4.19}$$

where $C_{ijnkl} = \partial C_{ijnk}/\partial y^l$. The expression (4.19) exhibits a remarkable property of $N_i{}^j$, namely that $N_i{}^j$ is constructible just from the tensor $g_{ij}(x,y)$ and its derivatives and does not depend explicitly on the fields $S^A(x)$. Hence, although the representation (3.14) of the tensor $(g_{ij}(x,y)$ leaves the tensors $S^A(x)$ ambiguous (see, for example, the scale transformations (2.61)), such indefiniteness does not appear in the tensor density (4.19).

It is worth noting that this last observation has a simple analogy in the usual Riemannian approach to general relativity. Thus, if one follows the tetradic approach[3] according to which the orthonormal tetrads $h^P_i(x)$ of a Riemannian metric tensor $a_{ij}(x)$ are regarded as genuine field variables, then the Euler-Lagrange derivative $E_P{}^i$ of the gravitational Lagrangian density with respect to h^P_i will explicitly depend on h^P_i. However, the gravitational field equations may easily be represented in such a form that $E_P{}^i$ occurs only in the combination $N_j{}^i \overset{\text{def}}{=} h^P_j E_P{}^i$. Since

$$N_j^{\ i} = h_j^P E^{mn}\, \partial a_{mn}/\partial h_i^P = 2E_j^{\ i},$$

where E^{mn} denotes the Euler-Lagrange derivative of the gravitational Lagrangian density with respect to a_{mn}, $N_j^{\ i}$ depends explicitly only on a_{mn} and its derivatives.

At this stage we can clarify the explicit form of $N_i^{\ j}$, as well as of the Euler-Lagrange derivative (4.12) itself, in the important case when the initial direction-dependent scalar density L is of the Finslerian form (3.86). Accordingly, it will be assumed in the remainder of this section that L_g of the form (4.6) is the osculating version of the Finslerian L given by Equation (3.86). Under these conditions, we can substitute Equations (3.105)–(3.107) in the expressions (4.17) and (4.18) for M^{ijk} and M^{ijkl} and, after that, substitute M^{ijk} and M^{ijkl} in (4.12). Since the identity $y^l \partial v_{ijkl}/\partial S^A = -\partial v_{ijk}/\partial S^A$ holds by virtue of the third of the Finslerian identities (3.87), the terms in M^{ijkl} which are proportional to y^i, y^j, y^k or y^l will be transferred to M^{ijk}. Also, any term in M^{ijk} which is proportional to y^i, y^j or y^k may be neglected because $y^i \partial v_{ijk}/\partial S^A = 0$ in accordance with the first identity in (3.87). Finally, the decomposition (3.98) is used in M^{ijk} taking into account Equation (3.103). This procedure yields after some straightforward calculations the following result.

PROPOSITION 1. In the case when the direction-dependent scalar density L is chosen to have the Finslerian form (3.86), it is legitimate to replace M^{ijk} and M^{ijkl} in the decomposition (4.12) of the Euler-Lagrange derivative by the following simplified tensor densities:

$$\begin{aligned} M^{*ijk} = & -\Psi^{*ijk} + \mathscr{A}^{(ijk)}(-2y^k_{mn}L^{*ij,mn} - 4y^k_m y^l_n(\partial L^{*ij,mn}/\partial y^l + \\ & + \tfrac{1}{2}J\delta^i_l T^{jmn}) - 4y^k_l L^{*ij,nl}{}_{|n} - 2y^r_l L^{*ij,ml} C_m{}^k{}_{r|n} y^n + \\ & + 2y^k_l \Pi^{*ij,l}) \end{aligned} \tag{4.20}$$

and

$$M^{*ijkl} = 2\mathscr{A}^{(ijkl)}(L^{*ij,mn} y^k_m y^l_n) \tag{4.21}$$

where

$$\begin{aligned} \Psi^{*ijk} = & \ \Omega^{ijk} + \mathscr{A}^{(ijk)}(2G_m{}^i \Pi^{*jk,m} + 2y^h M_h{}^i{}_{nm} L^{*jk,nm} - \\ & - 8G_n{}^i \Gamma_s{}^j{}_m L^{*ks,mn} - 2JT^{imn} G_m{}^j G_n{}^k) + \end{aligned} \tag{4.22}$$

and $\Pi^{*ij,m}$ denotes the part of $\Pi^{ij,m}$ which remains after neglecting terms proportional to y^i or y^j (Equation (3.115)), and T^{imn} is the tensor (3.101).

4.2. Application to Non-Gravitational Fields

In the preceding section we have developed a method for the Finslerian generalization of the gravitational field equations by exploiting the concept of osculation. In the present section it is shown that such an approach can consistently be applied in order to generalize equations of other physical fields.

Suppose we are given a field $W'(x)$ satisfying a set of differential equations. The Finslerian technique may be used to generalize these equations by introducing a field $W(x,y)$ dependent on both the point x^i and the tangent vector y^i; for example, we may replace in the equations Riemannian covariant derivatives by Cartan covariant ones[4]. After so doing, we may adopt the viewpoint that our x-dependent field is a manifestation of $W(x,y)$ through osculation along a vector field $y^i(x)$, and introduce the x-dependent field

$$W(x) = W(x, y(x)) \tag{4.23}$$

osculating to $W(x,y)$ *along* $y^i(x)$.

In order to formulate field equations suitable for describing such osculating fields we proceed in the same fashion as in the preceding section where we made use of the representation $F(x,y) = v(S^A(x), y)$ (Equation (3.5)) of the Finslerian metric function so as to treat $S^A(x)$ as gravitational field variables. Accordingly, we select the x-dependent fields $r^a(x)$, $a = 1, 2, \ldots$, from which the generalized field $W(x,y)$ can be constructed:

$$W(x,y) = q(r^a(x),\ S^A(x), y) \tag{4.24}$$

and treat them as the genuine field variables for the field $W(x)$ defined by (4.23).

The Lagrangian density which yields the field equations for $W(x)$ in terms of these $r^a(x)$ may be constructed in a straightforward way. Let the initial field $W'(x)$ have an associated Lagrangian density, say, of the first-order form

$$L_{W'} = L_{W'}(a_{ij}(x),\ W'(x),\ \mathrm{D}_i W'(x)) \tag{4.25}$$

where a_{ij} is a Riemannian metric tensor and D_i is the covariant derivative associated with a_{ij}. Then our purpose can be achieved in two steps. First, we replace in Equation (4.25) the field $W'(x)$ by its generalization $W(x,y)$, $a_{ij}(x)$ by the Finslerian metric tensor $g_{ij}(x,y)$, and D_i by the Cartan covariant derivative $|i$. This leads us to a direction-dependent scalar density of the form

$$\begin{aligned} &L(g_{ij}, \partial g_{ij}/\partial x^k, C_{ijk}, W(x,y), \partial W(x,y)/\partial x^i, \partial W(x,y)/\partial y^i) = \\ &\quad = L_{W'}(g_{ij}, W(x,y), W(x,y)_{|i}) \end{aligned} \tag{4.26}$$

which, after substituting (4.24) together with $g_{ij}(x,y) = v_{ij}(S^A(x), y)$ (Equation (3.14)), takes the form

$$L = L(y^i, S^A(x), \partial S^A(x)/\partial x^i, r^a(x), \partial r^a(x)/\partial x^i). \tag{4.27}$$

Second, we construct an osculating scalar density L_W, i.e. we substitute a vector field $y^i(x)$ for the directional variable y^i in (4.27), which yields the Lagrangian density

$$L_W = L_W(y^i(x), S^A(x), \partial S^A(x)/\partial x^i, r^a(x), \partial r^a(x)/\partial x^i) \tag{4.28}$$

(cf. Equation (4.6)) from which follow the field equations for $W(x)$ in terms of the Euler-Lagrange derivative E_a of L_W with respect to $r^a(x)$, the fields $S^A(x)$ being the gravitational field variables. In particular, the vacuum equations for the generalized field $W(x)$ will read

$$E_a = 0. \tag{4.29}$$

Let us apply this method to the electromagnetic field. The conventional (rescaled) Lagrangian density of the electromagnetic field is[5]

$$L = (-a)^{1/2} a^{im}(x) a^{jn}(x) F_{ij}(x) F_{mn}(x) \tag{4.30}$$

where

$$F_{ij}(x) = \partial A_j(x)/\partial x^i - \partial A_i(x)/\partial x^j \tag{4.31}$$

denotes the electromagnetic field strength tensor; $A_i(x)$ is the electromagnetic vector potential; $a^{ij}(x)$ is the Riemannian metric tensor of space-time; $a = \det(a_{ij})$. Suppose the equations for the direction-dependent generalization $A_i(x, y)$ have been formulated in some way. Then we obtain the direction-dependent density

$$L = J(x, y) g^{im}(x, y) g^{jn}(x, y) F_{ij}(x, y) F_{mn}(x, y) \tag{4.32}$$

where

$$\begin{aligned} F_{ij}(x, y) &\overset{\text{def}}{=} A_j(x, y)_{|i} - A_i(x, y)_{|j} = \partial A_j(x, y)/\partial x^i - \\ &- G_i^l(x, y)\, \partial A_j(x, y)/\partial y^l - \partial A_i(x, y)/\partial x^j + G_j^l(x, y)\, \partial A_i(x, y)/\partial y^l \end{aligned} \tag{4.33}$$

in accordance with the definition (1.43) of the Cartan covariant derivative. Further, $A_i(x, y)$ is to be written in the form (4.24),

$$A_i(x, y) = a_i(r^a(x), S^A(x), y) \tag{4.34}$$

where a_i is a covariant vector field. Substituting (4.34) together with (3.14) in Equation (4.32) yields a direction-dependent density of the form (4.27), and the subsequent replacement $y^i \to y^i(x)$ gives the desired generalized Lagrangian density L_A of the electromagnetic field in the form (4.28).

A comparatively short calculation yields the following expressions for the Euler-Lagrange derivatives of L_A with respect to r^a and S^A:

$$\frac{\delta L_A}{\delta r^a} = 4 \frac{\partial a_j}{\partial r^a} \mathrm{d}_k (J F^{kj}(x, y(x))) + 4 J F^{kj}(x, y(x)) \frac{\partial^2 a_j}{\partial y^m \, \partial r^a} y^m_k \tag{4.35}$$

and

$$\begin{aligned} \frac{\delta L_A}{\delta S^A} &= 4 \frac{\partial a_j}{\partial S^A} \mathrm{d}_k (J F^{kj}(x, y(x))) + 4 J F^{kj}(x, y(x)) \frac{\partial^2 a_j}{\partial y^m \, \partial S^A} y^m_k + \\ &+ \frac{\partial v_{ij}}{\partial S^A} M_A{}^{ij} + \frac{\partial v_{ijk}}{\partial S^A} M_A{}^{ijk} \end{aligned} \tag{4.36}$$

(cf. Equation (4.12)), where

$$M_A{}^{ij} = \mathrm{d}_m(\partial L_A/\partial g_{ij,m}) - \partial L_A/\partial g_{ij} \tag{4.37}$$

and

$$M_A{}^{ijk} = 4\mathscr{A}^{(ijk)}(Jy^m y^k_m F^{nj}(x, y(x))g^{li}\,\partial a_n/\partial y^l) \tag{4.38}$$

are tensors (cf. Equations (4.13)–(4.15)). If we introduce in the conventional way an electromagnetic field strength tensor $F_{(0)ij}$ by constructing it from the skew-symmetric derivatives of the osculating vector potential $A_i(x, y(x))$:

$$F_{(0)ij}(x) = \mathrm{d}_i(A_j(x, y(x))) - \mathrm{d}_j(A_i(x, y(x))), \tag{4.39}$$

then from Equation (4.33) we get

$$F_{ij}(x, y(x)) = F_{(0)ij} - y^l_i\,\partial a_j/\partial y^l + y^l_j\,\partial a_i/\partial y^l. \tag{4.40}$$

The above formulae are considerably simplified if the vector field $y^k(x)$ is *stationary*, that is

$$y^k_m = 0 \tag{4.41}$$

in which case the relation (4.35) reduces to merely

$$\frac{\delta L_A}{\delta r^a} = 4\frac{\partial a_j}{\partial r^a}\,\mathrm{d}_k(JF_{(0)}{}^{kj}). \tag{4.42}$$

Under these conditions, the generalized electromagnetic field equations in vacuo

$$\delta L_A/\delta r^a = 0 \tag{4.43}$$

will hold whenever the classical Maxwell equations

$$\mathrm{D}_k F_{(0)}{}^{kj} \equiv \frac{1}{J}\mathrm{d}_k(JF_{(0)}{}^{kj}) = 0 \tag{4.44}$$

are valid. However, the sets (4.43) and (4.44) do not seem to be equivalent in general.

The simplest way of choosing the direction-dependent generalization $A_i(x, y)$ is to use

$$A_i(x, y) = A_P(x)h^P_i(x, y) \tag{4.45}$$

where $h^P_i(x, y)$ is a covariant orthonormal n-tuple of the Finslerian metric tensor. In other woards, we assume that the contravariant orthonormal n-tuple $h^i_P(x, y)$ can be chosen such that the scalars

$$A_P = A_i(x, y)h^i_P(x, y) \tag{4.46}$$

are independent of the directional variable y^i. Also, the choice of $h^P_i(x, y)$ will naturally be stipulated by the condition that they can be constructed from the same collection of x-dependent fields $S^A(x)$ as can the Finslerian metric tensor

(Equation (3.14)), that is that there exists a representation of the form

$$h_i^P(x, y) = e_i^P(S^A(x), y). \tag{4.47}$$

Regarding the scalars $A_P(x)$ as a set of electromagnetic field variables, i.e. putting

$$\{r^a(x)\} = \{A_P(x)\}, \tag{4.48}$$

we find that the functions a_i in (4.34) are

$$a_i(A_P(x), S^A(x), y) = A_P(x)e_i^P(S^A(x), y). \tag{4.49}$$

Since Equations (4.47)–(4.49) entail

$$\partial a_i/\partial A_P = h_i^P, \tag{4.50}$$

the Euler-Lagrange derivative (4.35) takes the form

$$\frac{\delta L_A}{\delta A_P} = 4Jh_j^P(x, y(x))\mathrm{D}_k F_{(0)}{}^{kj} + 4Jy_k^m h_{jm}^P(x, y(x))F_{(0)}{}^{kj} \tag{4.51}$$

where $h_{jm}^P(x, y) = \partial h_j^P(x, y)/\partial y^m$.

From Equations (4.51) and (4.39) it may be concluded that the stationarity condition (4.41) reduces the present Finslerian generalization of the electromagnetic field equations to the usual Maxwell form written with respect to the osculating vector potential $A_i(x, y(x))$ and in terms of the osculating Riemannian metric tensor $a_{ij}(x) = g_{ij}(x, y(x))$. So, in the case when all gravitational effects are neglected (which implies, of course, that the derivative y_m^k must be zero), a light-wave in vacuo will be described by the usual equations in terms of the osculating metric tensor $a_{ij}(x)$. If gravitational effects are taken into account, the proposed Finslerian equations will differ from the Maxwell equations by the second term on the right-hand side of (4.51).

4.3. Derivation of the Finslerian Equations of Motion of Matter from the Gravitational Field Equations

Let L_M be a Lagrangian density used for describing matter (including non-gravitational fields) in the Finslerian approach based on the concept of osculation. Then we may in general assume that L_M is a function of the gravitational variables $S^A(x)$, field $y^i(x)$ and a set of tensors $Q^{aj\cdots}_{i\cdots}(x)$, $a = 1, 2, \ldots$, in terms of which matter is described. Let L_M be, for definiteness, of the first-order form

$$L_M = L_M(y^n, \mathrm{d}_m y^n, S^A, \mathrm{d}_m S^A, Q^a, \mathrm{d}_m Q^a), \tag{4.52}$$

so that the Lagrangian density

$$L_T = L_g + L_M, \tag{4.53}$$

where the gravitational Lagrangian density L_g is assumed to be of the form (4.6), will represent the total system of gravitational field plus matter.

The gravitational field equations in the presence of matter will read

$$E_{TA}{}^{i\ldots}_{j\ldots} = E_{gA}{}^{i\ldots}_{j\ldots} + E_{MA}{}^{i\ldots}_{j\ldots} = 0 \tag{4.54}$$

where E denotes the Euler-Lagrange derivative defined in accordance with Equation (3.26), and after substituting $E_{gA} = -E_{MA}$ in the Noether identity (3.25) written for L_g we obtain the following *equations of motion of matter*

$$\mathrm{d}_j N_{Mi}{}^j - E_{MA}{}^{n\ldots}_{j\ldots}\,\mathrm{d}_i S^{Aj\ldots}{}_{n\ldots} = -\mathrm{d}_j(E_{gi} y^j) - E_{gj}\,\mathrm{d}_i y^j. \tag{4.55}$$

Here, $N_{Mi}{}^j$ is the tensor density obtained by substituting E_{MA} for E_A in (3.28), and $E_{gi} = -\partial L_g/\partial y^i$ in accordance with (3.27) and (4.6).

It will be noted that L_M itself, as a density, will satisfy the Noether identity of the form (3.25), namely,

$$\begin{aligned} \mathrm{d}_j(N_i^{*j} - E_{Mi} y^j) = E_{MA}{}^{n\ldots}_{j\ldots}\,\mathrm{d}_i S^{Aj\ldots}{}_{n\ldots} + E_{Ma}{}^{n\ldots}_{j\ldots}\,\mathrm{d}_i Q^{aj\ldots}{}_{n\ldots} + \\ + E_{Mj}\,\mathrm{d}_i y^j \end{aligned} \tag{4.56}$$

where E_{Ma} denotes the Euler-Lagrange derivative of L_M with respect to Q^a; the symbol E_{Mi} appears as a result of the replacement of L by L_M in (3.27);

$$N_i^{*j} = N_{Mi}{}^j + Q^{an\ldots}{}_{i\ldots} E_{Man\ldots}{}^{j\ldots} + \cdots - Q^{aj\ldots}{}_{n\ldots} E_{Mai\ldots}{}^{n\ldots} - \cdots. \tag{4.57}$$

The reader can obtain different forms of the equations of motion by combining Equations (4.55) and (4.56).

The right-hand side of Equation (4.55) can be rewritten in a more elegant form when L_g satisfies the homogeneity condition (4.7). Indeed, since (4.7) implies the identity $y^i\,\partial L_g/\partial y^i = 0$ (see the Euler theorem (1.3) for $r = 0$), it is apparent that (4.55) may be expressed in the form

$$\mathrm{d}_j N_{Mi}{}^j - E_{MA}{}^{n\ldots}_{j\ldots}\,\mathrm{d}_i S^{Aj\ldots}{}_{n\ldots} = \frac{y^j}{M(x)}(\mathrm{d}_i(M(x)E_{gj}) - \mathrm{d}_j(M(x)E_{gi})) \tag{4.58}$$

where the density $M(x)$ of weight -1 is defined by the requirement

$$\mathrm{d}_j(y^j/M(x)) = 0 \tag{4.59}$$

so that this density is, in fact, the same $M(x)$ which appears in the Clebsch representation (4.68) of the field $y^i(x)$.

An incoherent fluid (a cloud of dust) presents the simplest case of a gravitational field source. In the Riemannian approach to general relativity such a source can be described by a Lagrangian density of the form

$$L = \tfrac{1}{2} n(x) S^2(x, q(x)) \tag{4.60}$$

where $S^2(x,q) = a_{mn}(x) q^m q^n$ denotes the squared Riemannian metric function; $q^i(x)$ is a vector field tangent to the streamlines of the fluid; $n(x)$ is the density of the fluid. The Einstein gravitational field equations with this kind of fluid as a source are well known to imply the continuity equation and that the streamlines of the fluid are Riemannian geodesics. The latter circumstance in turn makes it possible to conclude, by considering a tube of streamlines, that test bodies follow Riemannian geodesics[6].

This kind of gravitational field source can be directly generalized to the Finslerian case: *a Finslerian incoherent fluid* will be described by the Lagrangian density

$$L_M = \tfrac{1}{2}n(x)F^2(x, q(x)). \tag{4.61}$$

This definiton yields

$$E_{MA}{}^{i\ldots}_{j\ldots} = -n(x)FF_A{}^{i\ldots}_{j\ldots} \tag{4.62}$$

where F_A denotes the derivative of $F = v(S^A, y)$ with respect to S^A. Recalling the invariance identities (3.10), we get

$$N_{Mi}{}^j = -nq_iq^j. \tag{4.63}$$

On substituting Equations (4.62) and (4.63) in the left-hand side of the equations of motion (4.55), we obtain the expression

$$-\,\mathrm{d}_j(nq_iq^j) + \tfrac{1}{2}n(\partial v^2/\partial S^A)\,\mathrm{d}_iS^A$$

or, inserting $v^2 = F^2$ and recalling the definition (1.26) of the Finslerian Christoffel symbols γ_{imj},

$$-\,\mathrm{d}_j(nq_iq^j) + n\gamma_{imj}(x, q(x))q^m(x)q^j(x).$$

This can be directly rewritten as

$$-\,l_i\,\mathrm{d}_j(Nl^j) - N(x)(d_jl_i - \gamma_{imj}(x, l)l^m)l^j$$

where $l^i = l^i(x) \equiv q^i(x)/F(x, q(x))$ and $N(x) = n(x)F^2(x, q(x))$ or, in view of the definition (1.54) of the δ-derivative,

$$-\,l_i\,\mathrm{d}_j(Nl^j) - N\delta l_i/\delta s.$$

Therefore, the equations of motion take the following form:

$$l_i\,\mathrm{d}_j(N(x)l^j) + N(x)\delta l_i/\delta s = \mathrm{d}_j(E_{gi}y^j) + E_{gj}\,\mathrm{d}_iy^j. \tag{4.64}$$

An interesting implication follows from (4.64) in the particular case when the right-hand side of the equations of motion (4.55) vanishes:

$$\mathrm{d}_j(E_{gi}y^j) - E_{gj}\,\mathrm{d}_iy^j = 0. \tag{4.65}$$

Indeed, upon contracting Equation (4.64) with l^i and taking into account (4.65) together with the identity $l^i\delta l_i/\delta s \equiv 0$, we find that the continuity equation

$$\mathrm{d}_j(N(x)l^j) = 0 \tag{4.66}$$

holds. These equations reduce (4.64) to simply

$$\delta l^i/\delta s = 0. \tag{4.67}$$

Equation (4.67) says that the streamlines are Finslerian geodesics.

Concluding this section, it is noted that the replacement of Riemannian geodesics by Finslerian ones does not affect the validity of the (weak) equivalence principle, according to which the trajectory of a test body moving in a gravitational field is independent of the body's mass.

4.4. Significance of the Auxiliary Vector Field from the Viewpoint of the Clebsch Representation

The auxiliary role of the vector field $y^i(x)$ in the Finslerian gravitational Lagrangian density (4.6) can be eliminated if we invoke *the Clebsch representation of* $y^i(x)$:

$$y^i(x) = Me^{ijkl}P_jQ_kR_l \tag{4.68}$$

where the four-dimensional case is considered; $M(x)$ is a density of weight -1; e^{ijkl} denotes the permutation symbol; $P(x)$, $Q(x)$ and $R(x)$ are three independent scalars called the *Clebsch potentials*[7]; $P_i = \mathrm{d}_iP \equiv \partial P/\partial x^i$, etc. Indeed, on substituting (4.68) in the Lagrangian density L_g having the form (4.6), we may treat L_g as a function of the form

$$L_g = L_g(P_i, Q_i, R_i, S^A, \mathrm{d}_mS^A, \mathrm{d}^2_{mn}S^A) \tag{4.69}$$

where the homogeneity (4.7) of L_g permits us to suppress dependence on $M(x)$. By construction, the scalars P, Q, R in (4.69) are auxiliary quantities. It can be proved, however, that they may be re-interpreted as a set of field variables complementary to S^A without destroying the validity of the initial field equations written in terms of S^A.

Generally speaking, such a possibility occurs in the case when the invariance identities associated with the Lagrangian density imply that the Euler-Lagrange derivatives with respect to the auxiliary functions vanish whenever the field equations are satisfied. This possibility is of course a rare one. Nevertheless, it occurs when the role of the auxiliary quantities is played by a set of independent scalars $v^a(x)$, $a = 1, 2, \ldots,$ whose number is not greater than the dimension number N of the background manifold.

To prove this statement, let us consider the Lagrangian density, for definiteness, of the following first-order form

$$L = L(v^a, \mathrm{d}_iv^a, r_{ij}, \mathrm{d}_nr_{ij}), \tag{4.70}$$

where the Riemannian metric tensor $r_{ij}(x)$ is assumed to be the field variable. Then it can be readily verified that the Noether identities (3.25) written for the density (4.70) are of the form

$$(\mathrm{d}_iv^a)\delta L/\delta v^a = 2r_{ik}J\mathrm{D}_m(J^{-1}\delta L/\delta r_{km}) \tag{4.71}$$

where the symbol δ/δ denotes the Euler-Lagrange derivative; D_m stands for the

Riemannian covariant derivative associated with r_{ij}; $J^2 = |\det(r_{mn})|$. The identity (4.71) shows that the equality

$$(d_i v^a)\delta L/\delta v^a = 0 \tag{4.72}$$

will hold whenever the field equations $\delta L/\delta r_{km} = 0$ are satisfied. But since the scalars v^a are assumed to be independent, the covariant vectors $d_i v^a$ must be linearly independent. Therefore, (4.72) implies $\delta L/\delta v^a = 0$. The proof is complete. Naturally, this result retains its validity if arbitrary tensor variables are considered and any orders of derivatives are allowed in L. Thus, we get

PROPOSITION 2. If the role of the auxiliary functions in the Lagrangian density is played by a set of independent scalars whose number is not greater than the dimension number of the background manifold, then these scalars may be treated as additional field variables without affecting the validity of the initial set of field equations.

From this proposition it follows that we may treat the Lagrangian density (4.69) as a function of the field variables S^A, P, Q, and R not involving auxiliary functions at all. The independent scalars P, Q, R define *a special coordinate system* as follows

$$x^{*1} = P(x),\ x^{*2} = Q(x),\ x^{*3} = R(x),\ x^{*0} \textit{ is arbitrary}. \tag{4.73}$$

By the Clebsch representation (4.68),

$$y^n(x^*) = y^0(x^*)\delta^n_0, \tag{4.74}$$

irrespective of the particular choice of x^{*0}. Owing to the homogeneity (4.7) of L_g in y^i, the factor y^0 may be neglected when $y^n(x^*)$ is substituted in the field equations, so that we may put

$$y^n(x^*) = \delta^n_0 \tag{4.75}$$

without loss of generality. This leads to

PROPOSITION 3. Relative to the special coordinate system (4.73) defined by the Clebsch potentials of the vector field $y^i(x)$ one may write the equality (4.75) in the field equations associated with an arbitrary homogeneous Lagrangian density of the form (4.6).

Thus, the Lagrangian density of the form (4.69) contains information not only about the actual field variables $S^A(x)$ but also about the special coordinate system, which may be regarded as the characteristic feature of the Finslerian approach based on the concept of osculation.

The presence of the vector field in the theory can be conveniently used to determine the world time. In this respect, it is natural to call $y^i(x)$ *the vector field of the world time*. This field generates the temporal slicing of spacetime, *a slice* being defined (as a three-dimensional hypersurface) by the equation $x^{*0} = \text{const}$. On each slice one can choose intrinsically a coordinate system

x^{*a}, $a = 1, 2, 3$, that is, a triple (P, Q, R). Any vector tangent to a slice may be represented as $X_i = aP_i + bQ_i + cR_i$, where a, b, c are scalars, and any such vector will be orthonormal to y^i in view of the Clebsch representation (4.68) of the latter: $y^i X_i = 0$ (a consequence of the skew-symmetry of the tensor e^{ijkl}). The value of x^{*0} on a slice will just be that of the world time.

In connection with this, one should make several remarks. In the literature there is not as yet a common definition of the concept of time. Constructions often rest upon half-intuitive considerations. In the introduction to a collection of articles devoted to the problem of time (ed. by Gold, 1967), the situation has been characterized as follows:

"Introspective understanding of the flow of time is basic to all our physics, and yet it is not clear how this idea of time is derived or what status it ought to have in the description of the physical world" (p. vii).

"Nearly everyone who works in physics has come to the conclusion that an understanding of the nature of time is basic. Each physicist has developed a point of view about it. There are great divergences in these points of view which have not been well aired; the literature is disappointingly limited in this subject. I think we shall discover that there is no agreement regarding even the most basic matters. It amazes me that a concept like time can have such a profound tradition in the physical sciences in spite of the fact that it is regarded in such widely different ways" (p. 1).

Nevertheless, it would not be an exaggeration to say that a particular definition of time is necessary for a detailed study of any system of interacting bodies and fields, the Universe being the largest object of such a type. The necessity of introducing a unified concept of time has been emphasized in a book by Bondi (1960), where, in discussing the assumption of the existence of an omnipresent cosmic time, the author argues that

"the development of relativistic cosmology is impossible without such an assumption. Cosmic time is in any case necessary if the ordinary, but not perfect, cosmological principle is assumed. For if it is asserted that every fundamental observer sees a changing universe, but that it presents the same aspect to them all, then it must be possible for observer A to find a time t_A according to his clock at which he sees the universe in the same state as observer B sees it at a time t_B by his clock. The universe itself therefore acts as a synchronizing instrument which enables A and B (and hence all observers) to synchronize their clocks.... It can be shown from fairly simple geometrical arguments, with one or two reasonable assumptions, that a cosmic time must exist in any simple model" (p. 71).

It can also be argued that the real world is not an absolute empty space in which there float independent observers, each of which has his/her own individual reference standard for a determination of time. The real world, as well as any part of it, is primarily an energy-matter system, and time is one of its most important attributes, so that a clear-cut analytical definition of time is much needed.

The idea that all geometric and any other properties of real space-time are determined by the system of material bodies and fields filling space-time was the leit-motif in the construction of the general theory of relativity. It would seem that a consequence of this idea would be that Newtonian absolute time, which is an abstract concept and therefore unacceptable, should be replaced by a vector field which geometrically represents the flow of time and which is determined by an assembly of material bodies and fields through a set of field equations. This however did not happen simply for the reason that Riemannian geometry was chosen as the geometrical ansatz, and in this geometry there is only one primary geometric object, namely the field of the symmetric second-rank tensor. The needed definition is suggested by the Finslerian approach which in fact introduces, through the notion of osculation, the concept of world time as a physical object possessing a clear geometrical image described by a set of physical field equations on a par with the concept of metric. As a result, the field $y^i(x)$ is no longer regarded as an auxiliary quantity and acquires a meaning as real as the concept of metric.

In order to be able to speak about the solutions of this set, one should make use of the conditions relating the field $y^i(x)$ to gravitating material bodies which would play the part of boundary conditions. Such a relation arises in an obvious way in the case when the gravitational field is generated by a static body. In this case the world time evidently coincides with the direction of the proper time of the body which is given by the four-dimensional velocity vector of motion of the body. The static case will be treated in the next section.

4.5. Static Gravitational Field

The case of a static gravitational field plays an important part in general relativity because all gravitational effects in the solar system may be described with high accuracy by means of an approximation to the static spherically symmetric solution of the gravitational field equations[8]. Since in our approach we have at our disposal the special coordinate system x^{*i} obeying the remarkable property (4.74), a static gravitational field will be defined in the logically simplest way with respect to these x^{*i}. Accordingly, we introduce the following

DEFINITION. A gravitational field treated in the Finslerian approach is called *static* if the special coordinate system x^{*i} defined by the Clebsch potentials of the field $y^i(x)$ in accordance with Equation (4.73) is chosen such that the following two conditions are satisfied:

(i) The gravitational field variables $S^A(x)$ are independent of x^{*0}, i.e.

$$\partial S^A(x^*)/\partial x^{*0} = 0. \tag{4.76}$$

(ii) The mixed components of the osculating Riemannian metric tensor $a_{ij}(x) = g_{ij}(x, y(x))$ vanish, i.e.

$$a_{0b}(x^*) = 0, \tag{4.77}$$

where $b = 1, 2, 3$.

If these two conditions are satisfied, x^{*i} are called *static coordinates.*

It will be noted that the representation $g_{mn}(x, y) = v_{mn}(S^A(x), y)$ (Equation (3.14)) together with (4.74) entails

$$a_{mn}(x^*) = v_{mn}(S^A(x^*),\ y^0(x^*)\delta^k_0) = v_{mn}(S^A(x^*),\ \delta^k_0)$$

where in the last step we have made use of the homogeneity of the Finslerian metric tensor in y^k. Therefore, (4.76) implies that

$$\partial a_{mn}(x^*)/\partial x^{*0} = 0 \qquad (4.78)$$

irrespective of the form of the function $y^0(x^*)$.

To simplify matters, we shall stipulate that the field $y^i(x)$ satisfies the unit length condition

$$F^2(x, y(x)) \equiv a_{mn}(x) y^m(x) y^n(x) = 1 \qquad (4.79)$$

keeping in mind that the homogeneity condition (4.7) makes such a stipulation legitimate without loss of generality. With (4.74), the condition (4.79) takes the form

$$a_{00}(x^*)(y^0(x^*))^2 = 1$$

relative to the static coordinates x^{*i}, from which it follows, by virtue of (4.78), that the condition (4.79) implies $\partial y^0(x^*)/\partial x^{*0} = 0$. These observations may be summarized as follows:

$$\begin{aligned} &y^i(x^*) = y^0(x^{*0})\delta^i_0, \qquad a_{ij}(x^*) = a_{ij}(x^{*b}), \qquad y^0 = (a^{00})^{1/2}, \\ &y_0 = (a_{00})^{1/2}, \qquad a^{0b}(x^*) = a_{0b}(x^*) = 0. \end{aligned} \qquad (4.80)$$

The indices $a, b, c, \ldots$ refer to the spatial part of the static coordinates and so will run from 1 to 3.

Let us denote

$$a^k = y^m(x) \mathrm{D}_m y^k(x) \qquad (4.81)$$

where D_m stands for the Riemannian covariant derivative associated with $a_{mn}(x)$. Then, as a consequence of (4.80), the following series of important identities will be valid in the case of a static gravitational field relative to an arbitrary, not necessarily static, coordinate system x^i:

$$\mathrm{D}_m y^k = y_m a^k, \qquad y_k a^k = 0, \qquad \mathrm{D}_k y^k = 0, \qquad (4.82)$$

$$a_k = \mathrm{d}_k a(x), \qquad a(x) = a(x^*(x)), \qquad a(x^*) = -\ln(y_0(x^*)), \qquad (4.83)$$

$$y^n \mathrm{D}_n a^k = -y^k a^i a_i, \qquad y_k \mathrm{D}_n a^k = -y_n a^i a_i \qquad (4.84)$$

which can easily be verified directly. Here, $y^i = y^i(x)$ and $y_i = y_i(x) = a_{ij}(x) y^j(x)$ everywhere; $a_i = a_{ij} a^j$.

In the remainder of the present section we shall briefly describe the relationship between the Finslerian geodesics of a static gravitational field and the Riemannian geodesics related to the osculating Riemannian metric tensor

$a_{ij}(x)$. The coordinates x^i will be assumed to be static, although the star will be omitted. The indices $i, j, \ldots$ will range over 0, 1, 2, 3.

The motion of any test material point in the static world can be described by the velocity

$$V^n = \mathrm{d}x^n/\mathrm{d}x^0 \tag{4.85}$$

which does not involve any metric function. It is approapriate, therefore, to compare the different types of geodesics in the static world in terms of this velocity. By the definition (4.85), we have

$$V^0 = 1, \tag{4.86}$$

while the spatial velocity

$$V^a = \mathrm{d}x^a/\mathrm{d}x^0 = (\mathrm{d}x^a/\mathrm{d}t)/c, \tag{4.87}$$

where $t = x^0/c$, gives the ratio of the three-dimensional velocity as the latter is used in classical mechanics to the light velocity c. At the same time, a Finslerian metric function $F(x, y)$ gives rise to the four-dimensional velocity

$$x'^n = \mathrm{d}x^n/\mathrm{d}s. \tag{4.88}$$

Here, s is the Finslerian arc-length defined by $\mathrm{d}s = F(x, \mathrm{d}x)$, so that the vector (4.88) has unit Finslerian length:

$$F(x, x') = 1. \tag{4.89}$$

In contrast to V^n, the quantity x'^n is a four-dimensional vector. The relation

$$\mathrm{d}s/\mathrm{d}x^0 = F(x, \mathrm{d}x)/\mathrm{d}x^0 = F(x, \mathrm{d}x/\mathrm{d}x^0) = F(x, V)$$

enables us to express the velocity (4.88) in terms of the velocity (4.85):

$$x'^n = V^n/F(x, V). \tag{4.90}$$

In terms of the velocity V^n, the equation of Finslerian geodesics may be written as

$$\mathrm{d}V^n/\mathrm{d}x^0 = -\gamma_i{}^n{}_j(x, V)V^iV^j + V^n\mathrm{d}(\ln F(x, V))/\mathrm{d}x^0 \tag{4.91}$$

(Equation (1.25)). In view of (4.86), the following identity will hold:

$$\mathrm{d}(\ln F(x, V))/\mathrm{d}x^0 = \gamma_i{}^0{}_j(x, V)V^iV^j. \tag{4.92}$$

The relations (4.91) and (4.92) together with (4.86) imply that the three-dimensional velocity (4.87) will be governed by the equation

$$\begin{aligned} \mathrm{d}V^a/\mathrm{d}x^0 = & -\gamma_0{}^a{}_0(x, V) - 2\gamma_b{}^a{}_0(x, V)V^b - \gamma_b{}^a{}_c(x, V)V^bV^c + \\ & + V^a(\gamma_0{}^0{}_0(x, V) + 2\gamma_b{}^0{}_0(x, V)V^b + \gamma_b{}^0{}_c(x, V)V^bV^c). \end{aligned} \tag{4.93}$$

If in some region W of space-time we have

$$q \stackrel{\text{def}}{=} (\delta_{ab}V^aV^b)^{1/2} \ll 1 \tag{4.94}$$

we may employ the Taylor series expansion of the Finslerian metric tensor

$g_{ij}(x, V^n) \equiv g_{ij}(x, 1, V^a)$ with respect to V^a:

$$g_{ij}(x, V) = a_{ij}(x) + 2C_{(0)ija}(x)V^a + C_{(0)ijab}(x)V^aV^b + O(q^3). \tag{4.95}$$

Here, we have substituted the equality

$$g_{ij}(x, 1, 0) = a_{ij}(x)$$

which is obtained if we use the definition $a_{ij}(x) = g_{ij}(x, y(x))$ and take into account the first equality in (4.80) together with the homogeneity of the Finslerian metric tensor $g_{ij}(x, y)$ in y^n; and the following notation is adopted:

$$2C_{(0)ija} \stackrel{\text{def}}{=} \left. \frac{\partial g_{ij}(x, V)}{\partial V^a} \right|_{V^a = 0} = 2C_{ija}(x^n, 1, 0)$$

and

$$2C_{(0)ijab} \stackrel{\text{def}}{=} \left. \frac{\partial^2 g_{ij}(x, V)}{\partial V^a \partial V^b} \right|_{V^a = 0} = 2C_{ijab}(x^n, 1, 0)$$

where C_{ijk} is the Cartan torsion tensor and C_{ijkm} is its derivative (1.17). Because of the identities (1.16), (1.18) and $y^n(x) = y^0 \delta^n_0$ (Equation (4.74)) we have

$$C_{(0)0ij} = C_{(0)00ij} = 0, \; C_{(0)0ijm} = -C_{(0)ijm}. \tag{4.96}$$

With (4.96), the expansion (4.95) takes the form

$$g_{00}(x, V) = a_{00}(x) + O(q^3), \tag{4.97a}$$

$$g_{0a}(x, V) = O(q^2), \tag{4.97b}$$

$$g_{ab}(x, V) = a_{ab}(x) + 2C_{(0)abc}(x)V^c + C_{(0)abcd}(x)V^cV^d + O(q^3). \tag{4.97c}$$

The reciprocal of (4.97c) reads

$$\begin{aligned} g^{ab}(x, V) = a^{ab}(x) - 2C_{(0)}{}^{ab}{}_c(x)V^c - \\ - (C_{(0)}{}^{ab}{}_{cd} - 4C_{(0)}{}^{ae}{}_c C_{(0)}{}^b{}_{ed})V^cV^d + O(q^3), \end{aligned} \tag{4.98}$$

where the indices are raised by means of the tensor $a^{ab}(x)$, for example $C_{(0)}{}^b{}_{ed} = a^{bc}C_{(0)ced}$. In writing the third term on the right-hand side of Equation (4.98) we have made use of the relation

$$\begin{aligned} \partial C^{ab}{}_c(x, y)/\partial y^d \equiv \partial(g^{ae}(x, y)g^{bf}(x, y)C_{efc}(x, y))/\partial y^d = \\ = g^{ae}(x, y)g^{bf}(x, y)C_{efcd}(x, y) - \\ - 2(g^{ae}(x, y)C^{bf}{}_d(x, y) + C^{ae}{}_c(x, y)g^{bf}(x, y))C_{efcd}(x, y) \end{aligned}$$

which follows from the relation (1.20).

Substituting the expansion (4.97a)–(4.98) in the general expression (1.26) of the Finslerian Christoffel symbols $\gamma_m{}^i{}_n$, we find that

$$\begin{aligned} \gamma_0{}^a{}_0(x, V) = -\tfrac{1}{2}(a^{ab} - 2C_{(0)}{}^{ab}{}_c V^c - (C_{(0)}{}^{ab}{}_{cd} - \\ - 4C_{(0)}{}^{ae}{}_c C_{(0)}{}^b{}_{ed})V^cV^d)\, \partial a_{00}/\partial x^b + O(q^3), \end{aligned} \tag{4.99a}$$

$$\gamma_0{}^0{}_0(x, V) = 0, \qquad \gamma_b{}^a{}_0(x, V) = O(q^2), \qquad \gamma_b{}^a{}_c(x, V) = \{{}_b{}^a{}_c\}(x) + O(q),$$
$$\gamma_b{}^0{}_0(x, V) = \tfrac{1}{2}a^{00}\,\partial a_{00}/\partial x^b + O(q^3). \tag{4.99b}$$

With formulae (4.93) and (4.99), the expansion of the equations of Finslerian geodesics in the region W reads

$$\begin{aligned} \mathrm{d}V^a/\mathrm{d}x^0 = \tfrac{1}{2}(a^{ab} &- 2C_{(0)}{}^{ab}{}_c V^c - (C_{(0)}{}^{ab}{}_{cd} - \\ &- 4C_{(0)}{}^{ae}{}_c C_{(0)}{}^b{}_{ed})V^c V^d)\,\partial a_{00}/\partial x^b - \\ &- \{{}_b{}^a{}_c\}V^b V^c + 2\{{}_b{}^0{}_0\}V^a V^b + O(q^3). \end{aligned} \tag{4.100}$$

So far, no assumptions have been made about the intensity of the gravitational field, so that Equation (4.100) refers to slow motion in an arbitrarily high static gravitational field. If, however, the gravitational field is not strong (which is the case in our solar system), then in Equation (4.100) one can perform an expansion of the functions a^{ab}, a_{00}, $C_{(0)abc}, \ldots$, depending solely on x^a. If, in addition, the static gravitational field is assumed to be spherically symmetric, we should use an expansion of the form (4.193) and carry out expansions simultaneously in q and $U = B/r$. The particular form of the expansion will depend on whether the orders of smallness of q and U are comparable.

Characteristically, the Finslerian corrections are those in velocities, i.e. in q. These corrections enter into the equations for Finslerian geodesics in a way that involves the Finslerian tensor C_{ijk}. If all Finslerian corrections are ignored (the case of very low velocities) or if we formally put the Finslerian tensor C_{ijk} equal to zero (not wishing to go beyond the Riemannian concepts), the expansion (4.100) reduces to the expansion in U of the equations of Riemannian geodesics written with respect to the osculating Riemannian metric tensor $a_{ij}(x)$.

As a consequence of this, if one attempts to adopt the hypothesis of Finslerian geodesics rather than that of Riemannian geodesics, i.e. one postulates that test bodies in a gravitational field move along Finslerian geodesics, then, when the velocity of motion is low, we are unable to distinguish between the Finslerian and the Riemannian cases.

In the next section we shall attempt to overcome the problem of selecting a concrete Finslerian gravitational Lagrangian by the determination of the post-Newtonian parameters in the spherically symmetric static case. The main stages of this rather lengthy calculation will be marked by Propositions 4–8.

4.6. Reduction of the Gravitational Lagrangian Density to First-Order Form in the 1-Form Case

The general expressions (1.77) and (1.78) for the Finslerian curvature tensors $\tilde{K}_l{}^j{}_{hk}(x, y)$ and $K_l{}^j{}_{hk}(x, y)$ imply immediately that

$$\tilde{K}_l{}^j{}_{hk}(x, y(x)) = \mathrm{d}_k(\Gamma_l{}^j{}_h(x, y(x))) - \mathrm{d}_h(\Gamma_l{}^j{}_k(x, y(x))) + \\ + \Gamma_m{}^j{}_k(x, y(x))\Gamma_l{}^m{}_h(x, y(x)) - \\ - \Gamma_m{}^j{}_h(x, y(x))\Gamma_l{}^m{}_k(x, y(x)) \tag{4.101}$$

and

$$K_l{}^j{}_{hk}(x, y(x)) = \tilde{K}_l{}^j{}_{hk}(x, y(x)) - \Gamma_l{}^j{}_{hm}(x, y(x))y^m_k + \\ + \Gamma_l{}^j{}_{km}(x, y(x))y^m_h \tag{4.102}$$

where $\Gamma_l{}^j{}_{hm}(x, y) = \partial\Gamma_l{}^j{}_h(x, y)/\partial y^m$ is the tensor (1.48) and y^m_k denotes the covariant derivative (4.10) of the vector field $y^m(x)$. On substituting the relation

$$\Gamma_l{}^j{}_h(x, y(x)) = V_l{}^j{}_h(x, y(x)) + L_l{}^j{}_h(x)$$

(see Equation (2.116)), which is valid in any 1-form Finsler space, and taking into account the identities (2.89)–(2.91), we find after some simple calculations that the contractions

$$\tilde{K}(x, y(x)) \stackrel{\text{def}}{=} g^{lh}(x, y(x))\tilde{K}_l{}^k{}_{hk}(x,y(x)) \tag{4.103}$$

and

$$K(x, y(x)) \stackrel{\text{def}}{=} g^{lh}(x, y(x))K_l{}^k{}_{hk}(x, y(x)) \tag{4.104}$$

may be decomposed as follows:

$$\tilde{K}(x, y(x)) = \tilde{K}_1 + K_2, \tag{4.105}$$

$$K(x, y(x)) = K_1 + K_2 \tag{4.106}$$

where

$$\tilde{K}_1 = (M_{kmh} - V_{mkh})V^{hmk} + (V^{ik}{}_i - V^{ki}{}_i)(M^n{}_{nk} - C_A{}^B{}_B\,\mathrm{d}_k y^A) + \\ + V_m{}^i{}_i V^{hm}{}_h - V_A{}^k{}_B\,\mathrm{d}_k g^{AB} + V_A{}^i{}_i S^h_B\,\mathrm{d}_h g^{AB}, \tag{4.107}$$

$$K_1 = \tilde{K}_1 - (S^m_A\,\mathrm{d}_k y^A + y^n V_n{}^m{}_k)\Gamma^{hk}{}_{hm} + (S^m_A\,\mathrm{d}_h y^A + y^n V_n{}^m{}_h)\Gamma^{hk}{}_{km}, \tag{4.108}$$

$$K_2 = \frac{1}{J}\mathrm{d}_k(J(V^{ik}{}_i - V^{ki}{}_i)). \tag{4.109}$$

Here and in what follows, we use the same notation as in Section 4.1, namely y^i always denotes a vector field $y^i(x)$ and d_m stands for the total derivative defined by (4.8). In particular, the derivative $\mathrm{d}_k g^{AB}$ is to be understood as $\mathrm{d}_k(g^{AB}(x, y(x))$ or, by virtue of Equation (2.94),

$$\mathrm{d}_k g^{AB} = -2C^{AB}{}_D\,\mathrm{d}_k y^D \tag{4.110}$$

where $y^D = S^D_i(x)y^i(x)$. The symbol J stands for $J(x, y(x))$, where $J(x, y)$ is the Jacobian (2.96), so that, by (2.97) and (2.101),

$$\frac{1}{J}\mathrm{d}_k J = L_{j\ k}^{\ j} + C_{A\ B}^{\ B}\mathrm{d}_k y^A. \tag{4.111}$$

Using the relations (4.110) and (4.111) together with $\Gamma^{hk}{}_{hm} \equiv g^{hl}\,\partial\Gamma_{l\ h}^{\ k}/\partial y^m = \partial V^{hk}{}_h/\partial y^m + 2C_m{}^{lh}V_{l\ h}^{\ k}$ (see Equations (2.116) and (1.20)) and replacing $y^n V_{n\ k}^{\ m}$ by $w_k{}^m$ in accordance with (2.118), we can rewrite the expressions (4.107) and (4.108) in the following form which is more appropriate for our calculations:

$$\begin{aligned}\tilde{K}_1 = &-(V_{mkh} - M_{kmh})V^{hmk} + V_{m\ i}^{\ i}V^{hm}{}_h + (V^{ik}{}_i - V^{ki}{}_i)M^n{}_{nk} + \\ &+ (2C_A{}^{lh}V_{l\ h}^{\ k} - 2V_{B\ i}^{\ i}C_A{}^{Bk} - (V^{ik}{}_i - V^{ki}{}_i)C_{A\ B}^{\ B})\,\mathrm{d}_k y^A,\end{aligned} \tag{4.112}$$

$$\begin{aligned}K_1 = &-(V_{mkh} - M_{kmh} + 2C_{thk}w_m{}^t)V^{hmk} + V_{m\ i}^{\ i}V^{hm}{}_h + \\ &+ w_k{}^m(\Gamma^{ki}{}_{im} - \partial V^{hk}{}_h/\partial y^m) + (V^{ik}{}_i - V^{ki}{}_i)M^n{}_{nk} + \\ &+ (-S^m_A\,\partial V^{hk}{}_h/\partial y^m + \Gamma^{ki}{}_{iA} - 2V_{B\ i}^{\ i}C_A{}^{Bk} - (V^{ik}{}_i - V^{ki}{}_i)C_{A\ B}^{\ B})\,\mathrm{d}_k y^A.\end{aligned} \tag{4.113}$$

The Lagrangian densities $\tilde{L} = J\tilde{K}(x, y(x))$ and $L = JK(x, y(x))$ belong to the second-order type (4.6). However, formulae (4.105)–(4.109) show that each of these densities is the sum of two densities:

$$\tilde{L} = \tilde{L}_1 + L_2, \qquad L = L_1 + L_2$$

of which $\tilde{L}_1 = J\tilde{K}_1$ and $L_1 = JK_1$ do not involve the second-order derivatives of the gravitational field variables $S^A_n(x)$, and $L_2 = JK_2$ is an ordinary divergence. Since the Euler-Lagrange derivative of an ordinary divergence is known to vanish identically, we get

PROPOSITION 4. In the case of a 1-form Finsler space the second-order Finslerian gravitational Lagrangian densities $\tilde{L} = J\tilde{K}(x, y(x))$ and $L = JK(x, y(x))$ may be replaced by the first-order densities $\tilde{L}_1 = J\tilde{K}_1$ and $L_1 = JK_1$, respectively, without affecting the gravitational field equations.

It will be noted in this respect that the reduction of the gravitational Lagrangian density to a first-order expression is often used in formulating the Einstein gravitational field equations[9]. In the latter case, however, the divergence term involving all the second-order derivatives is not a density, so that the remaining part is not tensorial. The cirumstance that the Finslerian $\tilde{L}_1$ and L_1 are densities is advantageous in that it enables one to formulate conservation laws for the gravitational field in a strictly covariant way (see the succeeding section), something which is impossible in the Riemannian theory.

A mere glance at formulae (2.116)–(2.118) and (2.105) reveals that the Lagrangians (4.112) and (4.113) may be treated as functions of the scalar arguments $y^A(x)$, $\mathrm{d}_B y^A \equiv S^n_B\,\mathrm{d}_n y^A(x)$ and $M^A{}_{BC}$, so that

$$K_1 = K_1(y^A, \mathrm{d}_B y^A, M^A{}_{BC}) \tag{4.114}$$

and the same for $\tilde{K}_1$. The Euler-Lagrange derivative

$$E_A{}^i \stackrel{\text{def}}{=} \mathrm{d}_j(\partial L_1/\partial(\mathrm{d}_j S_i^A)) - \partial L_1/\partial S_i^A \tag{4.115}$$

can easily be written in terms of the derivatives of the expression (4.114) with respect to its arguments. Indeed, the definitions $y^A = S_i^A y^i$, $\mathrm{d}_B y^A = S_B^j (S_i^A \mathrm{d}_j y^i + y^i \mathrm{d}_j S_i^A)$ and $M^A{}_{BC} = S_B^i S_C^j (\mathrm{d}_i S_j^A - \mathrm{d}_j S_i^A)$ yield

$$\begin{aligned}
&\partial y^B/\partial S_i^A = y^i \delta_A^B, \qquad \partial y^A/\partial y^i = S_i^A, \qquad \partial \mathrm{d}_B y^C/\partial \mathrm{d}_j S_i^A = y^i S_B^j \delta_A^C,\\
&\partial \mathrm{d}_B y^C/\partial S_i^A = \delta_A^C S_B^j \mathrm{d}_j y^i - S_A^j S_B^i \mathrm{d}_j y^C, \qquad \partial \mathrm{d}_B y^A/\partial y^i = S_B^j \mathrm{d}_j S_i^A,\\
&\partial \mathrm{d}_B y^A/\partial \mathrm{d}_j y^i = S_B^j S_i^A, \qquad \partial M^B{}_{CD}/\partial \mathrm{d}_j S_i^A = 2\delta_A^B S_C^j S_D^i,\\
&\partial M^B{}_{CD}/\partial S_i^A = -2S_C^i M^B{}_{AD}
\end{aligned}$$

where we have made use of the equality

$$\partial S_B^n/\partial S_i^A = -S_B^i S_A^n$$

which is obtained by differentiating the relation $S_i^A S_A^n = \delta_i^n$ with respect to S_i^A. Therefore,

$$\partial K_1/\partial \mathrm{d}_j S_i^A = (\partial K_1/\partial \mathrm{d}_B y^A) S_B^j y^i + K_A{}^{ji}$$

where

$$K_A{}^{ij} = 2S_B^i S_C^j \partial K_1/\partial M^A{}_{BC} \equiv -K_A{}^{ji}. \tag{4.116}$$

Further,

$$\begin{aligned}
\partial K_1/\partial S_i^A = {} & y^i \partial K_1/\partial y^A + (\mathrm{d}_j y^i) S_B^j \partial K_1/\partial \mathrm{d}_B y^A - (\mathrm{d}_A y^C) S_B^i \partial K_1/\partial \mathrm{d}_B y^C -\\
& - K_B{}^{iD} M^B{}_{AD}.
\end{aligned}$$

The derivatives of J are given by Equations (2.100) and (2.101).

The above relations may be used to conclude that the Euler-Lagrange derivative (4.115) is equal to

$$E_A{}^i = \mathrm{d}_j(J K_A{}^{ji}) - J T_A{}^i \tag{4.117}$$

where

$$\begin{aligned}
T_A{}^i = {} & (S_A^i + y^i C_A{}^B{}_B) K_1 - K_B{}^{ni} M^B{}_{nA} -\\
& - S_B^i (\mathrm{d}_A y^D) \partial K_1/\partial \mathrm{d}_B y^D - \tfrac{1}{J} y^i E_A
\end{aligned} \tag{4.118}$$

and

$$E_A \stackrel{\text{def}}{=} \mathrm{d}_j(J S_B^j \partial \mathrm{K}_1/\partial \mathrm{d}_B y^A) - J \partial \mathrm{K}_1/\partial y^A. \tag{4.119}$$

If we take the Euler–Lagrange derivative of L_1 with respect to $y^i(x)$:

$$E_i \stackrel{\text{def}}{=} \mathrm{d}_j(\partial L_1/\partial \mathrm{d}_j y^i) - \partial L_1/\partial y^i, \tag{4.120}$$

we obtain

$$E_i = \mathrm{d}_j(JS^j_B S^A_i \,\partial K_1/\partial \mathrm{d}_B y^A) - S^A_i \,\partial L_1/\mathrm{d}y^A - \\ - JS^j_B(\mathrm{d}_j S^A_i)\partial K_1/\partial \mathrm{d}_B y^A \equiv S^A_i E_A,$$

i.e.

$$E_i = S^A_i E_A. \tag{4.121}$$

The vacuum gravitational field equations will read

$$\frac{1}{J}\,\mathrm{d}_j(JK_A{}^{ji}) = T_A{}^i. \tag{4.122}$$

It is seen from Equations (4.112), (4.113), and (2.117) that that part of $\tilde{K}_1$ or K_1 not involving the Finslerian tensor C_{ijk} is

$$K_{11} = -\tfrac{1}{2}M^{kmn}Z_{kmn} + M^k{}_{kn}M_m{}^{mn}. \tag{4.123}$$

This contains $y^i(x)$ only through the osculating Riemannian metric tensor $a_{ij}(x) = g_{ij}(x, y(x))$. From (4.123) it follows that the derivative

$$\tfrac{1}{2}K_{(1)A}{}^{ji} \stackrel{\mathrm{def}}{=} \partial K_{11}/\partial M^A{}_{ji}$$

is equal to

$$\tfrac{1}{2}K_{(1)A}{}^{ji} = -Z_A{}^{ji} - \tfrac{1}{2}(M_A{}^{ji} - M^{ji}{}_A + M^{ij}{}_A) + 2(S^j_A M^i - S^i_A M^j). \tag{4.124}$$

If we denote

$$T_{(1)A}{}^i = S^i_A K_{11} - K_{(1)B}{}^{ni}M^B{}_{nA} \tag{4.125}$$

and

$$E_{(1)A}{}^i = \frac{1}{J}\mathrm{d}_j(JK_{(1)A}{}^{ji}) - T_{(1)A}{}^i, \tag{4.126}$$

then it can be readily verified that $E_{(1)j}{}^i$ is equal to double the Einstein tensor associated with the osculating Riemannian metric tensor $a_{ij}(x) = g_{ij}(x, y(x))$, that is

$$E_{(1)i}{}^j = 2(A_i{}^j(x) - \tfrac{1}{2}\delta^j_i A(x)) \tag{4.127}$$

where $A(x) = A_i{}^i(x)$, and $A_i{}^j(x) = A_i{}^{kj}{}_k(x)$ is the contraction of the Riemannian curvature tensor $A_i{}^h{}_{jk}(x)$ associated with $a_{ij}(x)$. Indeed, in the Riemannian limit, where the Finslerian tensor C_{ijk} vanishes identically, the Finslerian curvature tensors $\tilde{K}_l{}^j{}_{hk}$ and $K_l{}^j{}_{hk}$ given by the expressions (1.77) and (1.78) each reduced to the Riemann curvature tensor. Therefore, those parts of the tensors (4.101) and (4.102) not involving C_{ijk} are each identical to $A_l{}^j{}_{hk}(x)$. It follows that the parts of the Finslerian Lagrangians (4.103) and (4.104) which do not involve C_{ijk} are both equal to $A(x)$ which is of course the gravitational Lagrangian in the Einstein formulation of general relativity. Since the difference between $JA(x)$ and JK_{11} is an ordinary divergence by the construction of K_{11}, the expression (4.126) is equal to the variational derivative

$$\delta \mathbf{A}(x)/\delta S_i^B = (\delta A(x)/\delta a_{mn})\delta a_{mn}/\partial S_i^B$$

as in the Riemannian approach where

$$\partial a_{mn}/\partial S_i^B = \delta_m^i a_{Bn} + \delta_n^i a_{Bm}$$

and

$$\delta A(x)/\delta a_{mn} = A^{mn}(x) - \tfrac{1}{2} a^{mn} A(x).$$

It is obvious from the general expressions of the Cartan connection coefficients that the curvature tensors $K_l{}^j{}_{hk}(x, y)$ and $\tilde{K}_l{}^j{}_{hk}(x, y(x))$ are positively homogeneous of degree zero in y^i (as is the Finslerian metric tensor), i.e. $K_l{}^j{}_{hk}(x, ty) = K_l{}^j{}_{hk}(x, y)$ for any positive number t, and $\tilde{K}_l{}^j{}_{hk}(x, t(x)y(x)) = \tilde{K}_l{}^j{}_{hk}(x, y(x))$ for any positive scalar $t(x)$. The property of homogeneity is transferred to the Lagrangians $\tilde{K}(x, y(x))$ and $K(x, y(x))$ and, then to the reduced Lagrangians $\tilde{K}_1$ and $\tilde{K}_1$, so that the identity

$$K_1(t(x)y_A, \, \mathrm{d}_B(t(x)y^A), \, M^A{}_{BC}) = K_1(y^A, \, \mathrm{d}_B y^A, M^A{}_{BC}), \tag{4.128}$$

and the same for $\tilde{K}_1$, holds. On differentiating Equation (4.128) with respect to t and $\mathrm{d}_n t$ and then putting $t = 1$, we get two identities

$$y^A(\partial K_1/\partial y^A) + (\mathrm{d}_B y^A)\partial K_1/\partial \mathrm{d}_B y^A = 0 \tag{4.129}$$

and

$$y^A \, \partial K_1/\partial \mathrm{d}_B y^A = 0 \tag{4.130}$$

(cf. the Euler theorem (1.3)), and the same for $\tilde{K}_1$. Obviously, the validity of Equations (4.129) and (4.130) may be verified directly from the expressions (4.107) and (4.108) for $\tilde{K}_1$ and K_1. If one wishes, one may use the homogeneity property (4.128) to specify the length of $y^i(x)$ (for example, by the condition (4.140)). The identities of the form (4.128)–(4.130) are also valid for the Lagrangian densities $\tilde{L}_1 = J\tilde{K}_1$ and $L_1 = JK_1$ because the Jacobian J is homogeneous of degree zero in y^i. The identities (4.129) may be rewritten equivalently as

$$y^i E_i = 0 \tag{4.131}$$

which follows from the definition (4.119) together with the relations (4.121) and (4.130).

An examination of the explicit form of the dependence of the Lagrangians $\tilde{K}_1$ and K_1 on y^A and $M^A{}_{BC}$ is the first necessary step in investigating the structure of the generalized gravitational field equations. Although this purpose may be achieved by straightforward calculations, such a procedure would be cumbersome and result in general in complicated expressions. Much simplification will arise in the particular case when the Finslerian metric function exhibits the property $C_i = 0$. Indeed, in this case we find from the expression (2.117) for $V_i{}^k{}_j$ that the contractions $V^{ki}{}_i$ and $V^{hk}{}_h$ have the following simple form

$$V^{ki}{}_i = -M^k, \qquad V^{hk}{}_h = M^k - 2C_m{}^{kh} w_h{}^m \tag{4.132}$$

where the notation $M_k = M^i{}_{ik}$ has been used. Therefore, the expression

$$(V^{ik}{}_i - V^{ki}{}_i)M^n{}_{nk} + V_m{}^i{}_i V^{hm}{}_h$$

entering Equations (4.112) and (4.113) will be equal to

$$2(M^k - C_m{}^{kh} w_h{}^m)M_k - M_m(M^m - 2C_n{}^{mh} w_h{}^m) = M^k M_k.$$

Further, the first equality in (4.132) reduces the contraction $\Gamma_i{}^k{}_k$ of the Cartan connection coefficients given by Equation (2.116) to an x-dependent object, namely $\Gamma_i{}^k{}_k = L_k{}^k{}_i(x)$, so that $\Gamma_k{}^i{}_{im} \equiv \partial\Gamma_i{}^k{}_k/\partial y^m = 0$. As a result of these simplifications, the condition $C_i = 0$ reduces the Lagrangians (4.112) and (4.113) to

$$\tilde{K}_1 = (M_{kmh} - V_{mkh})V^{hmk} + M^k M_k + 2(V_l{}^k{}_h C_A{}^{lh} + M_n C_A{}^{nk})\mathrm{d}_k y^A, \tag{4.133}$$

$$\begin{aligned} K_1 = (M_{kmh} - V_{mkh} - 2C_{thk} w_m{}^t)V^{hmk} - w_k{}^m \partial V^{hk}{}_h/\partial y^m + \\ + M^k M_k - (S_A^m \partial V^{hk}{}_h/\partial y^m - 2M_n C_A{}^{nk})\mathrm{d}_k y^A. \end{aligned} \tag{4.134}$$

Additional simplifications occur if the space is $S3$-like (Section 2.1). Let us consider the case where the $S3$-like condition (2.1) in the form (2.36) is valid simultaneously with $C_i = 0$. Then we have the handy identities (2.41) and (2.42) to be used below. To simplify the notation, we shall denote contractions with y^i by the index zero[10], for example

$$Z_{00m} = Z_{ijm} y^i y^j, \qquad M^{0mk} = M^{imk} y_i, \quad \text{etc.} \tag{4.135}$$

Let us examine the term $C_k{}^{ni} w_i{}^k$ that appears in Equation (4.132). The definition (2.118) of $w_i{}^k$ entails

$$C_k{}^{ni} w_i{}^k = C^{kni} Z_{i0k} - C_k{}^{ni} C_i{}^{km} Z_{00m}$$

or, in view of the identities (2.110) and (2.111),

$$C_k{}^{ni} w_i{}^k = C^{kni} M_{i0k} - C_k{}^{ni} C_i{}^{km} M_{00m}.$$

On substituting the identity (2.40) in this relation and taking into account the fact that $M_{000} \equiv 0$, we get the following result:

$$C_k{}^{ni} w_i{}^k = C^{kni} M_{i0k} + (N-2)M_{00}{}^n F^{-2}. \tag{4.136}$$

The first term in the final parenthesis in (4.133) can be evaluated in the same fashion. Indeed, it follows from (2.117) that

$$C^{nij} V_i{}^k{}_j = C^{nij} M_{ij}{}^k - 2C^{nij} C_i{}^k{}_m w_j{}^m + C^{nij} C_{ijm} w^{km}$$

where we have made use of the identities

$$C^{nij} M^k{}_{ij} = 0$$

(the tensor C^{nij} is symmetric, while $M^k{}_{ij}$ is a skew-symmetric in the indices i, j) and

$$C^{nij} Z_{ij}{}^k = C^{nij} M_{ij}{}^k$$

(see Equation (2.111)). From the definition (2.118) we obtain the expression

$$C^{nij} C_i{}^k{}_m w_j{}^m = C^{nij} C_i{}^k{}_m M_{j0}{}^m - C^{nij} C_i{}^k{}_m C_j{}^m{}_p M_{00}{}^p$$

in which (2.41) is to be substituted. By (2.40),

$$F^2 C^{nij} C_{ijm} w^{km} = -(N-2) h_m{}^n w^{km} = -(N-2) w^{kn}$$

because $h_m{}^n = \delta^n_m - y^n y_m F^{-2}$ and the circumstance that the definition (2.118) of $w_i{}^k$ implies that

$$y_m w_i{}^m = 0.$$

On bringing together these observations, we finally get

$$\begin{aligned} C^{nij} V_i{}^k{}_j = C^{nij} M_{ij}{}^k - 2C^{nij} C_i{}^k{}_m M_{j0}{}^m - \\ - (N-4) F^{-2} C^{nkp} M_{00p} - (N-2) F^{-2} Z^{k0n}. \end{aligned} \tag{4.137}$$

An essentially similar procedure enables us to clarify the explicit form of the first term in (4.134), namely, we have consecutively

$$\begin{aligned} (M_{kmh} - V_{mkh} - 2C_{thk} w_m{}^t) V^{hmk} &= -(Z_{mhk} + C_{hkt} w_m{}^t + C_{mht} w_k{}^t - \\ &\quad - C_{mkt} w_h{}^t) V^{hmk} = -Z_{mhk} (M^{mhk} + Z^{hkm}) - \\ &\quad - (C_{kht} w_m{}^t + C_{mht} w_k{}^t - C_{mkt} w_h{}^t)(M^{mhk} - \\ &\quad - C_r{}^{mk} w^{hr}) = -\tfrac{1}{2} M^{mhk} Z_{mhk} - 2C_{mht} M^{mhk} w_k{}^t + \\ &\quad + 2C_k{}^h{}_t C_r{}^{km} w_m{}^t w_h{}^r + (N-2) w_{ht} w^{ht} F^{-2} = \\ &= -\tfrac{1}{2} M^{mhk} Z_{mhk} - 2C_t{}^k{}_n M^{tnm} Z_{m0k} + \\ &\quad + 2M^{00t} M_t F^{-2} + 2C_k{}^{ht} C^{kmr} Z_{m0t} Z_{h0r} - \\ &\quad - 2(N-4) C^{tnm} M_{tn0} M_{00m} F^{-2} + \\ &\quad + (N^2 - 6N + 6) M^{00n} M_{00n} F^{-4} + \\ &\quad + (N-2) Z^{k0t} Z_{k0t} F^{-2} \end{aligned}$$

where in calculating the term $2C_k{}^h{}_t C_r{}^{km} w_m{}^t w_h{}^r$ the identity (2.42) is used. Similarly,

$$\begin{aligned} (M_{kmh} - V_{mkh}) V^{hmk} &= -\tfrac{1}{2} M^{mhk} Z_{mhk} - \\ &\quad - 2C_k{}^{ht} C^{kmr} Z_{m0t} Z_{h0r} + 2(N-4) C^{tnm} M_{tn0} M_{00m} F^{-2} - \\ &\quad - (N-2)(N-4) M^{00n} M_{00n} F^{-4} - (N-2) Z^{k0t} Z_{k0t} F^{-2}. \end{aligned}$$

The contraction $Z^{k0t}Z_{k0t}$ is given by (2.113).

By virtue of Equations (4.132) and (4.136), the explicit dependence of $V^{hk}{}_h$ on y^i can be expressed as follows

$$V^{hk}{}_h(x,y) = g^{kj}(x,y)(M_j(x) - 2C_{ij}{}^q(x,y)M^i{}_{pq}(x)y^p - \\ - 2(N-2)F^{-2}(x,y)M^p{}_{qj}(x,y)y_p y^q).$$

The term $\partial V^{hk}{}_h/\partial y^m$ in (4.134) is complicated because of the appearance of the derivative $\partial C_{ij}{}^k/\partial y^m$. Nevertheless, if the T-condition holds, this derivative may be expressed algebraically in terms of the tensor C_{ijk} in accordance with (2.37). Having done this, and taking the $S3$-like condition (2.36) together with $C_i = 0$ into account, straightforward calculations give the following result:

$$\begin{aligned} -w_k{}^m \partial V^{hk}{}_h/\partial y^m = & -2C^{mnt}M_{mn0}M_t + 2(N-1)M^t M_{00t}F^{-2} - \\ & - 2C^{trn}C_r{}^{mk}M_{tn0}M_{mk0} - 2C_{tn}{}^k M^{tnm}Z_{k0m} + \\ & + 4(N-1)C^{tnm}M_{tn0}M_{00m}F^{-2} - 2(N-2)M^{00n}M_{00n}F^{-4} + \\ & + NM^{mk0}M_{0mk}F^{-2} + \\ & + (N-2)(M^{mk0}M_{km0} + M^{mk0}M_{mk0} - M^{0mk}M_{0mk})F^{-2}. \end{aligned}$$

On substituting the above formulae in Equations (4.133) and (4.134) and cancelling similar terms, we eventually get

$$\begin{aligned} \tilde{K}_1 = & K_{11} + 2(N-4)C^{ntm}M_{nt0}M_{00m}F^{-2} - \\ & - (N-2)(N-4)M^{00n}M_{00n}F^{-4} - \\ & - \tfrac{1}{4}(N-2)(2M_{k0t}M^{k0t} + 2M_{k0t}M^{t0k} + M_{0kt}M^{0kt})F^{-2} - \\ & - 2C_h{}^{kt}C^{rhm}Z_{k0r}Z_{m0t} + 2(V_l{}^k{}_h C_A{}^{lh} + M_n C_A{}^{nk})\,\mathrm{d}_k y^A, \end{aligned} \tag{4.138}$$

$$\begin{aligned} K_1 = & K_{11} + 2M_n(C^{nkm}M_{k0m} + NM^{00n}F^{-2}) - C^{knt}M_{ntm}(M^m{}_{0k} + M_{k0}{}^m) + \\ & + 2(N+2)C^{ntm}M_{nt0}M_{00m}F^{-2} + (N^2 - 8N + 10)M^{00n}M_{00n}F^{-4} + \\ & + NM^{mk0}M_{0mk}F^{-2} + \tfrac{3}{2}(N-2)(M^{mk0}M_{km0} + \\ & + M^{mk0}M_{mk0} - \tfrac{1}{2}M^{0mk}M_{0mk})F^{-2} + \\ & + 2C_h{}^{kt}C^{rhm}(Z_{k0r}Z_{m0t} - M_{tk0}M_{rm0}) + \\ & + (-S_A^m\,\partial V^{hk}{}_h/\partial y^m + 2M_n C_A{}^{nk})\,\mathrm{d}_k y^A. \end{aligned} \tag{4.139}$$

Thus we have established the following result.

PROPOSITION 5. If $C_i = 0$, and the $S3$-like condition in the form (2.36) and the T-condition are valid, then the first-order Lagrangians $\tilde{K}_1$ and K_1 have the forms (4.138) and (4.139).

As is known from Section 2.3, the 1-form Berwald-Moór metric function satisfies these three conditions, so that (4.138) and (4.139) may be related to the case of this metric function.

It will be noted that if the Finslerian metric function possesses some invariance properties, they may be used to simplify the relations between the auxiliary vector field $y^i(x)$ and the field variables $S_i^A(x)$ in the field equations. For the 1-form Berwald-Moór metric function an important conclusion of this type can be drawn from Proposition 11 of Section 2.3, according to which the equality $y^A(x) = Q(x)$ (Equation (2.63)) may be assumed without loss of generality. However, the factor Q will, in fact, disappear in the field equations because of the homogeneity (4.128) of the Finslerian Lagrangians, so that we may put simply

$$y^A = 1. \tag{4.140}$$

Thus we have established:

PROPOSITION 6. If the Lagrangian is constructed from the 1-form Berwald-Moór metric function, one may put (4.140) in the field equations without loss of generality.

The equality (4.140) can be rewritten as

$$y^i(x) = S^i(x) \tag{4.141}$$

where S^i is the preferred vector field given by Equation (2.132). In the case of the 1-form Berwald-Moór metric function we obtain from (2.32) and (4.141) that

$$a_{ij}(x) \overset{\text{def}}{=} g_{ij}(x, y(x)) = y_i(x)y_j(x) - \frac{1}{N}\sum_{A=1}^{N} B_i^A B_j^A \tag{4.142}$$

where

$$B_i^A(x) = S_i^A(x) - S_i(x) \tag{4.143}$$

and

$$y_i(x) = S_i(x) = a_{ij}(x)S^j(x) = \frac{1}{N}\sum_{A=1}^{N} S_i^A(x). \tag{4.144}$$

It is seen from the above formulae that

$$y^i B_i^A = 0, \qquad y_A B_i^A = \frac{1}{N}\sum_{A=1}^{N} B_i^A = 0 \tag{4.145}$$

and that the unit length condition

$$a_{ij}(x)y^i(x)y^j(x) = 1 \tag{4.146}$$

holds when $y^A(x) = 1$.

In the remaining part of the present section we shall elucidate the structure of the gravitational field equations (in the static case) following from the Finslerian Lagrangian $\tilde{K}_1$ taking as an example the 1-form Berwald-Moór metric function, so that $\tilde{K}_1$ will be given by Equation (4.138) and the relations

(4.140)–(4.146) will be assumed valid. In this case, the following useful relations will hold in addition to Equations (4.82)–(4.84):

$$y^i M_{ijh} = y^i Z_{ijh} = y_j a_h - y_h a_j, \tag{4.147}$$

$$y^j M_{ijh} = y^j Z_{ijh} = y_i a_h, \tag{4.148}$$

$$y^j M_j = 0, \qquad y^n \mathrm{D}_n M^k = -y^k a^i M_i \tag{4.149}$$

where $y^i = y^i(x)$ wherever y^i occurs; $y_i = y_i(x) = a_{ij}(x) y^j(x)$; $M_j = M^i{}_{ij} = M_j(x)$. For example, Equation (4.147) is verified as follows. We have

$$y^i M_{ijh} = y_A M^A{}_{jh} = \frac{1}{N} \sum_{A=1}^{N} M^A{}_{jh}$$

where in the second step we have used the fact that (4.140) implies $y_A = 1/N$. Thus,

$$y^i M_{ijh} = \frac{1}{N} \sum_{A=1}^{N} (d_j S^A_h - \mathrm{d}_h S^A_j)$$

and taking into account the relation

$$y_h = y_A S^A_h = \frac{1}{N} \sum_{A=1}^{N} S^A_h$$

we obtain

$$y^i M_{ijh} = \mathrm{d}_j y_h - \mathrm{d}_h y_j \equiv \mathrm{D}_j y_h - \mathrm{D}_h y_j = y_j a_h - y_h a_j$$

where in the last step we have substituted the relation $\mathrm{D}_j y_h = y_j a_h$ given in (4.82)

In the four-dimensional case $N = 4$, which we restrict ourselves to in what follows, the second and third terms on the right-hand side of Equation (4.138) vanish because of the factor $(N-4)$. Moreover, in view of the identities (4.148) and $C_{ijk} y^i = 0$, the fifth term on the right-hand side of (4.138) will not contribute to the static values of the derivatives of $\tilde{K}_1$ with respect to its arguments, so that this term may also be neglected. As a result, the Lagrangian $\tilde{K}_1$ may be written in the following reduced form

$$\begin{aligned} \tilde{K}_1 = K_{11} - \tfrac{1}{2}(2M_{kpt} M^k{}_q{}^t + 2M_{kpt} M^t{}_q{}^k + M_{pkt} M_q{}^{kt}) y^p y^q F^{-2} + \\ + 2(V_l{}^k{}_h C_A{}^{lh} + M_n C_A{}^{nk}) \mathrm{d}_k y^A. \end{aligned} \tag{4.150}$$

Under these conditions, taking the derivatives of $\tilde{K}_1$ given by (4.150) and then substituting (4.140) together with Equations (4.82)–(4.84) and (4.147)–(4.149), we find after some simple calculations that the static values of these derivatives are (henceforth, calculations will be based solely on the Lagrangian $\tilde{K}_1$, and the wave will not be written)

$$K_A{}^{ji} = K_{(1)A}{}^{ji} - 2y_A(y^j a^i - y^i a^j), \tag{4.151}$$

$$\partial K_1/\partial \mathrm{d}_B y^A = 2C_A{}^{lh} M_{lh}{}^B + 2M_n C_A{}^{nB} - 4y^B a_A, \tag{4.152}$$

$$\begin{aligned} \partial K_1/\partial y^A &= \partial K_{11}/\mathrm{d}y^A + 4C_{Apq} q^p a^q = \\ &= C_{Apq}(-\tfrac{1}{2}M^p{}_{mn} M^{qmn} + M_k{}^{pn} M^{kq}{}_n + \\ &+ M^{kmp} M_{mk}{}^q - 2M^p M^q + 4a^p a^q) \end{aligned} \tag{4.153}$$

where $K_{(1)A}{}^{ji}$ is given by (4.124). The relation (4.137) has been used in calculating (4.152). It follows from Equations (4.151), (4.82), (4.84) and $y^A = 1$ that

$$\frac{1}{J}\mathrm{d}_j(JK_A{}^{ji}) = \frac{1}{J}\mathrm{d}_j(JK_{(1)A}{}^{ji}) + 2y_A a^i(\mathrm{D}_n a^n + a_n a^n). \tag{4.154}$$

The substitution of (4.151) together with $y^A = 1$ and $C_A{}^B{}_B = 0$ in (4.118) gives

$$\mathrm{T}_A{}^i = T_{(1)A}{}^i - 2S^i_A a_n a^n + 4y^i y_A a_n a^n + 4a^i a_A - y^i E_A, \tag{4.155}$$

where we have made use of the fact that the static value of the Lagrangian (4.150) equals $K_{11} - 2a_n a^n$ as a consequence of (4.147) and (4.148). The first term on the right-hand side of (4.155) is given by (4.125).

At this stage, let us write down the explicit form of the Euler-Lagrange derivative $JE_A{}^i$ in the static case. The substitution of (4.154) and (4.155) in (4.117) yields

$$E_A{}^i = E_{(1)A}{}^i + 2y_A y^i(\mathrm{D}_n a^n - a_n a^n) + 2S^i_A a_n a^n - 4a_A a^i + y^i E_A \tag{4.156}$$

where $E_{(1)A}{}^i$ is given by (4.126) or, in another form,

$$E^{ij} = E_{(1)}{}^{ij} + 2y^i y^j(\mathrm{D}_n a^n - a_n a^n) + 2a^{ij} a_n a^n - 4a^i a^j + y^j E^i. \tag{4.157}$$

A close inspection of the structure of the quantities (4.156) and (4.119) shows that the interesting relation

$$E_B = C_B{}^A{}_i E_A{}^i \tag{4.158}$$

exists in the static case considered here. This we now prove. In fact, it follows directly from (4.156) and the identities $C_{ijk} y^i = 0$ and $C_i{}^m{}_m = 0$ that

$$C_B{}^A{}_i E_A{}^i = C_B{}^A{}_i(E_{(1)A}{}^i - 4a_A a^i). \tag{4.159}$$

By virtue of Equations (4.125), (4.126) and $C_m{}^A{}_i S^i_A = C_m{}^i{}_i = 0$, we have

$$\begin{aligned} C_B{}^A{}_i E_{(1)A}{}^i &= C_B{}^A{}_i\left(\frac{1}{J}\mathrm{d}_j(JK_{(1)A}{}^{ji}) + K_{(1)C}{}^{ni} M^C{}_{nA}\right) = \\ &= C_B{}^A{}_D\left(\frac{1}{J}S^D_i\,\mathrm{d}_j(JK_{(1)A}{}^{ji}) + K_{(1)C}{}^{nD} M^C{}_{nA}\right) = \\ &= C_B{}^A{}_D\left(\frac{1}{J}\mathrm{d}_j(JK_{(1)A}{}^{jD}) - \tfrac{1}{2}K_{(1)A}{}^{ji} M^D{}_{ji} + K_{(1)C}{}^{nD} M^C{}_{nA}\right) \end{aligned} \tag{4.160}$$

where we have made use of the fact that

$$K_{(1)A}{}^{ji}\,\mathrm{d}_j S^D_i = \tfrac{1}{2}K_{(1)A}{}^{ji} M^D{}_{ji}$$

in accordance with our notation $M^D{}_{ji} = \mathrm{d}_j S^D_i - \mathrm{d}_i S^D_j$ (Equation (2.106)) and the skew-symmetry of the tensor $K_{(1)A}{}^{ji}$ in its superscripts. It can be readily verified that the expression (4.124) of the tensor $K_{(1)A}{}^{ji}$ implies the following equalities

$$C_B{}^A{}_D\left(\frac{1}{J}\mathrm{d}_j(JK_{(1)A}{}^{jD})\right) = C_B{}^A{}_D\left(\frac{1}{J}\mathrm{d}_j(J(2M_A{}^{Dj} + 2S^j_A M^D))\right), \tag{4.161}$$

$$-\tfrac{1}{2}K_{(1)}{}^{pji}M^q{}_{ji} + K_{(1)C}{}^{np}M^C{}_n{}^q = \tfrac{1}{2}M^p{}_{mn}M^{qmn} - M_k{}^{pn}M^{kq}{}_n - M^k{}_{mp}M^m{}_k{}^q + 2M^pM^q. \tag{4.162}$$

On the other hand, the first term in the definition (4.119) of E_B is equal to

$$\frac{1}{J}\mathrm{d}_j(J(2C_B{}^A{}_D M_A{}^{Dj} + 2C_{BD}{}^j M^D - 4y^j a_B)) \tag{4.163}$$

because of (4.152). Since the last term in the parenthesis in (4.163) may be omitted in the static case (indeed, relative to the static coordinates, $\mathrm{d}_j(Jy^j a_B) = \mathrm{d}_0(Jy^0 a_B) = 0$) and $C_B{}^A{}_D$ are constants ($C_B{}^A{}_D$ are functions of y^C only and are equal to 1 according to (4.140)), the expression (4.163) is identical to the right-hand side of (4.161). The remaining part of E_B defined by (4.119) is equal to $-\partial K_1/\partial y^B$, and comparison of (4.153) with (4.163) shows immediately that this part is equal to (4.162) plus $4C_{Bpq}a^p a^q$. Therefore, E_B is indeed identical to (4.159). The proof of the equality (4.158) is complete.

The first term on the right-hand side of the expression (4.157) for E^{ij} is symmetric in i,j in accordance with the definition (4.127). Therefore, all terms in (4.157) are symmetric in i,j except for the last one $y^j E^i$ which cannot be symmetric in i,j because this would contradict (4.131), $y^i E_i = 0$. Thus, *the tensor E^{ij} will be symmetric in its superscripts if and only if $E_A = 0$.*

It is apparent from Equation (4.158) that the condition $E_A = 0$ will be valid if the static gravitational field is generated by a perfect fluid whose energy-momentum tensor is of the form

$$T_M{}^{ij} = k((n + p)y^i(x)y^j(x) - pa^{ij}) \tag{4.164}$$

where $n = n(x)$ is the density of the fluid, $p = p(x)$ the pressure and k denotes a constant. Indeed, in this case the gravitational field equations $E^{ij} = T_M{}^{ij}$ will imply that $E_A = 0$ because of (4.158) and the identities $y^i C_{ijk} = 0$ and $C_i{}^m{}_m = 0$. In particular, such an implication occurs in the vacuum case. So, we get the following result.

PROPOSITION 7. In the case when the static gravitational field is generated by a perfect fluid, and in particular in vacuo, the tensor E^{ij} has the following restricted form

$$E^{ij} = E_{(1)}{}^{ij} + 2y^i y^j(D_n a^n - a_n a^n) + 2a^{ij}a_n a^n - 4a^i a^j \tag{4.165}$$

and therefore is symmetric in its superscripts.

By the definition (4.120), the quantity JE_i is nothing but the Euler-Lagrange derivative of the Lagrangian density $\tilde{L}_1 = J\tilde{K}_1$ with respect to $y^i(x)$. By virtue of the equality (4.121) the vanishing of E_A means that E_i vanishes also. So, although $y^i(x)$ has been treated as an auxiliary vector field (not as one of the actual field variables) and therefore E_i not postulated to be zero, under certain conditions the field equations demand the vanishing of E_i. Thus we get the following:

PROPOSITION 8. Under the conditions stated in Proposition 7, the Euler-Lagrange derivative of the Lagrangian density $\tilde{L}_1$ with respect to $y^i(x)$ vanishes as a consequence of the field equations.

It will be noted in this respect that, since the difference between $\tilde{L}_1$ and the initial Finslerian Lagrangian density L is an ordinary divergence, and the Euler-Lagrange derivative of an ordinary divergence is identically zero, *the equality*

$$JE_i = -\partial L/\partial y^i \tag{4.166}$$

will hold whenever the field equations are satisfied. Therefore, $E_i = 0$ will imply $\partial L/\partial y^i = 0$. Obviously, the latter equality cannot be treated as a condition imposed on the Finslerian Lagrangian density L because this would imply that L is independent of y^i and, therefore, would confine us to Riemannian geometry. Nevertheless, that $\partial L/\partial y^i = 0$ whenever the field equations are satisfied appears to be quite reasonable in the static case.

Among the various possible solutions of the static gravitational field equations, the case of the spherically symmmetric field is the simplest. Most of the relativistic gravitational effects observed in the solar system can be described by post-Newtonian approximations of the spherically symmetric static solution[11]. The case of interest to us may be defined as follows.

DEFINITION. A static gravitational field is called *spherically symmetric* if the static coordinates x^i can be chosen in such a way that the following two conditions are satisfied relative to these x^i:

(i) The spatial part a_{ab} of the osculating Riemannian metric tensor $a_{ij}(x) = g_{ij}(x, y(x))$ is proportional to the Kronecker symbol, that is

$$a_{ab} = -M^2\delta_{ab} \tag{4.167}$$

where M is a scalar function.

(ii) Both y_0 and M are functions of the radius r only, that is

$$y_0 = y_0(r), \qquad M = M(r) \tag{4.168}$$

where

$$r = (\delta_{ab}x^a x^b)^{1/2}. \tag{4.169}$$

Under these conditions, the static coordinates x^i will be called *isotropic*[12].

In the rest of this section, the coordinates x^i used will be assumed to be static and isotropic. By virtue of Equations (4.168) and (4.83), the entities $a = -\ln y_0$ and $a_n = d_n a$ will be functions of the radius (4.169). Let also $u(r) = \ln M$ and $u_n = \mathrm{d}_n u$, so that $u_0 = 0$. Then, it can be readily verified that the Riemannian Christoffel symbols $\{{}^{\,k}_{m\,n}\}$ constructed from the osculating Riemannian metric tensor a_{ij} are equal to

$$\{{}^{\,c}_{b\,a}\} = u_a \delta^c_b + u_b \delta^c_a - u^c a_{ab}, \qquad \{{}^{\,k}_{0\,0}\} = a_{00}\, a^k,\ \{{}^{\,0}_{0\,b}\} = \\ = -a_b,\ \{{}^{\,0}_{a\,b}\} = \{{}^{\,c}_{a\,0}\} = 0. \tag{4.170}$$

After substituting these formulae in the Riemannian Ricci tensor

$$A_{lh}(x) = \mathrm{d}_k \{{}^{\,k}_{l\,h}\} - \mathrm{d}_h \{{}^{\,k}_{l\,k}\} + \{{}^{\,k}_{m\,k}\}\{{}^{\,m}_{l\,h}\} - \{{}^{\,k}_{m\,h}\}\{{}^{\,m}_{l\,k}\}$$

we obtain that the tensor $E_{(1)i}{}^{j}$ given by (4.127) can be written as

$$E_{(1)0}{}^{0} = 4f\,\mathrm{D}_n a^n - 2(f^2 - 2f - 2q)a_n a^n, \tag{4.171}$$

$$E_{(1)ab} = 2(1 - \mathrm{f})\mathrm{D}_a a_b - 2(f^2 + q + 1)a_a a_b + \\ + 2((f - 1)\mathrm{D}_n a^n + (f + q)a_n a^n)a_{ab}, \tag{4.172}$$

$$E_{(1)i}{}^{i} = -4\mathrm{D}_n a^n + 2E_{(1)0}{}^{0} \tag{4.173}$$

and $E_{(1)0}{}^{a} = E_{(1)a}{}^{0} = 0$. Here,

$$f = u'/a', \qquad q = f'/a', \tag{4.174}$$

where $u' = \mathrm{d}u/\mathrm{d}r$, $\mathrm{a}' = \mathrm{d}a/\mathrm{d}r$ and $f' = \mathrm{d}f/\mathrm{d}r$.

From Equation (4.165) we have

$$E_0{}^{0} = E_{(1)0}{}^{0} + 2\mathrm{D}_n a^n, \tag{4.175}$$

$$E_{ab} = E_{(1)ab} + 2a_n a^n a_{ab} - 4a_a a_b, \tag{4.176}$$

$$E_i{}^{i} = E_{(1)i}{}^{i} + 2\mathrm{D}_n a^n + 2a_n a^n. \tag{4.177}$$

In the case of the Einstein vacuum equations $E_{(1)i}{}^{j} = 0$, the expressions (4.171) and (4.173) immediately yield

$$D_n a^n = 0, \qquad f^2 - 2f - 2q = 0. \tag{4.178}$$

On the other hand, our vacuum equations $E_i{}^{j} = 0$ lead, as a consequence of Equations (4.171), (4.173), (4.175) and (4.177), to the equalities

$$3\mathrm{D}_n a^n = a_n a^n, \qquad f^2 - \tfrac{8}{3}f - 2q - \tfrac{1}{3} = 0 \tag{4.179}$$

and not to (4.178). It is apparent that the set of two equations $E_0{}^{0} = 0$ and $E_i{}^{i} = 0$ is equivalent to (4.179). Therefore, *the equations* $\mathrm{E}_i{}^{j} = 0$ *are equivalent to the equations* (4.179) *plus*

$$E_{ab} \equiv (1 - \mathrm{f})\mathrm{D}_a a_b - (f^2 + q + 3)a_a a_b + ((f - 1)\mathrm{D}_n a^n + \\ + (f + q + 1)a_n a^n)a_{ab} = 0 \tag{4.180}$$

which is obtained after substituting (4.172) in (4.176).

We get from (4.167)–(4.169) that

$$n_a \overset{\text{def}}{=} \mathrm{d}_a r = x^a/r, \qquad a^{ab} n_a n_b = -M^{-2}, \qquad \mathrm{d}_b n_a = (\delta_{ab} - n_a n_b)/r$$

so that

$$\begin{aligned} &a_b = a' n_b, \qquad a_m a^m = -M^{-2} a'^2, \\ &\qquad \mathrm{d}_a a_b = \left(a'' - \frac{1}{r} a'\right) n_a n_b + \frac{1}{r} a' \delta_{ab} \end{aligned} \tag{4.181}$$

($a'' = \mathrm{d}a'/\mathrm{d}r$) from which we can conclude with the help of the explicit form (4.170) of the Christoffel symbols that

$$\begin{aligned} D_n a^n &\overset{\text{def}}{=} \mathrm{d}_n a^n + \left\{ {}^{\,n}_{n\ m} \right\} a^m = \mathrm{d}_n a^n - a_m a^m + 3u_m a^m = \\ &= \mathrm{d}_n a^n + (3f - 1) a_m a^m. \end{aligned}$$

Here, the first term is equal to

$$\begin{aligned} \mathrm{d}_n a^n = \mathrm{d}_a(a^{ab} a_b) &= -\delta^{ab} \mathrm{d}_a (M^{-2} a_b) = -2u_b a^b - M^{-2} \delta^{ab} d_a a_b = \\ &= -2f a_m a^m - M^{-2} \delta^{ab} d_a a_b \end{aligned}$$

so that using (4.181) we find after cancelling similar terms that

$$D_n a^n = -M^{-2}\left(a'' + \frac{2}{r} a' + (f-1) a'^2\right). \tag{4.182}$$

Let us solve the system (4.179)–4.180) in the post-Newtonian approximation. This means that we are to substitute the asymptotic expansion

$$\begin{aligned} &y_0 = 1 - Br^{-1} - Cr^{-2} + 0(\mathrm{r}^{-3}),\ M = 1 + Ar^{-1} + Dr^{-2} + \\ &\qquad + O(r^{-3}) \end{aligned} \tag{4.183}$$

in equations (4.179)–(4.180) in order to express the constants A,C, and D in terms of the constant B. We begin the expansion with unity because we assume, as usual, that the metric tensor a^{ij} will have the pseudo-Euclidean form $a^{00} = 1$ and $a^{ab} = -\delta^{ab}$ at infinity. Since $a = -\ln y_0$ and $u = \ln M$, we get from (4.183) that

$$\begin{aligned} a &= Br^{-1} + (C + \tfrac{1}{2}B^2) r^{-2} + O(r^{-3}), \\ u &= Ar^{-1} + (D - \tfrac{1}{2}A^2) r^{-2} + O(r^{-3}). \end{aligned} \tag{4.184}$$

So if we write

$$f = f_0 + O(r^{-1}), \qquad q = q_0 + O(r^{-1})$$

we find from (4.184) and the definition (4.174) of f and q that

$$f_0 = A/B, \qquad q_0 = \left(\frac{2(D-C)}{A} - A - B\right) A/B^2. \tag{4.185}$$

It will be noted that Equation (4.180), as well as Einstein's equations $E_{(1)ab} = 0$, imply $f_0 = 1$ as a result of the factor $(1 - f)$ of $\mathrm{D}_a a_b$. This statement follows from the observation that

$$a_a a_b = O(r^{-4}), \qquad \mathrm{D}_n a^n = O(r^{-4}), \tag{4.186}$$

whereas

$$\mathrm{D}_a a_b = O(r^{-3}), \tag{4.187}$$

which we now demonstrate, indeed from Equation (4.184) we have

$$a' = -B/r^2 + O(1/r^3), \qquad a'' = 2B/r^3 + O(1/r^4). \tag{4.188}$$

Since $a_b = a' n_b$ (see (4.181)) and $n_b = O(1)$, the quantity a_b is of the same order $O(1/r^2)$ as a', so that the first equality in (4.186) is correct. Next, on substituting (4.188) in (4.182), all terms of the order $O(1/r^3)$ are cancelled, the remaining expression being

$$D_n a^n = -(2C + B^2)/r^4 + (1 - f_0)B^2/r^4 + O(1/r^5) \tag{4.189}$$

so that the second equality in (4.186) is valid too. As to Equation (4.187), the definition of $\mathrm{D}_a a_b$ reads

$$\mathrm{D}_a a_b = \mathrm{d}_a a_b - \{{}_{acb}\} a^c.$$

It is seen from (4.170) that each term in $\{{}_{acb}\} a^c$ is proportional to $a'^2 = O(1/r^4)$. On the other hand, according to the last equality in (4.181), $\mathrm{d}_a a_b$ contains the third-order expression

$$\mathrm{a}'' n_a n_b + (\delta_{ab} - n_a n_b) a'/r = (3 n_a n_b - \delta_{ab}) B/r^3 + O(1/r^4).$$

Thus, Equation (4.187) is valid and the expansion of Equation (4.180), as well as the Einstein equations $E_{(1)ab} = 0$, reads

$$(1 - f_0)(3\mathrm{n}_a \mathrm{n}_b - \delta_{ab}) B/r^3 + O(1/r^4) = 0$$

which implies that $f_0 = 1$. The proof is complete.

The first equality in (4.185) shows that $f_0 = 1$ means

$$A = B. \tag{4.190}$$

Since

$$a_n a^n = -B^2/r^4 + O(1/r^5)$$

(see Equations (4.181) and (4.188)), the first equation in (4.179) implies, by virtue of (4.189) and $f_0 = 1$, that,

$$C = -B^2/3. \tag{4.191}$$

Next, substituting $f_0 = 1$ in the second equation in (4.179) we find that $q_0 = -1$, so that the second equality in (4.185) implies, in view of Equations (4.190) and (4.191), that

$$D = B^2/6. \tag{4.192}$$

Considering the square of (4.183), we obtain the expansion of the components $a_{00} = (y_0)^2$ and $a_{ab} = -M^2\delta_{ab}$ (see (4.80) and (4.167)) of the osculating Riemannian metric tensor a_{ij} in the following form

$$a_{00} = 1 - 2B/r + (B^2 - 2C)/r^2 + O(1/r^3),$$
$$a_{ab} = -\delta_{ab}(1 + 2A/r + (A^2 + 2D)/r^2) + O(1/r^3)$$

or, on substituting Equations (4.190)–(4.192),

$$\begin{aligned} a_{00} &= 1 - 2U + 2\beta U^2 + O(1/r^3), \\ a_{ab} &= -\delta_{ab}(1 + 2\gamma U + \tfrac{3}{2}\lambda U^2) + O(1/r^3), \end{aligned} \tag{4.193}$$

where

$$\gamma = 1, \qquad \beta = 5/6, \qquad \lambda = 8/9 \tag{4.194}$$

and the notation $U = B/r$ has been used.

It is noted for the purposes of comparison that Einstein's equations (4.178) imply, as a consequence of Equations (4.189) and $f_0 = 1$, that $C = -B^2/2$ and $D = B^2/4$, from which it directly follows that $\gamma = \beta = \lambda = 1$.

4.7. Conservation Laws for the Gravitational Field in the 1-Form Case

The 1-form approach developed in the preceding section has yielded vacuum gravitational field equations of the form (4.122). Since the tensors $K_A{}^{ji}$ entering the left-hand side of these equations are skew-symmetric in their superscripts (see (4.116)), we can draw the interesting conclusion that the vectors $T_A{}^i$ on the right-hand side of Equation (4.122) are conserved quantities; that is, the (weak) conservation law

$$\frac{1}{J}\mathrm{d}_i(JT_A{}^i) = 0 \tag{4.195}$$

will be valid whenever the gravitational field equations (4.122) are satisfied. The conservation law (4.195) is obviously covariant. Moreover, it is also integrable in the following sense. The N scalars

$$P_A = \int_V JT_A{}^i \mathrm{d}V_i \tag{4.196}$$

give a measure of the gravitational flux through any $(N-1)$-dimensional hypersurface V in such a way that if the hypersurface V is chosen to be closed, the scalars P_A vanish by virtue of the relation (4.195) and the Gauss theorem.

Such a conservation law has no analogue in the usual Riemannian approach to gravitational field theory[13]. On the other hand, a conservation law of this type does appear in any theory which uses a set of covariant vector fields

$S_i^a(x)$, $a = 1,2,\ldots$, as field variables, on the assumption that their derivatives enter the Lagrangian density L in the skew-symmetric combinations $F^a{}_{ij} = \mathrm{d}_i S_j^a - \mathrm{d}_j S_i^a$ only. This property is inherent in the theories of electromagnetic, Yang-Mills, and tetradic[14] gravitational fields, all these theories employing Lagrangian scalar densities of the form

$$L = L(a_{mn}(x),\ S_i^a(x),\ F^a{}_{ij}(x)) \tag{4.197}$$

in which the Riemannian metric tensor $a_{mn}(x)$ of space-time plays an auxiliary role. In this respect, the Lagrangian (4.114) used in the 1-form Finslerian approach to gravitational field theory appears to be quite similar to that of the Yang-Mills field, the vector field $y^i(x)$ together with its derivative $\mathrm{d}_j y^i(x)$ playing the role of auxiliary quantities in (4.114), analogous to $a_{mn}(x)$ in (4.197). By analogy with the treatment of the vector densities

$$t_a{}^i \overset{\text{def}}{=} \partial L/\partial S_i^a \tag{4.198}$$

as the conserved currents in Yang-Mills field theory[15], we call $T_A{}^i$ given by Equation (4.118) *the gravitational field currents.*

Another type of conservation law can be obtained by using the so-called Hamiltonian complex $H_i'^j$, which in the case of the density (4.197) is given by

$$H_i'^j = -\delta_i^j L + (\mathrm{d}_i S_n^a)\,\partial L/\partial \mathrm{d}_j S_n^a \tag{4.199}$$

in accordance with the definition (3.34). Introducing the tensor

$$N_a{}^{jn} \overset{\text{def}}{=} \partial L/\partial \mathrm{d}_j S_n^a = 2\partial L/\partial F^a{}_{jn} \tag{4.200}$$

we can rewrite (4.199) as

$$H_i'^j = -\delta_i^j L + N_a{}^{jn}\mathrm{d}_i S_n^a. \tag{4.201}$$

It is well known from the calculus of variations[16] that the divergence of the Hamiltonian complex is equal to the negative of the partial derivative of the Lagrangian density with respect to x^i whenever the Euler-Lagrange derivative of the Lagrangian density with respect to the field variables vanishes. In terms of the present notation this reads

$$\mathrm{d}_j H_i'^j = -(\partial L/\partial a_{mn})\mathrm{d}_i a_{mn} \tag{4.202}$$

whenever the field equations

$$\mathrm{d}_j N_a{}^{ji} = t_a{}^i \tag{4.203}$$

(cf. the field equations (4.122)) are satisfied.

Although the Hamiltonian complex (4.199) is not of tensorial nature, the conservation law (4.202) is, that is the relation (4.202) has the same form relative to any admissible coordinate system. If the tensor $a_{mn}(x)$ admits a local coordinate system x^i possessing the property $\mathrm{d}_i a_{mn} = 0$ (which assumes the vanishing of the Riemann curvature tensor associated with a_{mn}) then, relative to these x^i, Equation (4.202) becomes a genuine conservation law:

$$\mathrm{d}_j H_i'^j = 0. \tag{4.204}$$

Taking into account the evident identities

$$\mathrm{d}_i S_n^a = F^a{}_{in} + \mathrm{d}_n S_i^a$$

and

$$N_a{}^{jn} \mathrm{d}_n S_i^a = \mathrm{d}_n (S_i^a N_a{}^{jn}) - S_i^a \mathrm{d}_n N_a{}^{jn} = \mathrm{d}_n (S_i^a N_a{}^{jn}) + S_i^a t_a{}^j,$$

where in the last step we have used the field equations (4.203), we obtain the following relation between the Hamiltonian complex (4.201) and the conserved currents (4.198):

$$H_i'^j = -\delta_i^j L + N_a{}^{jn} F^a{}_{in} + \mathrm{d}_n (S_i^a N_a{}^{jn}) + S_i^a t_a{}^j. \tag{4.205}$$

On the other hand, the invariance identities (see Section 3.1) associated with the density (4.197) can easily be written in the form

$$H_i{}^j = t_a{}^j S_i^a + N_a{}^{nj} F^a_{ni} - \delta_i^j L \tag{4.206}$$

with

$$H_i{}^j = -2a_{ij} \partial L / \partial a_{jn} \tag{4.207}$$

(cf. Equation (3.31)). Comparing (4.205) with (4.206), we get the following result:

$$H_i'^j = H_i{}^j + \mathrm{d}_n (S_i^a N_a{}^{jn}). \tag{4.208}$$

The quantity (4.207) is obviously a tensor density. Therefore, we can conclude from (4.208) that the non-tensorial part of the Hamiltonian complex (4.205) is the ordinary divergence of the tensor density $S_i^a N_a{}^{jn}$. Since the latter is skew-symmetric in its superscripts, we have

$$\mathrm{d}_j \mathrm{d}_n (S_i^a N_a{}^{jn}) \equiv 0,$$

and the tensorial part $H_i{}^j$ of the Hamiltonian complex will satisfy the same conservation law (4.202) as the total Hamiltonian complex $H_i'^j$, i.e.

$$\mathrm{d}_j H_i{}^j = -(\partial L / \partial a_{mn}) \mathrm{d}_i a_{mn}. \tag{4.209}$$

It is $H_i{}^j$ that plays the role of the energy-momentum tensor density in the Maxwell electromagnetic theory, in which $t_a{}^j \equiv 0$, but $H_i{}^j \neq 0$. It is usually assumed in tetradic gravitational theories that the density L, Equation (4.197), does not involve any auxiliary functions, which results in $H_i{}^j \equiv 0$, while $t_a{}^j \neq 0$. In the case of the Yang-Mills field, $t_a{}^j$ and $H_i{}^j$ do not vanish identically. Since the Yang-Mills field is intermediate between the electromagnetic and gravitational fields, the example of this field explains clearly the difference between the conserved currents, the Hamiltonian complex, and the conventional energy-momentum tensor.

An essentially similar line of reasoning may be applied to the Hamiltonian complex $H_i^{\prime j}$ associated with the Finslerian Lagrangian K_1 given by Equation (4.144). The equality (4.208) will be replaced by

$$H_i^{\prime j} = H_i^{\ j} + \mathrm{d}_n(JK_i^{\ jn)} \tag{4.210}$$

where

$$H_i^{\ j} = -\delta_i^j L_1 + JK_A^{\ jn} M^A_{\ \ in} + JT_i^{\ j} + Jy^n(\mathrm{d}_i S_n^A)\,\partial K_1/\partial \mathrm{d}_j y^A \tag{4.211}$$

or, after substituting (4.118) and cancelling similar terms,

$$H_i^{\ j} = -y^j E_i - S_n^D(\mathrm{d}_i y^n)\, J\,\partial K_1/\partial \mathrm{d}_j y^D + y^j C_i{}^m{}_m L_1. \tag{4.212}$$

The conservation law takes the following form

$$\mathrm{d}_j H_i^{\ j} = -(\mathrm{d}_j y^n)\,\partial L_1/\partial y^n - (\mathrm{d}^2_{im} y^n)\,\partial L_1/\partial \mathrm{d}_m y^n \tag{4.213}$$

(instead of (4.209)), where the right-hand side represents the negative of the partial derivative of $L_1 = JK_1$ with respect to x^i.

Relative to the special coordinate system x^i with the property $\mathrm{d}_i y^n = 0$ (see Equation (4.75)), the relation (4.213) takes the form of a genuine conservation law:

$$\mathrm{d}_j H_i^{\ j} = 0 \tag{4.214}$$

which makes it possible to assign a measure of gravitational flux

$$H_i = \int_V H_i^{\ j} \mathrm{d}V_j \tag{4.215}$$

(cf. the definition (4.196)) to any three-dimensional hypersurface V. In contrast to (4.206) $H_i^{\ j}$ given by (4.212) is not a tensorial quantity, because of the non-tensorial second term on the right-hand side of (4.212). However, the measure (4.215) has an invariant sense since the special coordinate system is defined in a covariant way by means of a set of scalars (Equation (4.73)).

The general conclusion derivable from the analysis performed in this chapter is that the Finslerian generalization of gravitational field equations may develop rightfully. The obtained equations are such that they can in principle distinguish the Finslerian deviations of space-time geometry from the Riemannian pattern, It seems that the more generally the metric geometry is taken for describing space-time, the less decisive is the verdict of the conditions for the acceptability of theories implied by the known relativistic gravitational effects. Since such effects are scanty nowadays, they impose but inconsiderable limitations on the allowed extensions of the geometrical basis of space-time.

This notwithstanding, having attempted in this book to advocate the constructive possibilities of Finsler geometry, we are not entitled to end the analysis on this pessimistic note. Fresh prospects are opened up in two directions. First, we may venture to generalize the conventional Lorentzian special relativity by using Finslerian metrics. Such an attempt will be made in

Chapter 7. The second approach, which we shall try to realize in the next chapter, is to treat the two-storied structure of Finsler geometry described in Section 1.2 as introducing, gauge potentialities with the view of relating the Finslerian generalization of the Riemannian geometry of space-time to the internal symmetries of fundamental physical fields.

Problems

4.1. The vanishing of $N_i{}^j$ given by Equation (4.19) does not appear to imply in general that the tensor densities M^{ij}, M^{ijk}, and M^{ijkl} will vanish. However, if M^{ijk} and M^{ijkl} vanish, the equation $N_i{}^j = 0$ takes the form

$$(2\delta_i^m g_{jn} + 2y^m C_{ijn})M^{ij} = 0.$$

Prove that this equation entails $M^{ij} = 0$ in the case when g_{ij} is a Finslerian metric tensor.

4.2. Examine what simplifications in the formulae of Sections 3.4 and 4.1 will arise in the case of stationary $y^i(x)$, that is when $y_m^k = 0$.

4.3. Estimate the order of the second term on the right-hand side of the electromagnetic field equations (4.51) (this term represents Finslerian corrections) in the case of the static gravitational field near the Earth.

4.4. Prove that equation (4.65) holds identically in the case when the gravitational field is static.

4.5. Examine the post-Newtonian corrections in the equations of Finslerian goedesics for the spherically symmetric static gravitational field.

4.6. Following the technique developed in Sec. 4.6, consider the static gravitational field equations generated by choosing the Lagrangian K_1 instead of $\tilde{K}_1$ taken in the second part of Section 4.6.

4.7. So far we have used the concept of connection in the sense of a linear connection, that is in constructing the covariant derivative of a tensor the connection coefficients have been assumed to be independent of the tensor. However, it is possible to use the nonlinear covariant derivative $\mathscr{D}_j X^k = \partial X^k/\partial x^j + \Gamma_n{}^k{}_j(xX)X^n$ of any vector field $X^k(x)$ dependent solely on x^i, in place of the linear derivative $X^k_{|j} = \partial X^k/\partial x^j + \Gamma_n{}^k{}_j(x, y)X^n$ given by the Finslerian definition (1.42).

The theory of nonlinear connections is described in detail in Section 6.4 of the monograph by Rund (1959).

Proceeding in this way, one can avoid resorting to the concept of osculation though one would have to pay dearly for this: the field equations would become strongly nonlinear. Formulate the respective nonlinear generalizations of the equations for the scalar and electromagnetic fields.

4.8. Using solely the osculating Riemannian metric tensor $a_{ij}(x) \stackrel{\text{def}}{=} g_{ij}(x, y(x))$, one may construct scalar densities from the tensor a_{ij} and its derivatives. Let L_g be some scalar density constructible in such a way; for example, L_g can be taken to be the Einstein gravitational Lagrangian density $(-\det(a_{ij}))^{1/2}$ R where R denotes the complete contraction of the Riemann curvature tensor associated with a_{ij}. Prove that the vacuum field equations $E_{gAi\ldots}^{\ j\ldots} = E_{gi} = 0$ are equivalent to the Einstein-type equations $\delta L_g / \delta a_{mn} = 0$.

4.9. Formulate a Finslerian generalization of the electromagnetic field equations so that the equations imply that photons follow isotropic Finslerian geodesics.

4.10. Develop an alternative osculating approach based on the idea of replacing the covariant directional variable y_i which enters the contravariant Finslerian metric tensor $g^{mn}(x^k, y_i)$ by the gradient $\partial v / \partial x^i$ of a scalar field $v(x)$.

Notes

[1] The following example clearly illustrates the situation. Let the direction-dependent scalar density L be of the particular form $L = y^i Z_i(x)$, where $Z_i(x)$ denotes an arbitrary covariant vector density. Considering the integral $I(y)$, we can take y^i outside the integral sign because y^i are independent of x^i. Thus, $I(y) = y^i \int_G Z_i(x) \mathrm{d}^N x$. However, the integration of a covariant vector density over a region is not a coordinate-invariant operation. So, $\int_G Z_i(x) \mathrm{d}^N x$ and, hence, $I(y)$ fails to be of tensorial nature. For this reason, the tensorial meaning of the double integration $\int_G \int_H L \, \mathrm{d}^N x \, \mathrm{d}^N y$, where H is a region of the tangent space, would be even more obscure.

[2] Nevertheless, it is known that in a Finsler space of constant curvature and of dimension $N > 2$ the covariant divergence of some analog of the Riemannian Einstein tensor vanishes identically (Rund, 1962). The corresponding problem for Finsler spaces of arbitrary scalar curvature was discussed by Shibata (1978b). Since this circumstance refers to particular Finsler spaces, it does not alter our reasoning.

[3] See, e.g., Treder, 1971.

[4] The idea of substituting the Finslerian covariant derivative for the Riemannian one in a Finslerian generalization of equations for physical fields has been employed by many authors; see, e.g. Horvath and Moór, 1952; Horvath, 1958, 1963; Takano, 1964, 1982; Ikeda, 1981a,b; Ishikawa, 1981.

[5] See, e.g., Landau and Lifschitz, 1962, Section 33.

[6] See Synge, 1960, Section IV.4.

[7] See Rund (1976), where the Clebsch representations of many geometrical objects have been introduced and investigated. The representation (4.68) corresponds to the four-dimensional case $N = 4$. However, the conslusions reached in Section 4.4 may be directly generalized to any dimension number N by considering $N - 1$ independent scalars. Incidentally, the possibility examined in Proposition 2 occurs also in the case when a pseudo-Euclidean metric tensor $e_{ij} = \Sigma_{a=1}^{4} q_a \mathrm{d}_i v^a \mathrm{d}_j v^a$ with q_a being $+1$ or -1 plays the role of an auxiliary quantity in a Lagrangian density (as in bimetric gravitational theories and the field-theoretic approaches (Cavalleri and Spinelli, 1980) to gravity in flat space-time).

[8] See Synge, 1960.

[9] See, e.g., Landau and Lifschitz, 1962, Section 93.

[10] The reader should be careful not to confuse this notation with the time components of objects in the static case.

[11] See, e.g., Synge, 1960.

[12] We follow the terminology adopted in general relativity; see, e.g., Synge, 1960, Section VIII.5.

[13]The reader can find a systematic discussion of the status of the conservation laws in Einstein's general relativity in Trautman, 1962.
[14]See Treder, 1971.
[15]See, e.g., Konopleva and Popov (1981).
[16]See Lovelock and Rund, 1975, Chapter 6.

Chapter 5

Parametrical Representation of Physical Fields. The Relevance to Gauge Theories

5.1. Application of the Parametrical Representation of the Indicatrix

According to the definitions introduced in Section 1.1 Finsler geometry deals with those objects which depend on both the points x^i and the tangent vectors y^i, so that a first acquaintance with this geometry suggests that the Finslerian generalizations of physical field equations should be expressed in terms of (x,y)-dependent tensors. However, the use of the concept of the indicatrix (Section 1.2) suggests another possible approach, namely that one can develop a theory of fields dependent on the points x^i and on the $N-1$ scalars u^a playing the role of intrinsic coordinates on the indicatrix. The parametrical representation $l^i = t^i(x^n, u^a)$ (Equation (1.100)) of the indicatrix enables one to cast any (x,y)-dependent field $W(x,y)$ of degree-zero homogeneity with respect to y^i into an (x,u)-dependent form $w(x,u)$ merely by substituting $t^i(x,u)$ for y^i. The function $w(x,u)$ defined as

$$w(x,u) \stackrel{\text{def}}{=} W(x,t(x,u)) \tag{5.1}$$

will be called *the parametrical representation of the object* $W(x,y)$. In particular, the parametrical representation of the Finslerian metric tensor $g_{ij}(x,y)$ will read

$$b_{ij}(x,u) \stackrel{\text{def}}{=} g_{ij}(x,t(x,u)). \tag{5.2}$$

The symmetry of the Cartan torsion tensor (1.15) is reflected in b_{ij} as follows:

$$\frac{\partial b_{ij}}{\partial u^a}\frac{\partial u^a}{\partial y^k} - \frac{\partial b_{ik}}{\partial u^a}\frac{\partial u^a}{\partial y^j} = 0. \tag{5.3}$$

The fact that the parameters u^a are scalars gives the significant advantage that these parameters do not affect the transformation laws of the derivatives of tensorial objects with respect to x^i; that is, for any tensorial object $w(x,u)$, the derivative $\partial w/\partial x^i$ will obey the same transformation law for changes of the coordinates x^i, as in the case when the object w does not depend on u^a at all. In particular, the derivative

$$b_{ij,k} = \partial b_{ij}(x,u)/\partial x^k \tag{5.4}$$

will transform in the same way as the partial derivative of a Riemannian metric tensor, in contrast to $\partial g_{ij}(x,y)/\partial x^k$ whose transformation law is of the complicated form (3.36) because of the vector nature of the argument y^i.

Therefore, we may be sure that *the parametrical Christoffel symbols*

$$b_m{}^i{}_n(x,u) = \tfrac{1}{2}b^{ik}(b_{mk,n} + b_{nk,m} - b_{mn,k}) \tag{5.5}$$

are connection coefficients, i.e. their transormation law is (3.38). The covariant derivative of a vector field $w^i(x,u)$ may be defined by means of these coefficients in the conventional way:

$$\mathrm{D}_j^* w^i(x,u) \overset{\text{def}}{=} \partial w^i(x,u)/\partial x^j + b_k{}^i{}_j(x,u)w^k(x,u). \tag{5.6}$$

Under these conditions, the usual Riemannian construction

$$Q_i{}^j{}_{mn}(x,u) = \partial b_i{}^j{}_m/\partial x^n - \partial b_i{}^j{}_n/\partial x^m + b_k{}^j{}_n b_i{}^k{}_m - b_k{}^j{}_m b_i{}^k{}_n \tag{5.7}$$

yields a tensor which may be called *the parametrical curvature tensor*.

These observations open up the possibility of using conventional Riemannian techniques to formulate the equations which describe the dependence of fields $w(x,u)$ on x^i in terms of the parametrical metric tensor $b_{ij}(x,u)$ and Christoffel symbols $b_m{}^i{}_n(x,u)$. Such field equations may be based on the usual variational principles used in theoretical physics. For example, a vector field $w_i(x,u)$ may be described by the Lagrangian density

$$L = (|b|)^{1/2}(-b^{ij}b^{mn}(\mathrm{D}_i^* w_m)(D_j^* w_n) + M^2 b^{ij} w_i w_j) \tag{5.8}$$

($b = \det(b_{ij})$), so that the associated field equations will read

$$b^{ki}\mathrm{D}_k^*\mathrm{D}_i^* w_j(x,u) + M^2 w_j(x,u) = 0. \tag{5.9}$$

Such an approach is quite free from the difficulty arising in the theory of (x,y)-dependent fields where a variational principle for describing the dependence of such fields on x^i cannot be formulated in a straightforward way (see the beginning of Section 4.1). Of course one may, ignoring this point, formulate the equations merely by replacing Riemannian covariant derivatives by Cartan covariant derivatives (see the first part of Section 4.2). The relationship between such generalizations and parametrical equations of the type (5.9) may be found directly by using the identities

$$\frac{\partial W_j(x,y)}{\partial x^i} = \frac{\partial w_j(x,u)}{\partial x^i} + \frac{\partial u^a(x,y)}{\partial x^i}\frac{\partial w_j(x,u)}{\partial u^a},$$

$$\frac{\partial W_j(x,y)}{\partial y^k} = \frac{\partial u^a(x,y)}{\partial y^k}\frac{\partial w_j(x,u)}{\partial u^a}$$

which imply that, according to the definition (1.43) of the Cartan covariant derivative,

$$W_{j|i} = D_i^* w_j + u_{|i}^a \partial w_j / \partial u^a - (\Gamma_i{}^k{}_j - b_i{}^k{}_j) w_k$$

where $w_i(x, u) = W_i(x, t(x, u))$ is the parametrical representation of the field $W_i(x, y)$.

However, such an approach gives rise to the new trouble that the field equations so formulated are not invariant under arbitrary x-dependent transformations $u^a = u^a(x, \bar{u})$ of the parameters u^a, although the equations will be invariant under any transformations

$$u^a = u^a(\bar{u}^b) \tag{5.10}$$

independent of x^i. This difficulty will be completely overcome in the next section by introducing so-called gauge fields; in particular, the totally invariant generalizations of Equations (5.8) and (5.9) will be (5.63) and (5.64).

It will be noted that the treatment of fields as functions of x^n and u^a justifies our regarding (5.10) as the simplest transformation of u^a in the sense that it does not involve any coordinates x^i, whereas writing $y^i(\bar{y}^j)$ fails to be meaningful unless a fixed point x^i to which it refers is indicated. This circumstance is another manifestation of the scalar nature of the variables u^a. In general, (x, u)-dependent fields can be classified according to their transformation laws under the combined transformations

$$x^i = x^i(\bar{x}^j), \qquad u^a = u^a(\bar{u}^b), \tag{5.11}$$

where x^i and u^a, as well as $\bar{x}^i$ and $\bar{u}^a$, are regarded as entirely independent variables. That is, the following classification scheme may be proposed:

scalar-scalar field $w(x, u)$:

$$\bar{w}(\bar{x}, \bar{u}) = w(x(\bar{x}),\ u(\bar{u}));$$

scalar-vector field $w^a(x, u)$:

$$\bar{w}^a(\bar{x}, \bar{u}) = \frac{\partial \bar{u}^a}{\partial u^b} w^b(x(\bar{x}), u(\bar{u}));$$

vector-scalar field $w^i(x, u)$:

$$\bar{w}^i(\bar{x}, \bar{u}) = \frac{\partial \bar{x}^i}{\partial x^j} w^j(x(\bar{x}),\ u(\bar{u}));$$

vector-vector field $w^{ia}(x, u)$:

$$\bar{w}^{ia}(\bar{x}, \bar{u}) = \frac{\partial \bar{x}^i}{\partial x^j}\frac{\partial \bar{u}^a}{\partial u^b} w^{jb}(x(\bar{x}),\ u(\bar{u}));$$

etc.

The projection factors t_a^i and u_i^a of the indicatrix give examples of vector-vector fields as follows from their definition (1.101), so that

$$\bar{t}_a^i(\bar{x}, \bar{u}) = \frac{\partial \bar{x}^i}{\partial x^j}\frac{\partial u^b}{\partial \bar{u}^a} t_b^j(x(\bar{x}), u(\bar{u})). \tag{5.12}$$

Therefore, the induced metric tensor g^*_{ab}, g^{*ab} (see Equation (1.113)) of the indicatrix is a scalar-tensor field, while the orthonormal n-tuples $g^{*(a)}_{\ b}$, $g^{*\ b}_{(a)}$ defined by the relations (1.133) and (1.137) will transform as scalar-vector fields. On the other hand, the orthonormal n-tuples g^P_i, g^i_P (Equations (1.135) and (1.139)) of the Finslerian metric tensor are evidently vector-scalar fields. These basic tensors make it possible to interchange the three types of indices $i, a, (a)$, for example

$$w^a \to w^{(a)} = g^{*(a)}_{\ b} w^b, \quad \text{so that} \quad w^b = g^{*\ b}_{(a)} w^{(a)}; \tag{5.13}$$

$$w^{(a)} \to w^i = g^{\ i}_{(a)} w^{(a)}, \quad \text{so that} \quad w^{(a)} = g^{(a)}_{\ i} w^i; \tag{5.14}$$

$$w^i \to w^{(a)} = g^{(a)}_{\ i} w^i, \quad \text{so that} \quad w^i = g^{\ i}_{(a)} w^{(a)} - l^i l_j w^j \tag{5.15}$$

where in (5.15) we have made use of the last equality (1.140′). The lowering of indices can be defined in a natural way:

$$w_a = g^*_{ab} w^b, \qquad w_{(a)} = q^a w^{(a)}, \qquad w_i = b_{ij} w^j$$

where the constants q^a equaling $+1$ or -1 represent the signature of the metric tensor (see Equations (1.133) and (1.137)).

There exist intimate relationships between the differential operators acting on y^i and on u^a. For example, suppose we are given a scalar-vector field $w_a(x, u)$. Then, treating the indicatrix as a Riemannian space (Section 1.2) we may use the usual Riemannian definition of *the covariant derivative on the indicatrix*, to be denoted by D^*_b:

$$\mathrm{D}^*_b w_a \stackrel{\text{def}}{=} \partial w_a / \partial u^b - q^{*c}_{a\ b} w_c \tag{5.16}$$

where the notation $q^{*c}_{a\ b}$ is used for *the Christoffel symbols of the indicatrix*:

$$q^{*c}_{a\ b} = \tfrac{1}{2} g^{*cd} (\partial g^*_{ad} / \partial u^b + \partial g^*_{bd} / \partial u^a - \partial g^*_{ab} / \partial u^d). \tag{5.17}$$

If we represent the field w_a in the (x, y)-dependent form:

$$W_i(x, y) \stackrel{\text{def}}{=} g^{(a)}_i g^{*\ b}_{(a)} w_b = F(x, y) u^a_i(x, y) w_a(x, u(x, y)), \tag{5.18}$$

where we have used the relation (1.134), then the converse will read

$$w_a = t^i_a W_i \tag{5.19}$$

because of the identity $F u^b_j t^j_a = \delta^b_a$ (see Equation (1.102)). Substituting (5.19) in (5.16) yields

$$\mathrm{D}^*_b w_a = W_i \mathrm{D}^*_b t^i_a + t^i_a t^j_b W_i|_j, \tag{5.20}$$

where $|_j$ denotes the Minkowskian derivative (1.66), and

$$\mathrm{D}^*_b t^i_a \stackrel{\text{def}}{=} \partial t^i_a / \partial u^b - q^{*c}_{a\ b} t^i_c + F C_{n\ b}^{\ i} t^n_a \tag{5.21}$$

with $C_{n\ b}^{\ i} = C_{n\ m}^{\ i} t^m_b$.

The following proposition is valid.

PROPOSITION 1. *The definition* (5.21) *is equivalent to*

$$\mathrm{D}^*_b t^i_a = -l^i t^m_a t^n_b g_{mn}. \tag{5.22}$$

Proof. Differentiating Equation (1.110) with respect to u^c yields

$$\partial g^*_{ab}/\partial u^c = 2FC_{ijk}t^i_a t^j_b t^k_c + g_{ij}(t^j_b\, \partial t^i_a/\partial u^c + t^i_a\, \partial t^j_b/\partial u^c) \tag{5.23}$$

where, in writing the first term on the right-hand side, we have used the identity $\partial/\partial u^c = Ft^k_c\, \partial/\partial y^k$ ensuing from the definition (1.100). From (5.23) it follows that the Christoffel symbols (5.17) can be written, after lowering the index c, as

$$q^*_{acb} = FC_{ijk}t^i_a t^j_b t^k_c + g_{ij}t^j_c\, \partial t^i_a/\partial u^b. \tag{5.23'}$$

On contracting this last relation with u^c_n and taking into account the last identity in (1.102), we get

$$u^c_n q^*_{acb} = C_{ijn}t^i_a t^j_b + F^{-1}(g_{in} - l_i l_n)\, \partial t^i_a/\partial u^b \tag{5.24}$$

which is equivalent to (5.22) because of the definition (5.21), Equation (1.115) and the relation

$$l_i\, \partial t^i_a/\partial u^b = -t^i_a\, \partial l_i/\partial u^b = -t^i_a F t^k_b\, \partial l_i/\partial y^k = -t^i_a t^k_b h_{ik} = -g^*_{ab}$$

(see Equation (1.113)). The proof is complete[1].

Substituting (5.22) in (5.20) and taking into account the fact that the definition (5.18) together with the identity $l^i u^a_i = 0$ (see (1.102)) implies that $W_i l^i = 0$, we get

$$\mathrm{D}^*_b w_a = t^i_a t^j_b W_{i|j}. \tag{5.25}$$

The contraction of (5.25) with g^{*ab} yields

$$g^{*ab}\mathrm{D}^*_b w_a = g^{ij} W_{i|j} \tag{5.26}$$

where we have made use of the equality

$$g^{*ab} t^i_a t^j_b = h^{ij} \equiv g^{ij} - l^i l^j$$

derivable from Equation (1.115). The relation (5.26) says that *the covariant divergences of* w_a *and* W_i *are equal to each other.*

The action of the operator D^*_b may be extended to any tensor in accordance with

$$\mathrm{D}^*_b w_{cd}{}^{ij} = \partial w_{cd}{}^{ij}/\partial u^b - q^{*e}_{c\ b} w_{ed}{}^{ij} - q^{*e}_{d\ b} w_{ce}{}^{ij} + FC_n{}^i{}_b w_{cd}{}^{nj} + FC_n{}^j{}_b w_{cd}{}^{in},$$

$$\mathrm{D}^*_b w^{cd}{}_{ij} = \partial w^{cd}{}_{ij}/\partial u^b + q^{*c}_{e\ b} w^{ed}{}_{ij} + q^{*d}_{e\ b} w^{ce}{}_{ij} - FC_i{}^n{}_b w^{cd}{}_{nj} - FC_j{}^n{}_b w^{cd}{}_{in},$$

$$\mathrm{D}^*_b w = \partial w/\partial u^b.$$

It is clear from this definition that the operator D_b^* is such that, if we are given a tensor $w^{\cdots}_{\cdots}$ under the combined transformations (5.11), then $\mathrm{D}_b^* w^{\cdots}_{\cdots}$ is again a tensor. Moreover,

$$\mathrm{D}_b^* g^{*cd} = \mathrm{D}_b^* g^*_{cd} = \mathrm{D}_b^* b^{ij} = \mathrm{D}_b^* b_{ij} = 0.$$

The operator D_b^* exhibits all the properties common to linear covariant derivatives.

It will be noted that, given one parametrical representation $l^i = t^i(x, u)$ (Equation (1.100)) of the indicatrix, we may obtain an infinite number of representations by making arbitrary sufficiently smooth invertible transformations $u^a = u^a(x^n, \tilde{u}^b)$, under which $t^i(x, u) \rightarrow \tilde{t}^i(x, \tilde{u}) = t^i(x, u(x, \tilde{u}))$. Among these representations, a preference may be assigned to those which are constructed from the same set of tensors $S^A(x)$ which appear in the representation of the Finslerian metric tensor $g_{ij}(x, y) = v_{ij}(S^A(x), y)$ (Equation (3.14)). Accordingly, we introduce the following

DEFINITION. Let a Finslerian metric tensor be functionally represented by a relation of the form $g_{ij}(x, y) = v_{ij}(S^A(x), y)$. Then the parametrical representation $l^i = t^i(x, u)$ of the indicatrix is called *uniform* if there exists a vector field $T^i(S^A, u)$ such that

$$t^i(x, u) = T^i(S^A(x), u). \tag{5.27}$$

Since the representation $l^i = t^i(x, u)$ was defined in Section 1.2 as inverse to the representation $u^a = u^a(x, y)$ given by Equation (1.92), the definition (5.27) is tantamount to the requirement that there exist scalars $M^a(S^A, y)$ such that

$$u^a(x, y) = M^a(S^A(x), y). \tag{5.28}$$

The projection factors $t^i_a = \partial t^i / \partial u^a$,

$$U^a_i(x, y) \stackrel{\text{def}}{=} F(x, y) u^a_i(x, y) \tag{5.29}$$

and the parametrical metric tensor (5.2) will also be called uniform if they are expressed in terms of the uniform parametrical representation of the indicatrix. Accordingly, the uniform parametrical metric tensor is obtained by substituting (5.27), together with (3.14), in (5.2), which gives

$$b_{ij}(x, u) = P_{ij}(S^A(x), u) \tag{5.30}$$

where

$$P_{ij}(S^A, u) = v_{ij}(S^A, T(S^A, u)).$$

In particular, in a 1-form Finsler space the above definition states that there exist functions $M^a(y^A)$ such that the representation (5.28) is of the form

$$u^a = M^a(y^A) \tag{5.31}$$

where $y^A = S^A_i(x) y^i$ (the notation (2.88)). In this case,

$$t^i(x,u) = S^i_A(x)t^A(u). \tag{5.32}$$

In other words, the uniform 1-form entities are constructed in such a way that $l^A = t^A(u)$ is the parametrical representation of the indicatrix of the Minkowskian space given by the Minkowskian metric function $F_{\text{Mink}}(y^A)$ from which the 1-form Finslerian metric function is constructed in accordance with (2.87). Therefore, in the 1-form case, the uniform projection factors U^a_i will satisfy the relation

$$\partial_h U^a_A = (\partial U^a_A/\partial y^C)y^m \partial_h S^C_m \tag{5.33}$$

which is similar to (2.94), and the uniform metric tensor of the indicatrix will be independent of x^i:

$$g^*_{ab} = g^*_{ab}(u). \tag{5.34}$$

For example, by virtue of the representations (2.64) and (2.72) (which are uniform), the uniform projection factors related to the 1-form Berwald-Moór metric function may be written in the following form

$$t^i_a = \sum_{A=1}^{N} S^i_A C^A_a l^A, \qquad U^a_i = N\sum_{A=1}^{N} S^A_i C^a_A l_A. \tag{5.35}$$

At this stage we have described a handy way of casting the fields dependent on the points x^i and on the tangent vectors y^i into the form where the point u^a of the Finslerian indicatrix is used instead of y^i. Treating the indicatrix as a particular physical internal space leads directly to the appearance of gauge fields and tensors, as will be described in more detail in the succeeding section.

5.2. The Emergence of Gauge Fields

The geometrical significance of the parameters u^a lies in the fact that they serve as the coordinates of points on the indicatrix (see Section 1.2). As a consequence of this auxiliary role, u^a may be subjected to arbitrary (sufficiently smooth and invertible) transformations

$$u^a = u^a(x, \tilde{u}) \tag{5.36}$$

without affecting any of the geometrical constituents of a Finsler space. The transformations (5.36) generalize (5.10) in that they allow for a dependence on x^i (our change of notation in the present section, namely the writing of a wave instead of the bar used in (5.10), is motivated only by reasons of convenience). This generalization means that the transformations of u^a in the indicatrices supported by different points x^i are different. Naturally, such a generalization will not affect the conclusions made in the preceding section. However, as soon as the transformations of u^a are allowed to depend on x^i, the partial derivative (5.4) of the parametrically represented metric tensor $b_{ij}(x,u)$ with respect to x^k, as well as the covariant derivative (5.6), the parametrical Christoffel symbols

(5.5), and the curvature tensor (5.7), fail to be invariant under (5.36). In order to find the required invariant notions, we shall proceed as follows.

The simplest case of an (x, u)-dependent field which is tensorial under (5.36) is presented by a scalar-vector field $w_a(x, u)$, that is the field possessing the following two properties: firstly, each component of w_a is a scalar under general coordinate transformations

$$x^i = x^i(\bar{x}^j), \tag{5.37}$$

namely

$$\bar{w}_a(\bar{x}, u) = w_a(x, u), \tag{5.38}$$

and secondly, the action of (5.36) is reflected by a transormation of the form

$$\tilde{w}_a(x, \tilde{u}) = u_a^b w_b(x, u) \tag{5.39}$$

where we have used the notation

$$u_a^b = \partial u^b(x, \tilde{u})/\partial \tilde{u}^a. \tag{5.40}$$

In the case when the transformations (5.36) are taken to be independent of x^i, the object $w_a(x, u)$ is a typical example of a field defined on the space of points x^i which has tensor properties under the 'internal' transformations $u^a = u^a(\tilde{u})$ which are related in no way to the 'external' coordinates x^i. Following the terminology adopted in the theory of guage fields[2], we shall call the transformations (5.36) *local gauge transformations*, or simply *gauge transformations*, in contrast to the *global transformations* $u^a = u^a(\tilde{u})$[3].

It is obvious that the dependence of gauge transformations on x^i necessitates the introduction of additional fields, called gauge fields, in order to construct the covariant derivative of w_a with respect to x^i. To this end, let us denote by $w_{a,i}$ and $\tilde{w}_{a,i}$ the partial derivatives of $w_a(x, u)$ and $\tilde{w}_a(x, \tilde{u})$ with respect to x^i for constant u^a and $\tilde{u}^a$, respectively, i.e.

$$w_{a,i} = \partial w_a(x, u)/\partial x^i, \qquad \tilde{w}_{a,i} = \partial \tilde{w}_a(x, \tilde{u})/\partial x^i.$$

Using the notation

$$u_a{}^b{}_c = \partial u_b^a/\partial \tilde{u}^c, \qquad u_{,i}^a = \partial u^a(x, \tilde{u})/\partial x^i, \qquad u_{b,i}^a = \partial u_b^a(x, \tilde{u})/\partial x^i, \tag{5.41}$$

we obtain from (5.39):

$$\tilde{w}_{a,i} = u_a^b w_{b,i} + u_{a,i}^b w_b + u_a{}^b{}_c u_{,i}^c w_{bc} \tag{5.42}$$

where $w_{bc} = \partial w_b(x, u)/\partial u^c$. We see from (5.42) that the derivative $\tilde{w}_{a,i}$ involves, apart from the tensor term $u_a^b w_{b,i}$, two terms proportional to w_b and w_{bc}, respectively. Therefore, in order to construct an operator D_i such that $\mathrm{D}_i w_a$ is a vector under gauge transformations, we must compensate for the non-tensorial nature of these last two terms in (5.42) by introducing two relevant terms in $\mathrm{D}_i w_a$. Accordingly, we define the action of the covariant derivative D_i

on w_a as follows:

$$D_i w_a = w_{a,i} - A_a{}^b{}_i w_b + N_i^c w_{ac} \tag{5.43}$$

where $A_a{}^b{}_i(x,u)$ and $N_i^c(x,u)$ are called *gauge fields.*

We stipulate that these gauge fields transform as covariant vectors under general coordinate transformations (5.37):

$$\bar{A}_a{}^b{}_i(\bar{x},u) = \frac{\partial x^j}{\partial \bar{x}^i} A_a{}^b{}_j(x,u), \bar{N}_i^c(\bar{x},u) = \frac{\partial x^j}{\partial \bar{x}^i} N_j^c(x,u) \tag{5.44}$$

which, together with the assumption (5.38), ensures that the object

$$w_{ai} \overset{\text{def}}{=} D_i w_a \tag{5.45}$$

will be a covariant vector under (5.37). The appropriate transformation laws of our gauge fields under the guage transformations (5.36) can be readily found if we require that w_{ai} be a vector under (5.36), that is

$$\tilde{w}_{ai} = u_a^b w_{bi}, \tag{5.46}$$

where

$$\tilde{w}_{ai} = \tilde{w}_{a,i} - \tilde{A}_a{}^b{}_i \tilde{w}_b + \tilde{N}_i^c \tilde{w}_{ac} \tag{5.47}$$

in accordance with the definitions (5.43) and (5.45). Indeed, on substituting in (5.47) the relations (5.42) and (5.40) together with

$$\tilde{w}_{ac} \overset{\text{def}}{=} \partial \tilde{w}_a / \partial \tilde{u}^c = u_a{}^d{}_c w_d + u_a^b u_c^d w_{bd}, \tag{5.48}$$

which follows from (5.39), we get

$$\tilde{w}_{ai} = u_a^b (w_{b,i} - (u_c^d \tilde{A}_e{}^c{}_i - u_{e,i}^d - u_e{}^d{}_c \tilde{N}_i^c)\tilde{u}_b^e w_d + (u_c^d \tilde{N}_i^c + u_{,i}^d) w_{bd}), \tag{5.49}$$

where $\tilde{u}_b^e = \partial \tilde{u}^e / \partial u^b$ so that $\tilde{u}_b^e u_c^b = \delta_c^e$. Comparison of (5.49) with (5.43) shows that our requirement (5.46) will be satisfied if the following transformation laws are assigned to the gauge fields:

$$u_c^d \tilde{A}_b{}^c{}_i - u_{b,i}^d - u_b{}^d{}_c \tilde{N}_i^c = u_b^e A_e{}^d{}_i, \tag{5.50}$$

$$u_c^d \tilde{N}_i^c + u_{,i}^d = N_i^d. \tag{5.51}$$

The simplest way of associating tensors with the gauge fields introduced above consists in evaluating the commutator of the covariant derivative. In order to do this, we have to extend the definition of the operator D_i to the field w_{an} which is a vector under general coordinate transformations (5.37). For this purpose we may employ any object $D_i{}^k{}_n(x)$ dependent solely on x^i which transforms as connection coefficients under coordinate transformations (5.37) and define the action of D_i on w_{an} as follows:

$$D_i w_{an} = \partial w_{an} / \partial x^i - D_n{}^k{}_i w_{ak} - A_a{}^b{}_i w_{bn} + N_i^c \partial w_{an} / \partial u^c. \tag{5.52}$$

It is clear from this definition that the derivative

$$w_{ani} \overset{\text{def}}{=} \mathrm{D}_i w_{an},$$

as well as $\mathrm{D}_i w_a$ itself, is a covariant vector under gauge transformations (5.36):

$$\tilde{w}_{ani} = u_a^b w_{bni} \tag{5.53}$$

and, simultaneously, a tensor of covariant valence 2 under coordinate transformations (5.37):

$$\bar{w}_{ani}(\bar{x}, u) = \frac{\partial x^m}{\partial \bar{x}^n} \frac{\partial x^j}{\partial \bar{x}^i} w_{amj}(x, u).$$

Therefore, the commutator $\mathrm{D}_i w_{an} - \mathrm{D}_n w_{ai}$ will transform as a vector and as a skew-symmetric tensor under the transformations (5.36) and (5.37), respectively.

We may simplify matters by assuming that the connection coefficients $\mathrm{D}_i{}^k{}_n(x)$ are symmetrical in their subscripts, which means that they will disappear in the above commutator. Under these conditions, the substitution of (5.43) in (5.52) yields the following result after omitting all terms symmetrical in the indices i, n:

$$\mathrm{D}_i w_{an} - \mathrm{D}_n w_{ai} = -w_b F_a{}^b{}_{in} + w_{ac} N_{in}^c \tag{5.54}$$

where

$$F_a{}^b{}_{in} = \partial A_a{}^b{}_n / \partial x^i - \partial A_a{}^b{}_i / \partial x^n + N_i^c \partial A_a{}^b{}_n / \partial u^c - N_n^c \partial A_a{}^b{}_i / \partial u^c - \\ - A_a{}^c{}_i A_c{}^b{}_n + A_a{}^c{}_n A_c{}^b{}_i \tag{5.55}$$

and

$$N_{in}^c = \partial N_n^c / \partial x^i - \partial N_i^c / \partial x^n + N_i^e \partial N_n^c / \partial u^e - N_n^e \partial N_i^c / \partial u^e. \tag{5.56}$$

We call (5.55) and (5.56) *the gauge field strength tensors*, or simply *the gauge tensors*. They are skew-symmetric in the last two subscripts:

$$F_a{}^b{}_{in} = -F_a{}^b{}_{ni}, \qquad N_{in}^c = -N_{ni}^c. \tag{5.57}$$

The following proposition is valid.

PROPOSITION 2. *The objects* (5.55) *and* (5.56) *obey the following transformation laws:*

$$\bar{F}_a{}^b{}_{in}(\bar{x}, u) = \frac{\partial x^j}{\partial \bar{x}^i} \frac{\partial x^m}{\partial \bar{x}^n} F_a{}^b{}_{jm}(x, u), \tag{5.58}$$

$$\bar{N}_{in}^c(\bar{x}, u) = \frac{\partial x^j}{\partial \bar{x}^i} \frac{\partial x^m}{\partial \bar{x}^n} N_{jm}^c(x, u), \tag{5.59}$$

$$u_b^d \tilde{F}_a{}^b{}_{in}(x, \tilde{u}) - u_a{}^d{}_c \tilde{N}_{in}^c(x, \tilde{u}) = u_a^c F_c{}^d{}_{in}(x, u), \tag{5.60}$$

$$u_c^d \tilde{N}_{in}^c(x, \tilde{u}) = N_{in}^d(x, u). \tag{5.61}$$

Proof. If we make coordinate transformations (5.37), then it follows directly from the assumption (5.44) that the derivatives $\partial A_a{}^b{}_n/\partial u^e$ and $\partial N^c_i/\partial u^e$ transform like covariant vectors and that the skew-symmetrization of the partial derivatives of $A_a{}^b{}_n$ and N^c_n with respect to x^i yields skewsymmetric tensors. Thus, Equations (5.58) and (5.59) are valid. As to the transformation laws (5.60) and (5.61), we note that, owing to (5.53), the right-hand side of Equation (5.54) represents a vector under gauge transformations (5.36), i.e.

$$\tilde{w}_b \tilde{F}_a{}^b{}_{in} - \tilde{w}_{ac}\tilde{N}^c_{in} - u^c_a(w_d F_c{}^d{}_{in} - w_{ce}N^e_{in}) = 0,$$

which takes the following form after substituting (5.39) and (5.48):

$$(u^d_b \tilde{F}_a{}^b{}_{in} - u_a{}^d{}_c \tilde{N}^c_{in} - u^c_a F_c{}^d{}_{in})w_d - u^b_a(u^d_e \tilde{N}^e_{in} - N^d_{in})w_{bd} = 0. \tag{5.62}$$

Here, the functions w_d are entirely arbitrary except for the preassigned transformation laws (5.38) and (5.39) so that both w_d and $w_{bd} = \partial w_b/\partial u^d$ may take on essentially arbitrary values for any given value of the argument (x, u) independently of one another. The latter circumstance, together with the fact the coefficients of w_d and w_{bd} in (5.62) do not involve the functions w_d or their derivatives, implies the vanishing of these coefficients which yields (5.60) and (5.61). The proof is complete.

As regards the partial derivatives of gauge fields with respect to u^a, they are not tensors under (5.36). In particular, the transformation law of $\partial N^d_i/\partial u^e$ can be readily found directly from Equations (5.50) and (5.51). Indeed, differentiating (5.51) with respect to $\tilde{u}^b$ yields

$$u_b{}^d{}_c \tilde{N}^c_i + u^d_c\, \partial \tilde{N}^c_i/\partial \tilde{u}^b + u^d_{b,i} = u^e_b\, \partial N^d_i/\partial u^e.$$

If we add this relation to Equation (5.50), we obtain $u^d_c \tilde{N}^c_{ib} = u^e_b N^d_{ie}$, where

$$N^c_{ib} = \partial N^c_i/\partial u^b + A_b{}^c{}_i. \tag{5.63}$$

Thus, *the object* (5.63) *is a guage tensor.*

Transformation laws (5.58)–(5.59) indicate that the objects $F_a{}^b{}_{in}$ and N^a_{in} are tensors under general coordinate transformations (5.37). At the same time, Equation (5.61) shows that N^a_{in} represents a gauge tensor under gauge transformations (5.36), whereas $F_a{}^b{}_{in}$ is not a tensor under (5.36) because of the second term on the left-hand side of (5.60). Unless new geometric objects are introduced, it is impossible to separate out the tensor part in $F_a{}^b{}_{in}$. If, however, we introduce the Christoffel symbols $q^{*b}_a{}_c(x, u)$ of the internal space, such a separation can be readily made. In fact, using $q^{*b}_a{}_c$, we may construct the covariant derivative $\mathrm{D}_c w_a = w_{ac} - q^{*b}_a{}_c w_b$, define the object

$$F_{(1)a}{}^b{}_{in} \overset{\text{def}}{=} F_a{}^b{}_{in} - q^{*b}_a{}_c N^c_{in}, \tag{5.64}$$

and then rewrite the commutation relation (5.54) in the form

$$\mathrm{D}_i w_{an} - \mathrm{D}_n w_{ai} = -w_b F_{(1)a}{}^b{}_{in} + N^c_{in}\mathrm{D}_c w_a. \tag{5.65}$$

The relation (5.65) shows that *the object* (5.64) *is a gauge tensor*, i.e. $u^d_b \tilde{F}_{(1)a}{}^b{}_{in} = u^c_a F_{(1)c}{}^d{}_{in}$.

5.3. Finslerian Representation of Gauge Fields and Tensors

The first point to be made when we attempt to conceive in Finslerian terms the conclusion reached in the preceding sections is that the rules (5.13)–(5.15) may be used to convert any (x, u)-dependent tensor into a (x, y)-dependent tensor-scalar form which remains unaffected by transformations of the parameters u^a of the indicatrix. In particular, the Finslerian metric tensor $g_{ij}(x, y)$, the Cartan connection coefficients $\Gamma_i{}^j{}_k(x, y)$, and the curvature tensors $K_i{}^j{}_{mn}(x, y)$, $R_i{}^j{}_{mn}(x, y)$, as well as the Cartan covariant derivative of any (x, y)-dependent tensor, are all insensitive to such transformations.

Accordingly, given a scalar-vector field $w_a(x, u)$ obeying the vector transformation law (5.39), we may construct a vector-scalar field by means of the definition (5.18) which we may write as

$$W_i(x, y) = U^a_i(x, y) w_a(x, u(x, y)) \tag{5.66}$$

in terms of the notation (5.29). By Equation (1.102), U^a_i are connected with the projection factors t^i_a by the following reciprocity relations:

$$U^a_i t^i_b = \delta^a_b, \qquad U^a_i t^j_a = \delta^j_i - l_i l^j. \tag{5.67}$$

The usual procedure of Cartan covariant differentiation may now be applied to W_i which is an entirely (x, y)-dependent vector field. That is, the Cartan covariant derivative $W_{i|j}$ will be a tensor independent of transformations of u^a. The relation (5.66) implies

$$W_{i|j} = U^a_{i|j} w_a + U^a_i w_{a|j} \tag{5.68}$$

where

$$w_{a|j} \overset{\text{def}}{=} (w_a(x, u(x, y)))_{|j} = w_{a,j} + w_{ab}\, \partial u^b/\partial x^j - G_j{}^n u^b_n w_{ab} \tag{5.69}$$

in accordance with the definition (1.44) of the Cartan covariant derivative of a scalar function, and we have used the notation $u^b_n = \partial u^b(x, y)/\partial y^n$. The substitution of (5.69) in (5.68) yields

$$W_{i|j} = U^a_{i|j} w_a + U^a_i (w_{a,j} + u^b_{|j} w_{ab}), \tag{5.70}$$

where

$$u^b_{|j} = \partial u^b/\partial x^j - G_j{}^n u^b_n \tag{5.71}$$

is the Cartan covariant derivative of $u^a(x, y)$.

The reciprocity relations (5.67) make it possible to write Equation (5.70) in the following form

$$w_{a,j} + t^k_a U^b_{k|j} w_b + u^c_{|j} w_{ac} = W_{k|j} t^k_a. \tag{5.72}$$

By construction, $W_{k|j}$ is a gauge-invariant tensor. At the same time, t^k_a transforms like a covariant vector under gauge transformations (5.36) (see Equation (5.12)). So, *the left-hand side of Equation* (5.72) *is a covariant vector under gauge transformations* (5.36). Comparing (5.72) with definition (5.43) shows that the gauge fields are

$$A_a{}^b{}_i = -t^k_a U^b_{k|i} \equiv U^b_k t^k_{a|i}, \qquad N^a_i = u^a_{|i}. \tag{5.73}$$

Thus, we have established the following result.

PROPOSITION 3. The gauge field $A_a{}^b{}_i$ is constructed from the projection factors t^k_a and U^a_k of the indicatrix of a Finsler space in accordance with (5.73). The other gauge field N^a_i is just the Cartan covariant derivative of the scalars $u^a(x, y)$ which represent a coordinate system on the indicatrix.

By means of these gauge fields we may define the covariant derivative of a tensor of any type in the usual way, for example for a tensor of the form $v_a{}^b(x, u)$,

$$\mathrm{D}_i v_a{}^b = \partial v_a{}^b/\partial x^i - A_a{}^c{}_i v_c{}^b + A_c{}^b{}_i v_a{}^c + N^c_i \, \partial v_a{}^b/\partial u^c. \tag{5.74}$$

By construction,

$$\mathrm{D}_i g^*_{ab} = \mathrm{D}_i g^{*ab} = 0 \tag{5.74$'$}$$

and

$$A_a{}^b{}_i + A^b{}_{ai} = -g^*_{ac} g^{*cb}{}_{|i},$$

where $A^b{}_{ai} = g^{*bc} g^*_{ad} A_c{}^d{}_i$. Also,

$$\mathrm{D}_i b_{mn}(x, u) \overset{\text{def}}{=} \partial b_{mn}/\partial x^i - \Gamma_m{}^k{}_i b_{kn} - \Gamma_n{}^k{}_i b_{mk} + N^a_i \, \partial b_{mn}/\partial u^a = 0.$$

The next question concerns the Finslerian counterparts of the gauge tensors (5.55) and (5.56). It might be expected that they are expressible in terms of the Finslerian curvature tensors. Let us make the required calculations. In the preceding section, gauge fields have been defined as dependent on x^i and u^a, whereas the object $u^a_{|i}$ in Equation (5.73) is primarily dependent on x^i and y^i. However, since the covariant derivatives $u^a_{|i}(x, y)$ inherit from $u^a(x, y)$ a homogeneity of degree zero in y^n, we may use the parametrical representation (1.100) of the indicatrix:

$$l^k(x, y) = t^k(x, u) \tag{5.75}$$

to write

$$N^a_i(x, u) = u^a_{|i}(x, t(x, u)). \tag{5.76}$$

From (5.75) we obtain

$$l^k_{,i} = t^k_{,i} + t^k_a \partial u^a/\partial x^i \tag{5.77}$$

where $l^k_{,i} = \partial l^k(x,y)/\partial x^i$ and $t^k_{,i} = \partial t^k(x,u)/\partial x^i$. Therefore,

$$\partial N^c_n(x,u)/\partial x^i = \partial u^c_{|n}(x,y)/\partial x^i + Ft^k_{,i}\,\partial u^c_{|n}(x,y)/\partial y^k. \tag{5.78}$$

On substituting $t^k_{,i}$ from (5.77), taking into account the identity $l^k_{|i} = 0$ (see Equation (1.47)), and writing the first term on the right-hand side of (5.78) in terms of the Cartan covariant derivative

$$u^c_{|n|i} = \partial u^c_{|n}/\partial x^i - G_i^{\ k}\,\partial u^c_{|n}/\partial y^k - \Gamma_{i\ n}^{\ k} u^c_{|k},$$

we get the following result

$$\begin{aligned}\partial N^c_n(x,u)/\partial x^i &= u^c_{|n|i} - Ft^k_e u^e_{|i}\,\partial u^c_{|n}/\partial y^k + \Gamma_{n\ i}^{\ k} u^c_{|k} \equiv \\ &\equiv u^c_{|n|i} - N^d_i\,\partial N^c_n/\partial u^d + \Gamma_{n\ i}^{\ k} u^c_{|k}\end{aligned} \tag{5.79}$$

which evidently reduces the gauge tensor (5.56) to

$$N^c_{in} = u^c_{|n|i} - u^c_{|i|n}. \tag{5.80}$$

The right-hand side of Equation (5.80) is equal to $-Fu^c_j l^r K_{r\ ni}^{\ j}$ in accordance with the commutation rule (1.79). So, we get eventually

$$N^c_{in} = U^c_j l^r K_{r\ in}^{\ j}. \tag{5.81}$$

Similarly to (5.79), we may write

$$\begin{aligned}\partial A_{a\ n}^{\ b}(x,u)/\partial x^i = (A_{a\ n}^{\ b}(x,y))_{|i} - Ft^k_e N^e_i\,\partial A_{a\ n}^{\ b}(x,y)/\partial y^k + \\ + \Gamma_{n\ i}^{\ k} A_{a\ k}^{\ b}\end{aligned} \tag{5.82}$$

where

$$A_{a\ n}^{\ b}(x,u) = A_{a\ n}^{\ b}(x, t(x,u)).$$

From (5.82) it follows that the gauge tensor (5.55) is

$$F_{a\ in}^{\ b} = A_{a\ n|i}^{\ b} - A_{a\ i|n}^{\ b} - A_{a\ i}^{\ c} A_{c\ n}^{\ b} + A_{a\ n}^{\ c} A_{c\ i}^{\ b} \tag{5.83}$$

or, taking into account (5.73),

$$\begin{aligned}F_{a\ in}^{\ b} = -t^j_a(U^b_{j|n|i} - U^b_{j|i|n}) - t^j_{a|i} U^b_{j|n} + t^j_{a|n} U^b_{j|i} - \\ - t^j_a U^c_{j|i} t^k_c U^b_{k|n} + t^j_a U^c_{j|n} t^k_c U^b_{k|i}.\end{aligned} \tag{5.84}$$

By virtue of the reciprocity relations (5.67) the second and fourth, as well as the third and fifth, terms cancel on the right-hand side of Equation (5.84), which leaves us with

$$F_{a\ in}^{\ b} = t^j_a(U^b_{j|i|n} - U^b_{j|n|i}). \tag{5.85}$$

The gauge tensor (5.85) may be represented in other suggestive forms. Indeed, using the commutation formula (1.79) gives

$$F_{a\ in}^{\ b} = t^j_a U^b_k(R^k_{\ jin} + l^r K_{r\ in}^{\ m} U^c_j t^k_c|_m). \tag{5.86}$$

Here,

$$t^k_c|_m \overset{\text{def}}{=} (t^k_c(x, u(x, y)))|_m = F(u^e_m \, \partial t^k_c/\partial u^e + C_j{}^k{}_m t^j_c)$$

or, because of (5.22),

$$t^k_c|_m = U^e_m t^k_d q^{*d}_c{}_e - l^k t^i_c g_{im} \tag{5.87}$$

where $q^{*d}_c{}_e$ denote the Christoffel symbols (5.17) of the indicatrix. Therefore, (5.86) may be written as

$$F_a{}^b{}_{in} = t^j_a U^b_k (R^k{}_{jin} + l^r K_r{}^m{}_{in} q^{*k}_j{}_m) \tag{5.88}$$

where $q^{*k}_j{}_m = t^k_d U^c_j U^e_m q^{*d}_c{}_e$.

Thus, we arrive at the following conclusion.

PROPOSITION 4. The gauge tensors N^c_{in} and $F_a{}^b{}_{in}$ may be expressed in terms of the Finslerian curvature tensors $K_r{}^j{}_{in}$ and $R_r{}^j{}_{in}$ in accordance with (5.81) and (5.88). The vanishing of the Finslerian curvature tensor $K_r{}^j{}_{in}$ implies the vanishing of the gauge tensors.

Formula (5.88) clearly indicates the origin of the non-tensorial term $u_a{}^d{}_c \tilde{N}^c_{in}$ in the transformation law (5.60) of $F_a{}^b{}_{in}$ under gauge transformations, the tensor part (5.64) of $F_a{}^b{}_{in}$ being

$$F_{(1)a}{}^b{}_{in} = t^j_a U^b_k R^k{}_{jin} \tag{5.89}$$

which is proportional to the Cartan curvature tensor $R^k{}_{jin}$ of Finsler space. Owing to (1.82), $F_{(1)a}{}^b{}_{in} = -F_{(1)}{}^b{}_{ain}$.

Using the relation (5.77) and the definition (5.73), it follows that, similarly to the Finslerian identity $l^j{}_{|i} = 0$,

$$\mathrm{D}_i t^j(x, u) \overset{\text{def}}{=} \partial t^j/\partial x^i + \Gamma_n{}^j{}_i t^n + N^b_i \, \partial t^j/\partial u^b = 0. \tag{5.90}$$

Furthermore, the reader can readily verify that

$$\mathrm{D}_i t^j_a \overset{\text{def}}{=} \partial t^j_a/\partial x^i + \Gamma_n{}^j{}_i t^n_a - A_a{}^b{}_i t^j_b + N^b_i \, \partial t^j_a/\partial u^b = 0, \tag{5.91}$$

and also

$$\mathrm{D}_i U^a_j = 0, \qquad \mathrm{D}_i b_{mn} = 0, \qquad \mathrm{D}_i b^{mn} = 0,$$

where b_{mn} is the parametrical metric tensor (5.2). The identity (5.91) can be obtained by differentiating (5.90) with respect to u^a.

So, we have set forth the idea of a relationship between Finsler geometry and the guage-field concept and had no difficulties in finding the explicit relations between the associated gauge tensors and Finslerian curvature tensors, the principal relations being given by Propositions 3 and 4. The following section, which is largely of technical character, extends the guage ideas of the present section to the case of fields possessing internal spin 1/2, including the case when there also exists an external spin.

5.4. Gauge-Covariant Derivatives of Spinors and Isospinors

Up to now, we have only considered fields which behave as vectors or tensors under gauge transformations. However, the extension of the action of the gauge-covariant operator D_i to the important case of a two-component spinor can be readily accomplished by using techniques that have been developed in general relativity for such spinors.

It will be remembered in this vespect that the theory of two-component spinors $v^{\beta}(x)$, $v^{\beta'}(x)$ in four-dimensional pseudo-Riemannian space-time is developed[4] in terms of the tetrads $h_i^P(x)$ of a Riemannian metric tensor $a_{ij}(x)$ of signature $(+\ -\ -\ -)$ and the *the Pauli matrices* $\left\| s_P{}^{\alpha\beta'} \right\|$:

$$\left\| s_1{}^{\alpha\beta'} \right\| = \frac{1}{\sqrt{2}} \left\| \begin{matrix} 1 & 0 \\ 0 & -1 \end{matrix} \right\|, \quad \left\| s_2{}^{\alpha\beta'} \right\| = \frac{1}{\sqrt{2}} \left\| \begin{matrix} 0 & 1 \\ 1 & 0 \end{matrix} \right\|,$$

$$\left\| s_3{}^{\alpha\beta'} \right\| = \frac{1}{\sqrt{2}} \left\| \begin{matrix} 0 & i \\ -i & 0 \end{matrix} \right\|, \quad \left\| s_0{}^{\alpha\beta'} \right\| = \frac{1}{\sqrt{2}} \left\| \begin{matrix} 1 & 0 \\ 0 & 1 \end{matrix} \right\| \equiv \frac{1}{\sqrt{2}} \left\| \delta^{\alpha\beta'} \right\| \tag{5.92}$$

which are *Hermitian*:

$$(s_P{}^{\alpha\beta'})^{\mathscr{C}} = s_P{}^{\beta\alpha'}. \tag{5.93}$$

The superscript $\mathscr{C}$ will denote *complex conjugation*. Greek letters will be used to denot spinor indices which range over 1, 2. From these entities the following *spin tensors* are constructed:

$$s_i{}^{\alpha\beta'}(x) = h_i^P(x) s_P{}^{\alpha\beta'}. \tag{5.94}$$

They have the important properties

$$s_i{}^{\alpha\beta'} s^j{}_{\alpha\beta'} = \delta_i^j, \tag{5.95}$$

$$s_{i\alpha\beta'} s^{i\mu\nu'} = \delta_\alpha^\mu \delta_{\beta'}^{\nu'}. \tag{5.96}$$

Tensor indices $i, j, \ldots$ are raised and lowered, as usual, by means of the metric tensor, whereas spinor indices are raised and lowered by means of the so-called *spinor metric tensor* $e_{\alpha\beta}$, $e_{\alpha'\beta'}$, which has the properties

$$e_{\alpha\beta} = -e_{\beta\alpha} = e_{\alpha'\beta'} = -e_{\beta'\alpha'}, \qquad e_{\alpha\beta}e^{\nu\beta} = \delta^{\nu}_{\alpha},$$
$$e_{\alpha'\beta'}e^{\nu'\beta'} = \delta^{\nu'}_{\alpha'}, \qquad e_{12} = -e_{21} = -e^{12} = e^{21} = 1. \tag{5.97}$$

For example,

$$s_{P\alpha\beta'} = e_{\alpha\mu}e_{\beta'\nu'}s_P{}^{\mu\nu'}.$$

The raising and lowering of the tetrad indices P, Q is performed by using the constants $q_P = q^P$ which represent the signature of the metric tensor: $q_0 = -q_1 = -q_2 = -q_3 = 1$, for example

$$s^{P\alpha\beta'} = q_P s_P{}^{\alpha\beta'}, \qquad h_{Pi} = q^P h^P_i.$$

To avoid any confusion it is noted that, because of the skew-symmetry of $e_{\alpha\beta}$, one should arrange correctly the order of the factors in products such as

$$Z^\beta Y_\beta = e_{\alpha\beta}Z^\alpha Y^\beta = -Z_\beta Y^\beta,$$

in particular

$$v_\beta L^\beta{}_{\alpha j} = -L_{\beta\alpha j}v^\beta, \qquad v_\alpha e^{\alpha\beta} = -e^{\beta\alpha}v_\alpha.$$

By definition, two-component spinors are objects which under Lorentzian rotations of tetrads:

$$h^i_P \to \tilde{h}^i_P = L^Q_P h^i_Q, \qquad \sum_{P=0}^{3} q_P h^i_P h^j_P = \sum_{Q=0}^{3} q_Q \tilde{h}^i_Q \tilde{h}^j_Q \tag{5.98}$$

transform by means of the matrices $\|s^\beta_\alpha\|$ from the group $SL(2, C)$:

$$v^\beta \to \tilde{v}^\beta = s^\beta_\alpha v^\alpha, \qquad v^{\beta'} \to \tilde{v}^{\beta'} = s^{\beta'}_{\alpha'}v^{\alpha'}, \tag{5.99}$$

where $\det(s^\beta_\alpha) = 1$ and $s^{\beta'}_{\alpha'} = (s^\beta_\alpha)^{\mathscr{C}}$. These transformations leave the spinor metric tensor and the Pauli matrices invariant:

$$e^{\alpha\beta} \to \tilde{e}^{\alpha\beta} \overset{\text{def}}{=} s^\alpha_\mu s^\beta_\nu e^{\mu\nu} = e^{\alpha\beta}\det(s^\mu_\nu) = e^{\alpha\beta},$$

$$s_P{}^{\alpha\beta'} \to \tilde{s}_P{}^{\alpha\beta'} \overset{\text{def}}{=} L^Q_P s^\alpha_\mu s^{\beta'}_{\nu'} s_Q{}^{\mu\nu'} = s_P{}^{\alpha\beta'},$$

while the definition

$$V_P = \sqrt{2}s_{P\alpha\beta'}v^\alpha v^{\mathscr{C}\beta'} \tag{5.100}$$

yields a Lorentzian vector:

$$V_P \to \tilde{V}_P = L^Q_P V_Q. \tag{5.101}$$

If we allow an arbitrary dependence of the transformations (5.98)–(5.99) on x^i:

$$L^Q_P = L^Q_P(x), \qquad s^\beta_\alpha = s^\beta_\alpha(x),$$

the covariant derivative of x-dependent spinors can be defined in the usual linear way:

$$\mathrm{D}_j v^\beta(x) = \partial v^\beta/\partial x^j + L^\beta{}_{\alpha j} v^\alpha,$$
$$\mathrm{D}_j v^{\beta'}(x) = \partial v^{\beta'}/\partial x^j + L^{\beta'}{}_{\alpha' j} v^{\alpha'}, \tag{5.102}$$

$$\mathrm{D}_j v_\beta(x) = \partial v_\beta/\partial x^j - L^\alpha{}_{\beta j} v_\alpha,$$
$$\mathrm{D}_j v_{\beta'}(x) = \partial v_{\beta'}/\partial x^j - L^{\alpha'}{}_{\beta' j} v_{\alpha'}, \tag{5.103}$$

$$\mathrm{D}_j e_{\alpha\beta} = -L^\nu{}_{\alpha j} e_{\nu\beta} - L^\nu{}_{\beta j} e_{\alpha\nu} = e_{\beta\nu} L^\nu{}_{\alpha j} - e_{\alpha\nu} L^\nu{}_{\beta j} =$$
$$= L_{\beta\alpha j} - L_{\alpha\beta j}, \tag{5.104}$$

$$\mathrm{D}_j s_i^{\alpha\beta'}(x) = s_i^{\alpha\beta'}{}_{;j} + L^\alpha{}_{\nu j} s_i^{\nu\beta'} + L^{\beta'}{}_{\nu' j} s_i^{\alpha\nu'}. \tag{5.105}$$

Here, $L^\alpha{}_{\beta j} = L^\alpha{}_{\beta j}(x)$ and $L^{\alpha'}{}_{\beta' j} = (L^\alpha{}_{\beta j})^{\mathscr{C}}$, so that

$$\mathrm{D}_j v^{\mathscr{C}\beta'} = (\mathrm{D}_j v^\beta)^{\mathscr{C}}.$$

The semicolon denotes the Riemannian covariant derivative:

$$s_i^{\alpha\beta'}{}_{;j} = h_i^P(x)_{;j} s_P{}^{\alpha\beta'}.$$

The spinor connection coefficients $L^\alpha{}_{\beta j}$ should have a transformation law of the form

$$s_\nu^\beta(x)\tilde{L}^\alpha{}_{\beta j}(x) = s_\mu^\alpha(x) L^\mu{}_{\nu j}(x) - \partial s_\nu^\alpha(x)/\partial x^j \tag{5.106}$$

in order to guarantee that the operator D_j maintains spinor transformation properties:

$$\mathrm{D}_j \tilde{v}^\beta(x) = s_\alpha^\beta(x) \mathrm{D}_j v^\alpha(x).$$

From Equation (5.104) it is seen that $\mathrm{D}_j e_{\alpha\beta} = 0$ if and only if the coefficients $L_{\alpha\beta j}$ are symmetric in the first two indices:

$$L_{\alpha\beta j} = L_{\beta\alpha j}. \tag{5.107}$$

We shall say that the spinor connection coefficients $L^\alpha{}_{\beta j}$ are *metric* if

$$\mathrm{D}_j e_{\alpha\beta} = 0 \tag{5.108}$$

together with

$$\mathrm{D}_j s_i^{\alpha\beta'} = 0. \tag{5.109}$$

In the metric case we have, according to the definition (5.105),

$$s^{i\alpha\beta'}{}_{;j} + L^\alpha{}_{\nu j} s^{i\nu\beta'} + L^{\beta'}{}_{\nu' j} s^{i\alpha\nu'} = 0.$$

Contracting this equation with $s_{i\mu\beta'}$ yields, after taking into account the identities (5.96),

$$s^{i\alpha\beta'}{}_{;j} s_{i\mu\beta'} + 2L^\alpha{}_{\mu j} + \delta_\mu^\alpha L^{\beta'}{}_{\beta' j} = 0.$$

But from Equation (5.107) it follows that $L^{\beta'}{}_{\beta' j} = e^{\alpha'\beta'} L_{\alpha'\beta' j} = 0$. Therefore,

$$L^{\alpha}{}_{\beta j} = -\tfrac{1}{2} s^{i\alpha\nu'}{}_{;j} s_{i\beta\nu'} \equiv -\tfrac{1}{2} s_{P\beta\nu'} s_Q{}^{\alpha\nu'} R^{PQ}{}_{j}, \tag{5.110}$$

which in turn implies that

$$L^{\alpha'}{}_{\beta' j} = -\tfrac{1}{2} s^{i\nu\alpha'}{}_{;j} s_{i\nu\beta'} \equiv -\tfrac{1}{2} s_{P\nu\beta'} s_Q{}^{\nu\alpha'} R^{PQ}{}_{j}. \tag{5.111}$$

Here, $R^{PQ}{}_j$ are *the Ricci rotation coefficients*

$$R^{PQ}{}_{j} = h^{Q}_{i\,;j} h^{Pi}, \tag{5.112}$$

where $h^P_i = h^P_i(x)$. Since $h^Q_i h^{Pi}$ are constants (equal to $q_P \delta^Q_P$), and $a_{mn;j} \equiv 0$, these coefficients are skew-symmetric:

$$R^{PQ}{}_{j} = -R^{QP}{}_{j}.$$

Thus the metric spinor connection coefficients are determined uniquely. They have the form (5.110)–(5.111) which is called *the connection coefficients of Infeld and van der Waerden*[5].

This well-known method can be applied in more general situations provided that the condition that the covariant derivative of the metric tensor vanishes is fulfilled. For example, in order to define the covariant derivative of a spinor $v^\beta(x, y)$, where x^i and y^i are regarded as independent variables, it is sufficient to use a Finslerian tetrad $h^P_i(x, y)$ in order to construct the (x, y)-dependent spin tensors

$$s_i{}^{\alpha\beta'}(x, y) = h^P_i(x, y) s_P{}^{\alpha\beta'} \tag{5.113}$$

and replace everywhere the Riemannian covariant derivative, which in the previous formulae was denoted by the semicolon $;j$, by the Cartan covariant derivative $|j$. The result is

$$\mathrm{D}_j v^\beta(x, y) = v^\beta(x, y)_{|j} + L^\beta{}_{\alpha j}(x, y) v^\alpha(x, y), \tag{5.114}$$

where

$$v^\beta{}_{|j} = \partial v^\beta/\partial x^j - G_j{}^n\, \partial v^\beta/\partial y^n$$

in accordance with the definition (1.44), and $L^\alpha{}_{\beta j}(x, y)$ are given by Equation (5.110) in terms of the Finslerian Ricci rotation coefficients

$$R^{PQ}{}_j(x, y) \stackrel{\text{def}}{=} h^Q_i(x, y)_{|j} h^{Pi}(x, y) = -R^{QP}{}_j(x, y). \tag{5.115}$$

These definitions are such that, under spinor transformations (5.98)–(5.99) with the coefficients depending on x^i and y^i in an arbitrary way:

$$L^Q_P = L^Q_P(x, y), \qquad s^\beta_\alpha = s^\beta_\alpha(x, y),$$

the coefficients $L^\alpha{}_{\beta j}(x, y)$ transform as

$$s^\beta_\nu(x, y) \tilde{L}^\alpha{}_{\beta j}(x, y) = s^\alpha_\mu(x, y) L^\mu{}_{\nu j}(x, y) - s^\alpha_\nu(x, y)_{|j},$$

and

$$D_j\tilde{v}^\beta(x,y) = s^\beta_\alpha(x,y)D_jv^\alpha(x,y).$$

Otherwise, all the above formulae remain valid, in particular

$$\begin{aligned} D_j s_i{}^{\alpha\beta'}(x,y) &\stackrel{\text{def}}{=} s_i{}^{\alpha\beta'}(x,y)_{|j} + L^\alpha{}_{\nu j}(x,y)s_i{}^{\gamma\beta'}(x,y) + \\ &\quad + L^{\beta'}{}_{\nu' j}(x,y)s_i{}^{\alpha\nu'}(x,y) = 0. \end{aligned} \tag{5.116}$$

It will be noted that, in the Finslerian case, there exists a simple preferred choice of Finslerian tetrad $h_i^P(x,y)$, namely one where $h_i^0(x,y) = l_i \stackrel{\text{def}}{=} y_i/F(x,y)$. Since $l_{i|j} \equiv 0$, we get $R^{OP}{}_j(x,y) = 0$ in accordance with the definition (5.115), so that the Finslerian spinor connection coefficients may be expressed in terms of only three Pauli matrices:

$$L^\alpha{}_{\beta j}(x,y) = -\tfrac{1}{2}s_{(a)\beta\nu'}s_{(b)}{}^{\alpha\nu'}R^{ab}{}_j(x,y) \quad \text{if} \quad h_i^0(x,y) = l_i \tag{5.117}$$

where we have used the notation

$$s_{(a)}{}^{\alpha\beta'} = \delta^P_a s_P{}^{\alpha\beta'} \tag{5.118}$$

$(a, b = 1, 2, 3)$. Under these conditions, the contraction of (5.116) with l^i yields $0 = l^iD_js_i{}^{\alpha\beta'}(x,y) = D_j(l^is_i{}^{\alpha\beta'}(x,y)) = D_js_0{}^{\alpha\beta'}$, that is

$$D_js_0{}^{\alpha\beta'} = 0 \quad \text{if} \quad h_i^0(x,y) = l_i. \tag{5.119}$$

Turning now to the case which is of interest to us here, that is when spinors depend on x^i and u^a, we shall assume that the coefficients of our transformations

$$g^i_P \to \tilde{g}^i_P = L^Q_P g^i_Q, \quad \sum_{P=0}^{3} q_P g^i_P g^j_P = \sum_{Q=0}^{3} q_Q \tilde{g}^i_Q \tilde{g}^j_Q, \tag{5.120}$$

(cf. Equation (5.98)) and (5.99) depend on x^i and u^a:

$$L^Q_P = L^Q_P(x,u), \qquad s^\beta_\alpha = s^\beta_\alpha(x,u). \tag{5.121}$$

$g^i_P = g^i_P(x,u)$ is the orthonormal tetrad (1.139). In this case we can say that we deal with *spinor gauge transformations*, following the terminology introduced in Section 5.2. Accordingly, we put

$$s_i{}^{\alpha\beta'}(x,u) \stackrel{\text{def}}{=} g_i^P(x,u)s_P{}^{\alpha\beta'} \tag{5.122}$$

where the tetrad g_i^P is defined by Equation (1.135). *The spinor gauge connection coefficients* of the Infeld and van der Waerden type will read

$$L^\alpha{}_{\beta j}(x,u) = -\tfrac{1}{2}s_{P\beta\nu'}s_Q{}^{\alpha\nu'}R^{PQ}{}_j(x,u) \tag{5.123}$$

where

$$R^{PQ}{}_j(x,u) = g^{Pi}D_jg_i^Q \equiv -R^{QP}{}_j(x,u). \tag{5.124}$$

The gauge-covariant derivative D_j is written in terms of the Cartan connection coefficients, so that

$$\mathrm{D}_j g^P_i = \partial g^P_i/\partial x^j + N^c_j \partial g^P_i/\partial u^c - \Gamma_i{}^k{}_j g^P_k.$$

The skew-symmetry of the Ricci rotation coefficients (5.124) in P, Q entails

$$L_{\alpha\beta j}(x,u) = L_{\beta\alpha j}(x,u). \tag{5.125}$$

The coefficients (5.123) will evidently have a transformation law of the form

$$s^\beta_\nu(x,u)\tilde{L}^\alpha{}_{\beta j}(x,u) = s^\alpha_\mu(x,u)L^\mu{}_{\nu j}(x,u) - D_j s^\alpha_\nu(x,u),$$

where

$$\mathrm{D}_j s^\alpha_\nu(x,u) = \partial s^\alpha_\nu(x,u)/\partial x^j + N^c_j(x,u)\,\partial s^\alpha_\nu(x,u)/\partial u^c,$$

which ensures that the definition

$$\begin{aligned}\mathrm{D}_j v^\beta(x,u) = \partial v^\beta(x,u)/\partial x^j + N^c_j(x,u)\,\partial v^\beta(x,u)/\partial u^c + \\ + L^\beta{}_{\alpha j}(x,u)v^\alpha(x,u)\end{aligned} \tag{5.126}$$

will yield a gauge-covariant tensor, i.e.

$$\mathrm{D}_j \tilde{v}^\beta(x,u) = s^\beta_\alpha(x,u)D_j v^\alpha(x,u). \tag{5.127}$$

Also, the spinor gauge connection coefficients (5.123) are metric, that is

$$\begin{aligned}\mathrm{D}_j s_i{}^{\alpha\beta'}(x,u) &\overset{\text{def}}{=} \partial s_i{}^{\alpha\beta'}(x,u)/\partial x^j + N^c_j(x,u)\,\partial s_i{}^{\alpha\beta'}(x,u)/\partial u^c - \\ &\quad - \Gamma_i{}^k{}_j(x,t(x,u))s_k{}^{\alpha\beta'}(x,u) + L^\alpha{}_{\nu j}(x,u)s_i{}^{\nu\beta'}(x,u) + \\ &\quad + L^{\beta'}{}_{\nu' j}(x,u)s_i{}^{\alpha\nu'}(x,u) \equiv \\ &\equiv s_P{}^{\alpha\beta'}\,\mathrm{D}_j g^P_i(x,u) + L^\alpha{}_{\nu j}(x,u)s_i{}^{\nu\beta'}(x,u) + \\ &\quad + L^{\beta'}{}_{\nu' j}(x,u)s_i{}^{\alpha\nu'}(x,u) = 0\end{aligned} \tag{5.128}$$

and, as a consequence of (5.125),

$$\mathrm{D}_j \mathrm{e}_{\alpha\beta} = L_{\beta\alpha j}(x,u) - L_{\alpha\beta j}(x,u) = 0. \tag{5.129}$$

In particular, if the tetrad g^i_P is such that $g^i_0 = l^i \equiv t^i(x,u)$, the Ricci rotation coefficients (5.124) will reduce to $R^{0Q}{}_j(x,u)$ and

$$\begin{aligned}R^{ab}{}_j(x,u) &= g^{*(a)c}D_j g^{*(b)}_c \equiv \\ &\equiv g^{*(a)c}(\partial g^{*(b)}_c/\partial x^j - A_c{}^d{}_j g^{*(b)}_d + N^d_j\,\partial g^{*(b)}_c/\partial u^d),\end{aligned}$$

as a consequence of the metric relations (5.90)–(5.91) and the reciprocity relation (5.67); $g^{*(a)}_c$ are orthonormal triads of the indicatrix metric tensor (see Equations (1.133)–(1.140)). Therefore,

$$L^\alpha{}_{\beta j}(x,u) = -\tfrac{1}{2}s_{(a)\beta\nu'}s_{(b)}{}^{\alpha\nu'}R^{ab}{}_j(x,u) \quad \text{if} \quad g^i_0 = l^i \tag{5.130}$$

and

$$\mathrm{D}_j s_0{}^{\alpha\beta'} \overset{\text{def}}{=} L^\alpha{}_{\nu j}s_0{}^{\nu\beta'} + L^{\beta'}{}_{\nu' j} = 0 \quad \text{if} \quad g^i_0 = l^i. \tag{5.131}$$

Assuming $g^i_0 = l^i$, we can represent the spin tensors (5.122) as follows

$$s_i^{\alpha\beta'}(x,u) = U^a_i(x,u)s_a^{\alpha\beta'}(x,u) + t_i(x,u)s_0^{\alpha\beta'} \quad \text{if} \quad g^i_0 = l^i \tag{5.132}$$

where $t_i = b_{ij}t^j \equiv l_i(x,u)$ and

$$s_a^{\alpha\beta'}(x,u) \overset{\text{def}}{=} g^{*(b)}_a(x,u)s_{(b)}^{\alpha\beta'} = t^i_a(x,u)s_i^{\alpha\beta'}(x,u). \tag{5.133}$$

Inserting (5.132) in (5.128) and taking into account Equations (5.90)–(5.91), we find that the tensors (5.133) are such that

$$\begin{aligned} \mathrm{D}_j s_a^{\alpha\beta'}(x,u) \overset{\text{def}}{=} {} & \partial s_a^{\alpha\beta'}(x,u)/\partial x^j - A_a{}^b{}_j s_b^{\alpha\beta'}(x,u) + \\ & + N^c_j \partial s_a^{\alpha\beta'}(x,u)/\partial u^c + L^\alpha{}_{\nu j}(x,u)s_a^{\nu\beta'}(x,u) + \\ & + L^{\beta'}{}_{\nu' j}(x,u)s_a^{\alpha\nu'} = 0 \quad \text{if} \quad g^i_0 = l^i. \end{aligned} \tag{5.134}$$

Thus we have obtained the following result.

PROPOSITION 5. The spinor gauge connection coefficients (5.123) are metric, that is the identities (5.128) and (5.129) hold. In the case when the tetrads $g^i_p(x,u)$ of the parametrical metric tensor (5.2) are chosen such that $g^i_0 = l^i$, the connection coefficients reduce to (5.130) and two additional identities (5.131) and (5.134) hold.

Let us consider now the case of mixed tensors or spinors. As an example, suppose we are given a quantity $w^{i\beta}(x,u)$ where i is a vector index under general coordinate transformations $x^n = x^n(\bar{x}^m)$ and β is a spinor index under spinor gauge transformations (5.120)–(5.121). Then, choosing any quantities $D_k{}^i{}_j$ which are connection coefficients under the first type of transformation (e.g. $D_k{}^i{}_j$ may be the Cartan connection coefficients $\Gamma_k{}^i{}_j$, or the Christoffel symbols associated with the osculating Riemannian metric tensor), we can define the gauge-covariant derivative as follows

$$\mathrm{D}_j w^{i\beta} = \partial w^{i\beta}/\partial x^j + D_k{}^i{}_j w^{k\beta} + L^\beta{}_{\alpha j} w^{i\alpha} + N^c_j \partial w^{i\beta}/\partial u^c \tag{5.134'}$$

(cf. the definition (5.52)), so that a relation of the form (5.127) will hold with respect to the spinor index β.

Finally, if $\Phi^\beta(x,u)$ is a two-component spinor under spinor gauge transformations and, for each value of the index β, is a four-component space-time spinor with respect to a Riemannian metric tensor $a_{ij}(x)$ of signature $(+ - - -)$, then we can define

$$\begin{aligned} \mathrm{D}_j \Phi^\beta &= \partial \Phi^\beta/\partial x^j - A_j(x)\Phi^\beta + L^\beta{}_{\alpha j}(x,u)\Phi^\alpha + N^c_j(x,u)\partial\Phi^\beta/\partial u^c, \\ \mathrm{D}_j \bar{\Phi}^{\beta'} &= \partial\bar{\Phi}^{\beta'}/\partial x^j + \bar{\Phi}^{\beta'} A_j(x) + L^{\beta'}{}_{\alpha' j}(x,u)\bar{\Phi}^{\alpha'} + N^c_j(x,u)\partial\bar{\Phi}^{\beta'}/\partial u^c \end{aligned} \tag{5.135}$$

where A_j are the following 4×4 matrices

$$A_j(x) = -\tfrac{1}{8}R^{PQ}{}_j(x)(\gamma_P\gamma_Q - \gamma_Q\gamma_P). \tag{5.136}$$

Here, γ_P denotes *the Dirac matrices* which have the property that

$$\gamma_P \gamma_Q + \gamma_Q \gamma_P = 2a_{PQ} \tag{5.137}$$

where $||a_{PQ}||$ is the diagonal matrix with $a_{00} = -a_{11} = -a_{22} = -a_{33} = 1$. The bar over the spinor stands for the usual Dirac conjugation: $\bar{\Phi}^{\beta'} = (\Phi^\beta)^{\mathscr{H}} \gamma_0$, where the superscript $\mathscr{H}$ means the operation of Hermitian conjugation; $R^{PQ}{}_j(x)$ are the Riemannian Ricci rotation coefficients (5.112). The reason for choosing this A_j is that the covariant derivative of a four-component space-time spinor $\Phi(x)$, $\bar{\Phi}(x) = \Phi^{\mathscr{H}} \gamma_0$ can be defined as

$$\mathrm{D}_j \Phi = \partial \Phi / \partial x^j - A_j \Phi, \qquad \mathrm{D}_j \bar{\Phi} = \partial \bar{\Phi} / \partial x^j + \bar{\Phi} A_j. \tag{5.138}$$

Such a definition appears in the standard gauge approaches[6]. The identity

$$\gamma_R(\gamma_P \gamma_Q - \gamma_Q \gamma_P) - (\gamma_P \gamma_Q - \gamma_Q \gamma_P)\gamma_R = 4a_{RP}\gamma_Q - 4a_{RQ}\gamma_P,$$

which follows from (5.137), implies that

$$\mathrm{D}_j \gamma^i \overset{\text{def}}{=} \gamma^i_{;j} - A_j \gamma^i + \gamma^i A_j = 0 \tag{5.139}$$

where

$$\gamma^j = a^{jn}(x) h^P_n(x) \gamma_P.$$

Hence

$$\mathrm{D}_j(\bar{\Phi}\Phi) = \partial(\bar{\Phi}\Phi)/\partial x^j, \qquad \mathrm{D}_j(\bar{\Phi}\gamma^n \Phi) = (\mathrm{D}_j \bar{\Phi})\gamma^n \Phi + \bar{\Phi}\gamma^n \mathrm{D}_j \Phi. \tag{5.140}$$

The reader can readily verify that postulating Equations (5.138)–(5.140) leads unambiguously to the coefficients (5.136).

By construction, the derivative (5.135) will transform as a four-component spinor for any value of the index $\beta = 1, 2$. At the same time, spinor gauge transformations (which are given by the coefficients (5.121)) will act on $\mathrm{D}_j \Phi^\beta$ in accordance with the law (5.127). The corresponding simplest generalization of the Dirac equation will read

$$\begin{aligned} &-i\gamma^j(x) \mathrm{D}_j \Phi^\beta(x, u) + m\Phi^\beta(x, u) = 0, \\ &i(\mathrm{D}_j \bar{\Phi}^{\beta'}(x, u))\gamma^j(x) + m\bar{\Phi}^{\beta'}(x, u) = 0 \end{aligned} \tag{5.141}$$

where m is a real constant.

Further, it will be noted that for any two spinors v^β, u^β it is possible to define a spinor inner product

$$v_\beta u^\beta = e_{\alpha\beta} u^\alpha v^\beta$$

which does not vanish in general. However, the spinor length of any two-component spinor vanishes identically: $v_\beta v^\beta = 0$, because of $e_{\alpha\beta} = -e_{\beta\alpha}$. Therefore, from a spinor v^β, it is impossible to construct a nonvanishing bilinear form which is invariant under general $\mathrm{SL}(2, C)$-transformations – a fact which precludes the construction of a quadratic Lagrangian density for the purpose of formulating equations for (x, u)-dependent spinor fields. In particular, the product $e_{\alpha\beta} v^{\alpha} v^{\mathscr{C}\beta'}$, though nonvanishing in general, is not an invariant, because the transformation law of the complex-conjugated spinor $v^{\mathscr{C}\beta'}$ differs from that of v^{β}.

Also, the form

$$v^{\mathscr{C}} v \stackrel{\text{def}}{=} v^{\mathscr{C}1'} v^1 + v^{\mathscr{C}2'} v^2 \tag{5.142}$$

will not be invariant under SL(2, C). However, we may restrict ourselves to a subgroup of SL(2, C) such that this form will be invariant. Indeed, if $\tilde{v}^{\alpha} = U^{\alpha}_{\beta} v^{\beta}$, then

$$\tilde{v}^{\mathscr{C}} \tilde{v} = \sum_{\beta} v^{\mathscr{C}\alpha'} U^{\beta'}_{\alpha'} U^{\beta}_{\nu} v^{\nu} = \sum_{\beta,\alpha} v^{\mathscr{C}\alpha'} U^{\mathscr{T}\alpha'}_{\beta'} U^{\beta}_{\nu} v^{\nu}$$

where $U^{\alpha'}_{\beta'} = (U^{\alpha}_{\beta})^{\mathscr{C}}$ and the symbol $\mathscr{T}$ denotes the operation of transposition. So, $\tilde{v}^{\mathscr{C}} \tilde{v} = v^{\mathscr{C}} v$ if $\Sigma_{\beta} U^{\mathscr{T}\alpha'}_{\beta'} U^{\beta}_{\nu} = \gamma^{\alpha'}_{\nu}$, that is when the matrix $\| U^{\alpha}_{\beta} \|$ is *unitary*. Also, since these matrices are taken from SL(2, C), we must have $\det(U^{\alpha}_{\beta}) = 1$. Thus, the form (5.142) will remain invariant under the transformations which belong to the subgroup SU(2) of SL(2, C).

Taking into account the fact that the matrix $\| s_0{}^{\alpha\beta'} \|$ in (5.92) is proportional to the unit matrix $\| \delta^{\alpha\beta'} \|$, we can conclude that this matrix remains invariant under SU(2)-transformations:

$$\tilde{s}_0{}^{\alpha\beta'} \stackrel{\text{def}}{=} U^{\alpha}_{\nu} U^{\beta'}_{\mu'} s_0{}^{\nu\mu'} = s_0{}^{\alpha\beta'}. \tag{5.143}$$

This circumstance suggests that we should introduce the following *operation of conjugation* of two-component spinors:

$$\hat{v}_{\alpha} = t_{0\alpha\beta'} v^{\mathscr{C}\beta'}, \qquad v^{\mathscr{C}\beta'} = \hat{v}_{\alpha} t_0{}^{\alpha\beta'}, \tag{5.144}$$

where

$$t_{0\alpha\beta'} \stackrel{\text{def}}{=} \sqrt{2}\, s_{0\alpha\beta'} = \delta_{\alpha\beta'}, \qquad t_0{}^{\alpha\beta'} = \delta^{\alpha\beta'}, \tag{5.145}$$

in terms of which the quadratic form (5.142) is written as just

$$\hat{v} v \stackrel{\text{def}}{=} \hat{v}_{\beta} v^{\beta} = v^{\mathscr{C}1'} v^1 + v^{\mathscr{C}2'} v^2, \tag{5.146}$$

and for two different spinors,

$$\hat{u} v \stackrel{\text{def}}{=} \hat{u}_{\beta} v^{\beta} = u^{\mathscr{C}1'} v^1 + u^{\mathscr{C}2'} v^2 = (\hat{v} u)^{\mathscr{C}}. \tag{5.147}$$

Owing to the definition (5.145) and the identity (5.131), we have

$$\mathrm{D}_j t_0{}^{\alpha\beta'} = L^{\alpha}{}_{\nu j}(x, u) \delta^{\nu\beta'} + L^{\beta'}{}_{\mu' j}(x, u) \delta^{\alpha\mu'} = 0, \tag{5.148}$$

if $g^i_0 = l^i$.

The geometrical meaning of the subgroup SU(2) follows immediately from the property of invariance (5.143). Indeed, proper three-dimensional rotations must correspond to the elements of SL(2, C) such that $\tilde{V}_0 = V_0$ in Equation (5.101). However, according to Equations (5.100) and (5.145), this implies the invariance of $V_0 = t_{0\alpha\beta'} v^{\alpha} v^{\mathscr{C}\beta'}$, that is of the quadratic form (5.146). Thus, SU(2) is that subgroup of SL$(2, C)$ which corresponds to the subgroup of three-dimensional rotations of the Lorentz group. We shall call the two-component spinors which transform under SU(2)-transformations *isospinors*, bearing in mind that it is objects such as these that are elements of isotopic space as the latter concept is understood in the physics of elementary particles (see Sections 5.5 and 5.7). According to this definition, isospinor transformations are accompanied by rotations of triads g^i_a only, leaving the zero-vector g^i_0 of a tetrad g^i_P unchanged. In particular, *when considering* (x, u)*-dependent isospinors, we may simultaneously stipulate that* $g^i_0 = l^i$, *which makes it possible to exploit the reduced expressions* (5.130), (5.131), *and* (5.134).

Taking advantage of the property of invariance (5.143), we can avoid the use of primed indices in manipulations with isospinors if we put

$$t_{(a)\alpha}{}^{\nu} = \sqrt{2}\, s_{(a)\alpha\beta'}\, t_0{}^{\nu\beta'}. \tag{5.149}$$

In terms of this notation the connection coefficients (5.130) may be written as

$$L^{\alpha}{}_{\beta j}(x, u) = -\tfrac{1}{4} t_{(a)\beta\nu}\, t_{(b)}{}^{\alpha\nu} R^{ab}{}_{j}(x, u). \tag{5.150}$$

The covariant derivative of the conjugated spinors $\hat{v}^{\beta}$ will have the same form as that of the spinors v^{β} (Equation (5.126)). The object

$$V_a = t_{(a)\alpha\beta}\, v^{\alpha} \hat{v}^{\beta}, \tag{5.151}$$

which is the definition (5.100) when $P \doteq a$ written in terms of the notation (5.144) and (5.149), is a three-dimensional vector:

$$V_a \rightarrow \tilde{V}_a = R^b_a V_b, \quad \sum_{a=1}^{3} (V_a)^2 = \sum_{a=1}^{3} (\tilde{V}_a)^2$$

(cf. Equation (5.101)) under the SU(2)-restriction of the transformations (5.98)–(5.99). Because of Equations (5.134), (5.148) and (5.145)

$$\mathrm{D}_j t_a{}^{\alpha\beta} = 0 \tag{5.152}$$

where

$$t_a{}^{\alpha\beta} = t_{(b)}{}^{\alpha\beta} g^{*(b)}_{\ a}(x, u)$$

in accordance with the definition (5.133).

The simplest equation for an isospinor $v^{\beta}(x, u)$ may be assumed, by analogy with the Dirac equation, to have the form

$$i\sqrt{2}\, s^j{}_{\alpha\beta'}(x, u) \mathrm{D}_j v^{\alpha}(x, u) + q t_{0\alpha\beta'} v^{\alpha}(x, u) = 0 \tag{5.153}$$

or, in terms of the notation (5.149),

$$i t^j{}_{\alpha}{}^{\nu}(x, u) \mathrm{D}_j v^{\alpha}(x, u) + q v^{\nu}(x, u) = 0, \tag{5.154}$$

where q is a constant, and

$$s^j{}_{\alpha\beta'}(x, u) = g^*{}^j_{(a)}(x, u) s^{(a)}{}_{\alpha\beta'},$$
$$t^j{}_\alpha{}^\nu(x, u) = t^j_a(x, u) t^a{}_\alpha{}^\nu(x, u) \equiv g^*{}^j_{(a)}(x, u) t^{(a)}{}_\alpha{}^\nu. \tag{5.155}$$

Taking the complex conjugate of the left-hand side of (5.153), we get, because of (5.93),

$$-i\sqrt{2}\, s^j{}_{\beta\alpha'} \mathrm{D}_j v^{\mathscr{C}\alpha'} + q^{\mathscr{C}} t_{0\beta\alpha'} v^{\mathscr{C}\alpha'},$$

so that

$$t_{0\beta\nu'}(it^j{}_\alpha{}^\nu \mathrm{D}_j v^\alpha + qv^\nu)^{\mathscr{C}} = -it^j{}_\beta{}^\mu \mathrm{D}_j \hat{v}_\mu + q^{\mathscr{C}} \hat{v}_\beta \tag{5.156}$$

where we have used the definitions (5.144) and (5.145) together with Equation (5.148). The relation (5.156) shows that equation (5.154) implies

$$-it^j{}_\alpha{}^\nu \mathrm{D}_j \hat{v}_\nu + q^{\mathscr{C}} \hat{v}_\alpha = 0 \tag{5.157}$$

which may be regarded as the conjugated form of (5.153).

The constant q in equation (5.154) *must be complex*. In fact, contracting (5.154) with $\hat{v}_\nu$ yields the equation

$$i\hat{v}_\nu t^j{}_\alpha{}^\nu \mathrm{D}_j v^\alpha + q\hat{v}_\nu v^\nu = 0, \tag{5.158}$$

and contracting (5.157) with v^α yields the equation

$$-iv^\alpha t^j{}_\alpha{}^\nu \mathrm{D}_j \hat{v}_\nu + q^{\mathscr{C}} v^\alpha \hat{v}_\alpha = 0. \tag{5.159}$$

However,

$$(\hat{v}_\nu t^j{}_\alpha{}^\nu \mathrm{D}_j v^\alpha)^{\mathscr{C}} = v^\mu t^j{}_\mu{}^\nu D_j \hat{v}_\nu, \tag{5.160}$$

so that subtraction of (5.159) from (5.158) yields, after multiplying by $-i$,

$$\hat{v}_\nu t^j{}_\alpha{}^\nu \mathrm{D}_j v^\alpha + (\hat{v}_\nu t^j{}_\alpha{}^\nu \mathrm{D}_j v^\alpha)^{\mathscr{C}} - i(q - q^{\mathscr{C}})\hat{v}_\nu v^\nu = 0.$$

Here, the sum of the first two terms is real, while $\hat{v}_\nu v^\nu$ is also real because of the definition (5.146). Therefore, $i(q - q^{\mathscr{C}})$ must be a real constant, which is possible if and only if q has an imaginary part.

As in the case of Dirac spinors, we may define a *spinor current*

$$J^i \overset{\text{def}}{=} \hat{v}_\nu t^i{}_\alpha{}^\nu v^\alpha = (J^i)^{\mathscr{C}}. \tag{5.161}$$

If the spinors satisfy equations (5.154)–(5.157), then

$$\mathrm{D}_i J^i = \hat{v}_\nu v^\alpha \mathrm{D}_i t^i{}_\alpha{}^\nu.$$

However, because of the definition (5.155) and the identities (5.91) and (5.152),

$$\mathrm{D}_i t^j{}_\alpha{}^\nu = 0. \tag{5.162}$$

Therefore, the current (5.161) will be a conserved quantity:

$$\mathrm{D}_i J^i = 0$$

whenever equations (5.154)–(5.157) are satisfied.

Finally, following the technique described above, the gauge-covariant derivative D_a of a two-component spinor $v^\alpha(x, u)$ in the indicatrix can be defined as

$$D_a v^\alpha(x, u) = \partial v^\alpha(x, u)/\partial u^a + L^\alpha{}_{\beta a}(x, u) v^\beta(x, u), \tag{5.163}$$

and the same for the spinors $\Phi^\beta(x, u)$ entering Equations (5.135) and (5.141):

$$D_a \Phi^\alpha(x, u) = \partial \Phi^\alpha/\partial u^a + L^\alpha{}_{\beta a} \Phi^\beta. \tag{5.164}$$

Here,

$$L^\alpha{}_{\beta a}(x, u) = \tfrac{1}{2} s_{(c)}{}^{\alpha\gamma'} s_{(d)\beta\gamma'} R^{(c)(d)}{}_a(x, u) \tag{5.165}$$

are the spinor connection coefficients of the Infeld and van der Waerden type, and

$$R^{(c)(d)}{}_a = g^{*(c)b}(\partial g^{*(d)}_{\;b}/\partial u^a - q^{*e}_{b\;a} g^{*(d)}_{\;e}). \tag{5.166}$$

Here, $g^{*(c)}_{\;a}(x, u)$ is an orthonormal triad of the metric tensor of the indicatrix (see Equations (1.133) and (1.137)) and $q^{*e}_{b\;a}(x, u)$ are its Christoffel symbols (Equation (5.17)), so that (5.166) are the Ricci rotation coefficients of the indicatrix. Evidently,

$$D_a e_{\alpha\beta} = 0,$$

that is

$$L_{\alpha\beta a} = L_{\beta\alpha a}.$$

Also, the following proposition is valid.

PROPOSITION 6. *The spinor connection coefficients* (5.165) *are metric, that is*

$$\begin{aligned} D_b s_a{}^{\alpha\beta'}(x, u) \overset{\text{def}}{=} \partial s_a{}^{\alpha\beta'}(x, u)/\partial u^b - q^{*c}_{a\;b}(x, u) s_c{}^{\alpha\beta'}(x, u) + \\ + L^\alpha{}_{\nu b}(x, u) s_a{}^{\nu\beta'}(x, u) + L^{\beta'}{}_{\nu' b}(x, u) s_a{}^{\alpha\nu'}(x, u) = 0 \end{aligned} \tag{5.167}$$

where the spin tensors $s_a{}^{\alpha\beta'}(x, u)$ *are defined, as before, by* (5.133). *Also,*

$$D_b s_0{}^{\alpha\beta'} \overset{\text{def}}{=} L^\alpha{}_{\nu b}(x, u) s_0{}^{\nu\beta'} + L^{\beta'}{}_{\nu' b}(x, u) s_0{}^{\alpha\nu'} = 0. \tag{5.168}$$

Proof. Let us construct the spinor connection coefficients $L^\alpha{}_{\beta N}(x, z)$ of the Infeld and van der Waerden type in a four-dimensional Riemannian space having a Minkowskian coordinate system $z^P = (z^0, z^a = u^a)$ (Equation (1.95)) and a Riemannian metric tensor $g^*_{PQ}(x, z)$ which is such that $g^*_{00} = 1$, $g^*_{0a} = 0$, and $g^*_{ab}(x, z) = g^*_{ab}(x, u)$ (see Equations (1.111)–(1.113)). The capital Latin indices P, Q, R, M, N will range over 0, 1, 2, 3, and x^i will refer to any fixed point of the underlying manifold. Define

$$L^\alpha{}_{\beta N}(x, z) \overset{\text{def}}{=} \tfrac{1}{2} s_P{}^{\alpha\gamma'} s_{Q\beta\gamma'} R^{PQ}{}_N(x, z) \tag{5.169}$$

with

$$R^{PQ}{}_N = g^{*(P)M}(\partial g^{*(Q)}_{M}/\partial z^N - q^{*R}_{M\ N} g^{*(Q)}_{R}).$$

Here, $g^{*(P)}_{M}$ is an orthonormal tetrad of g^*_{PQ} which, because of the specific form of g^*_{PQ}, may be chosen so that

$$g^{*(0)}_{M} = \delta^0_M, \qquad g^{*(a)}_{0} = 0, \qquad g^{*(a)}_{b}(x, z) = g^{*(a)}_{b}(x, u).$$

Also, the Christoffel symbols $q^{*R}_{M\ N}(x, z)$ associated with g^*_{PQ} satisfy

$$q^{*0}_{M\ N} = q^{*R}_{0\ N} = q^{*R}_{M\ 0} = 0, \qquad q^{*b}_{a\ c}(x, z) = q^{*b}_{a\ c}(x, u).$$

These relations result in

$$R^{0Q}{}_N = R^{P0}{}_N = R^{PQ}{}_0 = 0, \qquad R^{ab}{}_c(x, z) = R^{(a)(b)}{}_c(x, u).$$

Hence, (5.165) is that part of the connection coefficients (5.169) which does not vanish identically. Denoting $c_P{}^{\alpha\beta'}(x, z) = g^{*(Q)}_{P}(x, z)s_Q{}^{\alpha\beta'}$, so that $c_a{}^{\alpha\beta'} = s_a{}^{\alpha\beta'}(x, u)$, whereas $c_0{}^{\alpha\beta'} = s_0{}^{\alpha\beta'}$ are constants, we have

$$\mathrm{D}_M c_P{}^{\alpha\beta'} \overset{\text{def}}{=} \partial c_P{}^{\alpha\beta'}/\partial z^M - q^{*Q}_{P\ M} c_Q{}^{\alpha\beta'} + L^\alpha{}_{\nu M} c_P{}^{\nu\beta'} + L^{\beta'}{}_{\nu' M} c_P{}^{\alpha\nu'} = 0.$$

For $M = 0$, each term of this Equation vanishes identically. Putting $M = a$ and $P = 0$ yields (5.168). For $P = a$ and $M = b$, we get (5.167). The proof is complete.

According to Equations (5.119), (5.134), (5.167) and (5.168), the spin tensors $s_a{}^{\alpha\beta'}(x, u)$, as well as the objects $t_a{}^{\alpha\beta}(x, u)$ (see Equation (5.152)), may be taken outside both operators D_j and D_a.

5.5. Linear Gauge Transformations. Finslerian Geometrization of Isotopic Invariance

In the preceding sections 5.2–5.4 the gauge transformations (5.36) were allowed to depend on $\tilde{u}^a$ in any nonlinear way. In the present section, we confine ourselves to *linear gauge transformations*

$$u^a = u^a_b(x)\tilde{u}^b \tag{5.170}$$

whick makes it possible to stipulate, without contradicting the transformation laws (5.38) and (5.39), that the above scalar-vector field w_a is independent of u^a:

$$w_a = w_a(x). \tag{5.171}$$

In such a case we should, firstly, omit the third terms on the right-hand sides of the derivatives (5.42) and (5.43) and, secondly, require that the gauge field $A_a{}^b{}_i$ is independent of u^a:

$$A_a{}^b{}_i = A_a{}^b{}_i(x). \tag{5.172}$$

Under these conditions, the gauge tensor (5.55) reduces to

$$F_{a\ in}^{\ b}(x) = \partial A_{a\ n}^{\ b}/\partial x^i - \partial A_{a\ i}^{\ b}/\partial x^n - A_{a\ i}^{\ c} A_{c\ n}^{\ b} + A_{a\ n}^{\ c} A_{c\ i}^{\ b} \tag{5.173}$$

and the transformation law (5.60) of $F_{a\ in}^{\ b}$ takes the exact tensor form

$$u_b^d \tilde{F}_{a\ in}^{\ b} = u_a^c F_{c\ in}^{\ d}; \tag{5.174}$$

the second gauge field N_i^c entirely disappears in our subsequent treatment.

In the recent gauge approach in physics gauge transformations are usually taken to be linear so that it is relations of the form (5.170)–(5.174) that are exploited[7]. An important example of such a theory is the Yang–Mills theory describing the isotopic invariance of physical fields by means of gauge fields. Isotopic invariance means the invariance of physical fields under rotations in an internal three-dimensional Euclidean space, the isotopic space[8]. Such an invariance may be incorporated in the assumed Finslerian metric structure of space-time; to do so[9] we have only to require that the indicatrix be a Euclidean space, which implies that it possesses a certain coordinate system $\{u^a\}$ relative to which the Christoffel symbols (5.17) of the indicatrix vanish identically:

$$q_{a\ c}^{*b} = 0. \tag{5.175}$$

The vanishing of the curvature of the indicatrix may also be formulated by saying that the Finsler space is $S3$-like with $S^* = -1$ (see Section 2.1 and also Proposition 3 in that Section; a striking example of a metric function with this property is the Berwald-Moór metric function described in detail in Section 2.3). By virtue of Equation (5.175), the relation (5.87) reduces to

$$t_c^k|_m = -l^k t_c^i g_{im} \tag{5.176}$$

which implies in turn that

$$U_k^b|_m = -l_k U_m^b \tag{5.177}$$

as a consequence of (5.67), (1.68) and $U_k^c l^k = 0$.

In the case when (5.175) is valid, the condition (5.172) may be reformulated in terms of the characteristic tensors of Finsler space. Indeed, from Equation (5.73) we have

$$U_{i|k}^b = -A_{a\ k}^{\ b} U_i^a. \tag{5.178}$$

Then, (5.177) and (5.178) together with $l_{j|k} = 0$ permit us to obtain directly the following commutation formula

$$U_i^b|_{h|k} - U_{i|k}^b|_h = -l_i U_{h|k}^b + A_{a\ k}^{\ b} U_i^a|_h + A_{a\ k}^{\ b}|_h U_i^a.$$

Here, the first and second terms on the right-hand side cancel because of (5.177) and (5.178), which leaves us with

$$U_i^b|_{h|k} - U_{i|k}^b|_h = A_{a\ k}^{\ b}|_h U_i^a \equiv F(\partial A_{a\ k}^{\ b}/\partial y^h - C_{k\ h}^{\ m} A_{a\ m}^{\ b}) U_i^a. \tag{5.179}$$

Comparison of (5.179) with the commutation rule (1.73) shows that

$$F\,\partial A_{a\ k}^{\ b}/\partial y^h = t_a^j U_i^b P_{j\ kh}^{\ i} \tag{5.180}$$

(we have used Equation (1.75)) or, equivalently,

$$\partial A_{a\ k}^{\ b}/\partial u^c = t_a^j U_i^b t_c^h P_{j\ kh}^{\ i}. \tag{5.181}$$

As a result of (5.67) and $l^j l_i P_{j\ kh}^{\ i} \equiv 0$, together with

$$l^j P_{jikh} = -l^j P_{ijkh} = C_{ikh|r}\,y^r$$

(see Equation (1.74)), the converse of (5.180) reads

$$F U_m^a t_b^n\,\partial A_{a\ k}^{\ b}/\partial y^h = P_{m\ kh}^{\ n} - l_m C^n{}_{kh|r}\,y^r + l^n C_{mkh|r}\,y^r. \tag{5.182}$$

Formulae (5.180)–(5.182) show that the vanishing of $\partial A_{a\ k}^{\ b}/\partial u^c$ is equivalent to the vanishing of the right-hand side of Equation (5.182).

So, we have established the following:

PROPOSITION 7. Let the indicatrices of a Finsler space be spaces of vanishing curvature and the parameters u^a used as the coordinates of points of the indicatrices be chosen such that the Christoffel symbols $q_{a\ c}^{*b}$ of the indicatrices vanish. Then the derivative of the gauge field $A_{a\ k}^{\ b}$ with respect to u^c is given by Equation (5.181), and the gauge field $A_{a\ k}^{\ b}$ is independent of u^c if and only if the Finsler space possesses the property

$$P_{mnkh} - l_m C_{nkh|r}\,y^r + l_n C_{mkh|r}\,y^r = 0. \tag{5.183}$$

Since $C_{ijk|m} = 0$ implies $C_{ijk|r}\,y^r = 0$ together with $P_{mnkh} = 0$ (see the general expression (1.74) of the tensor P_{mnkh}), condition (5.183) will certainly be satisfied in affinely-connected Finsler spaces, i.e. when $C_{ijk|m} = 0$ (see Equation (1.53)). Therefore, Proposition 7, together with Proposition 18 from Section 2.4, enables us to draw the conclusion that in a conformally flat 1-form Berwald-Moór Finsler space the gauge field $A_{a\ i}^{\ b}$ is independent of u^a. To calculate $U_{i|h}^a$ explicitly in such a space, it is convenient to take the Cartan covariant derivative of U_i^a in the form (1.70) and substitute (5.177) and

$$\partial_h U_i^a = S_i^A\,\partial_h U_A^a + U_A^a\,\partial_h S_i^A,$$

where $U_A^a(x, y) = S_A^i(x) U_i^a(x, y)$. This procedure yields:

$$U_{i|h}^a = S_i^A\,\partial_h U_A^a + U_A^a\,\partial_h S_i^A + F^{-1} G_h{}^n l_i U_n^a - G_{i\ h}^{*n} U_n^a. \tag{5.184}$$

Let us adhere, as in the last part of Section 2.4, to the condition (2.144) in order to specify the coordinate system x^i of our conformally flat 1-form Finsler space. Then it may be readily verified that the form (2.154) of the connection coefficients, together with Equation (2.145) and the identity $U_i^a y^i = 0$, leads to the cancelling of all terms but the first in (5.184):

$$U_{i|h}^a = S_i^A\,\partial_h U_A^a. \tag{5.185}$$

In the case when U^a_i are uniform (Equation (5.31)), it follows from (2.144) together with the fact that U^a_i are homogeneous of degree zero with respect to y^n, that $\partial_h U^a_A = 0$ so that (5.185) reduces to

$$U^a_{i|h} = 0. \tag{5.186}$$

It is natural to call the gauge fields $A_a{}^b{}_i$ and N^a_i *uniform* if the projection factors from which they are constructed in accordance with (5.73) are uniform. From Equations (5.73) and (5.186) it follows that, in the case under study, the uniform $A_a{}^b{}_i$ and therefore the gauge tensor (5.173) vanish:

$$A_a{}^b{}_i = 0, \quad F_a{}^b{}_{in} = 0. \tag{5.187}$$

At the same time, the second gauge field $N^a_i = \partial u^a/\partial x^i - G_i{}^n u^a_n$ will reduce to $\overset{\bullet}{N}{}^a_i = -G_i{}^n u^a_n$ in the uniform case ($\partial u^a/\partial x^i$ will vanish because of (2.144) and the fact that $u^a(y^A)$ are homogeneous functions of degree zero). For the Berwald-Moór metric function we have expressions (2.154) and (5.35) for $G_i{}^n$ and u^a_n, and also the expression (S.6) of the tensor $C_k{}^{ip}$ entering Equation (2.154). With these formulae, calculation of N^a_i yields the following simple result

$$N^a_i = -N\delta^A_i C^a_A K_i(x) \equiv -NK_p \sum_{A=1}^{N} C^a_A S^A_i S^p_A, \tag{5.188}$$

where C^a_A are the constants given by Equations (2.71) and (2.71′). We observe that

$$\partial N^a_i/\partial u^e = 0. \tag{5.189}$$

Thus, we have established the following result.

PROPOSITION 8. Let a Finsler space be a conformally flat 1-form Berwald-Moór space. If u^a are chosen such that the Christoffel symbols of the indicatrix vanish, then the gauge fields will be given by Equations (5.187)–(5.189) with respect to the coordinate system x^i possessing the property (2.144).

As is known [10], when one takes into account electromagnetic interactions the total isotopic invariance of strongly interacting fields turns out to be broken, there remaining only an invariance relative to rotations about a fixed axis. This can be readily explained from the geometrical standpoint suggested above. Namely, if we assume, as in Section 4.2, that the vector potential A_k of the electromagnetic field has the form (4.46), then the projection A_a of that field on the indicatrix:

$$A_a \overset{\text{def}}{=} g^{*(b)}{}_a\, g^i_{(b)} A_i = t^i_a A_i \tag{5.190}$$

(cf. Equation (5.18)) will be independent of u^a:

$$A_a = A_a(x).$$

This last relation means that in the indicatrix resting on some point x^i the electromagnetic field $A_a(x)$ takes on a constant value. Therefore, the internal space will be Euclidean with a fixed vector field specified in it and defined by Equation (5.190). It is this vector that represents the axis of rotation, about which the internal space is invariant.

The main results of this section are given by Propositions 7 and 8. In the succeeding section we shall present a particular example of nonlinear internal symmetries.

5.6. Example of Nonlinear Internal Symmetry

Generally speaking, the construction of the gauge-covariant derivative D_j performed in Sections 5.2–5.4 makes it possible to formulate systematically gauge-invariant equations for physical fields having various internal symmetries in isotopic space, without assuming that isotopic space (that is, the Finslerian indicatrix) is a flat space. The simplest case of such a field is the triplet of π-mesons which has isospin 1 and is a space-time pseudoscalar and, therefore, can be described by a (real) scalar-vector field $w_a(x, u)$ (up to a Jacobian multiplier). The two-component isospinors $v^\beta(x, u)$ treated in Section 5.4 will correspond to K-mesons because these mesons are known to have isospin $\frac{1}{2}$ and, like π-mesons, to be space-time pseudoscalars. Also, the nucleons which have isospin $\frac{1}{2}$ and simultaneously ordinary space-time spin $\frac{1}{2}$ can be represented by the objects $\Phi^\beta(x, u)$ (see Equations (5.135)–(5.141)).

We shall consider below the case in which the indicatrix is a space of constant curvature, for such a case is logically next in complexity to the case of the zero-curvature indicatrix treated in the preceding section. A rather broad class of corresponding Finslerian metric functions is given by Equation (2.9). However, in order to apply the theory to physical fields we must fix the choice of metric function by the requirement that the Finslerian metric tensor g_{ij} has the space-time signature $(+ - - -)$. The two relevant cases have been given in Problem 2.3, namely, the metric function (2.9) with $r^A = (1, 1, 1, 1)$ and $r^A = (1, -1, -1, -1)$. Let us also assume that the constant curvatures $K(x)$ of the indicatrices at any points x^i are positive. Then consideration of these two metric functions leads to only one possibility

$$F(x, y) = \left(\sum_{A=1}^{4} (S_i^A(x) y^i)^{f(x)} \right)^{1/f(x)}, \qquad 0 < f(x) < 1, \tag{5.191}$$

for which

$$K(x) = \frac{f^2(x)}{4(1 - f(x))}. \tag{5.192}$$

Indeed, as we have indicated in Problem 2.3, the metric functions (2.9) with $r^A = (1, 1, 1, 1)$ and $r^A = (1, -1, -1, -1)$ correspond to metric tensors of signature $(+ - - -)$ when (i) $r^A = (1, 1, 1, 1)$ and $f < 1$; (ii) $r^A = (1, -1, -1, -1)$

and $f > 1$. But the indicatrix curvature R defined by (1.125) is equal to $f^2/4(f-1)$ (Equation (2.16)), so that $R < 0$ in the case (i) and $R > 0$ in the case (ii). If the Finslerian metric tensor is taken to be positive-definite, this R will be equal to the actual curvature of the indicatrix. However, in the case of the signature $(+ - - -)$ the angular metric tensor $h_{ij} = g_{ij} - l_i l_j$, and hence the metric tensor of the indicatrix $g^*_{ab} = h_{ij} t^i_a t^j_b$ (Equation (1.113), will be negative-definite. Therefore, in the case under study, the proper curvature K of the indicatrix will be equal to $-R$ which is positive in case (i) and negative in case (ii). The positive-definite metric tensor of the indicatrix will be

$$r_{ab}(x, u) = -g^*_{ab}(x, u) \tag{5.193}$$

(cf. Equation (2.86)). We have thus obtained the following result.

PROPOSITION 9. The Finslerian metric function (5.191) has the property that the indicatrix is a three-dimensional space of positive constant curvature (5.192) and the Finslerian metric tensor $g_{ij}(x, y)$ has the space-time signature $(+ - - -)$.

The gauge transformations (5.36) act as coordinate transformations in the indicatrices and, therefore, will induce a tensor transformation law of (5.193):

$$\tilde{r}_{ab}(x, \tilde{u}) = r_{cd}(x, u(x, \tilde{u}))u^c_a(x, \tilde{u})u^d_b(x, \tilde{u}), \tag{5.194}$$

where the indices a, b, c, d are specified to range over $1, 2, 3$. If the transformations of the form (5.36) correspond to motions in the indicatrices, the metric tensor (5.193) will remain invariant, i.e. the transformation law (5.194) will reduce to merely

$$\tilde{r}_{ab}(x, \tilde{u}) = r_{ab}(x, u). \tag{5.195}$$

In the case under consideration there exist coordinate systems $\{u^a\}$ in the indicatrices such that

$$r_{ab}(x, u) = \delta_{ab} \pm \frac{u^a u^b}{K^{-2}(x) - \delta_{cd} u^c u^d} \tag{5.196}$$

where $||R^i_j||$ with i and j running from 1 to 4 is an orthogonal four-dimen-invariance transformations possessing the property (5.195) will have the following explicit form

$$u^a = R^a_b(x)\tilde{u}^b \pm R^a_4(x)(K^{-2}(x) - \delta_{cd}\tilde{u}^c\tilde{u}^d)^{1/2} \tag{5.197}$$

where $||\ R^i_j\ ||$ with i and j running from 1 to 4 is an orthogonal four-dimensional matrix. The last assertion follows from the circumstance that a three-dimensional space of positive constant curvature can be defined as the hypersurface

$$(u^1)^2 + (u^2)^2 + (u^3)^2 + (u^4)^2 = K^{-2}$$

of a four-dimensional Euclidean space. Substituting

$$u^4 = +(K^{-2} - \delta_{cd} u^c u^d)^{1/2}$$

in

$$(dl)^2 = (du^1)^2 + (du^2)^2 + (du^3)^2 + (du^4)^2$$

yields $(dl)^2 = r_{ab} du^a du^b$ with r_{ab} given by Equation (5.196). The group of transformations (5.197) represents an example of *a group of nonlinear internal symmetries*. The transformations (5.197) become linear only when $R^a_4 = 0$.

So, one may proceed to geometrize in Finslerian terms those nonlinear internal symmetries of physical fields which reflect the group of motions of a three-dimensional Riemannian space of positive constant curvature, to be denoted by V_3. It will be noted in this respect that such a symmetry has been used in the work by Meetz[11], where the triplet $p^a(x)$ of soft π-mesons has been treated as the coordinates of points in V_3 at any space-time point x^n, assuming at the same time that $p^a(x)$ are scalars under the coordinate transformations $x^i = x^i(\bar{x}^j)$, so that the gradients

$$p^a_i \stackrel{\text{def}}{=} \partial p^a(x)/\partial x^i \tag{5.198}$$

will be covariant vectors under changes of x^n. In formulating the basic equations, this author confined himself to a consideration of only x-independent transformations $p^a = p^a(\tilde{p}^b)$, which in the functional form will be written as

$$p^a(x) = p^a(\tilde{p}^b(x)). \tag{5.199}$$

The functions $p^a(\tilde{p}^b)$ are assumed to be invertible, twice differentiable, and otherwise arbitrary. The transformations (5.199) entail that the gradients (5.198) will behave as contravariant vectors:

$$p^a_i = \tilde{p}^b_i \, \partial p^a/\partial \tilde{p}^b. \tag{5.200}$$

Under these conditions, the Lagrangian density for π-mesons is written in the following nonlinear form

$$L_{\pi\pi} = \frac{1}{2} J(x) a^{ij}(x) r_{ab}(p(x)) p^a_i p^b_j \tag{5.201}$$

which is invariant under the nonlinear transformations

$$p^a(x) = R^a_b p^b(x) \pm R^a_4 (K^{-2} - \delta_{cd} p^c(x) p^d(x))^{1/2} \tag{5.202}$$

(see Equation (5.187)) with K, R^a_b and R^a_4 being constants. And further, the nucleon field equations are formulated in terms of the covariant derivative

$$D'_i \Phi^\beta = \left. \frac{\partial \Phi^\beta(x, p)}{\partial x^i} \right|_{p^b = p^b(x)} + \left. p^a_i(x) D_a \Phi^\beta(x, p) \right|_{p^b = p^b(x)}, \tag{5.203}$$

where $\Phi^\beta(x, p)$ is a four-component space-time spinor and a two-component isospinor; D_a denotes the spinor covariant derivative (5.164) in the space of p^a (see also Problem 5.6)).

Clearly, one can arrive at such constructions from purely geometrical considerations based on the Finslerian metric function (5.191), taking advantage of the parametrical concept of osculation (Section 5.7)). If, in so doing, we wish to go beyond the constraining assumption that the internal transformations should be taken in the form (5.199), which is not explicitly dependent on x^i, and consider the gauge transformations (5.197) with the functions K, R^a_b, and R^a_4 dependent on x^i in an arbitrary way, then we shall have to add in the definition (5.203) a new term proportional to Φ^β.

As for the interpretation of the parameters u^a as triplets of π-mesons, it is not unique. It would be natural, for example, to interpret geometrically the triplet, along with other physical fields, as a triplet of scalar fields $w^a(x)$ governed by the transformation law

$$w^a(x) = \frac{\partial u^a(x)}{\partial \tilde{u}^b(x)} \tilde{w}^b(x)$$

under the gauge transformations $u^a(x) = u^a(x, \tilde{u}^b(x))$. For linear internal transformations there is no difference between this transformation law and (5.199). They will be significantly different, however, in the case of nonlinear transformations, such as (5.197). Following the first method, one cannot add a mass term in the Lagrangian (5.201) for if we formally write down the term $m^2 p^a p^b r_{ab}$ it will not be a scalar, because the transformation law (5.199) is different from the vector one. For this reason, one can invariantly describe by this method only soft π-mesons, that is the case when the kinetic energy of a pion is much lower than its rest mass. In contrast, a second possibility, where π-mesons are treated as the components of a vector in isotopic space, admits of the ordinary scalar mass term $m^2 w^a(x) w^b(x) r_{ab}(x, u(x))$. It cannot be excluded that this second method may prove successful in constructing nonlinear models of the nucleon-meson interaction.

5.7. Use of the Parametrical Concept of Osculation

In considering (x, y)-dependent fields it is natural to construct a field theory by using the concept of osculation, that is, by substituting for the vector variable y^i a vector field $y^i(x)$ (Chapter 4). Similarly, if when developing a theory of fields dependent on x^i and u^a we wish that there should be a correspondence with the conventional theory of fields dependent solely on x^i, it is expedient to make use of *the parametrical concept of osculation*, according to which in (x, u)-dependent fields one should replace the parameters u^a by functions of the argument x^i, which will be written in the symbolic form

$$u^a \to r^a(x) \tag{5.204}$$

(we do not use the symbols $u^a(x)$ in order to avoid confusion with the notation used in preceding sections). These $r^a(x)$ will be assumed to be (sufficiently smooth) independent scalar functions. Gauge transformations will now be written in the parametrical form

$$r^a(x) = K^a(x, \tilde{r}^b(x)) \tag{5.205}$$

where the functions $K^a(x,\tilde{r})$ are assumed to be sufficiently smooth and such that $\det(\partial K^a/\partial \tilde{r}^b) \neq 0$. One should bear in mind that when replacing the parameters by scalar functions in accordance with (5.204) we should simultaneously replace the vectors l^i by vector fields $l^i(x)$, because l^i and u^a are algebraically related. This relation will be taken to be uniform (Equation (5.27)), so that we shall have

$$l^i(x) = T^i(S^A(x),\ r(x)). \tag{5.206}$$

The following notation will be used below:

$$r^a{}_{,i} = \partial r^a/\partial x^i, \qquad r^a_b = \frac{\partial K^a(x,\tilde{r})}{\partial \tilde{r}^b}\Bigg|_{\tilde{r}^b = \tilde{r}^b(x)},$$

$$r^a_n = (F(x,y)\partial u^a(x,y)/\partial y^n)\Big|_{l^i = l^i(x)},$$

$$T^k_a = t^k_a(x, r(x)), \qquad T^k_{ab} = (\partial t^k_a(x,u)/\partial u^b)\Big|_{u^c = r^c(x)},$$

and

$$l^n_i = \partial l^n(x)/\partial x^i + G_i{}^n(x, l(x))$$

(cf. the notation (4.10)).

Following this idea, the gauge fields $A_a{}^b{}_i(x,u)$ and $N^a_i(x,u)$ are assigned osculating values:

$$B_a{}^b{}_i(x) \stackrel{\text{def}}{=} A_a{}^b{}_i(x, r(x)), \tag{5.207}$$

$$M^a_i(x) \stackrel{\text{def}}{=} N^a_i(x, r(x)). \tag{5.208}$$

Owing to Equations (5.207)–(5.208) and (5.73),

$$M^a_i = r^a{}_{,i} - r^a_n l^n_i, \tag{5.209}$$

$$B_a{}^b{}_i = r^b_k \nabla_i T^k_a + l^n_i r^c_n r^b_k T^k_{ac}, \tag{5.210}$$

where ∇_i denotes the covariant derivative with respect to the osculating Cartan connection coefficients, so that

$$\nabla_i T^k_a = \partial T^k_a/\partial x^i + \Gamma_n{}^k{}_i(x, l(x)) T^n_a. \tag{5.211}$$

For these fields the relations (5.50) and (5.51), yield the following law of transformation under (5.205):

$$r^d_c \tilde{B}_b{}^c{}_i - \partial r^d_b/\partial x^i + u_b{}^d{}_c(x, \tilde{r}(x)) \tilde{r}^c_n l^n_i = r^e_b B_e{}^d{}_i, \tag{5.212}$$

$$r^d_c \tilde{M}^c_i + u^d_{;i}(x, l(x)) = M^d_i. \tag{5.213}$$

The gauge-covariance properties (5.60)–(5.61) of the gauge tensors $F_a{}^b{}_{in}$ and

N^c_{in} will evidently be retained in the osculating approach, so that we shall have

$$r^d_b \tilde{F}_a{}^b{}_{in}(x, \tilde{r}(x)) - u_a{}^d{}_c(x, \tilde{r}(x))\tilde{N}^c_{in}(x, \tilde{r}(x)) = r^c_a F_c{}^d{}_{in}(x, r(x)),$$

$$r^d_c \tilde{N}^c_{in}(x, \tilde{r}(x)) = N^d_{in}(x, r(x)).$$

The second terms on the right-hand sides of (5.209) and (5.210) are obviously tensors under gauge transformations (5.205). Therefore, the fields

$$M_{(1)i}{}^a = r^a_{,i}, \qquad B_{(1)a}{}^b{}_i = r^b_k \nabla_i T^k_a \tag{5.214}$$

will obey the same transformation laws (5.212)–(5.213) as M^a_i and $B_a{}^b{}_i$, so that in this respect the gauge fields (5.214) are interchangeable with (5.209)–(5.210). Also, the field $B_{(1)a}{}^b{}_i$ may be replaced by the simpler field

$$B_{(2)a}{}^b{}_i = r^b_k \mathrm{D}_i T^k_a, \tag{5.215}$$

where D_i stands for the Riemannian covariant derivative associated with the osculating Riemannian metric tensor, because the difference

$$B_{(1)a}{}^b{}_i - B_{(2)a}{}^b{}_i = (\Gamma_n{}^k{}_i(x, l(x)) - \{_n{}^k{}_i\})T^n_a r^b_k$$

is a gauge-covariant tensor.

For linear gauge transformations

$$r^a(x) = r^a_b(x)\tilde{r}^b(x) \tag{5.216}$$

Equation (5.212) reduces to

$$r^d_c \tilde{B}_b{}^c{}_i - \partial r^d_b/\partial x^i = r^e_b B_e{}^d{}_i. \tag{5.217}$$

This transformation law has exactly the form which defines a Yang-Mills field in the usual sense, i.e. an x-dependent field $B_a{}^b{}_i(x)$ which is such that, if some field $w^a(x)$ transforms under (5.216) as $\tilde{w}^a = r^a_b w^b$, then the definition $w^a_i = \partial w^a/\partial x^i + B_b{}^a{}_i w^b$ yields a gauge-covariant object: $\tilde{w}^a_i = r^a_b w^b_i$. The field $B_a{}^b{}_i(x)$ transforming according to (5.212) is in fact an extension of the Yang-Mills field to the case of arbitrary nonlinear gauge transformations; as before, we call it the Yang-Mills field. It should be mentioned that the transformation law of the form (5.217) holds also for nonlinear gauge transformations if the field $l^n(x)$ is stationary, that is when $l^n_i = 0$ (cf. Equation (4.41)). The Yang-Mills strength tensor will read

$$\mathscr{F}_a{}^b{}_{in} = \partial B_a{}^b{}_n/\partial x^i - \partial B_a{}^b{}_i/\partial x^n - B_a{}^c{}_i B_c{}^b{}_n + B_a{}^c{}_n B_c{}^b{}_i \tag{5.218}$$

(cf. Equation (5.173)). When this definition is applied to the gauge fields $B_{(1)a}{}^b{}_i$ and $B_{(2)a}{}^b{}_i$ as given by Equations (5.214) and (5.215), simple direct calculations yield, respectively,

$$\mathscr{F}_{(1)a}{}^b{}_{in} = T^j_a r^b_k(\tilde{K}^k{}_{jin} - (\nabla_i l_j)\nabla_n l^k + (\nabla_n l_j)\nabla_i l^k), \tag{5.219}$$

$$\mathscr{F}_{(2)a}{}^b{}_{in} = T^j_a r^b_k(A^k{}_{jin} - (\mathrm{D}_i l_j)D_n l^k + (\mathrm{D}_n l_j)D_i l^k). \tag{5.220}$$

In (5.219), $\tilde{K}^k{}_{jin}$ is the tensor (4.101), while $A^k{}_{jin}$ in (5.220) is the Riemann curvature tensor associated with an osculating Riemannian metric tensor.

As a result of these observations we obtain a unified geometric viewpoint on the gravitational and Yang–Mills fields. The Finslerian representation (5.207) of the Yang–Mills field establishes an algebraic relationship between the Yang–Mills field and the primary variables $S^A(x)$ and their first derivatives: according to Equations (5.207), (5.73), (5.27) and (5.29), there exist functions $\mathscr{B}_a{}^b{}_i(S^A, \mathrm{d}_n S^A, r^c)$ such that

$$B_a{}^b{}_i(x) = \mathscr{B}_a{}^b{}_i(S^A(x), \quad \mathrm{d}_n S^A(x), r^c(x)). \tag{5.221}$$

Therefore, as soon as the Finslerian metric function reflecting the internal symmetry of a certain class of physical fields is found, the functions $\mathscr{B}_a{}^b{}_i$ in Equation (5.221) will be known, and the variables $S^A(x)$ will describe both the gravitational and Yang–Mills fields. As regards the formulation of field equations for $S^A(x)$ in the context of the gauge approach, the present Finslerian technique offers the simple possibility of using the projection factors of the indicatrix to construct a Lagrangian W linear in a gauge strength tensor (as is done in the usual Einstein gravitational theory), for example,

$$W = g^{nh}(x, l(x)) r^a_h T^i_b F_{(1)a}{}^b{}_{in}, \tag{5.222}$$

whereas in the usual gauge approaches the simplest gauge field Lagrangian is quadratic in the gauge tensor because the indices b, i of different geometrical nature cannot be contracted. Finally, Equations (5.218)–(5.219) clearly illustrate the relationship between the curvature tensors associated with a metric tensor and the Yang–Mills strength tensors.

Summing up this chapter, we may conclude that we have had no serious difficulties in reinterpreting Finsler geometry in precise mathematical terms as providing a geometric basis for constructing theories for physical fields exhibiting internal symmetries. The reader may make additional steps in this direction in solving the problems appended to this chapter.

Problems

5.1. Verify directly that the Finslerian gauge fields (5.73) obey the transformation laws (5.50) and (5.51).

5.2. Prove that, for any gauge fields $A_a{}^b{}_i$ and N^a_i obeying the transformation laws (5.50) and (5.51), the transformation law of the derivative $\partial A_a{}^b{}_i/\partial u^e$ is such that the object

$$A_a{}^f{}_{ie} = \partial A_a{}^f{}_i/\partial u^e - q^{*b}_a{}_e A_b{}^f{}_i - \partial q^{*f}_a{}_e/\partial x^i + q^{*f}_c{}_e A_a{}^c{}_i -$$
$$- N^c_i \partial q^{*f}_a{}_e/\partial u^c - q^{*f}_a{}_c \partial N^c_i/\partial u^e$$

is a gauge tensor, that is $u^d_c \tilde{A}_a{}^c{}_{ie} = u^b_a u^g_e A_b{}^d{}_{ig}$.

5.3. Clarify the Finslerian nature of the gauge tensors N^a_{ie} (Equation (5.63)) and $A_a{}^f{}_{ie}$ (Problem (5.2).

5.4. Derive directly from the representation (5.88) of the gauge tensor $F_{a\ in}^{\ b}$ that in the case of the conformally flat 1-form Berwald-Moór metric function we have $F_{a\ in}^{\ b} = 0$. Find the explicit form of the second gauge tensor N_{in}^{c}, as given by Equation (5.81), in this case.

5.5. Each component of a Dirac spinor satisfies the Klein–Gordon equation $a^{ij}(x)\partial_i\partial_j\Phi + m^2\Phi = 0$. What equation is satisfied by the components of v^β subject to equation (5.154)?

5.6. In the work by Meetz (1969), nucleons are described by functions of the form $\Phi^\beta(x, p^a(x))$ which depend on the fields $p^a(x)$ of π-mesons. Φ^β is assumed to be a four-component space-time spinor under the changes $x^i = x^i(\bar{x}^j)$ and a two-component isospinor under the transformations (5.194) of p^a in the isospace of constant positive curvature. The interaction of nucleons with π-mesons is formulated in terms of a derivative of the type (5.203):

$$\mathrm{D}'_i\Phi^\beta(x, p(x)) = \partial'_i\Phi^\beta + (\partial_i p^a)\Phi^\beta{}_{\|a}.$$

Here, the derivative ∂'_i acts only on the explicit variable x^n; $\Phi^\beta{}_{\|a}$ is understood as the spinor derivative of the two-component isospinor in a curved isospace and is taken to be $\Phi_{\|a} = \partial\Phi/\partial p^a - \Omega_a\Phi$ with Ω_a being the 2×2 matrices

$$\Omega_a = ie_{abc}p^b\sqrt{2}s^c\,\frac{1}{2k}\,\frac{1}{k + (k^2 - p^2)^{1/2}}$$

where $p^2 = \delta_{ab}p^a p^b$ and s^c are the Pauli matrices (5.92); the curvature of the isospace is $K = 1/k$; the symbol e_{abc} denotes permutation. This definition of D'_i coincides with the covariant derivative used by Weinberg (1968) and corresponds to Gürsey's second nonlinear model (see Chang and Gürsey, 1967). Compare $\Phi^\beta{}_{\|a}$ with the covariant derivative $D_a\Phi^\beta(x, u)$ (Equation (5.164)) in the indicatrix.

5.7. Derive equations of the type (5.154) and (5.141), and also of the Klein–Gordon type, for a scalar-vector field $w^a(x, u)$ from a variational principle.

5.8. As we have argued at the beginning of Chapter 4, the action principle for the gravitational field equations in a Finslerian approach cannot be formulated in the usual way until the theory is developed in terms of objects which arc functions of independent variables x^i and y^i. The origin of this difficulty resides in the fact that y^i is a vector. As the shortest way of suppressing such a dependence, we have exploited in Chapter 4 the concept of osculation according to which a vector field $y^i(x)$ is to be substituted for the directional variable y^i. However, the ideas set forth in Sections 5.1–5.5 suggest an alternative approach, namely, one may use the parametrical representation $l^i = t^i(x, u)$ of the indicatrix in order to turn from the directional variable y^i to the scalars u^a and, then, treat x^i and u^a as independent variables. The role of the gravitational field variables will be played by the tensors $S^A(x)$ in the same fashion as in Chapter 4. Under these conditions, nothing prevents one from formulating the action principle in the obvious way $\delta\, Ld^4x = 0$. The gravitational Lagrangian density L will not involve any auxiliary functions, while it will contain three independent scalar parameters u^a. It may be expected that the gravitational field equations derivable in this parametrical approach will be very similar to those which appear in the osculating approach developed in Chapter 4, the distinction generally being merely that the field $t^i(x, u)$ is used instead of $y^i(x)$. Carry out the corresponding calculations.

5.9. The classical Yang–Mills gauge field equations are constructed on the basis of a Lagrangian which is quadratic in the gauge field strength tensor. Choosing the Lagrangian in the form of the square of the gauge field tensor $F_{(1)a}{}^{b}{}_{in}$ given by Equation (5.64) and treating the gauge fields $A_a{}^b{}_i$ and N_i^a as field variables, derive the corresponding generalized equations and clarify their Finslerian nature. Propose a generalization of the Palatini method to obtain field equations from the Lagrangian density $J(x, u)I(x, u)t_a^n(x, u)t_b^i(x, u)$ $F_{(1)}{}^{ab}{}_{in}(x, u)$ which is linear in the gauge field tensor.

5.10. Generalize the concept of SU(3)-invariance to the Finslerian case.

5.11. Introduce a gauge field in order to construct a derivative covariant under G-transformations (1.141).

Notes

[1] The reader acquinted with Section 8 of Chapter 5 of the monograph by Rund (1959) will not fail to notice that our Equation (5.24) is identical with the relation (5.8.8) of that monograph (with $k = 1$, $B_\alpha^i \to t_a^i$, $\gamma_{\alpha\beta} \to g_{ab}^*$ and $\gamma_{\alpha\beta}^\gamma \to q_a^{*c}{}_b$).
[2] See, e.g., Drechsler and Mayer (1977); Chaichian and Nelipa (1983).
[3] Gauge transformations of a type much more general than (5.36) are examined in recent papers by Rund (1980, 1981b, 1982). The equations of motion of an isotopic spin-particle in a classical gauge field are studied in Rund (1983).
[4] See, e.g., Kramer *et al.* (1980, Section 3.6); Treder (1971, Section 12); Penrose and Rindler (1984).
[5] Infeld and van der Waerden (1933).
[6] See, e.g., Utiyama (1956, p. 1604).
[7] Utiyama (1956); Drechsler and Mayer (1977); Konopleva and Popov (1981).
[8] See the standard testbooks on the theory of physical fields and particles, e.g., Källen (1964); Gasiorowicz (1966); Bogoljubov and Shirkov (1959).
[9] An attempt to interpret isotopic invariance in terms of geometric entities defined on space-time has been made by Horváth (1963). The construction is based on the introduction of a field of special 'λ-trieders', the group of Euclidean rotations of which is identified with the group of transformations of internal isotopic space. However, from the viewpoint of the geometrization of internal symmetries in terms of the geometry of the Finslerian indicatrix, the introduction of an auxiliary trieder appears to be superfluous so that we need not follow this method. A number of ideas concerning the possibility of developing gauge field theory with the help of Finsler geometry have been advanced by Takano (1964, 1982). However, this author does not use transformations of the indicatrix parameters, that is (5.36), as gauge transformations in his papers. The present section, as well as the chapter as a whole, follow ideas suggested in Asanov (1983b).
[10] See, e.g., the references in Note 8.
[11] Meetz (1969).

Part D

Additional Observations

Chapter 6

Classical Mechanics from the Finslerian Viewpoint

6.1. Parametrically Invariant Extension of the Lagrangian

Let a dynamical system of $N-1$ degrees of freedom be represented by a Lagrangian of the form

$$L_1 = L_1(x^a, V^a), \tag{6.1}$$

so that the trajectories of the system are the solutions of the Euler–Lagrange equations

$$\mathrm{d}p_{1a}/\mathrm{d}t - \partial L_1/\partial x^a = 0, \tag{6.2}$$

where $V^a = \mathrm{d}x^a/\mathrm{d}t$ is the velocity and

$$p_{1a} = \partial L_1/\partial V^a$$

is the momentum; the indices a, b run from 1 to $N-1$.

The form of Equations (6.2) depends, in general, on a particular choice of the parameter t, the latter being frequently identified in classical mechanics with the parameter of time. However, the trajectories of a dynamical system are entities of an objective nature and exist in reality as certain curves irrespective of any particular method of parametrization. For example, when considering an electric charge in an electromagnetic field we are given, in experiment, only its trajectories. The question of which parameter to use for an analytical representation of these curves – whether we use the time parameter in the Newtonian sense or the proper time of the charge, or any other parameter – is merely one of finding a suitable means to describe the charge motion.

The Euler-Lagrange equations (6.2) yield a stationary value of the integral $\int L_1\,\mathrm{d}t$. If we take any other admissible parameter q, so that $t = t(q)$ with $\mathrm{d}t/\mathrm{d}q > 0$, this same integral becomes $\int L_1\,(\mathrm{d}t/\mathrm{d}q)\mathrm{d}q$. Therefore, if we re-interpret the initial parameter t as an additional N-th coordinate:

$$x^N = t,$$

we go to a Lagrangian L of the form[1]

$$L(x^i, \dot{x}^i) = L_1(x^a, \dot{x}^a/\dot{x}^N)\dot{x}^N, \tag{6.3}$$

where $\dot{x}^i = \mathrm{d}x^i/\mathrm{d}q$ (cf. Equation (2.173′)).

The last Lagrangian possesses the important property that it is a homogeneous function of degree one with respect to the new velocities $\dot{x}^i$, i.e.

$$L(x, k\dot{x}) = kL(x, \dot{x}), \tag{6.4}$$

$k > 0$. Therefore, whichever parameter $\bar{q}$ with the property $\mathrm{d}q/\mathrm{d}\bar{q} > 0$ we take, the integral will be of the same form:

$$\int L(x, \mathrm{d}x^i/\mathrm{d}q)\mathrm{d}q = \int L(x, \mathrm{d}x^i/\mathrm{d}\bar{q})\mathrm{d}\bar{q}$$

where L is the same function (6.3). This means that the Euler-Lagrange equations associated with L:

$$\mathrm{d}(\partial L/\partial \dot{x}^i)/\mathrm{d}q - \partial L/\partial x^i = 0 \tag{6.5}$$

will retain their form under arbitrary replacements $q = q(\bar{q})$ of the parameter subject only to the condition that $\mathrm{d}q/\mathrm{d}\bar{q} > 0$. For these reasons, the homogeneous Lagrangian (6.3) may be called *the parametrically invariant extension of the initial Lagrangian* (6.1).

From (6.3) we have

$$L_a \stackrel{\text{def}}{=} \partial L/\partial \dot{x}^a = p_{1a},$$

$$L_N \stackrel{\text{def}}{=} \partial L/\partial \dot{x}^N = -H_1,$$

$$\partial L/\partial x^N = 0,$$

where

$$H_1 = V^a p_{1a} - L_1 \tag{6.6}$$

is the Hamiltonian function associated with L_1. The equations (6.5) take the form

$$\mathrm{d}p_{1a}/\mathrm{d}q - \dot{x}^N \partial L_1/\partial x^a = 0, \tag{6.7}$$

$$\mathrm{d}H_1/\mathrm{d}q = 0, \tag{6.8}$$

of which the latter means that the Hamiltonian H_1 is an integral of the equations under consideration.

The homogeneity condition (6.4) enables us to treat L as a Finslerian metric function so that Equations (6.5), and hence (6.7) and (6.8), represent Finslerian geodesics. Equations (6.7), being written in terms of the initial parameter t, are identical with the initial equations (6.2). Thus, we may summarize the situation as follows.

PROPOSITION 1. The trajectories of any dynamical system of $N - 1$ degrees of freedom given by a Lagrangian L_1 of the form (6.1) are the projections of the N-dimensional Finslerian geodesics associated with the parametrically extended Lagrangian L. The N-th component of these equations reads merely that the Hamiltonian function H_1 associated with L_1 is a constant along any geodesic.

6.2. The Hamilton-Jacobi Equation for Homogeneous Lagrangians

In classical mechanics an important role is played by the Hamilton–Jacobi method which is developed in terms of the Hamiltonian function. The construction of the Hamiltonian function offers the possibility of considering the velocities as functions of momenta. Namely, if the matrix having entries $\partial p_{1a}/\partial V^b$ is of maximum rank:

$$\operatorname{rank}(\partial p_{1a}/\partial V^b) = N-1,$$

then it may be stated on the basis of the theorem of the inverse function that there exists the inverse relation $V^a = V^a(x, p_{1b})$. Under these conditions, the Hamiltonian function $H_1(x^a, p_{1b})$ is defined by Equation (6.6), which entails, as can be readily verified, that

$$\partial H_1/\partial x^a = -\partial L_1/\partial x^a \tag{6.9}$$

and that the Euler-Lagrange equations (6.2) are equivalent to the following *canonical equations for the nonhomogeneous case*:

$$V^a = \partial H_1/\partial p_{1a}, \qquad \mathrm{d}p_{1a}/\mathrm{d}t = -\partial H_1/\partial x^a. \tag{6.10}$$

The Hamilton–Jacobi equation will read

$$\partial S/\partial t + H_1(x^a, \partial S/\partial x^b) = 0. \tag{6.11}$$

The significance of this partial differential equation is that for any solution $S(t, x^a)$ of Equation (6.11) the gradient $\partial S/\partial x^a$ represents a canonical momenta field; that is, if we denote $\partial S/\partial x^a = p_{1a}(t, x)$, then the $p_{1a}(t, x)$ will satisfy the canonical equations (6.10).

In the homogeneous case, however, the velocities $\dot{x}^i$, $i = 1, 2, \ldots, N$, cannot be expressed in terms of the derivatives L_i, because the Euler theorem (1.3) will imply the relation

$$\dot{x}^i L_i = L$$

which shows that

$$\operatorname{rank}(\partial L_i/\partial \dot{x}^j) < N.$$

This notwithstanding, the Hamilton–Jacobi theory for homogeneous Lagrangians can be developed just as well as for the nonhomogeneous Lagrangians if we define the momenta as

$$p_i \stackrel{\text{def}}{=} LL_i = p_i(x^k, \dot{x}^k), \tag{6.12}$$

i.e. in accordance with the Finslerian rule (1.6):

$$p_i(x, \dot{x}) = g_{ij}(x, \dot{x})\dot{x}^j \tag{6.13}$$

where

$$g_{ij}(x, \dot{x}) = \tfrac{1}{2}\partial^2 L^2(x, \dot{x})/\partial \dot{x}^i \partial \dot{x}^j$$

is the Finslerian metric tensor associated with L. To this end, it will be noted that in the case under consideration,

$$\partial p_i/\partial \dot{x}^j = g_{ij}$$

(Equation (1.10)) and, therefore,

$$\text{rank}\,(\partial p_i/\partial \dot{x}^j) = N$$

which ensures the existence of the inverse relation

$$\dot{x}^i = \dot{x}^i(x^k, p_j). \tag{6.14}$$

Since the matrix $||\partial \dot{x}^i/\partial p_j||$ is inverse to $||\partial p_i/\partial \dot{x}^j||$ and $\partial p_i/\partial \dot{x}^j = g_{ij}$, the relation (6.14) implies

$$\partial \dot{x}^i/\partial p_j = g^{ij}, \tag{6.15}$$

where g^{ij} is the tensor reciprocal to g_{ij}. Owing to the fact that the functions (6.12) are positively homogeneous of degree one in $\dot{x}^i$, the inverse relation (6.14) should have the same homogeneity property in p_j:

$$p_j\, \partial \dot{x}^i/\partial p_j = \dot{x}^i \tag{6.16}$$

or, by virtue of (6.15),

$$\dot{x}^i = g^{ij} p_j. \tag{6.17}$$

Using the last relation, we may define the Hamiltonian function H for a homogeneous Lagrangian Las follows:

$$H(x, p) = L(x, \dot{x}(x, p)) \equiv (g^{ij}(x, p)p_i p_j)^{1/2}, \tag{6.18}$$

which implies that

$$H\, \partial H/\partial p_i = \dot{x}^i, \tag{6.19}$$

$$\tfrac{1}{2}\partial^2 H^2(x, p)/\partial p_i\, \partial p_j = g^{ij}(x, p) \tag{6.20}$$

because of Equations (6.12) and (6.15).

Similarly to (6.9), we shall have

$$\partial H/\partial x^i = -\partial L/\partial x^i \tag{6.21}$$

which is a manifestation of the following general

PROPOSITION 2. *Let a Lagrangian L be constructed from some x-dependent fields* $S^A(x)$, *so that there exists a function* $Q(S^A, \dot{x})$ *such that* $L(x, \dot{x}) = Q(S^A(x), \dot{x})$ (*cf. Equation* (3.5)), *and hence* $H(x,p) = Z(S^A(x), p)$, *where* $Z(S^A, p) = Q(S^A, \dot{x}(S^A, p))$. *Then*

$$\partial Q/\partial S^A = -\partial Z/\partial S^A. \tag{6.22}$$

Proof. Differentiating the identity $g_{ij}g^{ik} = \delta^k_j$ with respect to S^A yields

$$g^{ik}\,\partial Q_{ij}/\partial S^A = -g_{ij}\,\partial Z^{ik}/\partial S^A, \tag{6.23}$$

where we have used the notation

$$Q_{ij} = \tfrac{1}{2}\partial^2 Q^2/\partial\dot{x}^i\,\partial\dot{x}^j, \qquad Z^{ik} = \tfrac{1}{2}\partial^2 Z^2/\partial p_i\,\partial p_k.$$

On contracting Equation (6.23) with $\dot{x}^j p_k$ and taking into account Equations (6.13) and (6.17), we obtain the relation

$$\dot{x}^i\dot{x}^j\,\partial Q_{ij}/\partial S^A = -p_i p_j\,\partial Z^{ij}/\partial S^A$$

which reads as (6.22). The proof is complete.

From Equations (6.5), (6.12) and (6.21) it follows that

$$\mathrm{d}(p_i/H)/\mathrm{dq} = -\partial H/\partial x^i. \tag{6.24}$$

Equations (6.19) and (6.24) represent *the canonical equations for the homogeneous case*. If we let the parameter be the Finslerian arc-length s, so that $\mathrm{d}s = L(x, \mathrm{d}x)$ and $L(x, \dot{x}) = H(x,p) = 1$, then the canonical equations take the form

$$\mathrm{d}x^i/\mathrm{d}s = \partial H/\partial p_i, \qquad \mathrm{d}p_i/\mathrm{d}s = -\partial H/\partial x^i, \tag{6.25}$$

which is similar to that of the nonhomogeneous case (Equation (6.10)).

Since in the homogeneous case the Lagrangian does not depend on any parameter t, it is natural to expect that *the Hamilton–Jacobi equation for the homogeneous case* can be written as just

$$H(x^i, \partial S/\partial x^j) = 1, \tag{6.26}$$

instead of (6.11). In fact, the following proposition is valid.

PROPOSITION 3. *For any solution* $S(x^i)$ *of the partial differential equation* (6.26) *the gradient*

$$p_i(x) \overset{\text{def}}{=} \partial S/\partial x^i \tag{6.27}$$

satisfies the canonical equations (6.25).

Proof. If we put $u^i(x) = \partial H(x, p(x))/\partial p_i(x)$, then, according to Equation (6.26), the vectors $p_i(x)$, and hence $u^i(x)$, will be of unit Finslerian length:

$H(x, p(x)) = L(x, u(x)) = 1$. Therefore, we may regard $u^i(x)$ as the field of the unit velocities $u^i = dx^i/ds$, so that the first canonical equation in (6.25) will be satisfied. In order to verify that our $p_i(x)$ given by (6.27) do satisfy the second of the canonical equations (6.25), we write

$$\frac{dp_i(x)}{ds} = \frac{\partial^2 S}{\partial x^n \partial x^i}\frac{dx^n}{ds} = \frac{\partial^2 S}{\partial x^n \partial x^i}\frac{\partial H}{\partial p_n}.$$

On the other hand, differentiating Equation (6.26) yields

$$\partial H/\partial x^i + (\partial H/\partial p_n)\partial^2 S/\partial x^n \partial x^i = 0.$$

Comparing the last two equations, we observe that our assertion is indeed valid. The proof is complete[2].

6.3. The Generalized Hamilton–Jacobi Theory Based on the Clebsch Representation of the Canonical Momenta Field

According to the Darboux theorem[3], any covariant vector field $y_i(x)$ of class C^1 can be represented in the following form:

$$y_i(x) = e\partial_i M + P_a \partial_i Q^a. \tag{6.28}$$

Here, $M(x)$, $P_a(x)$ and $Q^a(x)$ are scalar functions which are referred to as *the Clebsch potentials of* $y_i(x)$, and (6.28) as *the Clebsch representation of* $y_i(x)$ (cf. Equation (4.68)); $\partial_i = \partial/\partial x^i$; if the dimension number N is even, then $e = 0$ and the index a runs from 1 to $N/2$; if N is odd, then $e = 1$ and the index a ranges over $1, 2, \ldots, (N-1)/2$.

The above representation can be applied to the canonical momenta field $p_i(x)$ which satisfies the canonical equations (6.19) and (6.24) and is otherwise arbitrary[4]. Let for definiteness N be even and the scalars P_a, Q^a be functionally independent[5]. Then we have

$$p_i(x) = P_a Q^a{}_{,i}, \tag{6.29}$$

$$(\partial_h p_j - \partial_j p_h)p^h = Q^a{}_{,j} dP_a/dq - P_{a,j} dQ^a/dq, \tag{6.30}$$

where we have used the notation: $p^h = g^{hi}(x, p_n)p_i$ and

$$Q^a{}_{,j} = \partial_j Q^a, \qquad dQ^a/dq = p^i Q^a{}_{,i}, \ldots .$$

On the other hand, from Equation (6.24) we have

$$(\partial_h p_j - \partial_j p_h)p^h = -\tfrac{1}{2}\partial h^2/\partial x^j + H^{-1} p_j dH/dq, \tag{6.31}$$

where

$$h(x) \stackrel{\text{def}}{=} H(x, p(x)). \tag{6.32}$$

Since P_a, Q^a are assumed to be functionally independent, the coordinates x^i may be regarded as functions of P_a and Q^a, from which it follows that there exists a scalar function $\Phi(P_a, Q^a)$ such that

$$h(x) = \Phi(P_a(x), Q^a(x)). \tag{6.33}$$

Combining (6.30) with (6.31)–(6.33), we can readily obtain the relation

$$\Phi^{-1} Q^a{}_{,j}(\mathrm{d}(P_a/\Phi)/\mathrm{d}q + \partial\Phi/\partial Q^a) - P_{a,j}(\mathrm{d}Q^a/\mathrm{d}q - \Phi\partial\Phi/\partial P_a) = 0.$$

However, because the scalars Q^a, P_a are assumed to be functionally independent, the covariant vectors $Q^a{}_{,j}, P_{a,j}$ should be linearly independent, so that the last relation implies the equations

$$\mathrm{d}Q^a/\mathrm{d}q - \Phi\partial\Phi/\partial P_a = 0, \tag{6.34}$$

$$\mathrm{d}(P_a/\Phi)/\mathrm{d}q + \partial\Phi/\partial Q^a = 0 \tag{6.35}$$

which are called *the associated canonical equations.* Their structure is essentially similar to that of the primary canonical equations (6.19) and (6.24).

Moreover, Equations (6.34) and (6.35) imply in turn that

$$\begin{aligned}\mathrm{d}\Phi/\mathrm{d}q &= (\partial\Phi/\partial Q^a)\,\mathrm{d}Q^a/\mathrm{d}q + (\partial\Phi/\partial P_a)\,\mathrm{d}P_a/\mathrm{d}q = (\partial\Phi/\partial Q^a)\Phi\,\partial\Phi/\partial P_a + \\ &\quad + (\partial\Phi/\partial P_a)(-\Phi\partial\Phi/\partial Q^a + \Phi^{-1} P_a\,\mathrm{d}\Phi/\mathrm{d}q) = \\ &= \Phi^{-1} P_a(\partial\Phi/\partial P_a)\mathrm{d}\Phi/\mathrm{d}q,\end{aligned}$$

which yields the relation

$$P_a\partial\Phi/\partial P_a = \Phi(Q, P)$$

showing that *the function $\Phi(Q,P)$ must be homogeneous of degree one in P_a.*

Conversely, if the scalars P_a, Q^a satisfy the equation

$$H(x^n, P_a Q^a{}_{,i}) = \Phi(Q, P), \tag{6.36}$$

then, after differentiating this with respect to x^j and using the above formulae, it can be readily seen that the following relation should be valid for the field (6.29):

$$\begin{aligned}H(\mathrm{d}(p_j/H)/\mathrm{d}q + \partial H/\partial x^j) &= Q^a{}_{,j}\Phi(\mathrm{d}(P_a/\Phi)/\mathrm{d}q + \partial\Phi/\partial Q^a) - \\ &\quad - P_{a,j}(\mathrm{d}Q^a/\mathrm{d}q - \Phi\,\partial\Phi/\partial P_a)\end{aligned}$$

showing that the associated canonical equations entail the primary canonical equations (6.19) and (6.24).

Thus we get:

PROPOSITION 4. The canonical equations (6.19) and (6.24) are equivalent to equations (6.34) and (6.35) for the Clebsch potentials of the canonical momenta field.

In this context, Equation (6.36) may be regarded as *a generalized Hamilton–Jacobi equation for the Clebsch potentials.* In the case when the canonical momenta field $p_i(x)$ is taken to be a unit vector, we have $\Phi = H = 1$, and the associated canonical equations (6.34)–(6.35) read merely that the Clebsch potentials Q^a, P_a are the integrals of the equations of motion of the dynamical system.

The rotation of the congruence of the trajectories tangent to $p_i(x)$ can naturally be described by *the vorticity tensor*

$$\omega_{hj} \stackrel{\text{def}}{=} \partial(p_h/H)/\partial x^j - \partial(p_j/H)/\partial x^h \tag{6.37}$$

or, on substituting the Clebsch representation (6.29),

$$\omega_{hj} = Q^a{}_{,h}\partial(P_a/H)/\partial x^j - Q^a{}_{,j}\partial(P_a/H)/\partial x^h.$$

Simple straightforward calculations show that, along the direction of the canonical momenta field $p_i(x)$,

$$\mathrm{d}\omega_{hj}/\mathrm{d}q = -\omega_{il}\partial p^l/\partial x^h - \omega_{hl}\partial p^l/\partial x^j \tag{6.38}$$

which reads that *the Lie derivative of the vorticity tensor with respect to the canonical momenta field vanishes*.

In the four-dimensional case, we may consider, instead of the vorticity tensor, *the vorticity vector*[6]:

$$\omega^l = \tfrac{1}{2}(|g|)^{1/2} e^{lhkm}\omega_{hk}p_m$$

which is evidently orthogonal to p_m:

$$\omega^n p_n = 0;$$

the notation e^{lhkm} is used for the permutation symbol.

Thus we have seen that Finsler geometry may appropriately be used to geometrize classical mechanics, for all the basic dynamical concepts, such as the Lagrangian and Hamiltonian functions, the canonical and Hamilton–Jacobi equations, the Lagrangian brackets, etc., have Finslerian geometrical counterparts. A new step will be made in the succeeding chapter where we shall analyze the concepts of special relativity, thereby laying bare a complex of ideas for generalizing special relativity, which are in principle checkable experimentally.

Problems

6.1. Consider in Finslerian terms the holonomic dynamical system defined by the Lagrangian of the form

$$L_1 = \tfrac{1}{2}m_{ab}(x^c,t)V^aV^b - U(x^a,t).$$

6.2. Extend to the homogeneous case the concept of Poisson brackets.

6.3. Indicate the relationship between the vorticity tensor and the Lagrange brackets.

6.4. Indicate the multiple variational principle from which the associated canonical equations can be inferred.

6.5. The rescaled Lagrangian describing the motion of an electric charge in classical electromagnetic and gravitational fields has the homogeneous form

$$L = (a_{ij}(x)\dot{x}^i \dot{x}^j)^{1/2} + b_i(x)\dot{x}^i.$$

Derive the corresponding Hamilton–Jacobi equation.

6.6. Prove that the vorticity tensor satisfies the identity

$$p^h \omega_{hj} = 0.$$

6.7. Examine the Finslerian δ-parallel transportation

$$\delta h_i^P / \delta s \equiv \mathrm{d}h_i^P / \mathrm{d}s - G_i{}^k h_k^P = 0$$

(definition (1.55)) of a Finslerian covariant tetrad $h_i^P(x, \dot{x})$ along the trajectory of a test electric charge.

Notes

[1] This method of an additional coordinate was proposed by Rund (1966, p. 44).
[2] The definition (6.18) of the Hamiltonian function as well as the subsequent canonical equations and the Hamilton–Jacobi equation for homogeneous Lagrangians which we have treated in the present section were proposed by Rund; for more detail the reader is referred to the Hamilton–Jacobi theory developed in Rund (1966a,b, 1972).
[3] See, for example, Sternberg (1964, Section 6 of Chapter III).
[4] In what follows, we present the results obtained in Rund (1979), which are an extension to the homogeneous case of the results obtained in an earlier paper by that author (1977).
[5] The remaining possibility, namely that N is odd and/or the representation of the canonical momenta field in terms of scalars involving less than the highest number of independent scalars can be treated in a similar way; see Baumeister, 1978, 1979.
[6] Baumeister, 1979, Section 5.

Chapter 7

Finslerian Refinement of Special Relativity Theory

7.1. Allowance for the Dependence of Space-Time Scales on the Directions of Motion of Inertial Frames of Reference

A way of refining the ideas of special relativity theory by employing them in a Finslerian context arises if one interprets the dependence of the Finslerian metric tensor g_{ij} on the directional variable y^i as a dependence on the four-dimensional directions of motion of inertial frames of reference[1]. To emphasize the circumstance that the directional variable is strictly specified physically, we shall replace the symbol y^i by u^i throughout this section:

$$y^i \to u^i.$$

The vectors u^i will be used to indicate the directions of motion of inertial frames of reference, and will not be normalized in any sense. Pursuing our aims, we shall neglect the dependence of Finslerian objects on x^i.

The possibility of applying the relations of Riemannian geometry to the theory of relativity is based on the requirement that the Riemannian metric tensor of spacetime should be of space-time signature. Therefore, to be in agreement with the facts of special relativity we begin by setting forth the following:

Principle of space-time signature. The Finslerian metric tensor must have the space-time signature $(+ - - -)$ for any admissible vector u^i.

If this principle was not insisted upon, it would seem very likely that no Finslerian approach could even approximately be applied to the real world. Relativistic particles and fields exhibit Lorentz invariance as an exact symmetry to a high degree of accuracy, so that Lorentz invariance can not be neglected in any sense. On the other hand, a Finslerian approach consistent with the principle formulated above opens up ways of extending the special and, subsequently, general realtivity theories without contradicting the experimental facts established on the basis of the concepts of Riemannian geometry. The incontestable implication of the experimental facts found in special relativity is that the metric tensor, Riemannian or Finslerian, of space-time must have the signature $(+ - - -)$.

The mathematical formulation of the above principle requires that the set of four linearly independent covariant vectors $h_i^P(u)$ can be chosen to satisfy the relation

$$g_{ij}(u) = q_{PQ} h_i^P(u) h_j^Q(u) \tag{7.1}$$

with $q_{00} = -q_{11} = -q_{22} = -q_{33} = 1$, other q_{PQ} being zero. The tetrad indices P, Q ... will range over $0, 1, 2, 3$, and the tensor indices $i, j, \ldots$, over $1, 2, 3, 4$. The contravariant reciprocal of $h_i^P(u)$ will be denoted by $h_P^i(u)$, so that

$$h_i^P(u) h_Q^i(u) = \delta_Q^P, \qquad h_i^P(u) h_P^j(u) = \delta_i^j.$$

The entities $h_i^P(u)$ and $h_P^i(u)$ may be called *Finslerian tetrads* because, by definition (7.1), the orthonormality condition

$$g^{ij}(u) h_i^P(u) h_j^Q(u) = q^{PQ}, \qquad g_{ij}(u) h_P^i(u) h_Q^j(u) = q_{PQ} \tag{7.2}$$

holds, where $q^{PQ} = q_{PQ}$. Owing to the homogeneity of the Finslerian metric tensor $g_{ij}(u)$, the tetrads $h_i^P(u)$ and $h_P^i(u)$ may be assumed to be positively homogeneous of degree zero:

$$h_i^P(ku) = h_i^P(u), \qquad h_P^i(ku) = h_P^i(u), \tag{7.3}$$

where $k > 0$.

The tetrads obtain their dependence on the directions of motion of the inertial frames of reference from the Finslerian metric tensor. A change of the vector u^i indicating such a direction will result in a K-transformation of the Finslerian metric tensor: if the directions of two vectors u^i and u'^i are connected by the K-transformation (1.145):

$$l^n(u') = K_m^n l^m(u), \tag{7.4}$$

where $l^m(u) = u^m/F(u)$, then

$$g_{mn}(u') = K{*}_m^i K{*}_n^j g_{ij}(u), \qquad g^{mn}(u') = K_i^m K_j^n g_{mn}(u), \tag{7.5}$$

in accordance with Equations (1.148) and (1.149). Such a transformation may also be applied to the tetrads. Indeed, let a tetrad $h_i^P(u)$ satisfy the relation (7.1) for a fixed vector u^i. If we put

$$h_i^P(u') = K{*}_i^m h_m^P(u), \qquad h_P^i(u') = K_m^i h_P^m(u), \tag{7.6}$$

then comparison of the transformation law (7.6) with (7.5) shows immediately that $h_i^P(u')$ and $h_P^i(u')$ are tetrads associated with $g_{ij}(u')$, i.e. that the relation (7.1) holds for u'^i.

On the other hand, all the tetrads related to some fixed vector u^i are obtainable from any single $\bar{h}_P^m(u)$ by means of the Lorentz transformation $h_P^m(u) = L_P^Q \bar{h}_Q^m(u)$ acting on the tetrad indices. Therefore, the equality

$$h_P^m(u) = K_n^m L_P^Q \bar{h}_Q^n(u_1) \tag{7.7}$$

will connect any two Finslerian tetrads related to some u^i and u_1^i, provided that the directions of u^i and u_1^i can be connected through K-transformation, i.e. that a relation of the form (7.4) exists between u^i and u_1^i.

It follows from Equation (7.7) that the group of Lorentz transformations and the group of K-transformations have different geometrical meanings. Namely, the Lorentz group transforms into each other the members of the set of all tetrads related to a fixed direction of motion of an intertial frame of reference, whereas the K-transformations connect the tetrads related to different directions of motion of such frames. The two transformations act on different indices of the tetrads.

7.2. Finslerian Extension of the Special Principle of Relativity

The relativity of motion is one of the most fundamental physical ideas. The postulate of the relativity of motion in Newtonian mechanics, as expressed by Newton's first law (Galileo's law), is the basic axiom without which the construction of the perfect edifice of classical mechanics would be impossible.

The rejection of the absolute independent existence of space and time is the next step made by special relativity theory in the development of the principle of relativity. Special relativity theory asserts that space and time taken separately will not be physically real until a system of reference is specified. The pseudo-Euclidean metric tensor is the only concept which retains its absolute meaning: the form of this metric tensor is the same with respect to any inertial frame of reference. This condition determines the form of all the relations in special relativity theory.

The generalization of Riemannian metric concepts to the Finslerian case implies that the next step in the hierarchy of relativities should be made by dropping the absoluteness of the pseudo-Euclidean metric tensor; now, an absolute meaning may be ascribed only to the Finslerian metric tensor.

Proceeding from the transformation law (7.5) of the Finslerian metric tensor as the initial axiom for constructing kinematics, one may gradually derive all the kinematic relations. We shall follow this method, using the conventional techniques of special relativity theory. Naturally, we shall eventually come to the point at which the assertion "the space-time metric tensor is of pseudo-Euclidean form" is deprived of any specific physical meaning until we indicate the direction of motion of the inertial frame of reference. Since the use of any inertial frame of reference is thought to be equally justified in accordance with our general understanding of the principle of relativity, we arrive at the following:

Finslerian extention of the special principle of relativity. Kinematical experiments cannot serve to establish the existence of a preferred Riemannian metric tensor.

To avoid a confusion of concepts, we shall regard the notion of a frame of reference as representing a real physical object composed of basic material bodies and of physical instruments placed on these bodies, i.e. as a laboratory.

In the case of inertial frames of reference, the basic bodies are assumed to be moving inertially. It is emphasized that, when speaking of observations in some frame of reference, we imply that instruments are placed on the basic bodies of the frame of reference and are not transported so as to be placed on the basic bodies of other frames of reference. It is quite another matter that we may transport the instruments 'mentally'; the watershed between conventional special relativity theory and its Finslerian generalization may be expected to arise just from the difference in the readings of instruments transported from one inertial frame of reference to another, in reality and mentally.

An inertial frame of reference can be represented geometrically by means of its *proper system of reference*, that is by means of the tetrad $h^i_P(u)$ with the property that the zero-vector of the tetrad gives the four-dimentional direction of motion $l^i(u) = u^i/F(u)$ of the frame:

$$h^i_0(u) = l^i(u), \qquad h^0_i(u) = l_i(u) \tag{7.8}$$

where $l_i(u) = g_{ij}(u)l^j(u) \equiv \partial F(u)/\partial u^i$. A proper system of reference may be associated with any inertial frame of reference by the same prescription, the only ambiguity being that the spatial axes h^i_1, h^i_2, h^i_3 may be subjected to arbitrary three-dimensional Euclidean rotations and reflections. The proper systems of reference are orthonormal with respect to the Finslerian metric tensor. Simultaneoulsy, each proper system of reference is orthonormal with respect to the Riemannian metric intrinsic to itself. Generally speaking, however, the proper system of reference of one inertial frame of reference will not appear as orthonormal from the Riemannian viewpoint of another inertial frame of reference, the deformation of the Riemannian quantities being described by K-transformations of the form (7.6). It is this point that may be regarded as the essence of the generalization of special relativity theory implied by Finsler geometry.

Illustration. *Any n-tuple is orthonormal with respect to its own Euclidean metric tensor.* In an affine space V_2, two-dimensional for simplicity, we shall consider two pairs of vectors (v_1, v_2) and (w_1, w_2) assuming them to be arbitrary except for the condition that v_1 and w_1 are not collinear to v_2 and w_2, respectively. Each of the pairs may be considered orthogonal in the Euclidean sense. However, the Euclidean metric tensors with respect to which the pairs are orthogonal will in general be different:

The pair of vectors (v_1, v_2) is orthogonal with respect to the Euclidean metric tensor $v^{ij} = v^i_1 v^j_1 + v^i_2 v^j_2$. By means of Euclidean rotations one may obtain an infinite number of pairs of vectors orthogonal in the sense assigned by the pair (v_1, v_2)	The pair of vectors (w_1, w_2) is orthogonal with respect to the Euclidean metric tensor $w^{ij} = w^i_1 w^j_1 + w^i_2 w^j_2$. . . . . (w_1, w_2)

The centroaffine transformation K_i^j that connects (v_1, v_2) with (w_1, w_2) will also connect v^{ij} with w^{ij}. The set W of all non-collinear pairs of vectors from V_2 is broken into the direct sum of sets W_a, where W_a for each value of the index a is composed of all pairs of vectors orthogonal with respect to the same Euclidean metric tensor. Since the centroaffine transformations of V_2 involve four arbitrary parameters, W is a four-dimensional set of W_a.

7.3. Three Types of Velocities. The Fundamental Kinematic Relation

The following two statements concerning the law of composition of velocities are well known:

A_1: The parallelogram law of velocities is valid in Newtonian mechanics.

A_2: In special relativity theory three-dimensional relative velocities are added by means of the relativistic law of addition which differs essentially from the parallelogram law.

To clarify the rigorous geometrical meaning of these statements, let us consider some given inertial frame of reference β and denote by u_β^i the four-dimensional vector in the direction of which the frame β moves. The proper Lorentzian system of reference $h_i^P(u_\beta)$ will be chosen as the reference system in β, so that the relation $h_0^i(u_\beta) = l^i(u_\beta)$ (Equation (7.8)) will be assumed valid. Let two material points M_1 and M_2 move in the four-dimensional directions indicated by the vectors u_1^i and u_2^i, respectively. Denote

$$V_1^P = h_i^P(u_\beta)u_1^i/h_j^0(u_\beta)u_1^j, \qquad V_2^P = h_i^P(u_\beta)u_2^i/h_j^0(u_\beta)u_2^j \tag{7.9}$$

where

$$V_1^0 = V_2^0 = 1.$$

Then the motion of M_1 and M_2 relative to β will be represented by the three-dimensional velocities V_1^a and V_2^a, respectively, where $a = 1, 2, 3$. This observation suggests the introduction of the following

DEFINITION. The three-dimensional vector

$$V_{21}^a = V_2^a - V_1^a, \tag{7.10}$$

$a = 1, 2, 3$, will be called *the velocity of approach* of the material points M_1 and M_2 considered relative to the inertial frame of reference β.

This velocity may be obtained by considering in β the radius-vector$\mathbf{R}_{21}(x^0)$ directed from M_1 to M_2 and by putting

$$\mathbf{V}_{21} = \mathrm{d}\mathbf{R}_{21}/\mathrm{d}x^0$$

and it, as well as V_1^a, V_2^a and $\mathbf{R}_{21}$, rotates as a vector under Euclidean rotations of the spatial triad $h_i^a(u_\beta)$ of the proper Lorentzian system of reference $h_i^P(u_\beta)$ of the frame β. Therefore, in the case when three material points M_1, M_2, and M_3

are considred relative to the frame β, the velocities of approach will be added by means of the parallelogram law of velocities

$$V^a_{31} = V^a_{32} + V^a_{21},$$

because this is the law of composition of vectors which is effective in any three-dimensional Euclidean space.

Thus, apart from the relative velocity, the kinematics of special relativity involves another type of velocity, namely the velocity of approach. The law of addition of the latter is strictly the same as in Newtonian kinematics. As soon as we have separated the concept of the velocity of approach from the concept of the relative velocity, as the latter is used in special relativity, A_1 becomes a statement whose exact meaning must be specified in any proper geometrical theory. Indeed, what velocity do we deal with in A_1? If this is the velocity of approach, the truth of the statement A_1 will not be violated in generalizations of Newtonian kinematics as long as they are formulated in terms of affine geometry. So far as we assume condition A_1, we have to state that any generalization of Newtonian kinematics may only concern the concept of relative velocity.

Newtonian kinematics may be said to assume the identification of two logically different concepts of velocity, namely, the velocity of approach and the relative velocity. In this respect, special relativity is progressive because it clearly endows these two concepts with different content. This observation is instructive because the numerous troubles associated with the problem of aether[2] in prerelativistic physics ensued merely from the fact that the concept of the velocity of approach was intuitively confused with that of the relative velocity.

So, the quite legitimate question arises as to whether the concept of relative velocity, as used in special relativity theory, contains also the identification of different types of velocities. A rather simple examination shows that this is actually so; namely, it appears that the concept of relative velocity consists of two different concepts which we shall call the special relative velocity and the proper relative velocity and we denote them by V^a_S and V^a_P, respectively. It is this distinction that should be made in the Finslerian extension of special relativity theory. These relative velocities will obey different laws of addition, namely the velocities V^a_S will obey the conventional law of special relativity theory, whereas the law of addition of the velocities V^a_P will be defined by Finslerian K-transformations.

Indeed, in choosing the proper Lorentzian system of reference $h^P_i(u_\beta)$ as the reference system of the inertial frame of reference β, we have largely been guided by considerations of convenience, our physical assertion being only that such a choice was possible. If desired, we may proceed in β from the frame $h^P_i(u_\beta)$ to any other system of reference, orthonormal or not, which will give rise to the appropriate formal transformation of relations without affecting the physics. In particular, Lorentzian rotations may be used to replace the initial proper Lorentzian system of reference $h_i^P(u_\beta)$ by another reference systems $h_{1i}^{\ P}$ and $h_{2i}^{\ P}$ orthonormal in β:

$$h_{1i}^{\ P} = L_{1Q}^{\ P} h_i^Q(u_\beta), \qquad h_{2i}^{\ P} = L_{2Q}^{\ P} h_i^Q(u_\beta).$$

Let us constrain such a choice by the requirement

$$h_{10}^{\ i} = l^i(u_1), \qquad h_{20}^{\ i} = l^i(u_2), \tag{7.11}$$

which means that the tetrad $h_{10}^{\ i}$ related to the material point M_1 is obtained from $h_i^P(u_\beta)$ by means of a Lorentzian rotation such that the zero-vector $h_{10}^{\ i}$ of the tetrad is collinear to the vector u_1^i representing the four-dimentional direction of motion of the material point M_1; and the same for M_2. This leads us to the following

DEFINITION. The object

$$V_{S21}^a = h_{20}^{\ i} h_{1i}^{\ a} / h_{20}^{\ i} h_{1i}^{\ 0}, \tag{7.12}$$

$a = 1, 2, 3$, is called *the special relative velocity* of motion of the material point M_2 relative to the material point M_1, both M_1 and M_2 being considered with respect to the inertial frame of reference β.

Let us consider also a third material point M_3 moving in the direction of a vector u_3^i. Similarly to (7.11), we shall associate with M_3 the tetrad $h_{3i}^{\ P}$ possessing the property $h_{30}^{\ i} = l^i(u_3)$ with respect to the frame β. Then, owing to Equation (7.11) and the circumstance that the tetrads $h_{1i}^{\ P}$, $h_{2i}^{\ P}$ and $h_{3i}^{\ P}$ are obtainable from one another by suitable Lorentzian rotations, the function $V_{S31}^a(V_{S32}, V_{S21})$ must be essentially of the same form as in Riemannian special relativity theory. In other words, *the special relative velocities are added by means of the relativistic law* as the latter operates in the Riemannian formulation of special relativity theory.

This notwithstanding, we have not yet left the inertial frame of reference β which we fixed from the very beginning. The association with the inertially moving material point M_1 of the tetrad $h_{1i}^{\ P}$ in the frame β in such a way that $h_{10}^{\ i} = l^i(u_1)$, i.e. that M_1 is at rest with respect to $h_{1i}^{\ P}$, does not at all mean that we have a unique inertial frame of reference for M_1. As has been emphasized above, the choice of one or other reference system in each inertial frame of reference is dictated by convenience. In the present case, the only convenience of the choice of the tetrad $h_{1i}^{\ P}$ in the frame β is that the three-dimensional velocity of motion of M_1 vanishes with respect to this $h_{1i}^{\ P}$.

From the geometrical viewpoint, the essence of the problem is that, for the V_S-type relative velocities to be rigorously defined, it is sufficient to apply Lorentzian rotations in the *passive* sense only; that is, to consider the action of Lorentzian rotations as the rotations of a basis chosen in the affine space V_4, not as transformations among the vectors of V_4.

All the relations and laws of special relativity may be deduced step by step by an observer resting in a fixed inertial frame of reference and using Lorentzian rotations in the passive sense. In this case, of course, the observer will repeatedly be transferred 'mentally' to other inertial reference frames; however, "the mental Lorentz transformation in another inertial frame of reference" is nothing but a transition from the proper Lorentzian system of reference of an initial inertial frame of reference to another member of the set of tetrads associated with this inertial frame of reference. For example, the replacement of the proper Lorentzian system of reference of the inertial frame of reference β by the tetrad $h_{1}{}_{i}^{P}$ with the property $h_{1}{}_{0}^{i} = l^{i}(u_1)$ means that the observer remaining at rest in the frame β is "mentally transferred by means of a Lorentz transformation" from the initial frame β to another inertial frame of reference placed on the body M_1. This notwithstanding, the fact that the body M_1 has vanishing three-dimensional velocity with respect to $h_{1}{}_{i}^{P}$ does not in itself mean that the transition from β to the inertial frame of reference placed on M_1 is real, i.e. that instruments comoving with M_1 are used instead of instruments comoving with β.

The identification of "the mental Lorentz transition into another inertial reference frame" with a real transition into another inertial reference frame inevitably arises from the postulate that the primary metric structure of space-time is Riemannian. Riemannian geometry does not have the analytical means to make such an identification unnecessary. If, therefore, one is to elucidate the real state of affairs, one has to proceed from generalized metric concepts in order to discriminate analytically between the mental and real transitions. Only after this has been done in sufficient detail can the problem be referred to experiment as the higher authority.

Following our initial Finslerian approach, we may assert that the real transition from one inertial reference frame to another is achieved through the K-transformations which play the role of *active* kinematic transformations, because, by their definition (7.4), they transform the vectors of the tangent affine space into each other. Accordingly, the third type of velocity should be defined in terms of K-transformations.

To this end we shall consider two inertial frames of reference β_1 and β_2 moving in the direction of four-dimensional vectors u_1^i and u_2^i, respectively. The systems of reference $h_i^P(u_1)$ and $h_i^P(u_2)$ will be chosen in β_1 and β_2 to be proper Lorentzian. Therefore, K-transformations of the form (7.6) will entail that the relative components

$$R^P(1) \overset{\text{def}}{=} h_i^P(u_1)R^i, \qquad R^P(2) \overset{\text{def}}{=} h_i^P(u_2)R^i$$

of any vector R^i will be connected by the K-transformation

$$R^P(2) = R^Q(1)K^{*}{}_{Q}^{P}, \qquad R^P(1) = R^Q(2)K_Q^P, \tag{7.13}$$

which represents the transition from β_1 to β_2, and from β_2 to β_1. Here,

$$K_Q^P = K_m^n h_n^P(u_1)h_Q^m(u_1) = h_n^P(u_1)h_Q^n(u_2) \overset{\text{def}}{=} h_{(Q)}{}^{P}(u_2) \tag{7.14}$$

and

$$K_Q^P K^{*}{}_{R}^{Q} = \delta_R^P. \tag{7.15}$$

In particular,

$$u_2^P(1) = K_0^P, \qquad u_1^P(2) = K^{*P}_0, \qquad \delta_0^P = u_2^Q(1)K^{*P}_Q, \qquad \delta_0^P = u_1^Q(2)K_Q^P, \quad (7.16)$$

because $u_2^P(2) = u_1^P(1) = \delta_0^P$.

The motion of β_2 with respect to β_1 can be described by two three-dimensional velocities

$$U^a = K_0^a/K_0^0, \qquad U_a = K^{*0}_a/K^{*0}_0, \qquad (7.17)$$

both being defined relative to β_1 intrinsically, i.e. their definition does not assume a transition from the frame β_1 to another frame. Accordingly, we introduce the following

DEFINITION. The three-dimensional velocities (7.17) are called *the proper relative velocities* of motion of the inertial frame β_2 with respect to the inertial frame β_1.

The two velocities (7.17) have different physical and geometrical meanings. Indeed, let any source of light signals be at rest in β_2, for instance, at a point $R^a(2) = 0$. Then with respect to β_1 the events of the emission of light signals from such a source form a set of nonsimultaneous signals $R^P(1) = R^0(2)K_0^P$ with $R^0(2)$ arbitrary. Hence, $U^a = R^a(1)/R^0(1)$ Thus U^a is found by observing the motion of signal sources which are at rest in β_2.

U_a is of a different nature. Consider, with respect to β_2, a set of events: arbitrary $R^a(2)$ and fixed $R^0(2)$, for instance, $R^0(2) = 0$. The set may be visualized as a flash in β_2 at the proper time $R^0(2) = 0$. With respect to β_1 a set of nonsimultaneous signals $R^Q(1)$ will be observed such that

$$R^Q(1)K^{*0}_Q \equiv R^0(1)K^{*0}_0 + R^a(1)K^{*0}_a = 0.$$

Therefore,

$$U_a R^a(1)/R^0(1) = -1$$

which shows that at any moment $R^0(1)$ the vectors $R^a(1)$ form a three-dimensional plane of simultaneous events in β_1. The velocity U_a represents the direction of motion of this plane.

In this way, both velocities U^a and U_a can be observed directly in β_1. Therefore, observations of motions of different frames β_2 with respect to a fixed frame β_1 make it possible, in principle, to find in β_1 the relation

$$U_a = U_a(U^b) \qquad (7.18)$$

which may be called *the fundamental kinematic relation.*

7.4. Finslerian Kinematics

The fundamental kinematic relation (7.18) can be constrained by a set of conditions dictated by general physical considerations. First of all, we demand that

(i) $U_a(U^b = 0) = 0$

which reads that if β_2 is at rest relative to β_1 in the sense assigned by the velocity U^a then β_2 is at rest relative to β_1 in the sense assigned by the velocity U_a, too. A further condition is:

(ii) one-to-one correspondence and smoothness of the relation (7.18) between U_a and U^a.

Clearly, the fundamental kinematic relation can be easily rewritten in tensor form. In fact, the three-dimensional velocities (7.17) correspond to the four-dimensional ones

$$u^P = K^P_0, \qquad u_P = K^{*0}_P, \tag{7.19}$$

so that

$$U^a = u^a/u^0, \qquad U_a = u_a/u_0. \tag{7.20}$$

Substituting (7.20) in (7.18), contracting the result with u^a, and using the identity

$$u^P u_P \equiv u^0 u_0 + u^a u_a = 1 \tag{7.21}$$

ensuing from Equations (7.19) and (7.15), we get

$$1/u_0 = u^0 - u^a U_a$$

and hence the tensor form of the fundamental kinematic relation (7.18) will be

$$u_P(u^Q) = (u^0 - u^a U_a(u^b/u^0))^{-1}\, U_P(u^b/u^0), \tag{7.22}$$

where we have put $U_0 = 1$.

If the fundamental kinematic relation is known, the Finslerian metric tensor can easily be calculated in accordance with the following

PROPOSITION 1. The fundamental kinematic relation written in the tensor form (7.22) is identical with the Finslerian correspondence between contravariant and covariant vectors.

In fact, transforming $u^P(1)$ and $u_P(1)$ into β_2 by means of the general law (7.13) and taking into account Equations (7.19) and (7.15), we obtain $u^P(2) = \delta^P_0$ and $u_P(2) = \delta^0_P$, so that

$$u_P(2) = q_{PQ} u^Q(2), \tag{7.23}$$

where q_{PQ} are the constants that represent the signature of the metric tensor (Equation (7.1)). Since

$$u^P(2) = h^{(P}{}_Q{}^{)}(u(1))u^Q(1), \qquad u_P(2) = h^Q_{(P)}(u(1))u_Q(1)$$

in accordance with Equations (7.13) and (7.14), we can conclude from Equations (7.23) and (7.1) that

$$u_P(1) = g_{PQ}(u(1))u^Q(1)$$

actually holds. This relation can also be drawn directly from Equations (7.5) and (7.19). The proof of Proposition 1 is complete.

The next step will be to demand the condition of symmetry:

(iii) $K^m_n = K^n_m$

which is evidently identically with $K^P_Q = K^Q_P$. This condition essentially simplifies the kinematics because from (iii) it follows that the reverse proper relative velocities

$$U^{*a} = K^{*a}_0/K^{*0}_0, \qquad U^*_a = K^0_a/K^0_0, \tag{7.24}$$

describing the motion of β_1 with respect to β_2, can be determined in β_1 without leaving it. Although the last assertion could seem rather surprising at first glance, nevertheless a comparison of (7.17) with (7.24) shows immediately that the following result is valid.

PROPOSITION 2. The condition $K^P_0 = K^0_P$ (the condition $K^{*P}_0 = K^{*0}_P$) is a necessary and sufficient one for the velocity U^a (the velocity U_a) of motion of β_2 with respect to β_1 to be identical with the velocity U^*_a (with the velocity U^{*a}) of motion of β_1 with respect to β_2.

It has been tacitly assumed up to this point that $\det(K^m_n) \neq 0$. The physical meaning of this condition is that an inverse transformation exists for every K-transformation. It seems to be sufficient for kinematical considerations to assume that

(iv) $\det(K^n_m) \equiv \det(K^Q_P) = 1$.

Let us now consider inertial frames of reference β_2 with arbitrary admissible velocities U^a relative to a fixed β_1. Denoting by k_A and S^P_A the eigenvalues and eigenvectors of the matrix having entries $K^{*P}_Q(U^a)$, we have

$$K^{*P}_Q S^Q_A = k_A S^P_A, \tag{7.25}$$

$$K^Q_P S^A_Q = k^A S^A_P, \tag{7.26}$$

where the index A runs over $1, 2, 3, 4$,

$$k^A = 1/k_A, \tag{7.27}$$

and S^A_P are reciprocal to S^P_A, i.e.

$$S^A_P S^P_B = \delta^A_B. \tag{7.28}$$

Owing to the symmetry condition (iii) and the reality of K^m_n and K^P_Q, the eigenvalues k_A are real, and S^A_P and S^P_A can be regarded as real vectors. A sufficient smoothness of the dependence of S^A_P and S^P_A on U^a will be assumed below. On contracting Equation (7.25) with K_P^R, we get $S_A^R = k_A K_P^R S_A^P$ from which it follows, after comparing this with (7.26) and taking into account the symmetry condition (iii) together with Equation (7.27), that

$$S_P^A = C S_B^Q \delta^{AB} \delta_{PQ}, \tag{7.29}$$

where C is a nonvanishing factor and δ stands for the Kronecker symbol.

The eigenvectors have a clear physical meaning indicated by the following:

PROPOSITION 3. If a signal moves relative to β_1 in the direction of the three-dimensional velocity having the components $C_A^a = S_A^a/4S_A^0$ (alternatively, the components $C_a^A = S_a^A/S_0^A$), then the three-dimensional velocity of motion of the signal relative to β_2 has the same components.

In fact, let the signal move relative to β_1 in the four-dimensional direction $R^P(1) = S_A^P$ for some value of index A. Then, as a consequence of the transformation law (7.13) and the definition (7.25), we get $R^P(2) = k_A S_A^P$. Therefore,

$$R^a(1)/4R^0(1) = R^a(2)/4R^0(2) = S_A^a/4S_A^0 = C_A^a$$

which completes the proof.

In view of Proposition 3 we may call C_A^a and C_a^A *invariant velocities*. Under K-transformations, the length and the direction of each of the four three-dimensional velocities C_A^a, as well as C_a^A, remain invariant.

When $S_A^0 \to 0$, the corresponding invariant velocity C_A^a can run to infinity. Accordingly, the various kinematics obeying the conditions (i)–(iv) can be classified as follows.

(a) $S_A^0 \neq 0$ for any value of A.
(b) $S_A^0 = 0$ for only one value of A.
(c) $S_A^0 = 0$ for exactly two values of A.
(d) $S^0{}_A \neq 0$ for only one value of A.

In this respect, it will be noted that if $S_A^0 = 0$ for some value of A, then, in view of (7.29), $S_0^A = 0$ for the same value of A, and vice versa.

We shall be interested below only in the case (a)[3]. This case is the simplest and leads to the Berwald–Moór Finslerian metric function studied in Section 2.3. Since Equation (7.25) defines each of the eigenvectors S_A^P up to a nonvanishing factor, we may normalize them by demanding

(v) $S_A^0 = 1/4$.

PROPOSITION 4. *The conditions* (iii)–(v) *yield the following fundamental kinematic relation*:

$$u_P = \frac{1}{4} \sum_{A=1}^{4} \frac{S_P^A}{S_Q^A u^Q}. \tag{7.30}$$

Proof. By substituting $P = 0$ in Equations (7.25) and (7.26) and then dividing the result by $4S^0_A$ and S^A_0, respectively, we get

$$\dot{k}_A = 4C^Q_A u_Q, \qquad k^A = C^A_Q u^Q, \tag{7.31}$$

where $u_Q = K^{*0}_{Q}$, $u^Q = K^Q_0$, and

$$C^A_0 = 4C^0_A = 1. \tag{7.32}$$

The relations (7.27) and (7.31) yield a set of four equations

$$u_P = \sum_{A=1}^{4} \frac{S^A_0 S^0_A S^A_P}{S^A_Q u^Q}.$$

By virtue of the condition (v) and Equation (7.29), we can replace here $S^A_0 \cdot S^0_A$ by $C/16$ for any value of A, so that, after taking into account Equation (7.21), we obtain

$$C = 4 \tag{7.33}$$

and hence (7.30). The proof is complete.

From Equations (7.29) and (7.39) we may conclude that $S^A_0 = 1$, in addition to the condition (v). Therefore, according to the definition of the invariant velocities given in Proposition 3, we get

$$S^A_P = C^A_P, \qquad S^P_A = C^P_A. \tag{7.34}$$

In view of (7.20), the relation (7.30) entails the following expression for the function (7.18):

$$U_a(U^b) = \left(\sum_{B=1}^{4} \frac{1}{1 + C^B_b U^b} \right)^{-1} \sum_{A=1}^{4} \frac{C^A_a}{1 + C^A_b U^b}. \tag{7.35}$$

From Equations (7.28), (7.29), (7.33) and (7.34) it is obvious that the invariant velocities satisfy the relations

$$C^P_A C^A_Q = \delta^P_Q, \tag{7.36}$$

$$\sum_{A=1}^{4} C^A_a = \sum_{A=1}^{4} C^a_A = 0, \tag{7.37}$$

$$\sum_{A=1}^{4} C^A_P C^A_Q = 4\delta_{PQ}, \tag{7.38}$$

that is, that the quantities C^A_P appearing in our kinematic considerations satisfy all the relations which hold for the constants C^A_P as the latter appear in Section 2.3 in the construction of the parametrical representation of the indicatrix associated with the Berwald–Moór metric function. Therefore, reasoning as in Section 2.3 after Equation (2.70), we may conclude that the relations (7.36)–(7.38) define the quantities C^A_a as the constants given by the matrix (2.71), up to three-dimensional rotations dependent on U^a in an arbitrary way. However, wishing to restrict our consideration to the simplest case, we shall set forth the following condition.

(vi) The invariant velocities are independent of U^a.

Hence, the C_P^A will be the constants given by the matrix (2.71). Because of Equation (7.37), relation (7.35) entails that condition (i) holds. Also, condition (ii) follows from (7.35), if we take into account (7.36). Thus, we have arrived at:

PROPOSITION 5. The conditions (iii)–(vi) imply the conditions (i)–(ii).

Now the fundamental kinematic relation (7.30) can readily be integrated to yield the Finslerian metric function. Indeed, because the vectors u_P and u^Q entering (7.30) have unit Finslerian length in view of (7.21), and for unit vectors the Finslerian correspondence (1.9) holds, we have $u_P = \partial F(u)/\partial u^P$. Integration results in

$$F(u^P) = \left[\prod_{A=1}^{4} (C_P^A u^P) \right]^{1/4}, \tag{7.39}$$

that is in the Berwald–Moór metric function (Equation (2.30)). Thus we have established the following result.

PROPOSITION 6. The conditions (iii)–(vi) result in the Berwald–Moór metric function.

Concluding the above analysis, we think it of interest to note that one may arrive at the Berwald–Moór metric function by means of the following 'naive' kinematic reasoning: in the group of Lorentz transformations, only the subgroup leaving the two-dimensional space-time interval S^* invariant has a proper kinematic meaning; S^* can be generalized to four dimensions in two natural ways:

$$S^* = ((u^0)^2 - (u^1)^2)^{1/2} = ((u^0 - u^1)(u^0 + u^1))^{1/2}$$

$$\downarrow \qquad\qquad\qquad \downarrow$$

$$(a_{mn}(x) u^m u^n)^{1/2} \qquad \left[\prod_{A=1}^{4} (S_m^A(x) u^m) \right]^{1/4}$$

Riemannian way   Finslerian way

7.5. Proper Finslerian Kinematic Effects

Taking into account Equations (7.26), (7.28) and (7.32), and also the definition of the invariant velocities C_A^a and C_a^A given in Proposition 3 of the preceding section, we can represent the coefficients of K-transformations in terms of the invariant velocities as follows

$$K_P^Q(U) = \sum_{A=1}^{4} C_A^Q C_P^A k^A(U), \tag{7.40}$$

where

$$k^A(U) = C^A_Q U^Q u^0 \tag{7.41}$$

in accordance with Equations (7.31) and (7.20). Since

$$\prod_{A=1}^{4} k^A = 1$$

as a consequence of the condition (iv) of the preceding section, formula (7.41) entails

$$1/u^0 = F(U), \tag{7.42}$$

where

$$F(U) = \left[\prod_{A=1}^{4} (C^A_P U^P) \right]^{1/4} \tag{7.43}$$

is in fact the Finslerian metric function (7.39) expressed in terms of the argument U^P. Substituting in (7.40) the relation (7.41) with u^0 taken from (7.42), we explicitly obtain

$$K^Q_P(U) = \frac{U^R}{F(U)} \sum_{A=1}^{4} C^Q_A C^A_P C^A_R. \tag{7.44}$$

The group of such K-transformations is obviously Abelian and three-parametrical.

Putting in (7.44) $P = 0$ and $Q = a$, and then taking into account (7.36) (or putting in (7.44) $P = a$ and $Q = 0$ and using (7.38)), we obtain the relation

$$K^a_0(U)/K^0_0(U) = K^0_a(U)/K^0_0(U) = U^a, \tag{7.45}$$

which agrees with Equations (7.19)–(7.20). Also, from (7.44) with $Q = a$ and $P = b$, together with (7.36), it follows that

$$K^a_b(U)/K^0_0(U) = \delta^b_a + U^c r_b{}^a{}_c \tag{7.46}$$

where

$$r_b{}^a{}_c = \sum_{A=1}^{4} C^a_A C^A_b C^A_c \tag{7.47}$$

plays the role of the tensor of anisotropy of space-time. Comparison of (7.19) with (7.42) yields

$$K^0_0(U) = 1/F(U). \tag{7.48}$$

Formulae (7.45)–(7.48) should be compared with the coefficients of the usual special Lorentz transformations, which are

$$L^0_0(V) = (1 - V^2)^{-1/2}, \tag{7.49}$$

$$L^1_0(V)/L^0_0(V) = L^0_1(V)/L^0_0(V) = V, \tag{7.50}$$

$$L_1^1(V)/L_0^0(V) = 1, \tag{7.51}$$

where $V \equiv V_S$ is the special relative velocity.

In the case when the velocities are small with respect to the light velocity c, which has so far been assumed to be equal to unity, we find from Equation (7.49) that

$$L_0^0 = 1 + \tfrac{1}{2}(V/c)^2 + O(V/c)^4, \tag{7.52}$$

while the relations (7.50) and (7.51) will not require any expansion. On the other hand, if we expand the function (7.43) with respect to the parameter

$$q = (\delta_{ab} U^a U^b)^{1/2}/c \tag{7.53}$$

then we obtain, according to the identities (7.37) and (7.38),

$$F = 1 - \tfrac{1}{2}q^2 + O(q^3), \tag{7.54}$$

so that, in view of the relation (7.48), we shall have

$$K_0^0 = 1 + \tfrac{1}{2}q^2 + O(q^3).$$

The form of the last expansion deviates from that of the expansion (7.52) beginning with the third order term. The right-hand sides of Equations (7.45) and (7.46) do not require any expansion procedure. In (7.46), the corrections are of the first order with respect to q. However, in corrections for the velocities the coefficients K_b^a usually enter in the contractions with U^b, in which case one may expect that the Finslerian corrections for the velocities may begin only with terms which are of the second order with respect to q.

On the basis of these observations one could in principle begin to search for possible relativistic Finslerian 'post-Riemannian' kinematic effects. Let us illustrate this possibility by giving the following example. In the usual special relativity theory the reciprocity principle rigorously holds. This principle claims that, if we consider two inertially moving material points M_1 and M_2 with respect to some inertial frame of reference, then their relative velocities V_{S12}^a and V_{S21}^a satisfy the relation[4]

$$V_{S12}^a = -V_{S21}^a. \tag{7.55}$$

For the proper relative velocities U^a, however, such a relation does not in general hold. Instead, U_{12}^a and U_{21}^a will be connected by a relation of the form (7.35) (as a consequence of Proposition 2), the expansion of which yields

$$U_{12}^a = -U_{21}^a + cO(q^2) \tag{7.56}$$

because of the identity (7.37). If an observation is made in two laboratories which move at a relative velocity of $\sim 3\,\mathrm{km\,s^{-1}}$ (such a situation occurs in experiments made near the Earth), then, according to (7.56), corrections to the reciprocity principle will amount to $(3/300000)^2 c = 10^{-10}\,c = 3\,\mathrm{cm\,s^{-1}}$.

7.6. Finslerian Kinematics as a Consequence of the Equations of Motion of Matter

The formulation of the Finslerian equations of motion of matter in Section 4.3 has led in a simple case to the conclusion that test bodies in a gravitational field move along Finslerian geodesics. As a consequence of this, there arises a clear geometrical picture of inertial motion, which generalizes its Riemannian prototype. Namely, the inertial motion of a test body in a gravitational field gives rise to the Finslerian parallel transportation of a Finslerian tetrad representing the proper orthonormal system of reference of the body. From the viewpoint of the usual Riemannian notions, such transportation of the comoving tetrad causes its deformation. In the limiting case, when we restrict ourselves to a small four-dimensional vicinity W such that all gradients of the gravitational field variables can be neglected, we may consider in W test bodies with different velocities of motion, while the proper tetrads will then depend on four-dimensional directions of motion in accord with the relations derived in the previous Sections 7.1–7.5.

In particular, in the static gravitation field generated by a static body M, say by the Earth, we obtain the following picture that is completely consistent with the equations of the gravitational field and of the motion of matter. With respect to the static coordinates $x^P = (x^0, x^a)$, the auxiliary vector field $y^P(x)$ has the particular form $\delta^P_0 y^0(x)$ (Equation (4.74)), and the Finslerian metric tensor $g_{PQ}(x, y)$ gives rise to the osculating Riemannian metric tensor of the static gravitational field:

$$a_{PQ}(x) \stackrel{\text{def}}{=} g_{PQ}(x, y(x)) = g_{PQ}(x, \delta^P_0)$$

(see Section 4.5). In a sufficiently smal vicinity W of any fixed point C_0 beyond M we can consider a set of inertial frames of reference (laboratories), that is a set of sufficiently small test bodies with proper systems of reference (see Equation (7.8)). These systems of reference are orthonormal with respect to the Finslerian metric tensor. If a frame $h^{(P)}_Q$ in W refers to inertial motion in the four-dimensional direction $y^P(x)$ (the inertial frame of reference which is instantaneously at rest relative to the Earth), then with respect to this frame the metric tensor $a_{PQ}(C_0)$ will be of pseudo-Euclidean form. If, however, the four-dimensional directions of motion deviates from the direction y^P, this tensor will cease to look pseudo-Euclidean (it would remain pseudo-Euclidean in the usual Riemannian approach) and will be deformed in accordance with formulae (7.5).

Thus we have obtained an analytically consistent picture of Finslerian kinematics for the system gravitational field plus test bodies. Nevertheless, we shall not have a complete set of kinematical concepts unless we include in our considerations the electromagnetic field. Indeed, electromagnetic signals are generally used in kinematical considerations. Again, we have no right just to posutlate in one way or another the behaviour of light signals under kinematic transformations. All properties of light signals are contained in the electromagnetic field equations, so that the question here reduces to the formulation of the latter.

In Section 4.2, we have proposed one particular method of formulating the electromagnetic field equations in the Finslerian approach. According to this method, the most important part of the equations, namely, the first term on the left-hand side of Equation (4.51), is written in terms of the osculating Riemannian metric tensor in the conventional way. In the vicinity W, these equations will assume the usual Maxewell form $d_Q F^{PQ} = 0$. Therefore, the corresponding equation of the light wave front in W will be written in the form $a_{PQ} x^P x^Q = 0$. Relative to an inertial frame of reference instantaneously at rest with respect to the body M, the latter equation will read

$$(x^0)^2 - (x^1)^2 - (x^2)^2 - (x^3)^2 = 0, \tag{7.57}$$

while relative to another frame of reference moving with respect to M with a velocity U^a, this same front will be described by the equation

$$g_{PQ}(U) R^P R^Q = 0, \tag{7.58}$$

where g_{PQ} and a_{PQ}, as well as the coordinates R^P and x^P, are related by K-transformations (see Equations (7.5) and (7.13)).

The kinematic picture thus arrived at is in all respects similar to that of Section 7.2, with the exception of one new essential point, namely, the presence of the preferred Riemannian metric tensor given by the osculating a_{PQ}. It is important that the latter tensor arises not from the properties of kinematic transformations, as it does in the usual Riemannian approach, i.e. not in terms of invariants of kinematic transformations, but as a result of putting more reality into the physical model itself. By going beyond the limitations imposed by a proper kinematic picture, we are able to investigate not an 'absolutely empty space-time' filled with imaginary inertial observers, who do not generate any field and who are merely engaged in the exchange of signals, but a space-time with a material content which begets a preferred Riemannian metric tensor. The latter is the very Riemannian metric tensor in terms of which the behaviour of a light wave is primarily described. It may also be said that the presence of a preferred Riemannian metric tensor is explained by the presence of a preferred frame of reference, namely, that which is at rest relative to the gravitating body M. The necessity of going beyond the framework of the proper kinematic picture in developing the Finslerian theory is a point of principle – and far more important than for the usual Riemannian theory, in which the Riemannian metric tensor due to gravitating bodies and the tensor which appears as an invariant of kinematic transformations turn out to be identical. The proper Finslerian approach does not involve any Riemannian metric tensor invariant under kinematic transformations, the Finslerian metric function assuming the role of such an invariant.

Let us consider Equation (7.58) in more detail. The explicit expression for the tensor $g_{PQ}(U)$ is given, according to Equations (2.32) and (7.34), by the formula

$$g_{PQ}(U) = (2U_P U_Q - 4U_M U_N \sum_{A=1}^{4} C_P^A C_Q^A C_A^M C_A^N)/F^2(U) \tag{7.59}$$

where $U_0 = 1$, Because $C_0^A = 1$ (Equation (7.32)), from (7.59) it follows that

$$g_{00} = (2 - 4U_M U_N \sum_{A=1}^{4} C_A^M C_A^N)/F^2.$$

Substituting the relation

$$\sum_{A=1}^{4} C_A^M C_A^N = \tfrac{1}{4}\delta^{MN}$$

resulting from Equations (7.36) and (7.38), we get

$$g_{00} = (2 - \delta^{MN} U_M U_N)/F^2 = (1 - \delta^{ab} U_a U_b)/F^2.$$

Passing to the approximation of low velocities, i.e. $q \ll 1$, where the parameter q is given by Equation (7.53), and taking into account the expansion (7.54) and

$$U_a = -U^a + O(q^2)$$

(a consequence of (7.35) and (7.37); cf. Equation (7.56)), we find

$$g_{00}(U) = 1 + O(q^3). \tag{7.60}$$

Similarly, the other components of the tensor (7.59) may be evaluated to give

$$g_{0a}(U) = O(q^2) \tag{7.61}$$

and

$$g_{ab}(U) = -\delta_{ab} + 2r_{abe} U^e + O(q^2), \tag{7.62}$$

where we have denoted $r_{abe} = r_{ab}{}^e$, and $r_{ab}{}^e$ is the anisotropy tensor given by Equation (7.47). In view of Equations (7.29), (7.33) and (7.34), we may substitute $C_A^e = \frac{1}{4}C_e^A$ in (7.47) to obtain

$$r_{abe} = \tfrac{1}{4} \sum_{A=1}^{4} C_a^A C_b^A C_e^A. \tag{7.63}$$

In the lowest-order approximation in the parameter q, Equation (7.58) yields, after substituting (7.60)–(7.62),

$$c^2 - (\delta_{ab} - 2r_{abe} U^e) c^a c^b = 0, \tag{7.64}$$

where c^a are the components of the light velocity, and c^2 is the square of the universal constant. Comparing Equations (7.57) and (7.64) we see that, according to equation (7.64), the velocity of light will be different in different directions, the ellipsoid of the light velocities being defined by the Finslerian anisotropy tensor (7.63). Since the components of the latter tensor are of the order of unity (as is obvious from Equations (7.63) and (2.71)), the deviations of the velocity of light from the value of the universal constant c will be of the order of q. In order to detect such deviations, one can suggest that a Michelson-type interferometer be used, though not in a laboratory resting on the Earth (as in the classical experiments; see Whittaker (1960), Törnebohm (1970), Mandelshtam (1972)), but in a laboratory moving sufficiently rapidly with respect to the Earth. One would expect that the effect should be highly anisotropic. It is of the order of q, i.e. rather large, if we recall that in the classical experiments of Michelson and his followers instrumental accuracy was brought to a level at which effects of the second order in q^2 could be felt. Attempts were made at that time to detect the effects caused by the motion of the Earth around the Sun, which corresponds to $q^2 \simeq 10^{-8}$. When a laboratory moves relative to the Earth with a velocity of $\sim 3 \text{ km s}^{-1}$, which is in principle attainable at present, we have $q \simeq 10^{-5}$, so that the Finslerian effect proves to be sufficiently large to allow us to hope that it is observable in spite of all the difficulties that may arise due to the necessity of using a laboratory moving rapidly relative to the Earth.

Problems

7.1. Consider invariant velocities for the special Lorentz transformations.

7.2. Derive the law of addition of the proper relative velocities.

7.3. What should the active kinematic transformations of the electromagnet field be?

7.4. What new features can Finslerian kinematics be expected to introduce in the Doppler effect?

7.5. Can Finslerian kinematics be developed on the basis of an arbitrary Finslerian metric tensor $g_{ij}(x, y)$ of space-time signature $(+ - - -)$?

7.6. Extend the kinematic ideas of the present chapter to noninertial reference frames taking as an example the motion of test electric charge in the classical gravitational and electromagnetic fields.

Notes

[1] The present chapter is based on the results of the papers by Asanov (1977a, 1979b).

[2] See Whittaker (1960), Mandelshtam (1972).

[3] The cases (b)–(d) have not been studied in the literature and may furnish a subject for an original investigation.

[4] The principle of reciprocity has been investigated in a paper by Berzi and Gorini (1969), where it has been shown that postulating the principle of relativity, together with the homogeneity and the isotropy of space-time, leads unambiguously to the reciprocity principle.

Concluding Remark

If the reader who has fully imbibed the contents of this volume should suddenly forget its whole mathematical formalism, then there will appear a highly extraordinary amalgam of ideas in which many seemingly quite alien concepts are interwoven. What relation is there between the SU(2) group of symmpetry of physical fields and our attempt to introduce the concept of world time as a physical field or, for that matter, between these ideas and the readings of a Michelson interferometer in a laboratory moving relative to the Earth? What have these concepts to do with the possibility of defining the energy-momentum of the gravitational field in a covariant and integrable manner? Might all these questions be related to the idea of measuring the length of a vector by the Nth-root of and Nth-order form? It seem that, if one arms oneself with the techniques of Finsler geometry, these relations appear as if by themselves. The conscientious reader will also see many other new relations.

The Finslerian approach is nowhere found to be an alternative to the Riemannian general or special theory of relativity. As a matter of fact, it leads to their extension, it develops their methods, and it brings forth new concepts of space-time.

APPENDIX A

Direction-Dependent Connection and Curvature Forms

Finsler geometry rests essentially on the assumption of the homogeneity of basic geometrical objects with respect to tangent vectors and on the symmetry of the connection coefficients. It turns out, however, that one may develop the theory without these two assumptions and, generally speaking, without assuming any metric tensor. In what follows, the main results arrived at by Rund (1981a, Chapter 8) are described. The exposition is performed using the currently popular language of differential forms[1].

In the simplest case, a direction-dependent[2] p-form r can be represented locally in terms of the standard tangent space basis $\{\mathrm{d}x^k\}$, that is,

$$r = r_{i_1 \ldots i_p}(x, y)\, \mathrm{d}x^{i_1} \wedge \ldots . \wedge \mathrm{d}x^{i_p}$$

where the symbol $\wedge$ stands for the exterior product. In such a case, we shall say that the p-form r is an element of the space L^p. However, in general a direction-dependent $(p + q)$-form r requires a local representation of a more general type

$$r = r_{i_1 \ldots i_p m_1 \ldots m_q}(x, y) \mathrm{d}x^{i_1} \wedge \ldots \wedge \mathrm{d}x^{i_p} \wedge \mathrm{d}y^{m_1} \wedge \ldots \wedge \mathrm{d}y^{m_q}.$$

In the following, the symbol d will be used for the exterior differentiation operator, whose action is defined by

$$\mathrm{d}r = \frac{\partial r_{i_1 \ldots i_p m_1 \ldots m_q}}{\partial x^k} \mathrm{d}x^k \wedge \mathrm{d}x^{i_1} \wedge \ldots \wedge \mathrm{d}x^{i_p} \wedge \mathrm{d}y^{m_1} \wedge \ldots \wedge \mathrm{d}y^{m_q} +$$

$$+ \frac{\partial r_{i_1 \ldots i_p m_1 \ldots m_q}}{\partial y^k} \mathrm{d}y^k \wedge \mathrm{d}x^{i_1} \wedge \ldots \wedge \mathrm{d}x^{i_p} \wedge \mathrm{d}y^{m_1} \wedge \ldots \wedge \mathrm{d}y^{m_q}.$$

In particular, if $A(x, y)$ is a zero-form (tensor, scalar, etc.), then the operator d acts as an ordinary differential:

$$\mathrm{d}A = \frac{\partial A}{\partial x^k} \mathrm{d}x^k + \frac{\partial A}{\partial y^k} \mathrm{d}y^k.$$

We shall call a tensor which is m times contravariant and n times covariant, for brevity, a type (m, n) tensor.

The generalization which we wish to attain starts with the following.

DEFINITION. Suppose we are given a set of N^2 1-forms, to be denoted by $\beta_h{}^j(x, y)$. The $\beta_h{}^j$ are called *connection 1-forms*, if the following three conditions are satisfied:

(i) The forms are elements of the space L^1, that is they admit a representation of the form

$$\beta_h{}^j(x, y) = \Gamma_h{}^j{}_k(x, y)\,\mathrm{d}x^k. \tag{A.1}$$

(ii) The coefficients $\Gamma_h{}^j{}_k$ in (A.1) are of class C^3 in their 2N arguments (x^n, y^n).

(iii) Under coordinate transformations $x^i = x^i(\bar{x}^j)$ of class C^3 these 1-forms transform in accordance with

$$\mathrm{d}B^j_h = B^j_l \bar{\beta}_h{}^l - B^l_h \beta_l{}^j, \tag{A.2}$$

where $B^j_h = \partial x^j / \partial \bar{x}^h$.

The last condition (A.2) is evidently equivalent to the definition (3.38) written for the connection coefficients $\Gamma_h{}^j{}_k$.

Since the left-hand side of (A.2) is an exact differential, the exterior derivative of the right-hand side of (A.2) must vanish:

$$\mathrm{d}B^j_l \wedge \bar{\beta}_h{}^l + B^j_l\,\mathrm{d}\bar{\beta}_h{}^l - \mathrm{d}B^l_h \wedge \beta_l{}^j - B^l_h\,\mathrm{d}\beta_l{}^j = 0$$

or, on substituting (A.2),

$$B^j_l \bar{W}_h{}^l = B^l_h W_l{}^j, \tag{A.3}$$

where the 2-forms $W_h{}^j(x, y)$ are defined as follows

$$W_h{}^j = \mathrm{d}\beta_h{}^j + \beta_l{}^j \wedge \beta_h{}^l \tag{A.4}$$

and are obviously type (1, 1) forms.

The exterior covariant derivative of any type (1, 0) tensor field with local components $X^j(x, y)$ can be defined conventionally as

$$\mathrm{D}X^j \overset{\mathrm{def}}{=} \mathrm{d}X^j + \beta_h{}^j X^h. \tag{A.5}$$

The condition (A.2) ensures that $\mathrm{D}X^j$ are 1-forms of the type (1, 0) again, so that one may apply the definition (A.5) to $\mathrm{D}X^j$ itself, yielding

$$\mathrm{D}(\mathrm{D}X^j) \overset{\mathrm{def}}{=} \mathrm{d}(\mathrm{D}X^j) + \beta_l{}^j \wedge \mathrm{D}X^l = W_l{}^j X^l. \tag{A.6}$$

This observation, together with the transformation law (A.3), suggests that $W_h{}^j$ can be called *curvature 2-forms*. Moreover, it can be readily verified that the following *Bianchi identity* holds:

$$\mathrm{D}W_h{}^j \overset{\mathrm{def}}{=} \mathrm{d}W_h{}^j + \beta_l{}^j \wedge W_h{}^l - \beta_h{}^l \wedge W_l{}^j = 0. \tag{A.7}$$

Formally, the above relations do not differ from those of the standard theory of connections dependent solely on x^i. However, in the direction-dependent case the curvature 2-forms as defined by Equation (A.4) fail in general to be elements of the space L^2. Indeed, the substitution of (A.1) in (A.4) yields the following decomposition:

$$W_l^{\ j} = \left(\frac{\partial \Gamma_l{}^j{}_k}{\partial x^h} + \Gamma_m{}^j{}_h \Gamma_l{}^m{}_k\right) \mathrm{d}x^h \wedge \mathrm{d}x^k + \frac{\partial \Gamma_l{}^j{}_k}{\partial y^m} \mathrm{d}y^m \wedge \mathrm{d}x^k. \tag{A.8}$$

Replacing in Equation (A.8) the ordinary differential $\mathrm{d}y^m$, which is not tensorial, by the exterior covariant derivative

$$\mathrm{D}y^m \overset{\text{def}}{=} \mathrm{d}y^m + \beta_i{}^m(x, y) y^i = \mathrm{d}y^m + \Gamma_i{}^m{}_k(x, y) y^i \, \mathrm{d}x^k, \tag{A.9}$$

we find the decomposition

$$W_l^{\ j} = W_l^{*j} + \frac{\partial \Gamma_l{}^j{}_k}{\partial y^m} \mathrm{D}y^m \wedge \mathrm{d}x^k, \tag{A.10}$$

in which

$$W_l^{*j} = \left(\frac{\partial \Gamma_l{}^j{}_k}{\partial x^h} - \frac{\partial \Gamma_l{}^j{}_k}{\partial y^m} \Gamma_r{}^m{}_h y^r + \Gamma_m{}^j{}_h \Gamma_l{}^m{}_k\right) \mathrm{d}x^h \wedge \mathrm{d}x^k \tag{A.11}$$

represents the part of $W_l^{\ j}$ which is contained in L^2 and, therefore, admits a representation of the form

$$W_l^{*j} = -\frac{1}{2} K_l{}^j{}_{hk} \, \mathrm{d}x^h \wedge \mathrm{d}x^k, \tag{A.12}$$

where the object $K_l{}^j{}_{hk}$ may be called *the curvature tensor*. Comparing Equations (A.11) and (A.12) shows that

$$\begin{aligned} K_l{}^j{}_{hk}(x, y) = {} & \left(\frac{\partial \Gamma_l{}^j{}_h}{\partial x^k} - \frac{\partial \Gamma_l{}^j{}_h}{\partial y^m} \Gamma_r{}^m{}_k y^r\right) - \\ & - \left(\frac{\partial \Gamma_l{}^j{}_k}{\partial x^h} - \frac{\partial \Gamma_l{}^j{}_k}{\partial y^m} \Gamma_r{}^m{}_h y^r\right) + \Gamma_m{}^j{}_k \Gamma_l{}^m{}_h - \Gamma_m{}^j{}_h \Gamma_l{}^m{}_k. \end{aligned} \tag{A.13}$$

The Equations (A.10), augmented by Equations (A.11) and (A.12), being an explicit representation of the curvature 2-forms (A.4), may be called *the equations of structure of the first kind.*

The tensor (A.13) is remarkable for the generality of its definition: it involves, as particular cases, the curvature tensor associated with affine connection coefficients dependent solely on x^i, and the Finslerian K-curvature tensor (1.78). The reader will be justified in concluding that among the three Finslerian curvature tensors: Cartan's $R_j{}^i{}_{hk}$, Berwald's $H_j{}^i{}_{hk}$, and Rund's $K_j{}^i{}_{hk}$, each of which in its own way generalizes the Riemann curvature tensor to the Finslerian case, it is exactly the K-tensor that seems to be the most elementary Finslerian curvature tensor. Unlike the K-tensor, the Finslerian tensors $R_j{}^i{}_{hk}$ and $H_j{}^i{}_{hk}$ include in their definition, apart from the connection coefficients, the

Cartan torsion tensor, and it is for this reason alone that they have no analogs in generalizing the concept of connection.

It will be noted that if we have a stationary vector field $v^i(x)$, that is

$$\mathrm{d}v^j = -\Gamma_{l\ k}^{\ j}(x, v)v^l\,\mathrm{d}x^k,$$

and introduce the osculating affine connection coefficients

$$A_{l\ k}^{\ j}(x) \overset{\text{def}}{=} \Gamma_{l\ k}^{\ j}(x, v(x)),$$

then the osculating values of many direction-dependent objects will reduce to their affine prototypes studied in a book by Eisenhart (1927). In particular,

$$K_{l\ hk}^{\ j}(x, v) = L_{l\ hk}^{\ j}$$

where

$$L_{l\ hk}^{\ j} \overset{\text{def}}{=} \frac{\partial A_{l\ h}^{\ j}}{\partial x^k} - \frac{\partial A_{l\ k}^{\ j}}{\partial x^h} + A_{m\ k}^{\ j} A_{l\ h}^{\ m} - A_{m\ h}^{\ j} A_{l\ k}^{\ m}$$

denotes the affine curvature tensor associated with $A_{l\ h}^{\ j}$. Since the integrability conditions for a system

$$\mathrm{d}v^j = -A_{l\ k}^{\ j}(x)v^l\,\mathrm{d}x^k$$

read

$$v^l L_{l\ hk}^{\ j} = 0$$

(the implication of the identity $\mathrm{dd}v^j = 0$; cf. (S.34)), the identical vanishing of the K-tensor $K_{l\ hk}^{\ j}(x, y)$ will ensure the validity of these conditions.

The torsion 2-forms $W^j(x, y)$ are introduced in the usual way, namely, relative to a canonical basis $\{\mathrm{d}x^k\}$,

$$\begin{aligned} W^j \overset{\text{def}}{=} -\mathrm{D}(\mathrm{d}x^j) &= -\beta_h^{\ j} \wedge \mathrm{d}x^h = \Gamma_{h\ k}^{\ j}\,\mathrm{d}x^h \wedge \mathrm{d}x^k = \\ &= \tfrac{1}{2} S_{h\ k}^{\ j}\,\mathrm{d}x^h \wedge \mathrm{d}x^k \end{aligned} \tag{A.14}$$

where

$$S_{h\ k}^{\ j}(x, y) = \Gamma_{h\ k}^{\ j}(x, y) - \Gamma_{k\ h}^{\ j}(x, y) \tag{A.15}$$

is the direction-dependent type $(1, 2)$ *torsion tensor*. The torsion 2-forms thus obtained are elements of L^2 and, therefore, require no decomposition procedure. From Equations (A.14) and (A.15) it follows directly that

$$\mathrm{D}W^j = -W_h^{\ j} \wedge \mathrm{d}x^h. \tag{A.16}$$

The covariant derivative of any tensor field can be defined by analogy with Cartan's Finslerian rule (1.41). Namely, if $T_{h\dots}^{j\dots}(x, y)$ denote the local components of a type (p, q) tensor field, then the definition will read

$$T_{h\dots|k}^{j\dots} \overset{\text{def}}{=} \frac{\partial T_{h\dots}^{j\dots}}{\partial x^k} - \frac{\partial T_{h\dots}^{j\dots}}{\partial y^m} \Gamma_{r\ k}^{\ m} y^r + \Gamma_{l\ k}^{\ j} T_{h\dots}^{l\dots} - \Gamma_{h\ k}^{\ l} T_{l\dots}^{j\dots} + \dots. \tag{A.17}$$

Because of (A.9), applying the definition (A.5) of the exterior covariant derivative to $T_{h\ldots}^{j\ldots}$ yields

$$\begin{aligned} \mathrm{D}T_{h\ldots}^{j\ldots} &\overset{\text{def}}{=} \mathrm{d}T_{h\ldots}^{j\ldots} + \beta_l^{\ j} T_{k\ldots}^{l\ldots} - \beta_h^{\ l} T_{l\ldots}^{j\ldots} + \cdots = \\ &= T_{h\ldots|k}^{j\ldots}\,\mathrm{d}x^k + \frac{\partial T_{h\ldots}^{j\ldots}}{\partial y^m}\,\mathrm{D}y^m. \end{aligned} \tag{A.18}$$

However, because of Equation (A.14), the corresponding result for tensorial p-forms ($p \geqslant 1$) would be valid only if the torsion tensor vanishes.

From direction-dependent objects it is possible to construct new objects by differentiating not only with respect to the coordinates x^i but with respect to the tangent vectors y^i as well. The tensorial character of this last operation is obvious: the derivative of a tensor with respect to y^i is again a tensor. Keeping this in mind, one may determine the derivative of forms with respect to y^i. This operation will be denoted by ∂_{y^l}, or ∂_y for the sake of brevity. It can be elegantly defined axiomatically. Namely, the result of applying ∂_y to a p-form of type (m, n) is a p-form of type $(m, n + 1)$, and the action of ∂_y will be specified by the following requirements:

(a) For a pair of p-forms r and s,

$$\partial_y(r + s) = \partial_y r + \partial_y s;$$

(b) For a p-form r, and a q-form t,

$$\partial_y(r \wedge t) = (\partial_y r) \wedge t + r \wedge (\partial_y t),$$

irrespective of the values of p and q;

(c) For a zero-form $A(x, y)$,

$$\partial_{y^l} A = \partial A/\partial y^l;$$

(d) For any p-form r,

$$\partial_y(\mathrm{d}r) = \mathrm{d}(\partial_y r),$$

that is, the operator ∂_y commutes with the operator d of exterior differentiation.

Applying the rule (d) to $r = x^k$ and using the rule (c), we get

$$\partial_y(\mathrm{d}x^k) = \mathrm{d}(\partial_y x^k) = 0.$$

Similarly, if we put $r = y^m$,

$$\partial_{y^l}(\mathrm{d}y^m) = \mathrm{d}(\partial_{y^l} y^m) = \mathrm{d}(\delta_l^m) = 0.$$

Therefore, for any direction-dependent $(p + q)$-form r it follows from the above formulae that

$$\partial_{y^l} r = \frac{\partial r_{i_1 \ldots i_p m_1 \ldots m_q}}{\partial y^l}\,\mathrm{d}x^{i_1} \wedge \ldots \wedge \mathrm{d}x^{i_p} \wedge \mathrm{d}y^{m_1} \wedge \ldots \wedge \mathrm{d}y^{m_q},$$

so that the directional derivative of any differential form is uniquely determined by the rules (a)–(d).

In particular, the directional derivative of the 1-form (A.9) will be given by

$$\partial_{y^l}(\mathrm{D}y^m) = \beta_l{}^m + \beta_r{}^m{}_l y^r \tag{A.19}$$

where

$$\beta_r{}^m{}_l = \partial_{y^l}\beta_r{}^m = \frac{\partial \Gamma_r{}^m{}_k}{\partial y^l}\,\mathrm{d}x^k$$

are evidently 1-*forms of type* (1, 2):

$$B^j_l \beta_h{}^l{}_m = \beta_p{}^j{}_l B^p_h B^l_m$$

as may be found, e.g., by taking the directional derivative of the transformation law (A.2). Also, the definition (A.4) entails that

$$\begin{aligned} W_l{}^j{}_m \overset{\text{def}}{=} \partial_{y^m} W_l{}^j = \mathrm{d}\beta_l{}^j{}_m + \beta_p{}^j{}_m \wedge \beta_l{}^p + \beta_p{}^j \wedge \beta_l{}^p{}_m \equiv \\ \equiv \mathrm{D}\beta_l{}^j{}_m - \beta_l{}^j{}_p \wedge \beta_m{}^p, \end{aligned} \tag{A.20}$$

while from (A.10) it follows that

$$W^{*j}_l{}_m \overset{\text{def}}{=} \partial_{y^m} W^{*j}_l = \mathrm{D}\beta_l{}^j{}_m + \beta_l{}^j{}_p \wedge \beta_r{}^p{}_m y^r + \beta_l{}^j{}_{pm} \wedge \mathrm{D}y^p, \tag{A.21}$$

where

$$\beta_l{}^j{}_{pm} \overset{\text{def}}{=} \partial_{y^m}\beta_l{}^j{}_p = \beta_l{}^j{}_{mp}.$$

It is clear that the directional derivative of a tensorial p-form will not be tensorial unless the form is an element of L^p; this phenomenon is illustrated by the relations (A.19) and (A.20), where the term $\beta_l{}^m$ renders the expression non-tensorial. The form W^{*j}_l, however, is an element of L^2, so it is legitimate to call (A.20) *the equations of structure of the second kind.*

Equations (A.21) can be used to obtain the explicit form of $\mathrm{D}W^{*j}_h$. Indeed, from (A.10) and $\mathrm{D}W_h{}^j = 0$ (Equation (A.7)) we get

$$\begin{aligned} \mathrm{D}W^{*j}_l = \mathrm{D}(\beta_l{}^j{}_m \wedge \mathrm{D}y^m) = \mathrm{D}\beta_l{}^j{}_m \wedge \mathrm{D}y^m - \beta_l{}^j{}_m \wedge (W^{*m}_r y^r) + \\ + (\beta_l{}^j{}_m \wedge \beta_r{}^m{}_p y^r) \wedge \mathrm{D}y^p, \end{aligned}$$

so that after taking into account the following implication of (A.21):

$$W^{*j}_l{}_m \wedge \mathrm{D}y^m = \mathrm{D}\beta_l{}^j{}_m \wedge \mathrm{D}y^m + (\beta_l{}^j{}_m \wedge \beta_r{}^m{}_p y^r) \wedge \mathrm{D}y^p,$$

we get the desired *Bianchi identities for the curvature* 2-*forms* W^{*j}_l:

$$\mathrm{D}W^{*j}_l = W^{*j}_l{}_m \wedge \mathrm{D}y^m - \beta_l{}^j{}_m \wedge (W^{*m}_r y^r). \tag{A.22}$$

Up till now, no metric tensor has been used. Let us now introduce, in addition to the given connection 1-forms of the type (A.1), a direction-dependent tensor $g_{ij}(x, y)$, which will be assumed to be non-singular, symmetric, and of class C^2 in its 2N arguments (x^i, y^i), but otherwise arbitrary. According to (A.17), we can define the covariant derivative

$$g_{hk|j} = \frac{\partial g_{hk}}{\partial x^j} - 2C_{hkl}\Gamma_{m\,j}^{\,l} y^m - g_{lh}\Gamma_{k\,j}^{\,l} - g_{kl}\Gamma_{h\,j}^{\,l}, \tag{A.23}$$

where the tensor

$$C_{hkl} \stackrel{\text{def}}{=} \frac{1}{2}\frac{\partial g_{hk}}{\partial y^l} = C_{khl} \tag{A.24}$$

is a generalization of Cartan's Finslerian torsion tensor. If we require the connection to be *metric*, that is, if

$$g_{hk|j} = 0 \tag{A.25}$$

identically, then from (A.23) we directly obtain that

$$\Gamma_{h\,k}^{\,j} = \tfrac{1}{2}S_{h\,k}^{\,j} + Y_{h\,k}^{\,j}, \tag{A.26}$$

where

$$\begin{aligned} Y_{h\,k}^{\,j} = \gamma_{h\,k}^{\,j} - \tfrac{1}{2}g^{jm}(g_{hp}S_{m\,k}^{\,p} + g_{kp}S_{m\,h}^{\,p}) - g^{jm}(C_{hmp}\Gamma_{q\,k}^{\,p} + \\ + C_{kmp}\Gamma_{q\,h}^{\,p} - C_{hkp}\Gamma_{q\,m}^{\,p})y^q = Y_{k\,h}^{\,j} \end{aligned} \tag{A.27}$$

is in fact the symmetric part of the connection coefficients; $\gamma_{h\,k}^{\,j}$ denote the Christoffel symbols constructed from g_{hk} in accordance with the rule (1.26).

Conversely, if a relation of the form (A.26)–(A.27) holds for any skew-symmetric tensor $S_{h\,k}^{\,j}$, then forming the combination $\Gamma_{h\,k}^{\,m}g_{jm} + \Gamma_{j\,k}^{\,m}g_{hm}$ we deduce after simple calculations equation (A.25). Thus, *the condition* (A.26)–(A.27) *is necessary and sufficient for the connection to be metric.* The determination of the symmetric part of $\Gamma_{h\,k}^{\,j}$ from equations (A.25)–(A.26) assumes the possibility of solving the equations

$$g^{jq}(g_{hq}\delta_k^l + C_{mqh}\delta_k^l y^m - C_{mkh}y^m\delta_q^l + C_{kqh}y^l)\Gamma_{p\,l}^{\,h}y^p = P_k^{\,j} \tag{A.28}$$

with

$$P_k^{\,j} = (\gamma_{h\,k}^{\,j} + \tfrac{1}{2}S_{h\,k}^{\,j} - \tfrac{1}{2}g^{jm}(g_{hp}S_{m\,k}^{\,p} + g_{kp}S_{m\,h}^{\,p}))y^h$$

for $\Gamma_{p\,l}^{\,h}y^p$.

If we further specify the tensor g_{ij} as Finslerian, which means that we assume the tensor C_{hkl}, as defined by (A.24), to be completely symmetric in its subscripts and orthogonal to y^h in the sense $y^m C_{mkl} = 0$, then Equation (A.28) can easily be solved to give

$$\Gamma_{q\,k}^{\,j}y^q = (\delta_l^j\delta_k^h - C_{l\,k}^{\,j}y^h)P_h^{\,l}.$$

Substituting the latter in (A.26), we find the most general asymmetric coefficients of a connection which is metric relative to a given Finslerian metric tensor. If the connection is also symmetric, that is, $S_{h\,k}^{\,j} = 0$, we evidently obtain the Cartan connection coefficients (1.32).

Problems

A.1. Show that the equations of structure (A.10) entail that

$$(K_l{}^j{}_{hk} - S_h{}^j{}_{k|l} + S_m{}^j{}_l S_h{}^m{}_k)\,\mathrm{d}x^h \wedge \mathrm{d}x^k \wedge \mathrm{d}x^l = 0,$$

which says that the so-called *cyclic identity*, according to which the cyclic sum over the subscripts l, h, k of the tensor in parentheses vanishes identically, holds.

A.2. Expressing the Bianchi identities (A.22) in component form, prove the identity

$$\left(K_l{}^j{}_{hk|m} - K_l{}^j{}_{pm} S_h{}^p{}_k + \frac{\partial \Gamma_l{}^j{}_m}{\partial y^p} K_r{}^p{}_{hk}\, y^r\right) \mathrm{d}x^h \wedge \mathrm{d}x^k \wedge \mathrm{d}x^m = 0.$$

A.3. The difference of any two connections is evidently a tensor. Therefore, if in addition to some connection 1-forms $\beta_h{}^j$ under study, we have sufficiently simple connection 1-forms $p_h{}^j$, it is natural to perform a decomposition of the connection and curvature forms isolating terms wholly pertinent to the second connection. For example, in 1-form Finsler spaces there naturally exists the connection of absolute parallelism, as given by the coefficients (2.89). Similarly, when considering direction-dependent connections, as a simple way of introducing connection coefficients it would be natural to take the following

$$p_h{}^j = S^j_A(x, y)\, \mathrm{d}S^A_h(x, y),$$

where $S^A_h(x, y)$ is a direction-dependent covariant N-tuple, and $S^j_A(x, y)$ is its contravariant reciprocal. Then

$$\beta_h{}^j = p_h{}^j + L_h{}^j,$$

where

$$L_h{}^j = -S^j_A\, \mathrm{D}S^A_h.$$

Perform the corresponding decomposition of the curvature 2-forms.

Notes

[1] The basic notions of the method of differential forms are explained in Chapter 2 of Kramer *et al.* (1980). The method is exposed in detail in Lovelock and Rund (1975, Chapter 5). In the preceding part of the book, we have carried out our treatment in terms of local coordinates of the underlying manifold. The present Appendix shows the reader how to reformulate the theory by resorting to a coordinate-free description.

[2] Strictly speaking, since we do not assume any homogeneity of the objects with respect to the tangent vectors y^i, these objects depend not only on the directions of y^i but also on their magnitudes. Nonetheless, following Rund, we retain the term 'direction-dependent' as not likely to be misleading.

Appendix B

General Gauge Field Equations Associated with Curved Internal Space

B.1. Introduction

The ever growing interest in developing interaction models for physical fields exhibiting internal properties, and in developing various gauge-geometric generalizations of the theory of Yang–Mills fields (see Kikkawa (1983), Hehl (1981), Rund (1982, 1983)), has generated the idea that the conventional general relativistic concept of space-time as a totality of events should be extended in order that it incorporates the concept of an internal space formed of internal points z^P. It is high time, perhaps, to generalize conventional general relativity with the aim of attaching to the internal space the meaning of a physically real object, i.e., an object whose geometric and other propreties can be studied experimentally. This would imply, in particular, that the experimentally observable internal properties of physical fields (for example, the properties of symmetry with respect to one or another group of internal transformations) are to be treated as manifestations of such an internal, but no longer transcendental, space and that the geometric properties of the internal space, like those of the background space-time, are to be determined by material bodies and fields.

In the present Appendix B we refrain from discussing the philosophical problems which arise in such an External + Internal General Relativity, and limit ourselves to a detailed investigation into the gauge-geometric essence of the generalization. Namely, we shall be engaged in transferring the basic geometric methods of general relativity to the external + internal space. Accordingly, our initial ansatz is: the general gauge field equations should be formulated against an arbitrarily curved external + internal space in a coordinately and gauge-covariant way. Such generalized equations having been formulated, various particular cases, for instance the gauge field equations in a non-curved internal space associated with a certain group of linear internal transformations, can be treated as particular solutions, just as the space-times admitting a certain group of motions are considered in general relativity to be particular solutions of the Einstein equations (see Kramer *et al.* (1980)). It may be said somewhat tentatively that the present theory bears approximately the same relation to gauge theories based on postulating the linearity of internal transformations as conventional general relativity does to field theories in pseudo-Euclidean space.

In the following,[1] the dimension number of the internal space (of the fibre) may be any positive integer which will be denoted by M. The internal space will be taken in its logically simplest representation, namely as a totality of parameters z^P subject to the condition that they are of scalar nature under general transformations (1) of the background coordinates x^i, that is, the parameters z^P will obey the transformation law (4). The x-dependent generalization (2) of the internal transformations $z^P(z'^Q)$ will be treated as general gauge transformations. From the viewpoint of differential geometry the idea of the present Appendix is very simple: we adjoin the principle of general gauge covariance to the usual principle of covariance with respect to x^i. Guided by this *combined covariance principle*, we may stepwise formulate the generalized gauge- and coordinate-covariant equations and generalize all the basic concepts of general relativity and gauge field theory in strict gauge-geometric terms. Since the combined principle of covariance proves to be the only thing required when formulating all the generalized gauge-covariant equations, we may regard it as the (local) extended principle of relativity (cf. the interpretation of the principle of relativity in ordinary general relativity (Treder (1971)). This Appendix is organized as follows. In Section 2 we introduce a set of gauge fields which have to be used in constructing the gauge-covariant derivatives in x^i and z^P. Then, in Section 3 we calculate various commutators of these derivatives to arrive at explicit representations of the associated gauge tensors. Examining more complicated commutation relations in Section 4 enables us to directly find the identities satisfied by the gauge tensors. After these gauge-geometric preliminaries, we set forth the $(N + M)$-fold variational principle for all the gauge fields under study (Section 5) Namely, to get the equations for gauge fields, an appropriate contraction of the gauge tensor associated with a gauge field is taken to be the required Lagrangian. Lagrangians which are linear as well as quadratic in the gauge tensors are considered. The gauge field equations are written in terms of the gauge-covariant derivatives of the gauge tensors. Such a general variational principle arises from the treatment of gauge fields and metric tensors as independent field variables, which is in fact a generalization of the method proposed by Palatini (1919) for deriving the gravitational field equations in the Riemannian approach. Since no auxiliary fields or functions enter the variational principle, the approach is entirely self-contained. For the sake of generality, we do not constrain the metric tensors in Section 5 by symmetry conditions, so that the theories (Hlavatý (1957)) which generalize the Einstein equations by permitting the x-dependent metric tensor to be non-symmetric also lie within the scope of the present theory.

In Section 6 we supplement the equations of Section 5 by formulating gauge-covariant equations for the (x, z)-dependent generalizations of physical fields which, generally, carry not only external but also internal spins. Defining a set of generalized currents, we derive the respective conservation laws. In particular, we find the generalizations of the spin tensor and the energy-momentum tensor and show that they are conserved in conjunction with their internal counterparts. Remarkably, as is immediately seen from Eqs (47), (74), and (80), the gauge field B_i related to the internal densities enters the equations in the same minimal way as the usual electromagnetic vector potential enters the field equations of the classical theory. Apparently, nothing prevents one from identifying the physical electric charge with the gauge-geometric internal charge $-q$ (see the end of Section 2) in order to treat the fields of charged particles in a gauge-geometric way as fields possessing the property of being internal densities.

To a considerable degree, when developing the present theory, the author was motivated by the gauge ideas (Chapter 5) suggested by Finsler geometry which deals with geometrical objects dependent on the points x^i and the tangent vectors y^i, for it looks natural to treat the transformations $y^i(x, y')$ (Equation (106)) as gauge transformations associated with space-time. In Section 7 we introduce the parametrical representation of (x, y)-dependent tensor fields to indicate the place which the (x, y)-dependent approach occupies in the general parametrical theory developed in Sections 2–6. In essence, the (x, y)-dependent approach proves to be a particular case of the general parametrical theory, namely the case when the dimension number M of the internal space is assumed to equal the dimension number N of the background manifold. We show that the gauge-covariant equations arising from the theory of (x, y)-dependent fields are equivalent to their (x, z)-dependent counterparts. At the same time, the formal aspects of the parametrical approach are simpler than those of the (x, y)-represented theory because the parameters z^P are scalars in accordance with the definition (4), while the variables y^i are vectors and, therefore, their counterparts of the transformation law (5) will affect the argument y.

The extent to which any general theory is useful depends on success in elaborating particular cases. In the present theory the simplest possibility consists in stipulating that the gauge-covariant derivatives of the metric tensors should vanish. Under these metric conditions, many simplified relations can be inferred and, besides, the gauge fields can be expressed through the metric tensors and the relevant torsion tensors (Section 8). Finally, imposing the particular Finsler-type conditions on the internal metric tensor gives rise to a theory which may be called the Finslerian parametric gauge approach (Sections 9–10). Apparently, Finsler geometry plays the same role in displaying the curvative of the internal space as does Riemannian geometry for the background 'external' space-time. Concluding the Appendix, we find in Section 11 the gauge subgroup under which the Finslerian parametrical relations remain invariant and describe in Section 12 the linear-gauge limit of the present theory.

The author hopes that future application of the present External + Internal General Relativity to models of particle interactions will serve to unify current gauge and general relativistic ideas.

B.2. The Parametrical Representation

We shall consider tensorial fields dependent on the arguments (x^i, z^P), where x^i denote the coordinates of points of the background N-dimensional differentiable maniforld and z^P is a set of parameters (the indices $i, j, \ldots$ will range from 1 to N, while the indices P, Q, $R, \ldots$ range from 1 to M, where N and M may be any integers; $N=4$ in the usual physical context). A sufficiently general example of such a field is $w_n{}^m{}_P{}^Q(x, z)$. The purpose of the present section will be to develop a tensor calculus operative under two sets of transformations. They are the general coordinate transformations

$$x^i = x^i(\bar{x}), \qquad \bar{x}^i = \bar{x}^i(x), \tag{1}$$

and the general invertible transformations of the parameters:

$$z^P = z^P(x, z'^Q), \qquad z'^P = z'^P(x, z^Q). \tag{2}$$

The transformations (2) will be treated as *gauge transformations*, while (1) are the usual coordinate transformations of classical tensor calculus. It will be assumed that the transformations are arbitrary except for the requirement that they be sufficiently smooth and invertible. The notation

$$B_i^j = \partial x^j/\partial \bar{x}^i, \qquad B_{im}^j = \partial B_i^j/\partial \bar{x}^m, \qquad z_Q^P = \partial z^P/\partial z'^Q,$$
$$z_{QR}^P = \partial z_Q^P/\partial z'^R \tag{3}$$

will be used below and the summation convention will be assumed.

We shall stipulate that the arguments z^P be of scalar nature:

$$\bar{z}^P = z^P, \tag{4}$$

which means $\bar{z}^P(\bar{x}, \bar{z}') = z^P(x(\bar{x}), z')$, so that the transformation law of the field $w_n{}^m{}_P{}^Q$ under (1) will read

$$B_m^k \bar{w}_n{}^m{}_P{}^Q(\bar{x}, z) = B_n^i w_i{}^k{}_P{}^Q(x(\bar{x}), z) \tag{5}$$

which does not affect the argument z. Similarly, a tensor law is postulated for transformations (2), that is,

$$z_Q^R w'_n{}^m{}_P{}^Q(x, z') = z_P^S w_n{}^m{}_S{}^R(x, z(x, z')). \tag{6}$$

Under these conditions, the object $w_n{}^m{}_P{}^Q$ may be called a *parametrical gauge tensor*. Note that the indices P, $Q, \ldots$ and $i, j, \ldots$ of gauge tensors transform like scalar indices under the transformations (1) and (2), respectively.

In order to construct *the gauge-covariant derivative* D_i with respect to x^i, we introduce three *gauge fields* $L_m{}^i{}_n(x, z)$, $N_i^P(x, z)$, and $D_P{}^Q{}_i(x, z)$ subject to the transformation laws

$$B^t_{lm} = B^t_i \bar{L}_l{}^i{}_m - B^q_l B^p_m L_q{}^t{}_p, \qquad \bar{N}^P_i = B^j_i N^P_j, \qquad \bar{D}_P{}^Q{}_i = B^j_i D_P{}^Q{}_j \tag{7}$$

and

$$L'_m{}^i{}_n = L_m{}^i{}_n, \qquad \partial z^P/\partial x^i + N'^Q_i z^P_Q - N^P_i = 0, \tag{8}$$

$$\partial z^P_Q/\partial x^i + z^P_{QR} N'^R_i + z^R_Q D_R{}^P{}_i - z^P_R D'_Q{}^R{}_i = 0. \tag{9}$$

Here, $\bar{L}_l{}^i{}_m$ means $\bar{L}_l{}^i{}_m(\bar{x}, z)$, $D'_Q{}^R{}_i$ means $D'_Q{}^R{}_i(x, z')$, etc. The transformation laws (7) read that, under the coordinate transformations (1), the object $L_l{}^i{}_m$ transforms like connection coefficients, and N^P_i and $D_P{}^Q{}_i$ like covariant vectors. Notice that the field $-D_P{}^Q{}_i$ transforms like the derivative $\partial N^Q_i/\partial z^P$; in this respect, cf. the definition (26) of the gauge tensor $A_j{}^Q{}_R$. The first member of Equations (8) claims that the field $L_m{}^i{}_n$ must be a gauge scalar. With these transformation laws the sought derivative D_i may be defined as follows

$$\mathrm{D}_i w_n{}^m{}_P{}^Q = \mathrm{d}_i w_n{}^m{}_P{}^Q + L_k{}^m{}_i w_n{}^k{}_P{}^Q - L_n{}^k{}_i w_k{}^m{}_P{}^Q + D_R{}^Q{}_i w_n{}^m{}_P{}^R -$$
$$D_P{}^R{}_i w_n{}^m{}_R{}^Q, \tag{10}$$

where we have introduced the following notation for the sake of brevity:

$$\mathrm{d}_i = \partial_i + N^P_i \mathrm{d}_P, \qquad \partial_i = \partial/\partial x^i, \qquad \mathrm{d}_P = \partial/\partial z^P. \tag{11}$$

It can be readily verified that the definitions (5)–(11) entail the tensor transformation laws

$$B^m_r \overline{\mathrm{D}_i w_n{}^r{}_P{}^Q} = B^j_i B^k_n \mathrm{D}_j w_k{}^m{}_P{}^Q, \qquad z^R_Q (\mathrm{D}_i w_n{}^m{}_P{}^Q)' = z^S_P \mathrm{D}_i w_n{}^m{}_S{}^R$$

required of the gauge-covariant derivative. Clearly, the operator D_i will obey all the rules common to linear covariant derivatives, including the chain rule.

Similarly, by introducing the fourth gauge field $Q_P{}^S{}_R\ (x, z)$ through postulation of the transformation laws

$$z^P_{QR} + z^S_Q z^T_R Q_S{}^P{}_T - z^P_S Q'_Q{}^S{}_R = 0, \qquad \bar{Q}_P{}^S{}_R = Q_P{}^S{}_R, \tag{12}$$

we arrive at *the gauge-covariant derivative* S_R *with respect to* z^R:

$$\mathrm{S}_R w_n{}^m{}_P{}^Q = \mathrm{d}_R w_n{}^m{}_P{}^Q + Q_S{}^Q{}_R w_n{}^m{}_P{}^S - Q_P{}^S{}_R w_n{}^m{}_S{}^Q, \tag{13}$$

obtaining

$$B^m_r \mathrm{S}_R w_n{}^r{}_P{}^Q = B^k_n \mathrm{S}_R w_k{}^m{}_P{}^Q, \qquad z^M_Q (\mathrm{S}_R w_n{}^m{}_P{}^Q)' = z^T_R z^N_P \mathrm{S}_T w_n{}^m{}_N{}^M.$$

Three additional gauge fields $E_i(x, z)$, $B_i(x, z)$, and $H_P(x, z)$ are required so that tensor densities may be dealt with. Thus, denoted the Jacobians of the transformations (1) and (2) by

$$B = \det(B^i_j), \qquad Z = \det(z^P_Q), \tag{14}$$

we call the objects $W^{(+)}_n(x, z)$ and $W^{(+)m}(x, z)$ *external vector densities* if their transformation laws read

$$\bar{W}^{(+)}_{k} = BB^{r}_{k}W^{(+)}_{r}, \qquad (W^{(+)}_{n})' = W^{(+)}_{n}, \qquad B^{k}_{n}\bar{W}^{(+)n} = BW^{(+)k},$$
$$(W^{(+)n})' = W^{(+)n}, \tag{15}$$

while the objects $W^{+}_{R}(x,z)$ and $W^{+R}(x,z)$ will be called *internal vector densities* if

$$\bar{W}^{+}_{R} = W^{+}_{R}, \qquad (W^{+}_{R})' = Zz^{S}_{R}W^{+}_{S}, \qquad \bar{W}^{+R} = W^{+R},$$
$$(W^{+R})'z^{S}_{R} = ZW^{+S}. \tag{16}$$

Similarly, we obtain *external anti-densities* $W^{(-)}_{n}$, $W^{(-)n}$ and *internal anti-densities* W^{-}_{R}, W^{-R} by replacing in the above transformation laws the factors B and Z by $1/B$ and $1/Z$, respectively; for instance, $z^{R}_{S}(W^{-S})' = W^{-R}/Z$. Postulating then the transformation laws

$$\bar{E}_i + \frac{1}{B}\partial_i B = E_j B^j_i, \qquad E'_i = E_i, \qquad \bar{B}_i = B^j_i B_j, \qquad \bar{H}_R = H_R,$$
$$B'_i + \frac{1}{Z}\mathrm{d}'_i Z = B_i, \qquad H'_R + \frac{1}{Z}\mathrm{d}'_R Z = H_S z^S_R, \tag{17}$$

we extend the action of the gauge-covariant derivatives D_i and S_R to external and internal densities as follows:

$$\mathrm{D}_i W^{(+)}_{n} = (\mathrm{d}_i + E_i)W^{(+)}_{n} - L_{n}{}^{k}{}_{i}W^{(+)}_{k},$$
$$\mathrm{D}_i W^{(+)n} = (\mathrm{d}_i + E_i)W^{(+)n} + L_{k}{}^{n}{}_{i}W^{(+)k}, \tag{18}$$

$$\mathrm{D}_i W^{+}_{R} = (\mathrm{d}_i + B_i)W^{+}_{R} - D_{R}{}^{T}{}_{i}W^{+}_{T},$$
$$D_i W^{+R} = (\mathrm{d}_i + B_i)W^{+R} + D_{T}{}^{R}{}_{i}W^{+T} \tag{19}$$

and

$$\mathrm{S}_R W^{+}_{P} = (\mathrm{d}_R + H_R)W^{+}_{P} - Q_{P}{}^{T}{}_{R}W^{+}_{T},$$
$$S_R W^{+P} = (\mathrm{d}_R + H_R)W^{+P} + Q_{T}{}^{P}{}_{R}W_{T}, \tag{20}$$

which implies

$$\overline{\mathrm{D}_i W^{(+)}_{k}} = BB^{r}_{k}B^{j}_{i}D_j W^{(+)}_{r}, \qquad \overline{\mathrm{D}_i W^{+P}} = B^j_i D_j W^{+P},$$
$$z^{Q}_{P}(\mathrm{D}_i W^{+P})' = Z\mathrm{D}_i W^{+Q},$$
$$(S_R W^{+}_{P})' = Zz^{Q}_{R}z^{T}_{P}\mathrm{S}_Q W^{+}_{T}, \qquad \overline{\mathrm{S}_R W^{+}_{P}} = \mathrm{S}_R W^{+}_{P}, \text{ etc.}$$

The action of S_R on external densities will be defined in accordance with the earlier rule (13) because the Jacobian B in (14) does not depend on z^P. The gauge-covariant differentiation of anti-densities will involve the negatives of the gauge fields E_i, B_i, and H_P, as exemplified by

$$\begin{aligned} \mathrm{D}_i W_P^- &= (\mathrm{d}_i - B_i) W_P^- - D_P{}^Q{}_i W_Q^-, \\ \mathrm{S}_R W_P^- &= (\mathrm{d}_R - H_R) W_P^- - Q_P{}^T{}_R W_T^-, \end{aligned} \tag{21}$$

so that products of the type $(\mathrm{D}_i W_P^+)(\mathrm{D}_j W_Q^-)$ or $(\mathrm{S}_R W_P^+)(\mathrm{S}_Q W^{-T})$ will be tensors, not densities. Obviously, the definitions (18)–(21) may be extended directly to general densities of the weight (p, q), that is, densities of external weight p and internal weight q, where p and q may be any real numbers. Namely, if W is such a density, so that $\bar{W} = B^p W$ and $W' = Z^q W$, then we may put merely $\mathrm{D}_i W = (\mathrm{d}_i + pE_i + qB_i)W$, $\mathrm{S}_R W = (\mathrm{d}_R + qH_R)W$. These definitions suggest the attractive gauge-geometric idea of treating physical fields carrying electric charge e as internal densities of weight $-q$, for the gauge field B_i enters the field equations in the minimal $(\mathrm{d}_i - eB_i)$-way which is typical of the electromagnetic vector potential.

B.3. Associated Gauge Tensors

The transformation laws set forth in the preceding section were the only restrictions (except for the required smoothness) that had to be imposed on the gauge fields (connection coefficients). Given these general conditions, we are in a position to find the associated gauge tensors (curvature tensors). It proves convenient to seek for such tensors by evaluating respective commutators of the studied derivatives D_i and S_R. Indeed, simple direct calculations yield the following result:

$$\begin{aligned} (\mathrm{D}_j \mathrm{D}_i - \mathrm{D}_i \mathrm{D}_j) W_{nP} = (M_j{}^R{}_i \mathrm{S}_R + S_j{}^k{}_i \mathrm{D}_k) W_{nP} + L_n{}^k{}_{ji} W_{kP} + \\ + E_P{}^Q{}_{ji} W_{nQ}, \end{aligned} \tag{22}$$

$$(\mathrm{D}_j \mathrm{S}_R - \mathrm{S}_R \mathrm{D}_j) W_{nP} = A_j{}^Q{}_R \mathrm{S}_Q W_{nP} + B_n{}^k{}_{jR} W_{kP} - Q_P{}^T{}_{Rj} W_{nT}, \tag{23}$$

$$(\mathrm{S}_M \mathrm{S}_N - \mathrm{S}_N \mathrm{S}_M) W_{nP} = Y_M{}^Q{}_N \mathrm{S}_Q W_{nP} + Z_P{}^Q{}_{MN} W_{nQ} \tag{24}$$

with the *gauge tensors* given by the following table:

$$S_i{}^k{}_j = L_i{}^k{}_j - L_j{}^k{}_i, \qquad Y_P{}^T{}_R = Q_P{}^T{}_R - Q_R{}^T{}_P, \tag{25}$$

$$A_j{}^Q{}_R = -\mathrm{d}_R N_j^Q - D_R{}^Q{}_j, \qquad M_j{}^R{}_i = \mathrm{d}_j N_i^R - \mathrm{d}_i N_j^R, \tag{26}$$

$$L_n{}^k{}_{ij} = \mathrm{d}_j L_n{}^k{}_i - \mathrm{d}_i L_n{}^k{}_j - L_n{}^t{}_j L_t{}^k{}_i + L_n{}^t{}_i L_t{}^k{}_j, \qquad B_n{}^k{}_{jR} = \mathrm{d}_R L_n{}^k{}_j, \tag{27}$$

$$E_P{}^Q{}_{ij} = \mathrm{d}_j D_P{}^Q{}_i - \mathrm{d}_i D_P{}^Q{}_j - D_P{}^R{}_j D_R{}^Q{}_i + D_P{}^R{}_i D_R{}^Q{}_j + Q_P{}^Q{}_R M_i{}^R{}_j, \tag{28}$$

$$Q_P{}^T{}_{Rj} = \mathrm{d}_j Q_P{}^T{}_R + Q_P{}^T{}_S \mathrm{d}_R N_j^S - D_P{}^S{}_j Q_S{}^T{}_R + D_S{}^T{}_j Q_P{}^S{}_R - \mathrm{d}_R D_P{}^T{}_j, \tag{29}$$

$$Z_P{}^Q{}_{MN} = \mathrm{d}_N Q_P{}^Q{}_M - \mathrm{d}_M Q_P{}^Q{}_N - Q_P{}^S{}_N Q_S{}^Q{}_M + Q_P{}^S{}_M Q_S{}^Q{}_N. \tag{30}$$

We also present the following commutators to be referred to below

$$(\mathrm{D}_j\mathrm{D}_i - \mathrm{D}_i\mathrm{D}_j)W = (M_j{}^R{}_i\mathrm{S}_R + S_j{}^k{}_i\mathrm{D}_k)W, \tag{31}$$

$$(\mathrm{D}_j\mathrm{D}_i - \mathrm{D}_i\mathrm{D}_j)W_n = (M_j{}^R{}_i\mathrm{S}_R + S_j{}^k{}_i\mathrm{D}_k)W_n + L_n{}^k{}_{ji}W_k, \tag{32}$$

$$(\mathrm{D}_j\mathrm{D}_i - \mathrm{D}_i\mathrm{D}_j)W_{nm} = (M_j{}^R{}_i\mathrm{S}_R + S_j{}^k{}_i\mathrm{D}_k)W_{nm} + L_n{}^k{}_{ji}W_{km} + {} + L_m{}^k{}_{ji}W_{nk}, \tag{33}$$

$$(\mathrm{D}_j\mathrm{D}_i - \mathrm{D}_i\mathrm{D}_j)W_P = (M_j{}^R{}_i\mathrm{S}_R + S_j{}^k{}_i\mathrm{D}_k)W_P + E_P{}^Q{}_{ji}W_Q \tag{34}$$

$$(\mathrm{D}_j\mathrm{D}_i - \mathrm{D}_i\mathrm{D}_j)W_{PQ} = (M_j{}^R{}_i\mathrm{S}_R + S_j{}^k{}_i\mathrm{D}_k)W_{PQ} + E_P{}^R{}_{ji}W_{RQ} + {} + E_Q{}^R{}_{ji}W_{PR} \tag{35}$$

and

$$(\mathrm{D}_j\mathrm{S}_R - \mathrm{S}_R\mathrm{D}_j)W = A_j{}^Q{}_R\mathrm{S}_Q W, \qquad (\mathrm{D}_j\mathrm{S}_R - \mathrm{S}_R\mathrm{D}_j)W_P = = A_j{}^Q{}_R\mathrm{S}_Q W_P - Q_P{}^T{}_{Rj}W_T, \tag{36}$$

$$(\mathrm{D}_j\mathrm{S}_R - \mathrm{S}_R\mathrm{D}_j)W_{PQ} = A_j{}^T{}_R\mathrm{S}_T W_{PQ} - Q_P{}^T{}_{Rj}W_{TQ} - Q_Q{}^T{}_{Rj}W_{PT}, \tag{37}$$

$$(\mathrm{D}_j\mathrm{S}_R - \mathrm{S}_R\mathrm{D}_j)W_n = A_j{}^Q{}_R\mathrm{S}_Q W_n + B_n{}^k{}_{jR}W_k, \tag{38}$$

$$(\mathrm{D}_j\mathrm{S}_R - \mathrm{S}_R\mathrm{D}_j)W_{nm} = A_j{}^Q{}_R\mathrm{S}_Q W_{nm} + B_n{}^k{}_{jR}W_{km} + B_m{}^k{}_{jR}W_{nk}, \tag{39}$$

$$(\mathrm{S}_M\mathrm{S}_N - \mathrm{S}_N\mathrm{S}_M)W = Y_M{}^Q{}_N\mathrm{S}_Q W, \qquad (\mathrm{S}_M\mathrm{S}_N - \mathrm{S}_N\mathrm{S}_M)W_n = = Y_M{}^Q{}_N\mathrm{S}_Q W_n, \tag{40}$$

$$(\mathrm{S}_M\mathrm{S}_N - \mathrm{S}_N\mathrm{S}_M)W_P = Y_M{}^Q{}_N\mathrm{S}_Q W_P + Z_P{}^Q{}_{MN}W_Q, \tag{41}$$

$$(\mathrm{S}_M\mathrm{S}_N - \mathrm{S}_N\mathrm{S}_M)W_{PQ} = Y_M{}^R{}_N\mathrm{S}_R W_{PQ} + Z_P{}^R{}_{MN}W_{RQ} + Z_Q{}^R{}_{MN}W_{PR}. \tag{42}$$

The objects (25)–(30) are tensors in the indices $i,j,\ldots$ under (1), and in the indices $P,Q,\ldots$ under (2) (this assertion ensues from the circumstance that the tensor W_{nP} in the commutators (22)–(24) is arbitrary except for the prescribed transformation laws), for example, $z^R_Q E'_P{}^Q{}_{ij} = z^M_P E_M{}^R{}_{ij}$, $\bar{E}_P{}^Q{}_{ij} = B^m_i B^n_j E_P{}^Q{}_{mn}$. The tensors (25) are in fact the *torsions* associated with the gauge fields under study.

Additionally, we commute the derivatives for the case of densities. With the aid of the definitions (18)–(21) we obtain

$$(\mathrm{D}_j\mathrm{D}_i - \mathrm{D}_i\mathrm{D}_j)W^+ = (M_j{}^R{}_i\mathrm{S}_R + S_j{}^k{}_i\mathrm{D}_k + F_{ji})W^+, \tag{43}$$

$$(\mathrm{D}_j\mathrm{S}_R - \mathrm{S}_R\mathrm{D}_j)W^+ = (A_j{}^P{}_R\mathrm{S}_P + b_{jR})W^+, \qquad (\mathrm{S}_M\mathrm{S}_N - \mathrm{S}_N\mathrm{S}_M)W^+ = = (Y_M{}^P{}_N\mathrm{S}_P + H_{MN})W^+ \tag{44}$$

and

$$(\mathrm{D}_j\mathrm{D}_i - \mathrm{D}_i\mathrm{D}_j)W^{(+)} = (M_j{}^R{}_i\mathrm{S}_R + S_j{}^k{}_i\mathrm{D}_k + H_{ji})W^{(+)}, \tag{45}$$

$$(\mathrm{D}_j\mathrm{S}_R - \mathrm{S}_R\mathrm{D}_j)W^{(+)} = (A_j{}^P{}_R\mathrm{S}_P + e_{jR})W^{(+)}, \tag{46}$$

where the gauge tensors are

$$F_{ji} = \mathrm{d}_j B_i - \mathrm{d}_i B_j + H_R M_i{}^R{}_j, \qquad E_{ji} = \mathrm{d}_j E_i - \mathrm{d}_i E_j, \qquad e_{jP} = -\mathrm{d}_P E_j,$$
$$H_{PQ} = \mathrm{d}_P H_Q - \mathrm{d}_Q H_P, \qquad b_{jP} = \mathrm{d}_j H_P - \mathrm{d}_P B_j + H_R \,\mathrm{d}_P N_j^R. \tag{47}$$

So far, the examination of gauge fields and tensors has not required any metric tensors. However, raising and lowering the indices of the gauge tensors under study assumes the introduction of two metric tensors, $a_{ij}(x, z)$ and $g_{PQ}(x, z)$, of different tensorial type, namely,

$$\bar{a}_{ij} = B_i^m B_j^n a_{mn}, \qquad a'_{ij} = a_{ij}, \qquad \bar{g}_{PQ} = g_{PQ}, \qquad g'_{PQ} = z_P^R z_Q^S g_{RS}, \tag{48}$$

which may be called the *external and internal metric tensors*, respectively. In general, these tensors will not necessarily be symmetric, so that $a_{ij} \neq a_{ji}$ and $g_{PQ} \neq g_{QP}$ unless the contrary is stated. The metric tensors will be assumed to be nondegenerate, which makes it possible to introduce the reciprocal tensors a^{ij} and g^{PQ} by means of the reciprocity relations $a^{ji} a_{mi} = a^{ij} a_{im} = \delta^j_m$ and $g^{PQ} g_{RQ} = g^{QP} g_{QR} = \delta^P_R$. With these tensors, we may write $M_{iPj} = g_{PQ} M_i{}^Q{}_j$, $E_P{}^{Qij} = a^{mi} a^{nj} E_P{}^Q{}_{mn}$, etc., preserving the tensor status of each and every index. The covariant derivatives

$$\begin{aligned} D_k a_{ij} &= \mathrm{d}_k a_{ij} - L_i{}^m{}_k a_{mj} - L_j{}^m{}_k a_{im}, \\ \mathrm{D}_k g_{PQ} &= \mathrm{d}_k g_{PQ} - D_P{}^R{}_k g_{RQ} - D_Q{}^R{}_k g_{PR}, \\ \mathrm{S}_R a_{ij} &= \mathrm{d}_R a_{ij}, \qquad \mathrm{S}_R g_{PQ} = \mathrm{d}_R g_{PQ} - Q_P{}^S{}_R g_{SQ} - Q_Q{}^S{}_R g_{PS} \end{aligned} \tag{49}$$

are non-vanishing in general.

Denoting

$$J = |\det(a_{ij})|^{1/2}, \qquad K = |\det(g_{PQ})|^{1/2}, \tag{50}$$

we obtain objects which are external and internal densities, that is, $\bar{J} = BJ$, $J' = J$, $\bar{K} = K$, $K' = ZK$ (cf. the transformation laws (15)–(16)), so that we ought to write

$$\begin{aligned} \mathrm{D}_k J &= (\mathrm{d}_k + E_k) J, \qquad \mathrm{D}_k K = (\mathrm{d}_k + B_k) K, \qquad \mathrm{S}_R J = \mathrm{d}_R J, \\ \mathrm{S}_R K &= (\mathrm{d}_R + H_R) K \end{aligned} \tag{51}$$

in accordance with the definitions (18)–(20). On the other hand, from the obvious identities $2\mathrm{d}J = J a^{ij} \,\mathrm{d}a_{ij}$ and $2\mathrm{d}K = K g^{PQ} \,\mathrm{d}g_{PQ}$ it follows that $a^{ij} \partial_k a_{ij} = 2/J \,\partial_k J$ and $g^{PQ} \partial_k g_{PQ} = 2/K \,\partial_k K$ together with relations of the same form for the case of the derivatives d_R. Therefore, according to Equation (49), the definitions

$$\begin{aligned} \hat{\mathrm{D}}_k J &= (\mathrm{d}_k - L_m{}^m{}_k) J, \qquad \hat{\mathrm{D}}_k K = (\mathrm{d}_k - D_P{}^P{}_k) K, \\ \hat{\mathrm{S}}_R K &= (\mathrm{d}_R - Q_M{}^M{}_R) K \end{aligned} \tag{52}$$

will yield gauge-covariant objects too. By comparing the definitions (52) with

(51) we obtain the relations

$$\mathrm{D}_k J = \hat{\mathrm{D}}_k J + L_k J, \qquad \mathrm{D}_k K = \hat{\mathrm{D}}_k K + D_k K,$$
$$\mathrm{S}_R K = \hat{S}_R K + Q_R K \tag{53}$$

with the objects

$$L_k = E_k + L_m{}^m{}_k, \qquad D_k = B_k + D_P{}^P{}_k, \qquad Q_R = H_R + Q_P{}^P{}_R \tag{54}$$

which will obviously be gauge tensors. The definitions (54) supplement the list (47). Clearly, the definitions (51) and (52) will coincide only if the gauge tensors (54) vanish; such a vanishing, however, will not be postulated below.

B.4. Identities Satisfied by the Gauge Tensors

As is well known, the Riemann curvature tensor $r_i{}^j{}_{mn}(x)$ satisfies the cyclic identity $r_i{}^j{}_{mn} + r_n{}^j{}_{im} + r_m{}^j{}_{ni} \equiv 0$. This identity may be derived from the Riemannian commutation relation $(\mathrm{R}_j\mathrm{R}_i - \mathrm{R}_i\mathrm{R}_j)v_n = r_n{}^k{}_{ji}v_k$ (where R_i denotes the Riemannian covariant derivative and $v_n(x)$ is a covariant vector field) by putting v_n as the gradient of a scalar $v(x)$, commuting the derivative R_n with R_i and R_j, and then allowing for the identity $(\mathrm{R}_j\mathrm{R}_i - \mathrm{R}_i\mathrm{R}_j)v \equiv 0$. This method can be used directly in our present general theory.

Indeed, by putting $W_n = \mathrm{D}_n W$ in the commutator (32) we can transform the left-hand side of (32) in accordance with $\mathrm{D}_j\mathrm{D}_i\mathrm{D}_n W = \mathrm{D}_j\mathrm{D}_n\mathrm{D}_i W + \mathrm{D}_j(\mathrm{D}_i\mathrm{D}_n - \mathrm{D}_n\mathrm{D}_i)W = \mathrm{D}_n\mathrm{D}_j\mathrm{D}_i W + (\mathrm{D}_j\mathrm{D}_n - \mathrm{D}_n\mathrm{D}_j)\mathrm{D}_i W + \mathrm{D}_j(\mathrm{D}_i\mathrm{D}_n - \mathrm{D}_n\mathrm{D}_i)W$, or, taking into account the commutators (32) and (31), $\mathrm{D}_j\mathrm{D}_i\mathrm{D}_n W = \mathrm{D}_n\mathrm{D}_j\mathrm{D}_i W + M_j{}^Q{}_n\mathrm{S}_Q\mathrm{D}_i W + L_i{}^k{}_{jn}\mathrm{D}_k W + S_j{}^k{}_n\mathrm{D}_k\mathrm{D}_i W + \mathrm{D}_j(M_i{}^Q{}_n\mathrm{S}_Q W + S_i{}^k{}_n\mathrm{D}_k W)$. Substituting the last relation in the left-hand side of (32) and cancelling similar terms, we obtain an identity of the form $(\quad)^Q\mathrm{S}_Q W + (\quad)^k\mathrm{D}_k W = 0$, where the parentheses denote combinations of terms not involving W. Since W is an arbitrary scalar field, both parentheses must vanish identically. In this way we find the following cyclic identities:

$$\mathscr{A}_{(ijn)}(\mathrm{D}_i M_n{}^Q{}_j + A_i{}^Q{}_P M_n{}^P{}_j + M_i{}^Q{}_m S_n{}^m{}_j) = 0 \tag{55}$$

(the explicit form of $(\quad)^Q = 0$) and

$$\mathscr{A}_{(ijn)}(L_i{}^k{}_{jn} + \mathrm{D}_i S_n{}^k{}_j + S_i{}^k{}_m S_n{}^m{}_j) = 0 \tag{56}$$

(the explicit form of $(\quad)^k = 0$), where the symbol $\mathscr{A}_{(\;)}$ means the cyclic interchange of indices ($\mathscr{A}_{(ijn)}X_{ijn} = X_{ijn} + X_{nij} + X_{jni}$).

Repeating this procedure in the case of commutator (34) with W_P chosen to be $\mathrm{S}_P W$ and taking into account the commutators (40), (31), (32), and (38), we can readily find two additional identities

$$\begin{aligned} &E_P{}^Q{}_{ij} + \mathrm{D}_j A_i{}^Q{}_P - \mathrm{D}_i A_j{}^Q{}_P + \mathrm{S}_P M_j{}^Q{}_i + Y_P{}^Q{}_R M_j{}^R{}_i + A_m{}^Q{}_P S_i{}^m{}_j + \\ &\quad + A_i{}^R{}_P A_j{}^Q{}_R - A_j{}^R{}_P A_i{}^Q{}_R = 0 \end{aligned} \tag{57}$$

and

$$B_i{}^k{}_{jP} - B_j{}^k{}_{iP} + S_P S_j{}^k{}_i = 0. \tag{58}$$

The identity (57) expresses the gauge tensor $E_P{}^Q{}_{ij}$ in terms of more elementary tensors. Moreover, examining the second member of the commutators (36) with $W_P = S_P W$ yields

$$Q_R{}^T{}_{Pj} - Q_P{}^T{}_{Rj} + S_R A_j{}^T{}_P - S_P A_j{}^T{}_R + D_j Y_P{}^T{}_R + A_j{}^Q{}_R Y_Q{}^T{}_P + + A_j{}^Q{}_P Y_R{}^T{}_Q + A_j{}^T{}_Q Y_P{}^Q{}_R = 0. \tag{59}$$

In the case of commutator (41), the associated identity is obtainable from (56) by the formal change $L_i{}^k{}_{jn} \to Z_P{}^Q{}_{MN}$, $D_n \to S_N$, and $S_i{}^k{}_n \to Y_P{}^Q{}_N$, as becomes clear after comparing (31) and (32) with (40) and (41). Thus,

$$\mathscr{A}_{(PMN)}(Z_P{}^Q{}_{MN} + S_M Y_P{}^Q{}_N + Y_P{}^Q{}_R Y_N{}^R{}_M) = 0. \tag{60}$$

The derivation of the associated Bianchi-type identities can also be conveniently based on the direct use of the known commutators. Applying the obvious identity

$$\mathscr{A}_{(ijh)}(D_i(D_h D_j - D_j D_h) - (D_h D_j - D_j D_h)D_i) = 0 \tag{61}$$

to (32) and using the commutator (38) together with the cyclic identities (55) and (56), we obtain the following result:

$$\mathscr{A}_{(ijh)}(D_i L_n{}^k{}_{jh} + M_i{}^P{}_j B_n{}^k{}_{hP} + S_i{}^m{}_h L_n{}^k{}_{mj}) = 0. \tag{62}$$

Similarly, applying (61) to the second member of (36) and taking into account the identities (58) and (60), we obtain

$$\mathscr{A}_{(ijh)}(D_i E_P{}^Q{}_{jh} + S_i{}^m{}_h E_P{}^Q{}_{mj} + M_i{}^R{}_h Q_P{}^Q{}_{Rj}) = 0. \tag{63}$$

In turn, the identity (61) with the operator D replaced by the operator S can be applied to the commutator (41), giving

$$\mathscr{A}_{(RMN)}(S_R Z_P{}^Q{}_{MN} + Y_M{}^T{}_R Z_P{}^Q{}_{TN}) = 0. \tag{64}$$

Finally, we apply the identities $D_j(S_R D_m - D_m S_R) -$
$- (S_R D_m - D_m S_R)D_j + D_m(D_j S_R - S_R D_j) - (D_j S_R - S_{R,}D_j)D_m - S_R(D_j D_m -$

$- D_m D_j) + (D_j D_m - D_m D_j)S_R = 0$ and
$D_j(S_R S_P - S_P S_R) - (S_R S_P - S_P S_R)D_j +$
$+ S_P(D_j S_R - S_R D_j) - (D_j S_R - S_R D_j)S_P - S_R(D_j S_P - S_P D_j) + (D_j S_P -$
$- S_P D_j)S_R = 0$ to $D_n W$ and S_N W, obtaining four new identities:

$$D_m B_n{}^k{}_{jR} - D_j B_n{}^k{}_{mR} + A_j{}^T{}_R B_n{}^k{}_{mT} - A_n{}^T{}_R B_n{}^k{}_{jT} + S_j{}^i{}_m B_n{}^k{}_{iR} - S_R L_n{}^k{}_{jm} = 0, \tag{65}$$

$$D_j Q_N{}^T{}_{Rm} - D_m Q_N{}^T{}_{Rj} + A_m{}^M{}_R Q_N{}^T{}_{Mj} - A_j{}^M{}_R Q_N{}^T{}_{Mm} - S_R E_N{}^T{}_{jm} - - S_j{}^k{}_m Q_N{}^T{}_{Rk} + M_j{}^Q{}_m Z_N{}^T{}_{QR} = 0, \tag{66}$$

$$\mathrm{S}_R B_n{}^k{}_{jP} - \mathrm{S}_P B_n{}^k{}_{jR} + Y_P{}^T{}_R B_n{}^k{}_{jT} = 0, \tag{67}$$

$$\mathrm{D}_j Z_N{}^T{}_{PR} - Y_P{}^M{}_R Q_N{}^T{}_{Mj} + \mathrm{S}_P Q_N{}^T{}_{Rj} + A_j{}^M{}_R Z_N{}^T{}_{MP} + A_j{}^M{}_P Z_N{}^T{}_{RM} - \mathrm{S}_R Q_N{}^T{}_{Pj} = 0. \tag{68}$$

Eqs (62)–(68) are the sought for Bianchi-type identities involving the generalized curvature tensors (gauge tensors) under study.

B.5. Variational Principle for the Parametrical Gauge Fields

The observations in Section 2 suggested the introduction of seven independent gauge fields N_i^P, $L_m{}^i{}_n$, $D_P{}^Q{}_i$, $Q_P{}^R{}_S$, E_i, B_i, and H_P of different tensorial types. To formulate differential equations for these fields, the associated gauge tensors derived in Section 3 can be used in constructing appropriate Lagrangians. Since the studied fields are functions of both sets of arguments x^i and z^P, the variational principle should be formulated in the $(N + M)$-fold way

$$\delta \int LJK \, \mathrm{d}^N x \, \mathrm{d}^M z = 0, \tag{69}$$

where the densities J and K are given by the definition (50). Also, we may treat the metric tensors a_{ij} and g_{PQ}, which unavoidably enter the relevant Lagrangians L given by the contraction of gauge tensors, as additional field variables, obtaining in this way equations for the gauge fields as functions of the metric tensors. Accordingly, the Lagrangians will be constructed from conbinations of these nine fields, no relationship between them being postulated a priori. This in fact extends the method known in current general relativity theory as the Palatini (1919) method. According to this method, the connection coefficients and the metric tensor in the Riemannian gravitational Lagrangian are treated as independent field variables, the relationship between them being inferred by solving the equations which arise after varying the action integral in the connection coefficients. It should also be noted that in the general approach which we develop here the equations will be doubled in the sense that, because the fields involved depend on the points x^i as well as on the parameters z^P, we need two sets of equations to describe the dependence on two sets of arguments x^i and z^P.

Considering the gauge field N_i^P as an example, we have at our disposal the associated gauge tensors $M_i{}^P{}_j$ and $A_i{}^P{}_Q$ to construct two quadratic Lagrangians $L_{N1} = M_{iPj} M^{iPj}$ and $L_{N2} = A_{iPQ} A^{iPQ}$. Differentiating yields

$$\partial L_{N1}/\partial(\partial_m N_j^P) = 4M^m{}_P{}^j, \qquad \partial L_{N1}/\partial(\mathrm{d}_Q N_j^P) = 4N_k^Q M^k{}_P{}^j,$$
$$\partial L_{N1}/\partial N_j^P = 4M^j{}_Q{}^n \mathrm{d}_P N_n^Q, \qquad \partial L_{N2}/\partial(\mathrm{d}_Q N_j^P) = -2A^j{}_P{}^Q,$$
$$\partial L_{N2}/\partial D_P{}^Q{}_j = 2A^j{}_Q{}^P.$$

With these formulae, the associated Euler-Lagrange derivatives constructed in accordance with the rule $\delta\mathscr{L}/\delta N^P_j = \partial_m(\partial\mathscr{L}/\partial(\partial_m N^P_j)) + \mathrm{d}_Q(\partial\mathscr{L}/\partial(\mathrm{d}_Q N^P_j)) - \partial\mathscr{L}/\partial N^P_j$, where $\mathscr{L} = JKL$, can be calculated.

The complete list of the associated Lagrangians of the linear as well as quadratic forms in the gauge tensors is given by the following table:

$$L_{N1} = M_{iPj}M^{iPj}, \quad L_{N2} = A_{iPQ}A^{iPQ}, \quad L_{L1} = L^{nk}{}_{nk}, \quad L_{L2} = B_{nkjP}B^{nkjP},$$
$$L_{Q1} = Z^{PQ}{}_{PQ}, \quad L_{Q2} = Q_{PRSj}Q^{PRSj} = L_{D2}, \quad L_{L3} = L^{nkij},$$
$$L_{D3} = E_{PQij}E^{PQij}, \quad L_{Q3} = Z_{PQRS}Z^{PQRS}, \quad L_{B1} = F_{ij}F^{ij},$$
$$L_{B2} = b_{iP}b^{iP} = L_{H1}, \quad L_{H2} = H_{PQ}H^{PQ}, \quad L_{E1} = E_{ij}E^{ij}, \quad L_{E2} = e_{iP}e^{iP}.$$

Thus, the total Lagrangian for all the gauge fields under study is

$$L = n_1 L_{N1} + n_2 L_{N2} + l_1 L_{L1} + l_2 L_{L2} + q_1 L_{Q1} + q_2 L_{Q2} + l_3 L_{L3} + \\ + d_3 L_{D3} + q_3 L_{Q3} + b_1 L_{B1} + b_2 L_{B2} + h_2 L_{H2} + e_1 L_{E1} + e_2 L_{E2}, \tag{70}$$

where $n_1, n_2, l_1, \ldots$ are constants. The Euler-Lagrange derivatives thereof are found to have the following explicit forms:

$$\delta L/\delta L_n{}^k{}_i = -l(\mathrm{D}^*_m a_k{}^{inm} - \tfrac{1}{2}S_t{}^i{}_p a_k{}^{tnp}) + 2l_2 \mathrm{S}^*_R B^n{}_k{}^{iR} + \\ + 4l_3(\mathrm{D}^*_m L^n{}_k{}^{im} - \tfrac{1}{2}S_t{}^i{}_p L^n{}_k{}^{tp}), \tag{70$'$}$$

$$\delta L/\delta D_P{}^Q{}_i = 2n_2 A^i{}_Q{}^P - 2q_2 \mathrm{S}^*_R Q^P{}_Q{}^{Ri} + 4d_3(\mathrm{D}^*_m E^P{}_Q{}^{im} - \tfrac{1}{2}S_t{}^i{}_m E^P{}_Q{}^{tm}), \tag{70$''$}$$

$$\delta L/\delta Q_P{}^Q{}_R = -q_1(\mathrm{S}^*_N g_Q{}^{RPN} - \tfrac{1}{2}Y_N{}^R{}_M g_Q{}^{NPM}) + 2q_2(\mathrm{D}^*_j Q^P{}_Q{}^{Rj} + A_j{}^R{}_N Q^P{}_Q{}^{Nj}) + \\ + 2d_3 E^P{}_Q{}^{jm} M_j{}^R{}_m + 4q_3(\mathrm{S}^*_N Z^P{}_Q{}^{RN} - \tfrac{1}{2}Y_N{}^R{}_M Z^P{}_Q{}^{NM}), \tag{70$'''$}$$

and

$$\delta L/\delta E_i = 4e_1(\mathrm{D}^*_j E^{ji} + \tfrac{1}{2}S_k{}^i{}_j E^{kj}) - 2e_2 S^*_P e^{iP},$$
$$\delta L/\delta B_i = 4b_1(\mathrm{D}^*_j F^{ji} + \tfrac{1}{2}S_j{}^i{}_k F^{jk}) - 2b_2 S^*_Q b^{iQ},$$
$$\delta L/\delta H_P = 2b_1 F^{ij} M_i{}^P{}_j + 2b_2(\mathrm{D}^*_j b^{jP} + A_j{}^P{}_Q b^{jQ}) + 4h_2(S^*_R H^{RP} + \tfrac{1}{2}Y_Q{}^P{}_R H^{QR}),$$

together with

$$\delta L/\delta N^P_j = -4n_1(\mathrm{D}^*_m M^j{}_P{}^m - \tfrac{1}{2}S_t{}^j{}_q M^t{}_P{}^q + M^t{}_Q{}^j A_t{}^Q{}_P) - 2n_2(\mathrm{S}^*_R A^j{}_P{}^R - \\ - Y_Q{}^Q{}_R A^j{}_P{}^R) - l_1 a_k{}^{jni} B_n{}^k{}_{iP} - 2q_2 Q^R{}_M{}^{Nj} Z_R{}^M{}_{NP} + 4l_3 B_n{}^k{}_{mP} L^n{}_k{}^{jm} - \\ - 4d_3 Q_N{}^R{}_{Pm} E^N{}_R{}^{jm} - 4b_1 F^{ij} b_{iP} - 2b_2 b^{jQ} H_{PQ} - 4e_1 E^{kj} e_{kP} - \\ - Q_N{}^R{}_P \delta L/\delta D_N{}^R{}_j - H_P \delta L/\delta B_j. \tag{70$''''$}$$

Here, $\delta L/\delta$ means $(1/JK)\delta\mathscr{L}/\delta$; the notation $a_k{}^{mni} = \delta^m_k a^{ni} - \delta^i_k a^{nm}$, $g_Q{}^{RPN} = \delta^R_Q g^{PN} - \delta^N_Q g^{PR}$ and

$$\mathrm{D}^*_i = \mathrm{D}_i + N_i, \qquad \mathrm{S}^*_R = \mathrm{S}_R + V_R \tag{71}$$

with

$$N_i = \frac{1}{JK}\mathrm{d}_i(JK) - L_n{}^n{}_i - D_P{}^P{}_i + S_n{}^n{}_i - A_i{}^P{}_P,$$
$$V_R = \frac{1}{JK}\mathrm{d}_R(JK) - Q_P{}^P{}_R + Y_P{}^P{}_R. \tag{72}$$

being used. The objects (72) are obviously of tensor nature, namely, $\bar{N}_i = B_i^j N_j$, $N'_i = N_i$, $\bar{V}_R = V_R$, and $V'_R = z_R^P V_P$.

In principle, the theory could be extended even further by supplementing the metric tensors a_{ij} and g_{PQ}, whose types are respectively external and internal, by the mixed metric tensor $b_P{}^i(x,z)$, thereby assigning an internal image to any external tensor, for example, $W^i = b_P{}^i W^P$, $W_{PQ} = b_P{}^i b_Q{}^j W_{ij}$, etc. Obviously, the mixed metric tensor will give rise to new possible Lagrangians, for example, $b_P{}^i b_Q{}^j E^{PQ}{}_{ij}$.

B.6. General Gauge-Covariant Physical Field Equations

Now, guided by the combined principle of covariance, we proceed to elaborate the theory of the (x,z)-dependent generalizations of physical field equations. A sufficiently general example thereof is the case when the mixed contravariant internal anti-density $W^{-Pn}(x,z)$ couples with the mixed covariant internal density $W^{+}_{Pn}(x,z)$ (see the definitions (16), (19) and (20)). In this case the simplest gauge-invariant quadratic Lagrangian reads

$$L_W = L_{W1} + L_{W2} + k_3 W^{+}_{Pn} W^{-Pn} \tag{73}$$

with

$$L_{W1} = k_1 s^{ij}(\mathrm{D}_i W^{+}_{Pn})(\mathrm{D}_j W^{-Pn}) \equiv k_1 s^{ij}[(\partial_i + B_i + N_i^R d_R)W^{+}_{Pn} - D_P{}^Q{}_i W^{+}{}_{Qn} - - L_n{}^k{}_i W^{+}_{Pk}][(\partial_j - B_j + N_j^T \mathrm{d}_T)W^{-Pn} + D_T{}^P{}_j W^{-Tn} + L_m{}^n{}_j W^{-Pm}] \tag{74}$$

and

$$L_{W2} = k_2 t^{QR}(\mathrm{S}_Q W^{+}_{Pn})(\mathrm{S}_R W^{-Pn}) \equiv$$
$$\equiv k_2 t^{QR}[(\mathrm{d}_Q + H_Q)W^{+}_{Pn} - Q_P{}^T{}_Q W^{+}_{Tn}][(\mathrm{d}_R - H_R)W^{-Pn} + Q_M{}^P{}_R W^{-Mn}] \tag{75}$$

being scalars in the coordinate sense as well as in the gauge sense. The notation (11) together with $s^{ij} = (a^{ij} + a^{ji})/2$, $t^{QR} = (g^{QR} + g^{RQ})/2$ has been used, and k_1, k_2, k_3 are constants. According to the variational principle (69), we obtain the following result after simple calculations:

$$I^n{}_k{}^i \stackrel{\text{def}}{=} \delta L_W/\delta L_n{}^k{}_i = k_1(W^{+}_{Pk}\mathrm{D}^i W^{-Pn} - W^{-Pn}\mathrm{D}^i W^{+}_{Pk}), \tag{76}$$

$$I^P{}_Q{}^i \stackrel{\text{def}}{=} \delta L_W/\delta D_P{}^Q{}_i = k_1(W^{+}_{Qn}\mathrm{D}^i W^{-Pn} - W^{-Pn}\mathrm{D}^i W^{+}_{Qn}), \tag{77}$$

$$I^j_P \overset{\text{def}}{=} \delta L_W/\delta N^P_j = -k_1[(S_P W^+_{Qn})(D^j W^{-Qn}) + (S_P W^{-Qn})(D^j W^+_{Qn})] - \\ - Q_N{}^R{}_P I^N{}_R{}^j + H_P I^j, \tag{78}$$

$$I^P{}_R{}^T \overset{\text{def}}{=} \delta L_W/\delta Q_P{}^R{}_T = k_2(W^+_{Rn} S^T W^{-Pn} - W^{-Pn} S^T W^+_{Rn}), \tag{79}$$

$$I^i \overset{\text{def}}{=} -\delta L_W/\delta B_i = I^P{}_P{}^i = I^n{}_n{}^i, \qquad I^P \overset{\text{def}}{=} -\delta L_W/\delta H_P = I^R{}_R{}^P \tag{80}$$

and

$$\delta L_{W1}/\delta W^+_{Pn} = k_1 \square_x W^{-Pn}, \qquad \delta L_{W1}/\delta W^{-Pn} = k_1 \square_x W^+_{Pn},$$

$$\delta L_{W2}/\delta W^+_{Pn} = k_2 \square_z W^{-Pn}, \qquad \delta L_{W2}/\delta W^{-Pn} = k_2 \square_z W^+_{Pn},$$

so that

$$\delta L_W/\delta W^+_{Pn} = (k_1 \square_x + k_2 \square_z - k_3) W^{-Pn}, \tag{81}$$

$$\delta L_W/\delta W^{-Pn} = (k_1 \square_x + k_2 \square_z - k_3) W^+_{Pn}. \tag{82}$$

Here, $\delta L/\delta$ means $(1/JK)\ \delta\mathscr{L}/\delta$ with $\mathscr{L} = JKL$ and the designations $\square_x = D^*_j D^j$, $\square_z = S^*_R S^R$ mean the associated gauge-covariant D'Alembertians in x^i and z^P, respectively, where the operators D^*_j and S^*_R are given by the definition (71); $D^j = s^{ji} D_i$ and $S^R = t^{RQ} S_Q$.

Clearly, the structure of the right-hand sides of the Euler-Lagrange derivatives (76)–(82) will remain valid for fields W of other tensor types. For the case of tensors which are not densities, for example W^{Pn}, $W_P{}^Q$ or $W_i{}^{mn}$, we merely have to identify W^+ and W^- in the above formulae, which will in particular imply that the right-hand sides of the relations (80) will vanish in the metric case (96). On the other hand, if we consider fields which are external densities, for example the field $W^{(+)+}_{Pn}(x,z)$ coupled with $W^{(-)-Pn}(x,z)$ (see the definitions (15) and (18)), and construct the Lagrangian in accordance with the above rule (73)–(75), that is,

$$L_W = k_1 s^{ij}(D_i W^{(+)+}_{Pn})(D_j W^{(-)-Pn}) + k_2 t^{QR}(S_Q W^{(+)+}_{Pn})(S_R W^{(-)-Pn}) + \\ + k_3 W^{(+)+}_{Pn} W^{(-)-Pn},$$

then we should supplement the list (76)–(82) by the new member

$$I^i_{(E)} \overset{\text{def}}{=} -\delta L_W/\delta E_i = k_1(W^{(+)+}_{Pn} D^i W^{(-)-Pn} - W^{(-)-Pn} D^i W^{(+)+}_{Pn}) \tag{83}$$

which is similar to the objects (80).

The definitions (76)–(80) and (83) may be called *currents*, for they immediately generalize their usual field-theoretical prototypes. With the above formulae, we obtain directly the remarkable relation

$$D^*_i I^P{}_Q{}^i + S^*_T I^P{}_Q{}^T = (W^+_{Qn} \delta/\delta W^+_{Pn} - W^{-Pn} \delta/\delta W^{-Qn}) L_W, \tag{84}$$

where the obvious identity

$$(W^{+}_{Qn}\delta/\delta W^{+}_{Pn} - W^{-Pn}\delta/\delta W^{-Qn})W^{+}_{Rm}W^{-Rm} = 0 \tag{85}$$

has been used. In the case when the field equations

$$\delta L_W/\delta W^{+}_{Pn} = \delta L_W/\delta W^{-Pn} = 0 \tag{86}$$

hold, relation (84) reduces to the following *generalized conservation law for the currents* $I^P{}_Q{}^i$ *and* $I^P{}_Q{}^T$:

$$\mathrm{D}^*_i I^P{}_Q{}^i + \mathrm{S}^*_T I^P{}_Q{}^T = 0. \tag{87}$$

We observe that $I^P{}_Q{}^i$ and $I^P{}_Q{}^T$ are not conserved separately.

In the case of the current (76), the analogue of the relation (84) reads

$$\mathrm{D}^*_i I^n{}_k{}^i = (W^{+}_{Pk}\delta/\delta W^{+}_{Pn} - W^{-Pn}\delta/\delta W^{-Pk})L_{W1}. \tag{88}$$

At the same time, introducing the new current

$$I^n{}_k{}^T = k_2(W^{+}_{Pk}\mathrm{S}^T W^{-Pn} - W^{-Pn}\mathrm{S}^T W^{+}_{Pk}), \tag{89}$$

we shall have

$$\mathrm{S}^*_T I^n{}_k{}^T = (W^{+}_{Pk}\delta/\delta W^{+}_{Pn} - W^{-Pn}\delta/\delta W^{-Pk})L_{W2} \tag{90}$$

in addition to (88). Comparing the relations (88) and (90) and noting again the identity (85), we arrive at the additional conservation law

$$\mathrm{D}^*_i I^n{}_k{}^i + \mathrm{S}^*_T I^n{}_k{}^T = 0 \tag{91}$$

provided that the field equations (86) hold. The conservation law (91) is quite similar to (87). Contracting the indices P and Q in (87), or the indices n and k in (91), we obtain the conservation law for the currents (80):

$$\mathrm{D}^*_i I^i + \mathrm{S}^*_T I^T = 0. \tag{92}$$

Notice that $\mathrm{D}^*_i I^P{}_Q{}^i = \mathrm{D}^*_i I^n{}_k{}^i = \mathrm{D}^*_i I^i = 0$ if $k_2 = 0$.

Further, if we denote

$$S_{nk}{}^i = s_{nm}I^m{}_k{}^i - s_{km}I^m{}_n{}^i, \tag{93}$$

where $s_{nm} = (a_{nm} + a_{mn})/2$, then, on substituting the general representation (76) of the current $I^n{}_k{}^i$ in the right-hand side of (93) and comparing the result with the usual definition of the *spin tensor* (see, e.g., Equation (4.9) in Bogoljubov *et al.* (1959)), we arrive at the conclusion that the object (93) represents a generalization of the spin tensor for the parametrical (x, z)-dependent field W under study. Similarly, we introduce the *internal spin tensors*

$$S_{PQ}{}^i = t_{PR}I^R{}_Q{}^i - t_{QR}I^R{}_P{}^i, \qquad S_{PQ}{}^T = t_{PR}I^R{}_Q{}^T - t_{QR}I^R{}_P{}^T,$$
$$S_{nk}{}^T = s_{nm}I^m{}_k{}^T - s_{km}I^m{}_n{}^T, \tag{94}$$

where $t_{PR} = (g_{PR} + g_{RP})/2$. From the conservation law (91) it follows that the spin tensor ${S_{nk}}^i$ is conserved only in conjunction with the internal spin tensor ${S_{nk}}^T$.

In the particular case when the metric tensors are symmetric, that is,

$$s_{nm} = a_{nm}, \qquad t_{PQ} = g_{PQ}, \tag{95}$$

and the gauge-covariant derivatives are metric, that is,

$$\mathrm{D}_i a_{nm} = \mathrm{D}_i g_{PQ} = \mathrm{S}_R a_{nm} = \mathrm{S}_R g_{PQ} = 0 \tag{96}$$

(see Section 8), the expression (93) reduces to

$$\begin{aligned} {S_{nk}}^i = k_1(&W^+_{Pk}\mathrm{D}^i {W^{-P}}_n + {W^{-P}}_k \mathrm{D}^i W^+_{Pn} - W^+_{Pn}\mathrm{D}^i {W^{-P}}_k - \\ &- {W^{-P}}_n \mathrm{D}^i W^+_{Pk}), \end{aligned} \tag{97}$$

which is essentially of the same form as the usual field-theoretic spin tensor (Equation (4.9) of Bogoljubov *et al.* (1959)), and the spin conservation laws derived from (87) and (91) together with (94) are merely

$$\mathrm{D}^*_i {S_{PQ}}^i + \mathrm{S}^*_T {S_{PQ}}^T = 0, \qquad \mathrm{D}^*_i {S_{nk}}^i + \mathrm{S}^*_T {S_{nk}}^T = 0.$$

Notice that the superscripts + and − enter the right-hand side of the expression (97), as well as the initial Lagrangian (73)–(75), in a symmetric way, so that the spin tensor remains invariant under the replacement $+ \longleftrightarrow -$ (in contrast to the currents).

Finally, the usual definition of the *energy-momentum tensor*

$$T^{ij} = -\frac{2}{JK}\partial \mathscr{L}_W / \partial a_{ij} \tag{98}$$

is to be supplemented by the definition

$$T^{PQ} = -\frac{2}{JK}\partial \mathscr{L}_W / \partial g_{PQ} \tag{99}$$

which may tentatively be called the *internal energy-momentum tensor*. In the case of the Lagrangian (73)–(75), we obtain

$$T^{ij} = k_1[(\mathrm{D}^i W^+_{Pn})(\mathrm{D}^j W^{-Pn}) + (\mathrm{D}^j W^+_{Pn})(\mathrm{D}^i W^{-Pn})] - a^{ij} L_W \tag{100}$$

and

$$T^{PQ} = k_2[(\mathrm{S}^P W^+_{Rn})(\mathrm{S}^Q W^{-Rn}) + (\mathrm{S}^Q W^+_{Rn})(\mathrm{S}^P W^{-Rn})] - g^{PQ} L_W.$$

Introducing also the tensor

$$T^{iR} = k_2[(\mathrm{D}^i W^+_{Pn})(S^R W^{-Pn}) + (D^i W^{-Pn})(S^R W^+_{Pn})]$$

of partly internal meaning and assuming the symmetry and metric conditions (95)–(96) hold, we obtain the following conservation laws for the energy-momentum tensors after some simple calculations:

$$\begin{aligned} \mathrm{D}_j^* T^{ij} + \mathrm{S}_R^* T^{iR} = k_1[(\mathrm{D}^j W^{-Pn})(\mathrm{D}_j \mathrm{D}^i - \mathrm{D}^i \mathrm{D}_j) W^+_{Pn} + \\ + (\mathrm{D}^j W^+_{Pn})(\mathrm{D}_j \mathrm{D}^i - \mathrm{D}^i \mathrm{D}_j) W^{-Pn}] + \\ + k_2[(\mathrm{S}^R W^{-Pn})(\mathrm{S}_R \mathrm{D}^i - \mathrm{D}^i \mathrm{S}_R) W^+_{Pn} + \\ + (\mathrm{S}^R W^+_{Pn})(\mathrm{S}_R \mathrm{D}^i - \mathrm{D}^i \mathrm{S}_R) W^{-Pn}] \end{aligned} \tag{101}$$

and

$$\begin{aligned} \mathrm{S}_Q^* T^{RQ} + \frac{k_1}{k_2} \mathrm{D}_i^* T^{iR} = k_1[(\mathrm{D}^j W^{-Pn})(\mathrm{D}_j \mathrm{S}^R - \mathrm{S}^R \mathrm{D}_j) W^+_{Pn} + \\ + (\mathrm{D}^j W^+_{Pn})(\mathrm{D}_j \mathrm{S}^R - \mathrm{S}^R \mathrm{D}_j) W^{-Pn}] + \\ + k_2[(\mathrm{S}^Q W^{-Pn})(\mathrm{S}_Q \mathrm{S}^R - \mathrm{S}^R \mathrm{S}_Q) W^+_{Pn} + \\ + (\mathrm{S}^Q W^+_{Pn})(\mathrm{S}_Q \mathrm{S}^R - \mathrm{S}^R \mathrm{S}_Q) W^{-Pn}] \end{aligned} \tag{102}$$

provided that the field equations (86) are satisfied. The commutators of the covariant derivatives entering the right-hand sides of the conservation laws (101) and (102) may be expressed in terms of the gauge tensors by using the commutation relations established in Section 3. Because of the last term on the right-hand side of Equations (100) the internal part L_{W2} of the Lagrangian L_W will contribute to the energy-momentum of the system of physical fields unless the internal space is entirely flat.

It will also be noted that, treating the internal densities W^+ as the fields carrying the electric charge (see the end of Section 2), we are to interpret the current I^i entering Equations (80) and (92) as the gauge-geometric image of the *electric current*. As regards the external + internal parametrical spinor equations, they may also be formulated under these general conditions (see Section 5.4).

Hitherto, it has been tacitly implied that the examined fields $w_n{}^m{}_P{}^Q(x, z)$, $W^+_{Pn}(x, z)$, and $W^{-Pn}(x, z)$ are real, whereas the fields carrying electric charge are usually represented by complex fields (see, e.g., Bogoljubov *et al.* (1959)). However, the techniques developed can be applied directly to complex fields, the only modification being that the densities should be redefined as follows. A complex field $W(x, z)$ will be called *a complex (p, q)-density* if $\bar{W} = W \exp(ipB)$ and $W' = W \exp(iqZ)$; in turn, $\overline{W^{\mathscr{C}}} = W^{\mathscr{C}}$ and $(W^{\mathscr{C}})' = (W')^{\mathscr{C}}$, where the superscript $\mathscr{C}$ denotes complex conjugation. Accordingly, the aforesaid real fields W^+_{Pn} and W^{-Pn} will be replaced by the complex fields W^+_{Pn} ($p = 0$ and $q = 1$) and W^{-Pn} ($p = 0$ and $q = -1$), respectively. Agreement with the above postulated transformation laws (17) will be retained in dealing with complex fields if we replace the gauge fields E_n, B_n, and H_P by iE_n, iB_n, and iH_P in all the formulae of the present section [this assertion ensues from the relation $(\mathrm{d}'_n \exp(iZ)/\exp(iZ) = i(\mathrm{d}'_n Z)/Z$ and its counterparts for B and d'_P].

B.7. Parametrical Representation of the (x, y)-Dependent Gauge Fields Associated with Space-Time

On restricting ourselves to the particular case when the number of the parameters z^P is equal to the dimension number N of the background manifold, that is, $M = N$, and denoting by y^i the tangent vectors supported by x^i, let us assume that we are given a set of N sufficiently smooth functions

$$z^P = z^P(x, y) \tag{103}$$

obeying the following two conditions. First, they must be of scalar nature under the general coordinate transformations (1), that is,

$$\bar{z}^P(\bar{x}, \bar{y}) = z^P(x, y), \tag{104}$$

where $B^j_i \bar{y}^i = y^j$ because y^i are vectors under (1). Second, we require that $\operatorname{rank}(\partial z^P/\partial y^i) = N$, which makes it possible to express the tangent vectors y^i in terms of the scalar parameters z^P:

$$y^i = y^i(x, z). \tag{105}$$

We call (105) the *parametrical representation of the tangent vectors* y^i. The associated *projection factors* $z^P_i = \partial z^P/\partial y^i$ and $y^i_P = \partial y^i/\partial z^P$ will obviously be connected by the reciprocity relations $z^P_i y^i_Q = \delta^P_Q$ and $z^P_i y^j_P = \delta^j_i$. According to the assumption (104), the projection factors will be vectors under the general coordinate transformations (1), that is, $\bar{z}^P_i = B^j_i z^P_j$ and $B^j_i \bar{y}^i_P = y^j_P$. At the same time, the gauge transformations (2), when rewritten in the language of the tangent vectors y^i, will read

$$y^i = y^i(x, y'), \tag{106}$$

implying the transformation laws $z^P_i = z^P_Q z'^Q_i$ and $z^P_Q y^i_P = y'^i_Q$.

The transformations (103) and (105) may be treated as coordinate transformations in the tangent space of the background space (space-time in the usual physical context), namely, as transformations between the coordinates $\{y^i\}$ and $\{z^P\}$. Under the transition (105), the transforms of (x, y)-dependent objects tensorial under the gauge transformations (106) will consistently be defined in accordance with the tensor laws. For example, considering again the mixed metric tensor $w_n{}^m{}_P{}^Q(x, z)$ subject to the transformation laws (5) and (6), we may treat it as the parametrical representation

$$w_n{}^m{}_P{}^Q(x, z) = z^Q_j y^i_P w_n{}^m{}_{,i}{}^j(x, y(x, z)) \tag{107}$$

of the (x, y)-dependent field

$$w_n{}^m{}_{,i}{}^j(x, y) = z^P_i y^j_Q w_n{}^m{}_P{}^Q(x, z(x, y)). \tag{108}$$

Accordingly, the internal metric tensor g_{PQ} given by the transformation laws (48) may be regarded as the parametrical representation $g_{PQ}(x, z) = y^i_P y^j_Q g_{ij}(x, y(x, z))$ of the (x, y)-dependent prototype $g_{ij}(x, y)$. It should be noted that the tensors which are scalars under the gauge transformations (106) will be transformed into the parametrical representation by means of the scalar law, as is exemplified by the case of the external metric tensor a_{ij} given by (48): $a_{ij}(x, z) = a_{ij}(x, y(x, z))$. In the language of the variables x and y the transformation law (6) will be written as $y^k_j w'_n{}^m{}_{,i}{}^j = y^l_i w_n{}^m{}_{,l}{}^k$, where we have used the notation $y^k_j = \partial y^k/\partial y'^j$.

The actions of the gauge-covariant derivative D_i on (x, y)-dependent and (x, z)-dependent fields may naturally be connected with each other by postulating

$$\mathrm{D}_i w_n{}^m{}_{,r}{}^s = y^s_Q z^P_r \mathrm{D}_i w_n{}^m{}_P{}^Q, \tag{109}$$

where

$$\mathrm{D}_i w_n{}^m{}_{,r}{}^s = (\partial_i - A_i{}^j \partial/\partial y^j) w_n{}^m{}_{,r}{}^s + L_k{}^m{}_i w_n{}^k{}_{,r}{}^s - L_n{}^k{}_i w_k{}^m{}_{,r}{}^s + \\ + D_k{}^s{}_i w_n{}^m{}_{,r}{}^k - D_r{}^k{}_i w_n{}^m{}_{,k}{}^s.$$

From this it follows that the relationships between the gauge fields of the two representations are given by the identities

$$\begin{aligned} \mathrm{D}_i z^P &\stackrel{\text{def}}{=} \partial z^P/\partial x^i - A_i{}^j z^P_j - N^P_i = 0, \\ \mathrm{D}_i y^j &\stackrel{\text{def}}{=} \partial y^j/\partial x^i + A_i{}^j + N^P_i y^n_P = 0, \end{aligned} \tag{110}$$

$$\mathrm{D}_i z^Q_n \stackrel{\text{def}}{=} \partial z^Q_n/\partial x^i - A_i{}^j \partial z^Q_n/\partial y^j - D_n{}^m{}_i z^Q_m + D_P{}^Q{}_i z^P_n = 0, \tag{111}$$

$$\mathrm{D}_i y^n_P \stackrel{\text{def}}{=} \partial y^n_P/\partial x^i + N^R_i \partial y^n_P/\partial z^R + D_m{}^n{}_i y^m_P - D_P{}^Q{}_i y^n_Q = 0. \tag{112}$$

In Equation (110), z^P means $z^P(x, y)$ and y^j means $y^j(x, z)$.

As regards the gauge-covariant derivative S_i with respect to y^i, which reads $\mathrm{S}_i w_n{}^m{}_{,r}{}^s = \partial w_n{}^m{}_{,r}{}^s/\partial y^i + Q_k{}^s{}_i w_n{}^m{}_{,r}{}^k - Q_r{}^k{}_i w_n{}^m{}_{,k}{}^s$, we may write

$$\mathrm{S}_i w_n{}^m{}_{,r}{}^s = z^R_i y^s_Q z^P_r \mathrm{S}_R w_n{}^m{}_P{}^Q \tag{113}$$

similarly to Equation (109), obtaining

$$\begin{aligned} \mathrm{S}_R y^m_P &\stackrel{\text{def}}{=} \partial y^m_P/\partial z^R + Q_n{}^m{}_R y^n_P - Q_P{}^T{}_R y^n_T = 0, \\ \mathrm{S}_R z^P_n &\stackrel{\text{def}}{=} \partial z^P_n/\partial y^i - Q_n{}^m{}_i z^P_m + Q_R{}^P{}_i z^R_m = 0, \end{aligned} \tag{114}$$

where the notation $Q_n{}^m{}_R = y^k_R Q_n{}^m{}_k$ and $Q_R{}^P{}_i = z^T_i Q_R{}^P{}_T$ is used.

Finally, the external and internal densities introduced in Section 2 can be readily put in the (x, y)-representation in a similar way. Indeed, because the Jacobian B given by the definition (14) is dependent solely on the argument x and therefore is the same in both representations, we have simply $W^{(+)}(x, z) = W^{(+)}(x, y(x, z))$, so that $E_i(x, z) = E_i(x, y(x, z))$. Next, since the internal densities $W^+(x, z)$ and $W^+(x, y)$ are densities in z and y, respectively, they relate to each other as follows: $W^+(x,z) = YW^+(x,y(x,z))$, where $Y = \det(y^i_P)$. Postulating $\mathrm{D}_i W^+(x,z) = Y\mathrm{D}_i W^+(x,y)$ and $\mathrm{S}_R W^+(x,z) = y^i_R Y S_i W^+(x,y)$, we obtain

$$\begin{aligned} \mathrm{D}_i Y &\stackrel{\text{def}}{=} \mathrm{d}_i Y - YB_i(x,y) + YB_i(x,z) = 0, \\ \mathrm{S}_R Y &\stackrel{\text{def}}{=} \partial Y/\partial z^R - Yy^i_R H_i(x,y) + YH_R(x,z) = 0. \end{aligned} \tag{115}$$

Notice that the notation d_i given by Equation (11) may be retained in both representations, remembering that $\mathrm{d}_i = \partial_i - A_i{}^j \partial/\partial y^j$ and $\mathrm{d}_i = \partial_i + N^P_i \,\partial/\partial z^P$ when the operator d_i is applied to (x,y)-dependent objects and to (x,y)-dependent objects, respectively.

The remarkable identities (107), (108), (111), (112), (114), and (115) lead directly to the conclusion that the relationships between the gauge tensors of both representations are given by the tensor law operative in the gauge-tensorial indices P, $Q, \ldots$, i.e., $y^m_Q z^R_n A_j{}^Q{}_R(x,z) = A_j{}^m{}_n(x,y)$, $y^n_R M_j{}^R{}_i(x,z) = M_j{}^n{}_i(x,y)$, $E_m{}^n{}_{ij}(x,y) = y^n_Q z^P_m E_P{}^Q{}_{ij}(x,z)$, $y^m_T z^P_n z^R_k Q_P{}^T{}_{Rj}(x,z) = Q_n{}^m{}_{kj}(x,y)$, $z^P_i z^Q_j H_{PQ}(x,z) = H_{ij}(x,y)$, $F_{ij}(x,z) = F_{ij}(x,y)$, etc. To prove this statement in all rigour, it is sufficient to compare the commutators of the gauge-covariant derivatives in the two representations and make allowance for the aforementioned identities. The conclusions made in this section make it clear also that the gauge-covariant equations derivable in the (x,y)-representation and in the related (x,z)-representation will everywhere be equivalent.

B.8. Implications of Metric Conditions

The condition that the derivative D_i be metric with respect to the tensor a_{nm}, that is,

$$\mathrm{D}_i a_{nm} = 0, \tag{116}$$

yields the relation

$$L_{nmji} + L_{mnji} + M_j{}^R{}_i \mathrm{S}_R a_{nm} = 0 \tag{117}$$

which is obtained after writing the commutator (33) for the case of a_{nm}. Another metric condition

$$\mathrm{S}_R a_{nm} = 0 \tag{118}$$

reads merely that the tensor a_{nm} should be independent of z^R, that is,

$$\partial a_{nm}/\partial z^R = 0 \tag{119}$$

in accordance with the definition (13) of the covariant derivative S_R. Under the condition (118), the above relation (117) reduces to $L_{nmji} = -L_{mnji}$ which represents an immediate generalization of the well-known identity satisfied by the Riemann curvature tensor. Simultaneously postulating the metric conditions (116) and (118) entails an additional relation $B_{nmjP} = -B_{mnjP}$ as should be clear from the commutation rule (39).

Similarly, if we require that

$$\mathrm{D}_i g_{PQ} = 0, \tag{120}$$

then the commutation rule (35), when applied to g_{PQ}, will give $E_{PQji} + E_{QPji} + M_j{}^R{}_i \mathrm{S}_R g_{PQ} = 0$, while postulating the metric condition

$$\mathrm{S}_R g_{PQ} \equiv \mathrm{d}_R g_{PQ} - Q_P{}^T{}_R g_{TQ} - Q_Q{}^T{}_R g_{PT} = 0 \tag{121}$$

yields $Z_{PQMN} = -Z_{QPMN}$ in accordance with Equation (42). When the metric conditions (120) and (121) are postulated simultaneously, we obtain the relations $E_{PQji} = -E_{QPji}$ and $Q_{PRNj} = -Q_{RPNj}$ as a consequence of Equations (35) and (37). Finally, the conditions

$$\mathrm{D}_i J = 0, \qquad \mathrm{S}_R J = 0, \qquad \mathrm{D}_i K = 0, \qquad \mathrm{S}_R K = 0 \tag{122}$$

yield in turn the relations $M_j{}^R{}_i \mathrm{S}_R K + KF_{ji} = 0$, etc., obtainable by putting $W^{(+)} = J$ and $W^+ = K$ in the commutators (43)–(46). In particular, if all four conditions (122) are postulated simultaneously, then all the gauge tensors (47) vanish.

Up till now we have developed the theory without assuming that the metric tensors are symmetric. Henceforth, we shall drop such generality in order to be able to explicitly resolve the equations for the gauge fields. With the symmetry conditions

$$a_{ij}(x, z) = a_{ji}(x, z), \qquad g_{PQ}(x, z) = g_{QP}(x, z), \tag{123}$$

and not assuming (118) and (121), the equations (116) and (120) may be rewritten as follows:

$$\mathrm{d}_i a_{mn} - L_{mni} - L_{nmi} = 0 \tag{124}$$

and

$$\mathrm{d}_i g_{PQ} - D_{PQi} - D_{QPi} = 0, \tag{125}$$

respectively; the notation

$$L_{mni} = a_{nk} L_m{}^k{}_i, \qquad D_{PQi} = g_{QR} D_P{}^R{}_i \tag{126}$$

being used. We observe that (124) and (125) are equations for just the symmetric parts of the coefficients (126) in the indices m,n and P,Q, while the skew-symmetric parts

$$V_{mni} \stackrel{\text{def}}{=} L_{mni} - L_{nmi}, \qquad W_{PQi} \stackrel{\text{def}}{=} D_{PQi} - D_{QPi} \tag{127}$$

may be given arbitrarily. So, we get

$$2L_m{}^i{}_n = a^{ik}\,\mathrm{d}_n a_{mk} + V_m{}^i{}_n \equiv a^{ik}(\partial_n a_{mk} + N_n^P \mathrm{d}_P a_{mk}) + V_m{}^i{}_n \tag{128}$$

and

$$2D_P{}^Q{}_i = g^{QR}\,\mathrm{d}_i g_{PR} + W_P{}^Q{}_i \equiv g^{QR}(\partial_i g_{PR} + N_i^S\,\mathrm{d}_S g_{PR}) + W_P{}^Q{}_i. \tag{129}$$

However, the objects $V_m{}^i{}_n$ and $W_P{}^Q{}_i$ given by the definition (127) fail to be gauge tensors, which leads us to select the tensorial parts from them. To this end we recall the definition (25) of the gauge torsion tensor $S_m{}^i{}_n$, obtaining the following relation from the metric condition (124):

$$2S_{min} = \mathrm{d}_n a_{mi} - \mathrm{d}_m a_{ni} + V_{min} - V_{nim}. \tag{130}$$

The inverse reads

$$V_{inm} = \mathrm{d}_i a_{nm} - \mathrm{d}_n a_{mi} + S_{min} + S_{imn} - S_{mni}, \tag{131}$$

as may be verified directly by substituting (130) in (131). Next, with the gauge tensor $A_j{}^Q{}_R$ given by the definition (26), the definition (127) entails $W_{PQi} = g_{PR}\,\mathrm{d}_Q N_i^R - g_{QR}\,\mathrm{d}_P N_i^R + T_{PQi}$, where

$$T_{PQi} = A_{iPQ} - A_{iQP} \tag{132}$$

is obviously a gauge tensor. Thus, the expressions (128) and (129) take the forms

$$\begin{aligned} L_m{}^i{}_n = \tfrac{1}{2}a^{ik}(\partial_n a_{mk} + \partial_m a_{nk} - \partial_k a_{mn} + N_n^P\,\mathrm{d}_P a_{mk} + \\ + N_m^P \mathrm{d}_P a_{nk} - N_k^P\,\mathrm{d}_P a_{mn} + S_{nmk} + S_{mnk} + S_{mkn}) \end{aligned} \tag{133}$$

and

$$D_P{}^Q{}_i = \tfrac{1}{2}[g^{QR}(\partial_i g_{PR} + N_i^S\,\mathrm{d}_S g_{PR} + g_{PS}\,\mathrm{d}_R N_i^S) - \mathrm{d}_P N_i^Q + T_P{}^Q{}_i]. \tag{134}$$

Similarly, from the metric condition (121) it follows that

$$Q_P{}^T{}_R = \tfrac{1}{2}g^{TS}(\mathrm{d}_R g_{SP} + \mathrm{d}_P g_{SR} - \mathrm{d}_S g_{PR} + Y_{RPS} + Y_{PRS} + Y_{PSR}). \tag{135}$$

Clearly, the torsion tensors $S_m{}^i{}_n$, $T_P{}^Q{}_i$, and $Y_P{}^Q{}_R$ represent degrees of freedom in (133)–(135) which may be chosen independently of the metric conditions.

B.9. Specification of the Internal Metric Tensor

The simplest possibility for specifying the metric tensor g_{PQ} is to set forth the requirement that a field $f_Q(x, z)$ exists such that

$$g_{PQ} = \mathrm{S}_P f_Q \equiv \mathrm{d}_P f_Q - Q_Q{}^R{}_P f_R, \tag{136}$$

assuming the field f_Q to possess the scalar-vector law of transformations, that is,

$$\bar{f}_Q = f_Q, \qquad f'_Q = z_Q^P f_P. \tag{137}$$

As long as the metric condition (121) is assumed to hold, the requirement (136) entails the relation $f_P Z_Q{}^P{}_{RS} + Y_{RQS} = 0$, as is clear from the commutation rule

(41). Further, by assuming the field f_P to be such that

$$\mathrm{D}_i f_P \equiv \partial_i f_P + N_i^Q \, \mathrm{d}_Q f_P - D_P{}^Q{}_i f_Q = 0 \tag{138}$$

and employing the metric condition (120), we obtain another relation

$$f_P Q_R{}^P{}_{Mj} - A_{jRM} = 0 \tag{139}$$

with the aid of the second member of the commutators (36), while from the commutator (34) it follows that

$$f_Q E_P{}^Q{}_{ji} + M_{jPi} = 0. \tag{140}$$

The next stage in the specification of the metric tensor g_{PQ} consists in stipulating that the field f_P should possess a generating function, that is, that a scalar field $f(x, z)$ exists such that

$$f_P = \mathrm{S}_P f \equiv \mathrm{d}_P f. \tag{141}$$

With the conditions (136) and (141), we have $g_{PQ} = \mathrm{d}_P \mathrm{d}_Q f - Q_Q{}^R{}_P f_R$, so that the symmetry condition (123) will entail $f_R Y_P{}^R{}_Q = 0$. If we assume simultaneously that

$$\mathrm{D}_i f \equiv \partial_i f + N_i^P f_P = 0, \tag{142}$$

then the commutation rules (36) and (31) enable us to conclude that

$$f_P A_i{}^P{}_Q = 0, \qquad f_P M_i{}^P{}_j = 0. \tag{143}$$

With the aid of the first member of Equations (143) and the definition of the gauge tensor $A_j{}^Q{}_R$ (Equation (26)), we can reduce the condition (138) to the following differential equation for the gauge field N_i^Q alone:

$$\partial_i f_P + N_i^Q \, \mathrm{d}_Q f_P + f_Q \, \mathrm{d}_P N_i^Q = 0. \tag{144}$$

However, equation (144) is obtainable by differentiating equation (142) with respect z^P, so that Equation (144) is a concomitant of Equation (142). Thus, under the above conditions, the gauge field N_i^P, and hence the gauge field $D_P{}^Q{}_i$ given by the expression (134), is primarily described by the sole algebraic equation

$$\partial_i f + N_i^P f_P = 0. \tag{145}$$

B.10. Transition to the Parametrical Finslerian Limit

In the two previous sections we have introduced a set of particular conditions which give rise to a number of interesting special types of gauge field. It seems, nevertheless, that the level so attained is still rather general and not simple. It is only natural, therefore, to seek further simplifying conditions with a view to reducing our gauge fields to particular cases which are simple and yet still sufficiently nontrivial. The relations arrived at the end of the preceding section suggest the consideration of the case where the gauge fields and tensors are expressible in terms of the external metric tensor a_{ij} and a single scalar function $f(x, z)$ generating the internal metric tensor g_{PQ}, as well as in terms of the torsion tensors $S_m{}^i{}_n$, $Y_P{}^Q{}_R$, and $T_P{}^Q{}_i$ which enter the equations as entirely self-contained objects depending on nothing. Of course, our relations could be simplified further by assuming the torsion tensors to be zero. However, we shall retain them for two reasons. First, this is a comparatively easy task. Second, present-day studies largely deal with theories generalizing general relativity to the case involving torsion Hehl (1981). Also, the external metric tensor a_{ij} could be a priori assumed to be Riemannian, that is, the condition (119) may be postulated, but we shall again refrain from so doing for we are not impelled to by in any way serious mathematical difficulties.

The form of the appropriate additional conditions is not difficult to perceive. To arrive at the possibility of solving the equation (145) for the gauge field N_i^P in an explicit form, it seems natural to constrain the generating function $f(x, z)$ in the definition (141) by the homogeneity condition

$$f(x, kz) = k^r f(x, z), \qquad k > 0, \tag{146}$$

where the degree r may be an arbitrary integer, except for $r = 0$ and $r = 1$. From this it follows that

$$f_P z^P = rf, \qquad f_{PQ} z^Q = (r-1) f_P, \qquad z^R \mathrm{d}_R f_{PQ} = (r-2) f_{PQ} \tag{147}$$

because of Euler's theorem on homogeneous functions (these are obtainable by differentiating the homogeneity condition with respect to k and then taking $k = 1$). The notation $f_{PQ} = \mathrm{d}_Q f_P$ has been used. Substituting the first two members of the identities (147) in equation (145) and taking account of the fact that $\partial_i z^P \equiv 0$, we arrive at the equation

$$z^P \left(\frac{1}{r} \partial_i f_P + \frac{1}{r-1} f_{PQ} N_i^Q \right) = 0$$

which obviously admits the solution

$$f_{PQ} N_i^Q = \frac{1-r}{r} \partial_i f_P. \tag{148}$$

As long as

$$\det(f_{PQ}) \neq 0, \tag{149}$$

Equation (148) can be solved for N_i^P in the explicit form

$$N_i^P = \frac{1-r}{r} f^{PQ} \partial_i f_Q, \tag{150}$$

where the notation f^{PQ} is used for the inverse of f_{PQ}, so that $f^{PQ} f_{RQ} = \delta_R^P$. The solution (150) is then to be substituted in the expressions (133) and (134) for the gauge fields $L_m{}^i{}_n$ and $D_P{}^Q{}_i$. However, because of the definition (136), the gauge field $Q_P{}^M{}_R$ will still be involved in the set of differential equations obtainable in this way, unless we set forth the following new condition to suppress the dependence of the metric tensor g_{PQ} on $Q_P{}^M{}_R$:

$$f_{PQ} = g_{PQ}, \tag{151}$$

that is,

$$g_{PQ} = \partial^2 f/\partial z^P \, \partial z^Q \equiv g_{QP}. \tag{152}$$

Denoting further

$$C_{PQR} = \tfrac{1}{2}\partial^3 f/\partial z^P \, \partial z^Q \, \partial z^R, \tag{153}$$

we shall have

$$C_{PQR} = \tfrac{1}{2}\mathrm{d}_R \, g_{PQ}, \tag{154}$$

while differentiating the reciprocity relation

$$g^{PQ} g_{RQ} = \delta^P_R \tag{155}$$

with respect to z^M, we obtain

$$-\ \mathrm{d}_M g^{PQ} = C_M{}^{PQ} \equiv C_{MRS}\, g^{PR} g^{QS}. \tag{156}$$

Clearly, the object C_{PQR} is symmetric in all subscripts and homogeneous of degree r-3 in z, while the metric tensor g_{PQ} is symmetric and homogeneous of degree r-2 in z.

With the equality (151), the solution (150) becomes

$$N^P_i = \frac{1-r}{r} g^{PQ} \, \partial_i f_Q. \tag{157}$$

In view of the relations (152) and (156), differentiating the solution (157) with respect to z yields

$$\mathrm{d}_P N^Q_i = \frac{1-r}{r}(-2C_P{}^{QR}\, \partial_i f_R + g^{QR}\, \partial_i g_{RP}) \equiv g^{QR} g_{PS}\, \mathrm{d}_R N^S_i, \tag{158}$$

which reduces the expression (134) to merely

$$D_P{}^Q{}_i = \tfrac{1}{2} g^{QR}\, \partial_i g_{PR} + N^S_i C_S{}^Q{}_P + \tfrac{1}{2} T_P{}^Q{}_i. \tag{159}$$

Thus, we have explicitly expressed the gauge field $D_P{}^Q{}_i$ in terms of the set $\{f, T_P{}^Q{}_i\}$. Also, the symmetry of the object C_{PQR} reduces (135) to

$$Q_P{}^M{}_R = C_P{}^M{}_R + q_P{}^M{}_R, \qquad q_P{}^M{}_R = \tfrac{1}{2}(Y_{RP}{}^M + Y_{PR}{}^M + Y_P{}^M{}_R). \tag{160}$$

Finally, the relations (147) and (151), when taken in conjunction with the metric conditions (138) and (120), imply $\mathrm{D}_j z^P \equiv \partial_j z^P + N^P_j + D_Q{}^P{}_j z^Q = 0$ or, because of $\partial_j z^P \equiv 0$,

$$N_j^P = -D_Q{}^P{}_j z^Q. \tag{161}$$

In view of the definition (136), the condition (151) is tantamount to

$$f_M Q_P^M{}_R = 0, \tag{162}$$

which can be conveniently restricted by assuming

$$r = 2. \tag{163}$$

Indeed, with the equality (163), it follows from Equations (146)–(147) together with (151)–(155) that

$$f_P z^P = 2f, \qquad z_P \stackrel{\text{def}}{=} g_{PQ} z^Q = f_P, \qquad f^P \stackrel{\text{def}}{=} g^{PQ} f_Q = z^P, \tag{164}$$

$g_{PQ} z^P z^Q = 2f$, $g_{PQ} = d_Q f_P$ and

$$z^P C_{PQR} = z^Q C_{PQR} = z^R C_{PQR} = 0. \tag{165}$$

In view of the identities (165) and the expressions (160), the condition (162) will hold if and only if $z_M(Y_{RP}{}^M + Y_{PR}{}^M) = 0$, which in turn entails

$$z^P Q_P{}^M{}_R = z^R Q_P{}^M{}_R = z_M Q_P{}^M{}_R = 0. \tag{166}$$

With the homogeneity condition (163), the field N_i^P given by Equation (157) appears to be positively homogeneous of degree one in z, that is, $N_i^P(x, kz) = kN_i^P(x, z)$, $k > 0$, which entails

$$z^Q \, \mathrm{d}_Q N_i^P = N_i^P \tag{167}$$

because of the Euler theorem on homogeneous functions. On comparing the relations (167) and (161), it follows from the definition (26) that $z^Q A_j{}^P{}_Q = 0$ in addition to $z_P A_j{}^P{}_Q = 0$ (see the first member of Equations (143)). Because of the definition (132), the two last identities imply $z^P T_P{}^Q{}_i = z_Q T_P{}^Q{}_i = 0$. If we additionally stipulate that the torsion tensor $T_P{}^Q{}_i$ be positively homogeneous of degree zero in z, that is, $T_P{}^Q{}_i(x, kz) = T_P{}^Q{}_i(x, z)$, $k > 0$, then from the expression (134) it follows that the gauge field $D_P{}^Q{}_i$ is homogeneous too, implying

$$z^R \, \mathrm{d}_R D_P{}^Q{}_i = 0. \tag{168}$$

Using Equations (168) and (167) together with (166), the contraction of the definition (29) with z^R is found to be zero: $z^R Q_P{}^Q{}_{Rj} = 0$. Assuming the condition (163) to hold and putting

$$f = F^2/2, \tag{169}$$

it is legitimate to call $F(x, z)$ the *parametrical Finslerian metric function*, and $g_{PQ}(x, z)$ the *parametrical Finslerian metric tensor*.

B.11. Proper Finslerian Gauge Transformations

It should be noted that the condition (151), and hence all the subsequent relations, are obviously not invariant under the general gauge transformations (2), though all the relations are tensorial when considered under the general coordinate transformations (1). The question arises as to whether it is possible to specify a subgroup of gauge transformations such that all the Finslerian relations (162)–(169) become invariant.

The conclusion that such a subgroup actually exists may readily be arrived at by carefully comparing the transformation law (12) of the gauge field ${Q_P}^M{}_R$ with the starting-condition (162) of the Finslerian reduction performed in the preceding section. With this aim, we note that the transformation law (12) implies that a gauge transformation $z^P = z^P(x, z')$ will leave the condition (162) invariant if and only if

$$f_P\, \partial^2 z^P / \partial z'^Q\, \partial z'^R = 0. \tag{170}$$

Now, we may assert that, *given any field* $f_P(x, z)$ *possessing the scalar-vector transformation law* (137), *the condition* (170) *selects a subgroup from the general group of all invertible transformations of the form* (2). This assertion ensues from the chain rule

$$f_P \frac{\partial^2 z^P}{\partial z'^R\, \partial z'^Q} = f_P \frac{\partial z''^S}{\partial z'^Q} \frac{\partial^2 z^P}{\partial z''^S\, \partial z'^R} + f''_P \frac{\partial^2 z''^P}{\partial z'^R\, \partial z'^Q}, \tag{171}$$

where we have substituted the equality $f_N\, \partial_z{}^N / \partial z''^P = f''_P$ following from the transformation law of f_P. Indeed, the relation (171) indicates that, given two gauge transformations $z^P(x, z')$ and $z^P(x, z'')$ satisfying the condition (170), the connecting transformation $z''^P(x, z')$ will also satisfy a condition of the type (170), that is, $f''_P\, \partial^2 z''^P / \partial z'^Q\, \partial z'^R = 0$. The proof is complete.

Additionally, it is appropriate to constrain the gauge transformations by the homogeneity condition

$$z^P(x, kz') = kz^P(x, z'), \qquad k > 0 \tag{172}$$

in order that they preserve the homogeneity conditions (such conditions are crucial in Finslerian approaches). Assuming f_P in the condition (170) to be the gradient (141) of a scalar f obeying the homogeneity property (146) with $r = 2$, so that $f_P = z_P$ in accordance with the second member of Equations (164), we call the subgroup formed by all (sufficiently smooth) invertible transformations $z^P = z^P(x, z')$ satisfying the conditions (172) and

$$z_P\, \partial^2 z^P / \partial z'^Q\, \partial z'^R = 0 \tag{173}$$

the *group of Finslerian gauge transformations*. The reader may readily verify that, given any scalar f, the definition (152) yields an object which transforms as a tensor under the gauge transformations obeying the condition (170) with $f_P = \partial f / \partial z^P$, that is, it possesses the transformation law $g'_{PQ} = z^R_P z^S_Q g_{RS}$ (cf.

the fourth member of Equations (48) and also Section 1.3. At the same time, $C'_{PQR} = z^M_P z^N_Q z^S_R C_{MNS} + \frac{1}{2}(z^M_{PR} z^N_Q + z^M_P z^N_{QR}) g_{MN}$, which is not a tensor law. However, contracting the last relation with z'^R and taking into account the identities $z'^R z^P_R = z^P$ and $z'^R z^M_{PR} = 0$ ensuing from the homogeneity requirement (172), we obtain $C'_{PQR} z'^R = z^M_P z^N_Q C_{MNS} z^S$ which shows that the Finslerian identity (165) is an invariant of Finslerian gauge transformations. The expression (160) of the gauge field $Q_P{}^M{}_R$ will also be invariant under such transformations, for Equation (160) was derived under the sole condition (152).

As regards the particular solution (157), which takes the form

$$N^P_i = -\tfrac{1}{2} g^{PQ} \partial_i z_Q \tag{174}$$

in the Finslerian case (163), the reader may readily and directly derive the following assertion: *given a Finslerian gauge transformation, the object* (174) *satisfies the gauge transformation law ascribed to the gauge field* N^P_i *by the second member of Equations* (8) *if and only if the gauge transformation obeys the condition*

$$z_Q \, \partial z^Q_P / \partial x^i = z^R_P g_{RS} \, \partial z^S / \partial x^i \tag{175}$$

(examine the transformation law of the right-hand side of (174) and then compare the result with the required transformation law (8)). While the condition (173) restricts the degree of dependence of the gauge transformations on the argument z^P, the condition (175) restricts the degree of dependence of the gauge transformations on the argument x^i. Notice that the condition (175) is satisfied by conformal-type transformations $z^P = r(x) z^P(z')$ with arbitrary functions $r(x)$ and $z^P(z')$.

It is also clear that under the conditions (172), (173), and (175) the Finslerian relation (159) will obey the transformation law (9) required of the gauge field $D_P{}^Q{}_i$. This assertion ensues, for example, from the relation $D_P{}^Q{}_i = -A_i{}^Q{}_P - \mathrm{d}_P N^Q_i$ (Equation (26)) because $A_i{}^Q{}_P$ is a gauge tensor under general gauge transformations, hence in particular under Finslerian gauge transformations. We call the gauge transformations obeying the conditions (172), (173), and (175) *proper Finslerian gauge transformations.*

The observations of this section may be summarized as follows: *the parametrical Finslerian relations* (146)–(169) *with* $r = 2$ *are gauge-covariant if and only if the gauge transformations are taken to be proper Finslerian.* This assertion offers the possibility of transfering Finslerian methods to the present parametrical gauge approach.

B.12. Flat Internal Space

The internal metric tensor $g_{PQ}(x, z)$ may be called *flat* if the tensor is independent of z^P relative to some coordinate system $\{z^P\}$:

$$g_{PQ} = g_{PQ}(x). \tag{176}$$

Clearly, transitions to other systems $\{z^P\}$ will in general destroy the condition (176) unless the transitions are restricted to *linear gauge transformations*

$$z^P = z^P_Q(x)z'^Q, \tag{177}$$

where the coefficients $z^P_Q(x)$ may be arbitrary except for the nonsingularity condition

$$\det(z^P_Q(x)) \neq 0. \tag{178}$$

In the linear case (177), we shall have

$$z^P_{QR} = 0 \tag{179}$$

(see the notation (3)), which reduces the transformation law (9) of the gauge field $D_P{}^Q{}_i(x, z)$ to the form

$$\partial z^P_Q(x)/\partial x^i + z^R_Q(x)D_R{}^P{}_i - z^P_R(x)D'_Q{}^R{}_i = 0, \tag{180}$$

not involving the gauge field N^P_i. According to (180), the assumption that the field $D_P{}^Q{}_i$ be independent of z^P, that is,

$$D_P{}^Q{}_i = D_P{}^Q{}_i(x) \tag{181}$$

will be of an invariant nature under the linear gauge transformations (177)–(178). The condition (181) entails

$$\mathrm{d}_j D_P{}^Q{}_i = \partial_j D_P{}^Q{}_i, \qquad \mathrm{d}_R D_P{}^Q{}_i = 0. \tag{182}$$

In turn, substituting (177) in Equation (8) yields

$$z'^Q\, \partial z^P_Q(x)/\partial x^i + N'^Q_i z^P_Q(x) - N^P_i = 0. \tag{183}$$

Obviously, this transformation law does not enable one to assign an invariant meaning to the condition $N^P_i = N^P_i(x)$, while the linear condition

$$N^P_i = K^P_i(x) + N_Q{}^P{}_i(x)z^Q \tag{184}$$

will be invariant under linear gauge transformations, even if we put $K^P_i = 0$ so as to restrict ourselves to the form

$$N^P_i = N_Q{}^P{}_i(x)z^Q. \tag{185}$$

In other words, the derivative $N_Q{}^P{}_i \overset{\text{def}}{=} \partial N^P_i/\partial z^Q$ may be postulated to be independent of z^P, in analogy with the condition (181). Similarly to (180), the transformation law of the field $N_Q{}^P{}_i(x)$ will read

$$\partial z^P_Q(x)/\partial x^i - z^R_Q(x)N_R{}^P{}_i(x) + z^P_R(x)N'_Q{}^R{}_i(x) = 0.$$

As regards the gauge field $Q_P{}^R{}_T$, the condition (179) reduces the transformation law (12) to the tensor law

$$z^S_Q(x)z^T_R(x)Q_S{}^P{}_T - z^P_S(x)Q'_Q{}^S{}_R = 0,$$

so that the conditions $Q_S{}^P{}_T = Q_S{}^P{}_T(x)$ or $Q_S{}^P{}_T = 0$ will be invariant under linear gauge transformations.

Putting

$$Q_S{}^P{}_T = 0 \tag{186}$$

and using the relations (182), it follows from the definitions (28)–(30) that

$$Q_P{}^T{}_{Rj} = 0, \qquad Z_P{}^Q{}_{MN} = 0 \tag{187}$$

and

$$E_P{}^Q{}_{ij} = \partial_j D_P{}^Q{}_i - \partial_i D_P{}^Q{}_j - D_P{}^R{}_j D_R{}^Q{}_i + D_P{}^R{}_i D_R{}^Q{}_j \equiv E_P{}^Q{}_{ij}(x), \tag{188}$$

while the substitution of (181) and (185) in (26) yields

$$A_j{}^Q{}_R = -N_R{}^Q{}_j(x) - D_R{}^Q{}_j(x), \qquad M_j{}^P{}_i = M_Q{}^P{}_{ji}(x) z^Q, \tag{189}$$

where

$$M_Q{}^P{}_{ji}(x) = \partial_j N_Q{}^P{}_i - \partial_i N_Q{}^P{}_j + N_Q{}^T{}_j N_T{}^P{}_i - N_Q{}^T{}_i N_T{}^R{}_j. \tag{190}$$

If we drop the torsion tensors (25):

$$S_i{}^k{}_j = 0, \qquad Y_P{}^T{}_R = 0 \tag{191}$$

and then assume that the metric conditions (116) and (121) hold, we arrive at the relations $\mathrm{D}_m a_k{}^{inm} = 0$, $\mathrm{S}_N g_Q{}^{RPN} = 0$, and $V_R = 0$ (see Equations (72)). In this case the Euler–Lagrange derivatives (70′)–(70‴) are reduced to

$$\delta L/\delta L_n{}^k{}_i = -l_1 N_m a_k{}^{inm} + 2l_2 \mathrm{S}_R B^n{}_k{}^{iR} + 4l_3(\mathrm{D}_m + N_m)L^n{}_k{}^{im}, \tag{192}$$

$$\delta L/\delta D_P{}^Q{}_i = 2n_2 A^i{}_Q{}^P + 4d_3(\mathrm{D}_m + N_m)E^P{}_Q{}^{im}, \tag{193}$$

$$\delta L/\delta Q_P{}^Q{}_R = 2d_3 E^P{}_Q{}^{jm} M_j{}^R{}_m, \tag{194}$$

and the relation (70⁗) becomes

$$\begin{aligned}\delta L/\delta N_j^P = {} & -4n_1((D_m + N_m)M^j{}_P{}^m + M^t{}_Q{}^j A_t{}^Q{}_P) - l_1 a_k{}^{jni} B_n{}^k{}_{ip} + \\ & + 4l_3 B_n{}^k{}_{mP} L^n{}_k{}^{jm},\end{aligned} \tag{195}$$

where the relations (186)–(189) have been used and gauge fields associated with densities have been neglected. If we also assume that the external metric tensor a_{ij} is independent of z^P, that is,

$$a_{ij} = a_{ij}(x), \tag{196}$$

then from the relation (133) we obtain

$$L_m{}^i{}_n = L_m{}^i{}_n(x), \tag{197}$$

which entails

$$B_n{}^k{}_{jR} = 0 \tag{198}$$

because of the definition (27). Under these conditions, the relations (192) and (195) are simplified and become just

$$\delta L/\delta L_n{}^k{}_i = -l_1 N_m a_k{}^{inm} + 4l_3(\mathrm{D}_m + N_m)L^n{}_k{}^{im}, \tag{199}$$

$$\delta L/\delta N_j^P = -4n_1((\mathrm{D}_m + N_m)M^j{}_P{}^m + M^t{}_Q{}^j A_t{}^Q{}_P), \tag{200}$$

and the field N_m given by the definition (72) is reduced to

$$N_m = -A_m{}^P{}_P(x). \tag{201}$$

The simplest way to satisfy the gauge equations $\delta L/\delta Q_P{}^Q{}_R = 0$ and $\delta L/\delta N_j^P = 0$ for the case of relations (194) and (200) is to put

$$M_j{}^R{}_m = 0. \tag{202}$$

Because $M_j{}^R{}_m$ is the gauge tensor associated with the gauge field N_i^P in accordance with the definition (26), condition (202) will be satisfied if we assume that, among the coordinate systems $\{z^P\}$ connected by linear gauge transformations (177), a system $\{z^P\}$ exists such that

$$N_i^P = 0, \tag{203}$$

and hence $N_Q{}^P{}_i = 0$. This reduces expressions (192) and (193) to

$$\delta L/\delta L_n{}^k{}_i = -l_1 D_P{}^P{}_m a_k{}^{inm} + 4l_3(\mathrm{D}_m + D_P{}^P{}_m)L^n{}_k{}^{im} \tag{204}$$

and

$$\delta L/\delta D_P{}^Q{}_i = -2n_2 D_Q{}^P{}_i + 4d_3(\mathrm{D}_m + D_R{}^R{}_m)E^P{}_Q{}^{im}, \tag{205}$$

where the fact that

$$N_m = D_P{}^P{}_m(x) \tag{206}$$

has been taken into account, because of Equations (189), (201), and (203).

The equations under consideration may be simplified even further if we restrict the linear gauge transformations (177) to *rotational* linear gauge transformations defined by the condition

$$z_Q^P(x) z_S^R(x) \delta_{PR} = \delta_{QS}. \tag{207}$$

At this point it will be noted that any metric tensor $g_{PQ}(x)$ may be diagonalized (locally) by some linear gauge transformation $z^P = z^{*P}_Q(x) z'^Q$:

$$g_{PQ}(x) z^{*P}_R(x) z^{*Q}_S(x) = q_R \delta_{RS},$$

where q_R (equal to $+1$ or -1) represents the signature of the tensor g_{PQ}. Our definition (207) corresponds to the case when the tensor $g_{PQ}(x)$ is positively definite, so that $q_R = 1$; other possible sets of q_R can be treated similarly. So, we take

$$g_{PQ} = \delta_{PQ}, \tag{208}$$

obtaining from the metric condition (125) the relation

$$D_{PQi} = -D_{QPi} \tag{209}$$

and hence

$$N_m = D_P{}^P{}_m = 0 \tag{210}$$

in accordance with the result (206). Thus, the Euler-Lagrange derivatives (204) and (205) become

$$\delta L/\delta L_{n\ i}^{\ k} = 4l_3 \mathrm{D}_m L^{n\ \ im}_{\ k}, \tag{211}$$

$$\delta L/\delta D_P{}^Q{}_i = -2n_2 D_Q{}^P{}_i + 4d_3 \mathrm{D}_m E^P{}_Q{}^{im}. \tag{212}$$

For dimension $M = 3$ we may use the permutation symbol $e^{PQR} = e_{PQR}$ with $e_{123} = 1$, so that $e_{PQR}e^{PST} = \delta^S_Q \delta^T_R - \delta^S_R \delta^T_Q$, to introduce the field $A^P_i(x)$ by means of the definition

$$A^P_i = \tfrac{1}{2} e^{PQR} D_{QRi}, \tag{213}$$

obtaining from the relation (209) the expression

$$D_{PQi} = e_{PQR} A^R_i. \tag{214}$$

Substituting (214) in (188), we find

$$F^P{}_{ij} \overset{\text{def}}{=} -\tfrac{1}{2} e^{PQR} E_{QRij} = \partial_i A^P_j - \partial_j A^P_i - e^P{}_{QR} A^Q_i A^R_j \tag{215}$$

and from (212) we obtain

$$\delta L/\delta A^P_i \overset{\text{def}}{=} e^Q{}_{RP}(\delta L/\delta D_R{}^Q{}_i) = -2n_2 A^i_P - 8d_3 \mathrm{D}_m F_P{}^{im}, \tag{216}$$

where the action of the gauge-covariant derivative D_m on the gauge tensor $F^P{}_{ij}$ is given by the classical rule

$$\mathrm{D}_m F^P{}_{ij} = \partial_m F^P{}_{ij} - L_i{}^k{}_m F^P{}_{kj} - L_j{}^k{}_m F^P{}_{ik} + e^P{}_{QT} A^Q_m F^T{}_{ij}. \tag{217}$$

The inverse of the representation (215) reads $E_{QRij} = -e_{QRP} F^P{}_{ij}$.

Problems

B.1. Examine the case when the internal space is of constant curvature.

B.2. Consulting the Appendix A, derive the equations of structure for the present general parametrical gauge theory.

B.3. Derive the gauge-covariant equations of geodesics.

B.4. What should be the gauge-covariant generalization of the Hamilton-Jacobi equation?

B.5. Find the gauge-covariant generalization of the Lorentz force.

B.6. Is it possible to assign the mass to the gauge fields in a gauge-covariant way?

Note

[1] This Appendix extends the work by Asanov (1985).

Solutions of Problems

Chapter 1

1.1. The required fifth condition reads

$$\partial g_{ij}/\partial y^k - \partial g_{ik}/\partial y^j = 0. \tag{S.1}$$

It can be inferred from these five conditions that g_{ij} is a Finslerian metric tensor. Indeed, contracting Equation (S.1) with y^k and using the identity $y^k\, \partial g_{ij}/\partial y^k = 0$ (which follows from the second condition via the Euler theorem (1.3) with $r = 0$), we get $y^k\, \partial g_{ik}/\partial y^j = 0$ or identically

$$g_{ij} - \partial y_i/\partial y^j = 0 \tag{S.2}$$

where $y_i \overset{\text{def}}{=} g_{ij} y^j$. Since the first condition requires that the tensor g_{ij} be symmetric, (S.2) yields

$$\partial y_i/\partial y^j - \partial y_j/\partial y^i = 0,$$

which is well known to be the necessary and sufficient condition for the local existence of a function $Q(x, y)$ such that $y_i = \partial Q/\partial y^i$. It remains to contract (S.2) with $y^i y^j$, yielding

$$y^i y^j g_{ij} - 2Q = 0,$$

where we have made use of the fact that $y_i(x, y^n)$ is, by definition, homogeneous of degree one with respect to y^n. This implies that, first, $y^j\, \partial y_i/\partial y^j = y_i$ and, second, Q is homogeneous of degree two with respect to y^i, so that $2Q = y^i\, \partial Q/\partial y^i$. Thus, if we write $F = (2Q)^{1/2} \equiv (y^i y^j g_{ij})^{1/2}$, we obtain a scalar function F which is homogeneous of degree one with respect to y^n and, as a consequence of our third and fourth conditions, satisfies the conditions (1.1) and $F > 0$. On substituting $y_i = \partial Q/\partial y^i$ and then $Q = F^2/2$ in (3.2), we get (1.5), which completes the proof.

1.2. The first relation follows from $l^m l_m = 1$ and Equations (1.39), (1.40):

$$l_{k;j} l^k = (g_{km} l^m)_{;j} l^k = l^m{}_{;j} l_m = -l_{m;j} l^m.$$

Taking into account the symmetry of the connection coefficients in its subscripts, we get

$$l^j(\partial l_k/\partial x^j - \partial l_j/\partial x^k) = l^j(l_{k;j} - l_{j;k}) = 0.$$

1.3. Just as we have established the relationship (1.122) between the tensor S_{ijmn} and the curvature tensor R^*_{abcd} of the indicatrix, so we may relate the Minkowskian covariant derivative $(F^{-2} S_{ijmn})|_k$ to the covariant derivative of R^*_{abcd}. To this end, let us denote by D_T and D^*_T the operators of the Riemannian covariant derivatives with respect to z^T associated with the tensors g_{PQ} and g^*_{PQ}, respectively, and introduce the designations

$$R_{PQRS,T} = D_T R_{PQRS}, \qquad R^*_{PQRS,T} = D^*_T R^*_{PQRS}$$

By construction, $D_k^{\text{Mink}}(F^{-2} S_{ijmn})$ and $R_{PQRS,T}$ represent components of the same tensor with respect to different coordinate systems $\{y^i\}$ and $\{z^P\}$ connected by the transformation (1.106). Therefore, in terms of the derivative $|_k = \mathrm{FD}_k^{\mathrm{Mink}}$,

$$(F^{-2}S_{ijmn})|_k = Fz_i^P z_j^Q z_m^R z_n^S z_k^T R_{PQRS,T} \tag{S.3}$$

which is similar to Equation (1.120). If we denote by $q_P{}^Q{}_R$ and $q_P^*{}^Q{}_R$ the Riemannian Christoffel symbols constructed in accordance with the definition (1.58) from the tensors g_{PQ} and g_{PQ}^* and with respect to the variables z^P, we find from Equations (1.109)–(1.112) that

$$q_a{}^b{}_c = q_a^*{}^b{}_c, \qquad q_0{}^b{}_c = \delta_c^b, \qquad q_a{}^0{}_b = -g_{ab}^*, \qquad q_0{}^0{}_0 = 1, \qquad q_0{}^a{}_0 = q_0{}^0{}_a = 0$$

from which it follows directly that

$$\begin{aligned} &R_{abcd,e} = e^{2z^0} R^*_{abcd,e}, \qquad R_{abcd,0} = -2R_{abcd}, \\ &R_{0bcd,e} = -R_{b0cd,e} = -R_{ebcd}, \qquad R_{ab0d,e} = -R_{abd0,e} = -R_{abed}, \end{aligned} \tag{S.4}$$

while other components of $R_{PQRS,T}$ vanish as a consequence of the fact that the components of $R^*_{PQRS,T}$ where the index zero occurs vanish identically. Substituting the Equalities (S.4) in (S.3) and taking into account the relation (1.121), together with $z_i^0 = l_i F$ and $F^{-2}|_k = -2Fl_k$ (see Equation (1.68)), we find after some calculations that the desired result is

$$\begin{aligned} &S_{ijmn}|_k + l_i S_{kjmn} + l_j S_{ikmn} + l_m S_{ijkn} + l_n S_{ijmk} = \\ &\quad = F^5 u_i^a u_j^b u_m^c u_n^d u_k^e \mathrm{D}_e^* R^*_{abcd}. \end{aligned}$$

1.4. Direct procedure.

1.5. The Riemann curvature tensor of a two-dimensional Riemannian space has the particular form (1.130). Applying this observation to the tangent Minkowskian space yields, for $N = 2$,

$$S_{ijmn} = C(g_{im}g_{jn} - g_{in}g_{jm}),$$

where C is a scalar. Therefore, $S^i{}_{jin} = Cg_{jn}$. On the other hand, Equations (1.16) and (1.62) entail that $y^n S_{ijmn} = 0$. So, $O = y^n S^i{}_{jin} = Cy_j$, which is possible only if $C = 0$. The proof is complete.

1.6. A Riemannian space of constant curvature is characterized by a condition of the form (1.130) with $R^* = \text{const}$. Therefore, as in the preceding Problem 1.5, we get $S_{ijmn} = 0$ identically, that is the tangent Minkowskian space is Euclidean.

If, additionally, all conditions imposed on the Finslerian metric function in Section 1.1 of the monograph by Rund (1959) are fulfilled, and $N > 2$, then, according to the theorem by Brickell (1967), the identical vanishing of the tensor S_{ijm} should mean that the Finsler space is reduced to a Riemannian one.

1.7. See Watanabe and Ikeda (1981).

1.8. Given an extended frame of reference represented by a basic congruence β, denote by $u^i(x)$ a field of vectors which are tangent to the lines of β and are unit: $F(x, u) = 1$. Then, it follows from Equations (1.36)–(1.40) that the sought for decomposition reads

$$u_{i;j} = u_j \dot{u}_i + \omega_{ij} + \sigma_{ij} + \tfrac{1}{3}\theta h_{ij}.$$

Here, the δ-differentiation is performed in the direction of $u^i(x)$, and

$$\dot{u}_j \overset{\text{def}}{=} u_{j;i}u^i = u^i \mathrm{D}_i u_j,$$

$$2\omega_{ij} \overset{\text{def}}{=} u_{i;j} - u_{j;i} + \dot{u}_j u_i - \dot{u}_i u_j = \mathrm{D}_i u_j - D_j u_i + \dot{u}_j u_i - \dot{u}_i u_j,$$

$$2\sigma_{ij} \overset{\text{def}}{=} u_{i;j} + u_{j;i} - \dot{u}_i u_j - \dot{u}_j u_i - \tfrac{2}{3}\theta h_{ij} = \mathrm{D}_j u_i + \mathrm{D}_i u_j - \dot{u}_i u_j - \dot{u}_j u_i - \tfrac{2}{3}\theta h_{ij} - 2C_{ijm}\dot{u}^m,$$

$$\theta \overset{\text{def}}{=} u^i{}_{;i} = \mathrm{D}_i u^i - C_i{}^i{}_m \dot{u}^m$$

are, respectively, Finslerian acceleration, rotation, shear, and expansion of an extended frame of reference; D_i denotes the Riemannian covariant derivative associated with the osculating

Riemannian metric tensor $a_{ij}(x) = g_{ij}(x, u(x))$, and $u^n = a^{nm} u_m$. We observe that the Finslerian acceleration and rotation are always identical to their Riemannian counterparts:

$$\dot{u}^{(F)}_j = \dot{u}^{(R)}_j, \qquad \omega^F_{ij} = \omega^{(R)}_{ij},$$

whereas the shear and expansion are affected by the Finslerian generalization:

$$\sigma^{(F)}_{ij} = \sigma^{(R)}_{ij} - 2C_{ijm}\dot{u}^m, \qquad \theta^{(F)} = \theta^{(R)} - C^{i}_{i\,m}\dot{u}^m.$$

In particular,

$$\theta^{(F+)} = \theta^{(R)} \quad \text{if} \quad C^{i}_{i\,m} = 0$$

(the case of the Berwald-Moór metric function).

If we have an extended Finslerian inertial frame of reference, that is, if $\dot{u}^i = 0$, then $u^n{}_{;i} = \mathrm{D}_i u^n$ in accordance with Equation (1.90) (incidentally, this implication will remain in the more general case when the field $u^n(x)$ is recurrent, that is when there exists a vector field $K_i(x)$ such that $\mathrm{D}_i u^n = K_i u^n$). Therefore, an arbitrary congruence of Finslerian geodesics proves to be a congruence of Riemannian geodesics with respect to the osculating Riemannian metric tensor associated with this congruence.

1.9. Commuting the indices j, k in Equation (1.159) shows that the identity $g_{mn} y^m_j y^n_{ki} = g_{mn} y^m_k y^n_{ji}$ holds in the case of metric G-transformations, from which the assertion ensues. As a consequence of this fact, the covariant Minkowskian divergence $\mathrm{D}^{\mathrm{Mink}}_i X^i \equiv J^{-1}(x,y)\mathrm{d}[J(x,y)\,X^i(x,y)]/\mathrm{d}y^i$, where $J = |\det(g_{ij})|^{1/2}$, will be invariant under metric G-transformations, yielding the identities

$$\frac{\partial}{\partial y'^i}\left[\frac{\partial y'^i}{\partial y^k}\det\left(\frac{\partial y^m}{\partial y'^n}\right)\right] = 0, \qquad \frac{\partial}{\partial y^i}\left[\frac{\partial y^i}{\partial y'^k}\det\left(\frac{\partial y'^n}{\partial y^m}\right)\right] = 0. \tag{S.5}$$

1.10. Immediate generalizations of the Klein-Gordon equation from (pseudo-) Euclidean space to the tangent Minkowskian space of a Finsler space can be formulated as follows. Since the tangent Minkowskian spaces are Riemannian, we can construct a scalar field Lagrangian density in the usual Riemannian way:

$$L_w = (g^{ij}(x,y) w_i w_j - M^2 w^2) J(x,y) \tag{S.6}$$

to derive the linear equation for the scalar field $w(x,y)$:

$$\frac{1}{J(x,y)}\frac{\mathrm{d}}{\mathrm{d}y^i}(J(x,y) g^{ij}(x,y) w_j) + M^2 w = 0. \tag{S.7}$$

Here, $w_i = \partial w(x,y)/\partial y^i$ and M is apositive constant which plays the role of the scalar field mass in a physical context; $J(x,y) = |\det(g_{ij}(x,y))|^{1/2}$. Alternatively, we can take the Lagrangian density

$$L_v = (g^{ij}(x, v_m) v_i v_j - M^2 v^2) J(x, v^n), \tag{S.8}$$

where $v_i = \partial v(x,y)/\partial y^i$, $v^m = g^{mn}(x, v_k) v_n$ and $J(x, v^m) = |\det(g_{ij}(x, v^m))|^{1/2}$, to obtain the essentially nonlinear equation for the scalar field $v(x,y)$:

$$E_1 + E_2 + M^2 v = 0 \tag{S.9}$$

where

$$E_1 = \frac{1}{J(x,v^m)}\frac{\mathrm{d}}{\mathrm{d}y^i}\left[J(x,v^m) g^{ij}(x,v_n) v_j\right],$$

$$E_2 = \frac{1}{J(x,v^m)}\frac{\mathrm{d}}{\mathrm{d}y^i}\left[\frac{1}{2} L_v \partial \ln J(x,v^m)/\partial v_i\right]$$

(cf. Equations (S.43)–(S.47)). Equation (S.7) can be rewritten in the explicit covariant form

$$g^{ij}(x,y)\mathrm{D}^{\mathrm{Mink}}_i \mathrm{D}^{\mathrm{Mink}}_j w + M^2 w = 0,$$

while converting equation (S.9) into a similar form would require a nonlinear covariant derivative (cf. Problem 4.6) to be introduced in the tangent Minkowskian space. Both linear and nonlinear Finslerian equations written above reduce to the same Klein-Gordon equation in the Riemannian limit.

As regards transformation properties, the nonlinear equation (S.9) as well as the equations

$$E_1 + M^2 v = 0 \tag{S.10}$$

and

$$E_2 = 0 \tag{S.11}$$

(but not the linear equation (S.7)) are invariant under metric G-transformations, that is if a function $v(x, y)$ is a solution of any of the Equations (S.9), (S.10) or (S.11), then the function $v'(x, y) = v(x, Z^*(x, y))$, where Z^{*m} are functions which realize metric G-transformations according to the definitions (1.141)–(1.144) and (1.156)–(1.157), also represents a solution. This assertion can be verified directly by making use of the identities (S.5). In particular, noting that in the massless case $M = 0$ each of the Equations (S.9), (S.10) or (S.11) admit the plane-wave linear solution

$$v(x, y) = C(x) \sin(k_m(x) y^m) \tag{S.12}$$

with the wave vector satisfying the Finslerian dispersion relation

$$g^{ij}(x, k_m(x)) k_i(x) k_j(x) = 0 \tag{S.13}$$

(cf Equation (S.50)) and $C(x)$ being a scalar, we obtain an infinite set of solutions

$$v(x, y) = C(x) \sin(k_m(x) Z^{*m}(x, y)) \tag{S.14}$$

which represent nonlinear waves in general. The equation (S.10) with a nonvanishing mass M admits solutions of the form (S.12) and (S.14) with the dispersion relation

$$g^{ij}(x, k_m(x)) k_i(x) k_j(x) = M^2. \tag{S.15}$$

Using the Berwald-Moór metric function (see Section 2.3), the Equation (S.9) reduces to (S.10) because of the identity (2.38). As a consequence of the nonlinearity of the equations (S.9)–(S.11), their solutions will not obey the principle of superposition. In the case of the linear equation (S.7) the principle of superposition will evidently hold, while plane-wave solutions of the form (S.12) will not, generally speaking, exist.

The Lagrangian density (S.8) vanishes when

$$g^{ij}(x, v_m) v_i v_j = M^2 v^2. \tag{S.16}$$

Making use of the replacement $v = e^{MS}$ transforms (S.16) to

$$g^{ij}\left(x, \frac{\partial S(x, y)}{\partial y^m}\right) \frac{\partial S(x, y)}{\partial y^i} \frac{\partial S(x, y)}{\partial y^j} = 1. \tag{S.17}$$

Comparing Equation (S.17) with the definitions (6.18) and (6.26) immediately shows that in every tangent space (i.e. at a fixed point x^i) Equation (S.17) is nothing but the Finslerian Hamilton-Jacobi equation having the form of the background space Hamilton-Jacobi equation

$$g^{ij}\left(\frac{\partial S(x)}{\partial x^m}\right) \frac{\partial S(x)}{\partial x^i} \frac{\partial S(x)}{\partial x^j} = 1,$$

that is of the case when the Finsler space is flat (with respect to x^i). According to Rund's Hamilton-Jacobi theory discussed in Section 6.2, solutions $S(x, y)$ of Equation (S.17) will be such that, at every fixed point x^i, the function $S(y)$ will give rise to a congruence of Finslerian geodesics with momenta field $p_i(y) = \partial S(y)/\partial y^i$. The associated field of unit velocities will be $u^i(y) \stackrel{\text{def}}{=} p_j(y) g^{ij}(p_n(y))$. In the case of solution (S.14) we obtain $u^i(y) = k^n y^i_n(y)$, where $k^i = g^{ij}(k_m) k_j$ and y^i_n are the entities which have been introduced in Section 1.3. In the particular case (S.12) we get merely $u^i = k^i$. When $M = 0$, the Equation (S.16) gives rise to congruences of isotropic geodesics.

In the massless case equation (S.10) becomes

$$\frac{\mathrm{d}}{\mathrm{d}y^i}(J(x, v^m)g^{ij}(x, v_n)v_j) = 0 \tag{S.18}$$

which may be regarded as the Finslerian generalization of the D'Alembert equation. Apart from the wave solutions (S.12)–(S.14), Equation (S.18) has the solution

$$v(x, y) = C(x)F^{2-N}(x, y),$$

where F is the Finslerian metric function. Taking F to be of the form

$$F(y) = \{(y^0)^f + e[(y^1)^f + \cdots + (y^{N-1})^f]\}^{1/f} \quad \text{with} \quad e = 1 \quad \text{or} \quad e = -1$$

[cf. the definition (2.9); the point x^i is meant to be fixed], we can also seek for the static fundamental solution $v = v(Y)$ of (S.18), where $Y = [(y^1)^f + \cdots + (y^{N-1})^f]^{1/f}$ is in fact the generalized radius. Upon substituting $v(Y)$ in (S.18), we obtain $v = C/Y^{N-3}$, so that in the four-dimensional case $N = 4$ we have $v = C/Y$ in strict conformity with the fundamental solution C/r of the usual three-dimensional Euclidean Laplace equation.

Similarly, for the electromagnetic field vector potential $A_i(x, y)$ we can define the Lagrangian density in two natural ways, namely: in the form

$$L = J(x, y)g^{im}(x, y)g^{jn}(x, y)F_{ij}F_{mn}$$

($F_{ij} = \partial A_j/\partial y^i - \partial A_i/\partial y^j$) which implies covariant equations in the tangent Minkowskian space regarded as a Riemannian space, or in the essentially nonlinear form

$$L = J(x, A)g^{im}(x, A)g^{jn}(x, A)F_{ij}F_{mn}$$

(cf. Equation (S.48)) which yields equations which are invariant under metric G-transformations provided that the potential $A_i(x, y)$ transforms under such transformations as a vector, i.e. $A'_i(x, y') = y_i^m A_m(x, y)$ (cf. Equation (1.156)).

If we modify Equation (S.7) as follows

$$\frac{1}{J(x, y)}\frac{\mathrm{d}}{\mathrm{d}y^i}[J(x, y)g^{ij}(x, y)w_j] + \frac{N-2}{4(N-1)}SF^{-2}w = 0, \tag{S.19}$$

where $SF^{-2} \equiv S^{ij}{}_{ij}F^{-2}$ is the complete contraction of the curvature tensor (1.60) of the tangent Minkowskian space, we get the conformally-invariant scalar field equation; that is for any (sufficiently smooth) nonvanishing function $Q(x, y)$, the substitution $g_{ij} = Q^2 f_{ij}$ and $w = Q^{(2-N)/2}q$ retains the form of the Equation (S.19):

$$\frac{1}{A(x, y)}\frac{\mathrm{d}}{\mathrm{d}y^i}[A(x, y)f^{ij}(x, y)q_j] + \frac{N-2}{4(N-1)}B(x, y)q = 0, \tag{S.20}$$

where $A = |\det(f_{ij})|^{1/2}$, $q_j = \partial q/\partial y^j$, and B denotes the complete contraction of the Riemann curvature tensor constructed out of f_{ij} with respect to y^i at a fixed x^i. Recalling relation (1.109), putting $Q = \exp z^0$ together with $p(x, z^0, u^a) = q(x, y(x, z^0, u^a))$, and taking into account Equations (1.111) and (1.112), we reduce the Equation (S.20) to

$$\left(\frac{\partial}{\partial z^0}\frac{\partial}{\partial z^0} + \Delta_u + \frac{N-2}{4(N-1)}R^*(u)\right)p = 0. \tag{S.21}$$

Here and in the sequel, we neglect an explicit indication of the dependence on x^i; $R^* = R^{*ab}{}_{ab}$ with R^*_{abcd} being the curvature tensor (1.116) of the indicatrix; Δ_u denotes the Laplace operator associated with the indicatrix:

$$\Delta_u p = \frac{1}{I(u)}\frac{\partial}{\partial u^a}\left[I(u)g^{*ab}(u)\frac{\partial p}{\partial u^b}\right]$$

where $I = |\det(g^*_{ab})|^{1/2}$. The Equation (S.21) admits solutions of the form $p = b(u)\exp(-ik_0 z^0)$, where $b(u)$ represents a solution of the eigenvalue problem

$$\left(\Delta_u + \frac{N-2}{4(N-1)}R^*\right)b = (k_0)^2 b. \tag{S.22}$$

Proceeding to a Dirac-type spinor equation, we shall restrict ourselves to the four-dimensional case $N = 4$ and assume that the Finslerian metric tensor g_{ij} has the signature of space-time so that $g_{ij} = \Sigma^3_{P=0} q_P h^P_i h^P_j$, where $q_0 = -q_1 = -q_2 = -q_3 = 1$ and h^P_i is a Finslerian tetrad (cf. Problem 2.3). Under these conditions, the Dirac-type spinor equation in the tangent Minkowskian space can be formulated as follows

$$-iF\gamma^n \nabla_n \Phi + M\Phi = 0 \tag{S.23}$$

(cf. Equation (5.141)). Here, $\gamma^n \equiv \gamma^n(x,y) = h^n_P(x,y)\gamma^P$, where $\gamma^P = q_P\gamma_P$ and γ_P are the 4×4 Dirac matrices; Φ is a four-component spinor; the operator $\nabla_n = \partial/\partial y^n - A_n$ involves the spinor connection coefficients

$$A_n = -\tfrac{1}{8}(\gamma_P\gamma_Q - \gamma_Q\gamma_P)h^{Pi}(x,y)\mathrm{D}^{\mathrm{Mink}}_n h^Q_i(x,y)$$

(cf. Equation (5.136)). Notice that the operator ∇_n is conformally-invariant in the sense that the substitution $g_{ij} = Q^2\bar{g}_{ij}$ and $h^P_i = Q\bar{h}^P_i$, together with $\Phi = Q^{-3/2}\Psi$, results in the relation $\nabla_n\Phi = Q^{-3/2}\bar{\nabla}_n\psi$. Therefore, on passing to the coordinates z^P in accordance with Equation (1.95) and then making the conformal transformation with $Q = \exp z^0$, we get Equation (S.23) in the form

$$-i\left[\gamma^0\frac{\partial}{\partial z^0} + \gamma^a\left(\frac{\partial}{\partial u^a} - A_a\right)\right]\psi + M\psi = 0, \tag{S.24}$$

where $\psi = \Phi e^{3z_0/2}$, $\gamma^0 = \gamma^{P=0}$, and $\gamma^a = g^{*(a)}_{(b)}\gamma^{(b)}$ with $\gamma^{(b)} = \gamma^{P=b}$ and $g^{*a}_{(b)}$ is given by Equation (1.133). Putting $\psi = \lambda\exp(-ik_0 z^0)$ leads to the eigenvalue problem

$$-i\gamma^a(\partial/\partial u^a - A_a)\lambda = (k_0 - M)\lambda.$$

1.11. Taking $l^1 = (\sin\psi\sin\theta\cos\varphi)^{2/f}$, $l^2 = (\sin\psi\sin\theta\sin\varphi)^{2/f}$, $l^3 = (\sin\psi\cos\theta)^{2/f}$ and $l^4 = (\cos\psi)^{2/f}$, direct calculations show for the case of F_1 that the nonvanishing components of the tensor (1.110) are

$$g^*_{\psi\psi} = \frac{4(f-1)}{f^2}, \qquad g^*_{\theta\theta} = \frac{4(f-1)}{f^2}\sin^2\psi, \qquad g^*_{\varphi\varphi} = \frac{4(f-1)}{f^2}\sin^2\psi\sin^2\theta.$$

For the case of F_2, the function $\sin\psi$ in these formulae is to be replaced by $\operatorname{sh}\psi$.

1.12. For the case of F_1, the sought for transformations are $y^A = [R^A_B\cdot(y'^B)^{f/2}]^{2/f}$, where $||R^A_B||$ denotes the matrix of four-dimensional rotations. Similarly, for the case of F_2, we can take $y^A = [L^A_B\cdot(y'^B)^{f/2}]^{2/f}$ with $||L^A_B||$ being the matrix of Lorentz transformations. From the last expression we obtain

$$y^A_B \stackrel{\mathrm{def}}{=} \partial y^A/\partial y'^B = L^A_B\cdot(y'^B/y^A)^{f/2-1} \equiv L^A_B\cdot(l'^B/l^A)^{f/2-1}$$

(no summation over the indices A and B). It can be readily verified directly that $y_A y^A_{BC} \equiv 0$.

Chapter 2

2.1. The statement can be verified directly by substituting Equations (2.1)–(2.3) in the corresponding Gauss-Codazzi equations. (Matsumoto, 1971, p. 204).

2.2. The T-condition assumes the vanishing of the scalar

$$T \stackrel{\mathrm{def}}{=} T^{mn}{}_{mn} = F(\partial C^i/\partial y^i + C^i C_i)$$

(see Equation (2.27)). But from (2.13) and (2.14) it follows that

$$C_A = \frac{2-f}{2F}\left(Nl_A - \frac{1}{l^A}\right), \qquad C^A = \frac{2-f}{2F(f-1)}\left(Nl^A - \frac{1}{l_A}\right)$$

so that

$$FT = F^2(\partial C^A/\partial y^A + C^A C_A) = \frac{(2-f)f}{4(f-1)}\left[N^2 - 2N + \sum_{A=1}^{N} \frac{1}{l^A l_A}\right]. \tag{S.25}$$

Owing to (2.11), the last term in brackets in (S.25) is not a constant unless $f = 0$. Therefore, FT is not a constant for the examined metric functions. Notice that $C_A = 0$ for $r^A = (1, \ldots, 1)$ and $l^A = N^{-f}$.

From Equations (2.11)–(2.14) we find consecutively

$$\begin{aligned} C^E{}_{BD} &= \frac{2-f}{2F}\left(h_B{}^E l_D + h_D{}^E l_B + \frac{1}{f-1} l^E h_{BD} + l^E l_B l_D - \delta^E_{BD}(l^B)^{-1}\right), \\ C_D{}^{EN} &= \frac{2-f}{2(f-1)F}(h_D{}^E l^N + h_D{}^N l^E + (f-1)l_D h^{EN} + l^E l^N l_D - \delta^{EN}_D (l_D)^{-1}), \end{aligned} \tag{S.26}$$

$$\begin{aligned} C^{ri}{}_m C_{rn}{}^j = \frac{2-f}{2F}\Bigg\{&\left(l_m C_n{}^{ij} + l_n C_m{}^{ij} + \frac{1}{f-1} l^i C_{mn}{}^j + \frac{1}{f-1} l^j C_{mn}{}^i\right) + \\ &+ \frac{f-2}{2(f-1)F}\Bigg[h_m{}^i h_n{}^j + \frac{1}{f-1} l^i l^j h_{mn} + (f-1) l_m l_n h^{ij} + l^i l_m h_n{}^j + \\ &+ l^j l_n h_m{}^i + l^i l_n h_m{}^j + l^j l_m h_n{}^i + l^i l^j l_m l_n - \sum_{A=1}^{N} S^A_m S^A_n S^i_A S^j_A \frac{1}{l_A l^A}\Bigg]\Bigg\}, \end{aligned}$$

$$\begin{aligned} C_{ABCD} = \frac{2-f}{2F}\Bigg[&2C_{ABC} l_D + 2C_{ADC} l_B + 2C_{BCD} l_A + 2C_{ABD} l_C + \\ &+ \frac{1}{F}(h_{AB} h_{DC} + h_{BD} h_{AC} + h_{AD} h_{BC}) + \frac{f-3}{F}(l_A l_B h_{CD} + l_A l_D h_{BC} + \\ &+ l_B l_D h_{AC} + l_C l_D h_{AB} + l_C l_B h_{AD} + l_C l_A h_{BD}) + \frac{(f-3)(f-1)}{F} l_A l_B l_C l_D + \\ &+ \frac{(1-f)(f-3)}{F} \delta_{ABCD} l_A (l^A)^{-3}\Bigg]. \end{aligned}$$

The substitution of these formulae in the definition (2.26) yields the following result:

$$\begin{aligned} T_{mnij} = \frac{f}{2}\Big[&(l_m C_{nij} + l_n C_{mij} + l_i C_{mnj} + l_j C_{mni}) + \\ &+ \frac{f(2-f)}{4(f-1)F}(h_{mn} h_{ij} + h_{mi} h_{nj} + h_{mj} h_{ni}) + \frac{f(f-2)}{4F}(l_m l_n h_{ij} + \\ &+ l_m l_i h_{nj} + l_m l_j h_{ni} + l_n l_i h_{mj} + l_n l_j h_{mi} + l_i l_j h_{mn}) - \\ &- \frac{f(f-1)(2-f)}{4F} l_m l_n l_i l_j + \frac{f(f-1)(2-f)}{4F} \sum_{A=1}^{N} S^A_m S^A_n S^A_i S^A_j l_A (l^A)^{-3}\Big]. \end{aligned} \tag{S.27}$$

We see that the right-hand side of (S.27) is proportional to f, so that $T_{mnij} = 0$ if $f = 0$ (Proposition 8 of Chapter 2).

2.3. According to Equation (2.12), we can represent the metric tensor as follows

$$g_{AB} = l_A l_B + h_{AB}$$

where

$$h_{AB} = (f-1)m_{AB}, \qquad m_{AB} = \delta_{AB} l_A (l^A)^{-1} - l_A l_B.$$

Using Equation (2.11), we get

$$m_{AB} = t_A t_B (\delta_{AB} r_A q_A - r_A r_B),$$

where we have used the notation $r_A = r^A$ and

$$t_A = l_A / r^A \equiv (l^A)^{f-1}, \qquad q_A = (t_A)^{f/(f-1)} \equiv (l^A)^{-f}.$$

The elucidation of the signature of g_{AB} is thus reduced to diagonalizing the quadratic form

$$(Z)^2 = Z^A Z^B m_{AB} \equiv x^A x^B (\delta_{AB} r_A q_A - r_A r_B).$$

Here, we denote $x^A = Z^A \cdot t_A$ (not summing over the index A).

Let us consider, for example, the four-dimensional case with

$$r^A = (1, -1, -1, -1).$$

Denoting the members of x^A as follows

$$x^1 = x, x^2 = y, x^3 = z, x^4 = u,$$

we get

$$\begin{aligned}(Z)^2 &= q_1 x^2 - q_2 y^2 - q_3 z^2 - q_4 u^2 - (x - y - z - u)^2 \equiv \\ &\equiv -((1-q_1)x - y - z - u)^2/(1-q_1) - \left(1 + q_2 - \frac{1}{1-q_1}\right)y^2 + \\ &+ \frac{2q_1}{1-q_1} y(z+u) - \left(1 + q_3 - \frac{1}{1-q_1}\right)z^2 - \left(1 + q_4 - \frac{1}{1-q_1}\right)u^2 + \\ &+ \frac{2q_1}{1-q_1} zu.\end{aligned}$$

Complementing the sum of the second and third terms on the right-hand side to a complete square, we obtain the desired result:

$$\begin{aligned}(Z)^2 = &-((1-q_1)x - y - z - u)^2/(1-q_1) - \\ &-(Q_2 y - q_1 z - q_1 u)^2/Q_2(1-q_1) - \\ &-(\sqrt{Q_2 Q_3 - (q_1)^2}\, z - \sqrt{Q_2 Q_4 - (q_1)^2}\, u)^2/Q_2(1-q_1),\end{aligned}$$

where $Q_A = q_A - q_1 - q_1 q_A$. This means that the metric tensor can be represented in the following form

$$g_{AB} = \begin{cases} h_A^0 h_B^0 - h_A^1 h_B^1 - h_A^2 h_B^2 - h_A^3 h_B^3 & \text{if } f > 1 \\ h_A^0 h_B^0 + h_A^1 h_B^1 + h_A^2 h_B^2 + h_A^3 h_B^3 & \text{if } f < 1 \end{cases}$$

with the orthonormal tetrad h_A^P given by

$$h_A^0 = l_A,$$

$$\left(\frac{1-q_1}{|f-1|}\right)^{1/2} h_A^1 = \{(1-q_1)l_1, l_2, l_3, l_4\},$$

$$\left(\frac{(1-q_1)Q_2}{|f-1|}\right)^{1/2} h_A^2 = \{0, -Q_2 l_2, q_1 l_3, q_1 l_4\},$$

$$\left(\frac{(1-q_1)Q_2}{|f-1|}\right)^{1/2} h_A^3 = \{0, 0, -(Q_2Q_3 - (q_1)^2)^{1/2} l_3, \quad (Q_2Q_4 - (q_1)^2)^{1/2} l_4\}.$$

In the above formulae it has been assumed that $l^A > 0$ (if l^2, l^3 or l^4 is equal to zero, then $\det(g_{AB}) = 0$). The equality $F(l) = 1$ reads for the case under study that

$$(l^1)^f - (l^2)^f - (l^3)^f - (l^4)^f = 1$$

which implies that $l^1 > 1$. Therefore, $q_1 \overset{\text{def}}{=} (l^1)^{-f} < 1$ and

$$1 - q_1 > 0.$$

Denoting $s^A = 1/q_A \equiv (l^A)^f$, we get

$$s^1 - s^2 - s^3 - s^4 = 1$$

so that

$$Q_2 \overset{\text{def}}{=} q_2 - q_1 - q_1 q_2 = q_1 q_2 (s^1 - s^2 - 1) = q_1 q_2 (s^3 + s^4) > 0$$

and further

$$Q_2 Q_3 - (q_1)^2 = q_1 q_2 (s^3 + s^4) q_1 q_3 (s^2 + s^4) - (q_1)^2 =$$
$$= (q_1)^2 q_2 q_3 ((s^3 + s^4)(s^2 + s^4) - s^2 s^3) > 0.$$

Similarly, $Q_2 Q_4 - (q_1)^2 > 0$. Care should be taken that t_A and l_1 be positive, while l_2, l_3, l_4 are negative.

In the case

$$r^A = (1, 1, 1, 1)$$

we take $0 < l^A < 1$ which entails that $t_A = l_A > 0$ and $q_A > 1$. Denoting

$$P_A = q_1 q_A - q_1 - q_A,$$

we have $P_2 > 0$ and

$$P_2 P_3 - (q_1)^2 > 0, \qquad P_2 P_4 - (q_1)^2 > 0.$$

Carrying out the required calculations, we get

$$g_{AB} = \begin{cases} h_A^0 h_B^0 + h_A^1 h_B^1 + h_A^2 h_B^2 + h_A^3 h_B^3 & \text{if } f > 1 \\ h_A^0 h_B^0 - h_A^1 h_B^1 - h_A^2 h_B^2 - h_A^3 h_B^3 & \text{if } f < 1 \end{cases}$$

with the orthonormal tetrad h_A^P given by

$$h_A^0 = l_A,$$

$$\left(\frac{q_1 - 1}{|f-1|}\right)^{1/2} h_A^1 = \{(q_1 - 1) l_1, -l_2, -l_3, -l_4\},$$

$$\left(\frac{(q_1 - 1)P_2}{|f-1|}\right)^{1/2} h_A^2 = \{0, P_2 l_2, -q_1 l_3, -q_1 l_4\},$$

$$\left(\frac{(q_1 - 1)P_2}{|f-1|}\right)^{1/2} h_A^3 = \{0, 0, (P_2 P_3 - (q_1)^2)^{1/2} l_3, -(P_2 P_4 - (q_1)^2)^{1/2} l_4\}.$$

(Asanov, 1984b).

2.4. We have

$$Y_A \overset{\text{def}}{=} S_A^i F\, \partial F/\partial y^i = e_\beta l_A^\beta F^2/f^\beta, \qquad A \in I_\beta. \tag{S.28}$$

The quantity Y_A may naturally be treated as a function of the arguments x^n and Y^A. Accordingly, the scalars $g_{AB} = \partial Y_A/\partial Y^B$ will be related to the metric tensor g_{ij} as follows:

$$g_{ij} = S_i^A S_j^B g_{AB}.$$

The examination of the structure of g_{AB} may be divided into two different cases.

(I) $A, B \in I_\beta$ for some value of β. In this case, we obtain from (S.28) that

$$g_{AB} = 2l_A l_B + e_\beta(g^\beta_{AB} - 2l^\beta_A l^\beta_B)F^2/(f^\beta)^2,$$

where $l_A = S^i_A l_i$, or, in view of (S.28),

$$g_{AB} = 2l_A l_B + \frac{1}{e_\beta}(Q^{\beta CD}_{\ \ AB}(x) l_C l_D - 2l_A l_B)$$

(no summation over β). Here, $Q^{\beta CD}_{\ \ AB}$ is defined by the relation

$$g^\beta_{AB} = Q^{\beta CD}_{\ \ AB}(x) l^\beta_C l^\beta_D.$$

(II) $A \in I_\beta$ and $B \in I_{\beta'}$, with $\beta \neq \beta'$. In this case,

$$g_{AB} = 2l_A l_B.$$

2.5. The same procedure as used in Section 2.2 may be applied to Euler's identities

$$l^k v^{ij}{}_k(x, l) = r v^{ij}(x, l), \qquad l^m l^n v^{ij}{}_{mn}(x, l) = r(r-1) v^{ij}(x, l),$$

where $v^{ij}{}_k(x, y) = \partial v^{ij}(x, y)/\partial y^k$, $v^{ij}{}_{mn}(x, y) = \partial v^{ij}{}_m(x, y)/\partial y^n$ and $g^{ij}(x, y) = v^{ij}(x, l(x, y))$, yielding

$$r l^i l_k = l_j v^{ij}{}_k(x, l), \qquad r l_k = l_i l_j v^{ij}{}_k(x, l),$$

$$r g_{mn} + r(r-1) l_m l_n = l_i l_j v^{ij}{}_{mn}(x, l).$$

Contrary to the equivalence of the representations (2.17) and (2.19), the homogeneity of g^{ij} in l^n does not imply the homogeneity of g_{ij} in l_n in the general case.

2.6. The tensor (2.19) will be a Finslerian metric tensor if and only if relation (S.1) holds. Differentiation of (2.19) with respect to y^k yields

$$\partial g_{mn}/\partial y^k = 2F^{-1} l_i l_p l_q (Q^{ij}_{mn} Q^{pq}_{jk} - Q^{pq}_{mn}\delta^i_k) = 0$$

so that the sought for necessary and sufficient condition reads

$$\mathscr{A}^{(ipq)}(Q^{ij}_{mn} Q^{pq}_{jk} - Q^{ij}_{mk} Q^{pq}_{jn} - Q^{pq}_{mn}\delta^i_k + Q^{pq}_{mk}\delta^i_n) = 0,$$

where $\mathscr{A}^{(ipq)}$ denotes the symmetrization operator.

2.7. With the expressions (2.32) and (2.33) for the Berwald-Moór metric tensor, the given equality can be written as

$$2WZ_A - N(Z_A)^2 = q/N, \tag{S.29}$$

where $Z_A = l_{1A} \cdot l^A_2$ (no summation over A), $W = l_{1A} l^A_2$ and

$$l_{1A} = g_{AB}(x, y_1) y^B_1/F(x, y_1), \qquad l^A_2 = y^A_2/F(x, y_2).$$

Equation (S.29) shows that Z_A are the same for any value of A. Therefore, $Z_A = W/N$, which in turn implies that $l_{1A} = W/Nl^A_2$ or, because of (2.33), $l_{1A} = Wl_{2A}$. Since l_{1A} and l_{2A} are unit vectors, we obtain $W = 1$, and further $Z = 1/N$ and $q = 1$.

2.8. Direct procedure.

2.9. In particular, it turns out that in a 1-form Finsler space with the property $C_k{}^k{}_i = 0$ the parameter of the Finslerian arc-length is projective relative to a coordinate system possessing the property (2.143).

In the Randers space the parameter of the Riemannian arc-length will be projective relative to the coordinate system possessing the property $\left\{{i \atop i\ m}\right\} = 0$. The projectively invariant tensor $B_i{}^m{}_{jn}$ (Equation (4.8.23) of Rund (1959)) is found to be

$$B_i{}^m{}_{jn} = \tfrac{1}{2}(a_{ij}F_n{}^m + a_{jn}F_i{}^m + a_{ni}F_j{}^m)a^{-1} - \tfrac{1}{2}u_i u_j F_n{}^m + u_j u_n F_i{}^m +$$
$$+ u_n u_i F_j{}^m + \dot{x}^k F_k{}^m(u_i a_{jn} + u_j a_{ni} + u_n a_{ij})a^{-3} + \tfrac{3}{2}\dot{x}^k F_k{}^m u_i u_j u_n a^{-6}.$$

The fundamental significance of this tensor is that its vanishing is a necessary and sufficient condition for the existence of a parametrization of curves such that geodesics can be described by equations of the form

$$\frac{d\dot{x}^m}{dt} + N_i{}^m{}_j(x)\dot{x}^i\dot{x}^j = 0 \tag{S.30}$$

where the coefficients $N_i{}^m{}_j$ do not depend on the velocities. However, the vanishing of the above tensor $B_i{}^m{}_{jn}$ implies $F_{ij} = 0$. In fact,

$$B_i{}^m \overset{\text{def}}{=} B_i{}^m{}_{jn}a^{jn} = \frac{N+1}{2}(F_i{}^m - u_i\dot{x}^k F_k{}^m a^{-2})a^{-1}.$$

If $B_i{}^m = 0$, then

$$a^2 F_i{}^m = u_i \dot{x}^k F_k{}^m.$$

Contracting this with u_m yields $F_i{}^m u_m = 0$, so that $F_i{}^m = 0$. Thus, in the classical electromagnetic theory a representation of the form (S.30) is impossible.

2.10. From the representation (2.35) of the Berwald-Moór metric tensor it follows that

$$Q^{ip}_{kh}(x) \overset{\text{def}}{=} \tfrac{1}{2}\partial^2(F^2 g^{ip})/\partial y^k\,\partial y^h =$$
$$= \delta^i_k\delta^p_h + \delta^i_h\delta^p_k - N\sum_{A=1}^{N} S^A_k S^A_h S^i_A S^p_A.$$

With this formula, Equation (2.155) shows that the connection coefficients are

$$G_k{}^i{}_h = NK_p \sum_{A=1}^{N} S^A_k S^A_h S^i_A S^p_A \equiv G_k{}^i{}_h(x) = \Gamma_k{}^i{}_h(x).$$

The substitution of these connection coefficients in the definition (1.78) of the tensor $K_j{}^i{}_{hk}$ yields the following simple result:

$$K_j{}^i{}_{hk}(x) = N\sum_{A=1}^{N} S^A_j S^i_A S^p_A(K_{pk}S^A_h - K_{ph}S^A_k), \tag{S.31}$$

where $K_{pk} = \partial K_p/\partial x^k$. After that, it follows from Equation (1.80) together with (S.31) and the representation (2.14) (with $f = 0$) of the Cartan torsion tensor that

$$R_{jihk} = l_j R_{ihk} - l_i R_{jhk} \tag{S.32}$$

where

$$R_{ihk} = N\sum_{A=1}^{N} l^A S^A_i S^p_A(K_{ph}S^A_k - K_{pk}S^A_h).$$

(cf. Equation (2.162)).

2.11. The required additional condition reads $C_{ikj|h} = 0$. This equality is tantamount to the requirement that $\partial G_h{}^i{}_j/\partial y^k = 0$. The answer to the second question of the present problem is in negative (the reader is referred to Rund (1959, p. 136)).

2.12. For the Randers metric, $M_i = (b_i - (b/F)u_i)t^{1/2}/2F$ and $FM = -(f^2 r + b^2 + 2Fb)/4F^2$. For the Kropina metric, $M_i = (2a_{ij}y^j/b - 2a^2 b_i/b^2)b/2a^2$ and $FM = b^3/2ra^4$.

2.13. Consider the implications of $S^{mn}{}_{mn} = 0$ and $T^{mn}{}_{mn} = 0$. (Matsumoto, 1974).

2.14. Given any vector field $X^i(x, y)$ and some fixed field of n-tuples $h^P_i(x, y)$ (which need not necessarily be orthonormal with respect to the Finslerian metric tensor g_{mn}), we put $X^P = h^P_i X^i$

and define the covariant derivative $\mathscr{A}_i$ for the case of absolute parallelism as follows:

$$X^P_{|i} = X^n_{|i} h^P_n + h^P_{n|i} X^n \overset{\text{def}}{=} h^P_n \mathscr{A}_i X^n$$

which yields

$$\mathscr{A}_i X^n = X^n_{|i} + h^P_{m|i} h^n_P X^m \equiv \partial X^n/\partial x^i - G_i^{\ m} \partial X^n/\partial y^m + B_m{}^n{}_i X^m$$

(h^n_P are reciprocal to h^P_n), where

$$B_m{}^n{}_i = \Gamma_m{}^n{}_i + h^P_{m|i} h^n_P \equiv h^n_P(\partial h^P_m/\partial x^i - G_i^{\ k} \partial h^P_m/\partial y^k)$$

are the connection coefficients of absolute parallelism. Note that $B_m{}^n{}_i \neq B_i{}^n{}_m$ in general. The action of $\mathscr{A}_i$ may be extended to any tensor in accordance with the usual rule:

$$\mathscr{A}_i T_m{}^j{}_n = \partial T_m{}^j{}_n/\partial x^i - G_i^{\ k} \partial T_m{}^j{}_n/\partial y^k + B_s{}^j{}_i T_m{}^s{}_n - B_m{}^s{}_i T_s{}^j{}_n -$$
$$- B_n{}^s{}_i T_m{}^j{}_s.$$

From the viewpoint of absolute parallelism the fundamental property of the covariant derivative $\mathscr{A}_i$ is that $\mathscr{A}_i h^P_n = \mathscr{A}_i h^n_P = 0$. Therefore, $\mathscr{A}_i T_m{}^j{}_n = 0$ if the field $T_m{}^j{}_n$ is parallel in the sense of absolute parallelism, i.e. if the projections $T_Q{}^P{}_R \overset{\text{def}}{=} h^P_j h^m_Q h^n_R T_m{}^j{}_n$ are constants. The analogue of the Minkowskian derivative (1.64) in the tangent spaces will read

$$\mathscr{M}_i X^n \overset{\text{def}}{=} \mathrm{D}_i^{\text{Mink}} X^n + X^m h^n_P \mathrm{D}_i^{\text{Mink}} h^P_m = \partial X^n/\partial y^i + D_k{}^n{}_i X^k$$

where $D_k{}^n{}_i = h^n_P \partial h^P_k/\partial y^i \neq D_i{}^n{}_k$ is a tensor; $\mathscr{M}_i h^P_n = \mathscr{M}_i h^n_P = 0$, so that $\mathscr{M}_i T_m{}^j{}_n = 0$ in the case of an absolutely parallel tensor $T_m{}^j{}_n$.

Evaluating the commutator of $\mathscr{A}_h$ and $\mathscr{A}_k$ yields the following result after straightforward calculations:

$$(\mathscr{A}_k \mathscr{A}_h - \mathscr{A}_h \mathscr{A}_k) X^i = \mathscr{R}_j{}^i{}_{hk} X^j - K_r{}^j{}_{hk} y^r \mathscr{M}_j X^i + (B_k{}^m{}_h - B_h{}^m{}_k) \mathscr{A}_m X^i$$

(cf. Equation (1.79)), where

$$\mathscr{R}_j{}^i{}_{hk} = \partial B_j{}^i{}_h/\partial x^k - G_k^{\ m} \partial B_j{}^i{}_h/\partial y^m - \partial B_j{}^i{}_k/\partial x^h + G_h^{\ m} \partial B_j{}^i{}_k/\partial y^m +$$
$$+ B_m{}^i{}_k B_j{}^m{}_h - B_m{}^i{}_h B_j{}^m{}_k + D_j{}^i{}_l K_r{}^l{}_{hk} y^r$$

may be regarded as the analogue of the Cartan curvature tensor $R_j{}^i{}_{hk}$. Let us assume that X^P are constants. Then $\mathscr{A}_j X^i = \mathscr{M}_j X^i = 0$, and from the above commutation relation it immediately follows that $\mathscr{R}_j{}^i{}_{hk} = 0$. Similarly, we find

$$(\mathscr{A}_k \mathscr{M}_h - \mathscr{M}_h \mathscr{A}_k) X^i = -\mathscr{P}_j{}^i{}_{kh} X^j + (G_h{}^j{}_k - B_h{}^j{}_k) \mathscr{M}_j X^i + D_k{}^j{}_h \mathscr{A}_j X^i$$

(cf. Equation (1.73)), where

$$\mathscr{P}_j{}^i{}_{kh} = -\mathscr{A}_k D_j{}^i{}_h + \partial B_j{}^i{}_k/\partial y^h + (G_k{}^m{}_h - B_h{}^m{}_k) D_j{}^n{}_m$$

is the second curvature tensor associated with absolute parallelism. Applying the above reasoning to the latter commutator enables one to conclude that $\mathscr{P}_j{}^i{}_{kh} = 0$. Finally,

$$\mathscr{S}_j{}^i{}_{hk} \overset{\text{def}}{=} \partial D_j{}^i{}_h/\partial x^k - \partial D_j{}^i{}_k/\partial x^h + D_m{}^i{}_k D_j{}^m{}_h - D_m{}^i{}_h D_j{}^m{}_k = 0.$$

If we assume additionally that $h^P_i(x, y)$ are orthonormal with respect to the Finslerian metric tensor g_{mn}, we shall have $\mathscr{A}_i g_{mn} = \mathscr{M}_i g_{mn} = 0$. (Cf. Holland and Philippidis, 1984).

Chapter 3

3.1. For a scalar $S(y^n)$, the invariance identities read

$$\bar{S}(y^n b^k_n) = S(y^k),$$

and differentiating with respect to b^k_n yields, after putting $\bar{x}^i = x^i$,

$$y^n \, \partial S/\partial y^k = 0$$

which is possible only if $\partial S/\partial y^k = 0$.

If we have a vector $S^i(y^n)$, then the invariance identities

$$B^i_m \bar{S}^m(y^n b^k_n) - S^i(y^k)$$

will give, after differentiating with respect to B^j_m and putting $\bar{x}^i = x^i$, the following relation

$$\delta^i_j S^m - y^m \, \partial S^i/\partial y^j = 0,$$

which entails, after contracting the indices i and j, that

$$S^m = S y^m,$$

where $S \overset{\text{def}}{=} \partial S^i/\partial y^i$ is a scalar of the form $S(y)$ which, therefore, should be a constant.

Let now $S^{ij}(y^n)$ be a tensor. In this case, the invariance identities

$$B^k_i B^h_j \bar{S}^{ij}(b^n_m y^m) = S^{kh}(y^n)$$

yield, in a similar fashion, that

$$\delta^k_q S^{ph} + \delta^h_p S^{kp} = y^p \, \partial S^{kh}/\partial y^q.$$

Contracting here the indices k and q, we get

$$N S^{ph} + S^{hp} = S^h y^p$$

where $S^h \overset{\text{def}}{=} \partial S^{qh}/\partial y^q$ is a vector of the form $S^h(y)$ and, therefore, $S^h = S y^h$ in accordance with the preceding result. Thus,

$$N S^{ph} + S^{hp} = S y^p y^h$$

which shows that the tensor S^{hp} must be symmetric, from which it follows that $(N + 1)S^{ph} = S y^p y^h$, etc.

3.2. By hypothesis,

$$F(x, y) = v(S^A_i(x), y^n)$$

with v a scalar function. Let us write

$$w(S^A_i, y^B) = v(S^A_i, S^n_B y^B).$$

Since w, y^B are scalars, the invariance identities for w will read

$$\bar{w}(B^j_i \bar{S}^A_j, \bar{y}^B) = w(S^A_i, y^B)$$

from which it follows after differentiating with respect to B^j_i that $\partial w/\partial S^A_i = 0$. Thus,

$$F(x, y) = w(S^A_i(x) y^i)$$

which agrees with the definition of a 1-form Finsler space given in Section 2.4.

3.3. In the case under study, the invariance identities take the form

$$a_{im} \, \partial v/\partial a_{jm} + a_{mi} \, \partial v/\partial a_{mj} = y_i y^j/F$$

or

$$2F a_{jn}(a_{im} \, \partial v/\partial a_{jm} + a_{mi} \, \partial v/\partial a_{mj}) = y_i y^j a_{jn}.$$

Since the left-hand side of this equation is symmetric in i, n we get

$$y_i y^j a_{jn} - y_n y^j a_{ji} = 0,$$

which evidently requires the proportionality

$$y_i = qa_{ij}y^j \tag{S.33}$$

where q is a scalar. Differentiating (S.33) with respect to y^j and then contracting the result with y^i, we obtain the relation

$$y_j = qa_{ji}y^i + y^i y^n a_{in}\,\partial q/\partial y^j$$

which is consistent with (S.33) if and only if the scalar q is independent of y^i. Thus, Equation (S.33) will read

$$y_i = q(x)a_{ij}(x)y^j$$

from which it follows that the Finslerian metric tensor $g_{ij} = \partial y_i/\partial y^j$ (Equation (1.10)) is equal to $q(x)a_{ij}(x)$ and therefore is Riemannian. (Rund, 1979, Section 5).

3.4. Direct procedure.

3.5. Direct procedure.

3.6. Let there be a tensor, say $T_l{}'_m$, of the form

$$T_l{}'_m = T_l{}'_m(g_{hk}, C_{hkr}, g_{hk,r})$$

Then the reasoning used in Section 3.2 when deriving Equation (3.41) may be applied to our tensor $T_l{}'_m$ to conclude that $\partial T_l{}'_m/\partial g_{jk,i}$ is equal to the right-hand side of (3.41), the first term being neglected. This in turn implies that the analog of Equation (3.44) will obviously read $\partial T_l{}'_m/\partial g_{jk,i} = 0$.

3.7. Direct procedure.

3.8. This statement can be proved as follows:

$$\begin{aligned}\Pi^{kh,lm}K_{hksm} &= (L^{kh,lm} + C_i{}^k{}_j y^h L^{ij,lm})K_{hksm} = \\ &= L^{kh,lm}(-K_{khsm} - 2C_{hkq}K_i{}^q{}_{sm}y^i) + C_i{}^k{}_j y^h L^{ij,lm}K_{hksm} = \\ &= -K_{hksm}(L^{kh,lm} + C_i{}^k{}_j y^h L^{ij,lm}) = -\Pi^{kh,lm}K_{hksm}.\end{aligned}$$

(Rund and Beare, 1972, p. 53).

3.9. Take the derivative $\partial/\partial S_i^j$ of the transformation law (3.38) of the connection coefficients.

3.10. Direct procedure.

Chapter 4

4.1. Contracting the equation with $C_m{}^n{}_k$ yields $2C_{ijk}M^{ij} = 0$ which reduces the equation to $2M^{mj}g_{jn} = 0$.

4.2. It is well known from the theory of differential equations that the necessary integrability condition for systems of type $y_m^k = 0$ reads $\mathrm{d}_n y_m^k - \mathrm{d}_m y_n^k = 0$ which, as can easily be seen, may be rewritten as

$$y^i(x)K_{ijhk}(x, y(x)) = 0 \tag{S.34}$$

in view of the definition (1.78) of the tensor K_{ijhk}. In such a case, Equation (1.83) will entail the skew-symmetry property

$$K_{ijhk}(x, y(x)) = -K_{jihk}(x, y(x)).$$

Moreover, it follows from formula (4.2.27a) of Rund (1959) that (S.34) implies

$$y^j(x)K_{ijhk}(x, y(x)) = y^h(x)K_{ijhk}(x, y(x)) = y^k(x)K_{ijhk}(x, y(x)) = 0.$$

This results in turn in $K'^{ij} = K^{ij}$ (see Equation (3.126)) and $K^{hk} = K^{kh}$ (contract the cyclic identity (1.84) with g^{ij}). Under these conditions, substituting Equations (3.98)–(3.100) and (3.81) in (3.132) gives the following simple expression for the fundamental tensor density Π^{ij}:

$$\Pi^{ij} = -JK^{ij} + \tfrac{1}{2}g^{ij}JK - \tfrac{1}{4}C_r(y^iK^{rj} + y^jK^{ri})J \tag{S.35}$$

for any $y^i(x)$ satisfying the condition (S.34). In addition,

$$C^{ijn}{}_{|n|m} - C^{ijn}{}_{|m|n})y^m = 0 \tag{S.36}$$

as follows from (S.14) together with the general commutation formula (1.79). It will also be noted that condition (S.34) implies that the tensor density Ψ^{*ijk} given by Equation (3.139) vanishes identically:

$$\Psi^{*ijk} = 0 \tag{S.37}$$

In the case of stationary $y^k(x)$, we have $M^{ij} = E^{ij}$ (see Equation (4.16)) and $M^{*ijk} = M^{*ijkl} = 0$ (see Equations (4.20), (4.21) and (S.37)). Therefore, the Euler–Lagrange derivative (4.12) and the tensor density (4.19) become merely

$$E_{Ap\ldots}^{q\ldots} = E^{ij}\,\partial v_{ij}/\partial S^{Ap\ldots}_{q\ldots}, \qquad N_i{}^j = (2\delta_i^m g_{jn} + 2y^m C_{ijn})E^{ij}.$$

The maximum number of simplifications in the structure of E^{ij} will arise when the integrability condition (S.34) is taken in conjunction with the choice of the Finslerian metric function in the Berwald–Moór form (see the end of Section 3.4). Indeed, $C_i = 0$ will reduce (S.35) to merely

$$\Pi^{ij} = -JK^{ij} + \tfrac{1}{2}JKg^{ij} \tag{S.38}$$

and (3.144) will become

$$L^{ij,mn}{}_{|m|n} = -JC^{ijn}{}_{|n|k}y^k \tag{S.39}$$

as a consequence of (S.36). Therefore, the substitution of Equations (S.38), (S.39), and (3.145) in the definition (3.78) of E^{ij} gives the following result: if the integrability condition (S.34) and the identities (3.140) specific to the Berwald–Moór metric function are satisfied, then

$$J^{-1}E^{ij} = K^{ij} - \tfrac{1}{2}Kg^{ij} - C^{ijn}{}_{|n|k}y^k + l^iZ^j + l^jZ^i - l^il^jZ.$$

4.3. Using Equations (4.10) and (4.80)–(4.83) we get, with respect to static coordinates, $y^k_m = y_m a^k - C_m{}^k{}_q a^q$. The quantities y_m and $C_m{}^k{}_q$, as well as h^P_j and h^P_{jm} in (4.51), are of the order of unity, while the gravitational acceleration a^k is of the order of r_g/R^2, where r_g is the gravitational radius of the Earth, which is $r_g \simeq 0.9$ cm, and R is the geometrical radius of the Earth, which is $R \simeq 6 \times 10^8$ cm. Therefore, the coefficients of $F_{(0)}{}^{kj}$ in the last term of Equation (4.51) are of the order $10^{-18}\,\text{cm}^{-1}$. Since these coefficients are so small, the Finslerian corrections to the conventional Maxwell equations appear to be beyond present-day experimental possibilities.

4.4. The proof is based on the homogeneity condition (4.7) for the gravitational Lagrangian density L_g. Namely, this condition entails the identity $y^iE_{gi} = 0$ (Equation (4.131)). In the static case, with respect to static coordinates we have $y^i = y^0\delta^i_0$ (see Equation (4.80)) and, therefore, $E_{g0} = 0$, so that the second term on the left-hand side of our equation (4.65) vanishes. The vanishing of the first term in equation (4.65) is verified as follows: $\mathrm{d}_j(E_{gi}y^j) = \mathrm{d}_0(E_{gi}y^0) = 0$.

4.5. Near the Earth the gravitational field has the order $U \simeq 10^{-9}$, which follows from the fact that in the Newtonian approximation, in which Equation (4.100) has the form

$$\mathrm{d}V^a/\mathrm{d}x^0 = -Ur^a/r^2,$$

the acceleration

$$a^a \overset{\text{def}}{=} \mathrm{d}v^a/\mathrm{d}t,$$

where $v^a = cV^a$ and $\mathrm{d}t = \mathrm{d}x^0/c$, equals the free fall acceleration $\simeq 10\,\text{m}\,\text{s}^{-2}$. Macroscopic bodies used in gravimetric experiments on the Earth's surface move rather slowly, the curvature of their trajectories being measured only for velocities lower than $1\,\text{m}\,\text{s}^{-1}$, so that

$$q \ll U.$$

The post-Newtonian corrections to U will be much greater than those in q. Therefore, in real gravimetric experiments near the Earth it is immaterial whether we say that the geometric prototype of freely falling bodies is represented by Riemannian or Finslerian geodesics.

The motion of the planets (from Mercury to Mars) around the Sun is described by

$$q^2 \simeq U$$

(and is equal to $\sim 10^{-8}$). In this case, the substitution of the expansion (4.193) in the equations (4.100) shows that

$$a^a = -\frac{cU}{r^2}[cr^a + 2S^a{}_{be}r^b v^e + O(U)], \tag{S.40}$$

where the Finslerian constants $S^a{}_{be} = \delta_{bd} S^{ad}{}_e$ are defined by the expansion

$$C_{(0)}{}^{ab}{}_e = S^{ab}{}_e + O(U).$$

The presence of the second, purely Finslerian, term in the brackets on the right-hand side of Equation (S.40) leads to the angular momentum $\mathbf{M} = m[\mathbf{r}\,\mathbf{a}]$ of a planet failing to be a conserved quantity. Indeed, from (S.40) it follows that

$$\frac{\mathrm{d}\mathbf{M}}{\mathrm{d}t} = -\frac{2mcU}{r^2}[\mathbf{r}\,\mathbf{S}] \tag{S.41}$$

where

$$S^a = S^a{}_{be} r^b v^e. \tag{S.42}$$

If in the observation data on the motion of some planet one should find a change in angular momentum which is inexplicable in terms of the Newtonian attraction of other planets (similar to the 'residual' displacement of the perihelion of the orbit of Mercury, for the explanation of which the relativistic theory of gravitation has been employed), then it would in principle be possible to estimate the Finslerian tensor $S^a{}_{bc}$ using Equations (S.41) and (S.42). (See also Coley (1982); Aringazin and Asanov (1985)).

4.6. See Asanov (1983a).

4.7. Given a Finslerian metric tensor $g^{mn}(x, y_i)$ and a scalar field $w(x)$, we can construct the associated Hamilton–Jacobi function

$$H(x, w_i) \overset{\text{def}}{=} (g^{mn}(x, w_i) w_m w_n)^{1/2} \tag{S.43}$$

(see the definition (6.18)), where $w_i = \partial w/\partial x^i$, and then naturally propose the Lagrangian density in the form

$$L_w = (H^2 - M^2 w^2) J, \tag{S.44}$$

where $J = |\det(g_{mn}(x, w^i))|^{1/2}$, $w^i = g^{ij}(x, w_n) w_j$, and M is a positive constant which plays the role of the mass of the scalar field. The corresponding Euler–Lagrange equations are found to be

$$\frac{1}{J(x, w^i)} \frac{\mathrm{d}}{\mathrm{d}x^m}[J(x, w^j) w^m + \tfrac{1}{2} L_w \, \partial \ln J / \partial w_m] + M^2 w = 0 \tag{S.45}$$

(cf. Equations (S.6)–(S.7)). The reader can readily verify that, if $\partial J/\partial w^n = 0$ (that is, $C_k{}^k{}_n = 0$, as for the Berwald–Moór metric function), this nonlinear generalization of the Klein–Gordon equation can be rewritten as

$$\mathscr{D}_m w^m + M^2 w = 0, \tag{S.46}$$

where

$$\mathscr{D}_m w^m = \mathrm{d}w^m/\mathrm{d}x^m + \Gamma_m{}^k{}_k(x, w^i) w^m \tag{S.47}$$

and $\Gamma_m{}^k{}_k$ is a contraction of the Cartan connection coefficients (1.32).

Similarly, for an electromagnetic field described by a vector potential $A_i(x)$, we can take

$$L_A = J(x, A)g^{im}(x, A)g^{jn}(x, A)F_{ij}F_{mn}, \tag{S.48}$$

where $F_{ij} = \partial A_j/\partial x^i - \partial A_i/\partial x^j \equiv \mathscr{D}_i A_j - \mathscr{D}_j A_i$. The associated nonlinear equations for the electromagnetic field will be

$$\frac{1}{J(x, A)} \frac{\mathrm{d}}{\mathrm{d}x^m}(J(x, A)F^{mn}) + \tfrac{1}{4}C^{ikn}(x, A)F_i^{\ m}F_{km} - \tfrac{1}{4}L_A \,\partial \ln J/\partial A_n = 0,$$

where $F_i^{\ m} = F_{ij}g^{jm}(x, A)$ and $F^{mn} = F_i^{\ n}g^{im}(x, A)$. If $C_k^{\ k}{}_n = 0$, this equation takes the following quasi-Maxwellian form:

$$\mathscr{D}_m F^{mn} + \tfrac{1}{4}C^{ikn}(x, A)F_i^{\ m}F_{km} = 0. \tag{S.49}$$

4.8. Considering the relation (3.14), we obtain $E_{gA}{}_{i\ldots}^{j\ldots} = E^{*mn}\,\partial v_{mn}(S^B(x), y(x))/\partial S^{Ai\ldots}_{\ j\ldots}(x)$ and $E_{gi} = 2E^{*mn}C_{mni}(x, y(x))$, where $E^{*mn} = \delta L_g/\delta a_{mn}$. Substituting the first relation in Equation (3.28) and taking into account the invariance identity (3.15), we obtain the following result: $E_i^{\ j} = E^{*mn}(\delta^j_m g_{in} + \delta^j_n g_{im} + 2y^j C_{mni}) \equiv 2g_{in}E^{*jn} + y^j E_{gi}$, from which the assertion ensues. Notice that in the case under study the right-hand side of the equations of motion of matter (4.55) vanishes identically.

4.9. Clearly, our purpose can only be achieved through some sort of nonlinear generalization of the electromagnetic field equations. However, the nonlinear generalization (S.48)–(S.49) does not help in solving our problem. An appropriate idea is suggested by the nonlinear equations (S.43)–(S.47). Indeed, it will easily be noted that the simplified case of the equation (S.45):

$$\frac{1}{J(w^i)} \frac{\mathrm{d}}{\mathrm{d}x^m}(J(w^j)g^{mn}(w_k)w_n) + M^2 w = 0$$

admits wave solutions $w \sim \sin(k_n x^n)$ with the wave vector obeying the Finslerian dispersion relation

$$g^{mn}(k_i)k_m k_n = M^2 \tag{S.50}$$

(cf. Equation (S.15)). Starting with this observation, we may proceed as follows. Given the vector potential $A_i(x)$ of the electromagnetic field and a tetrad $h^i_P(x)$ (for example, of an osculating Riemannian metric tensor), we can take the Lagrangian density to be of the form

$$L = J(x) \sum_{P=0}^{3} H^2(x, A_{Pi}), \tag{S.51}$$

(cf. Equations (S.43) and (S.44)), where $A_{Pi} = \partial A_P/\partial x^i$, $A_P = A_i h^i_P$, and $J = \det(h^P_i)$. The Euler-Lagrange derivative associated with (S.51) will be $E^j \stackrel{\text{def}}{=} \mathrm{d}(\partial L/\partial(\partial A_j/\partial x^i))/\mathrm{d}x^i - \partial L/\partial A_j = 2h^j_P \mathrm{d}J^{Pi}/\mathrm{d}x^i$, where $J^{Pi} = J(x)g^{mi}(x, A_{Pk})A_{Pm}$. Thus, the vacum electromagnetic field equations will be

$$\mathrm{d}J^{Pi}/\mathrm{d}x^i = 0. \tag{S.52}$$

In order to derive from the equations (S.52) the trajectory of a photon, we resort to the geometrical optics approximation, that is, we substitute in (S.52) the representation $A_i(x) = a_i(x)\sin S(x)$ neglecting all derivatives except for the first ones of S. In the case under study this standard procedure yields the following result

$$g^{ij}\left(x, \frac{\partial S}{\partial x^n}\right)\frac{\partial S}{\partial x^i}\frac{\partial S}{\partial x^j} = 0. \tag{S.53}$$

What we have thus obtained is in fact the Finslerian Hamilton–Jacobi equation (see Equation (6.26)) corresponding to isotropic Finslerian geodesics. Notice that, on neglecting the dependence of $g^{ij}(x, S_n)$ on x, the equation (S.53) admits plane one-wave solutions $S \sim \sin(k_n x^n)$ with Finslerian massless dispersion relation $g^{mn}(k_i)k_m k_n = 0$ (cf. Equations (S.13) and (S.50)) for k_i.

Chapter 5

5.1. Direct procedure.

5.2. Considering the gauge covariant operator D_i introduced in Section 5.2 and the operator D^*_b of covariant differentiation in the indicatrix which was used in Section 5.1, we can obtain the following relation after simple direct calculations:

$$\mathrm{D}^*_e \mathrm{D}_i w_a - \mathrm{D}_i \mathrm{D}^*_e w_a = -A_a{}^f{}_{ie} w_f + N^c_{ie} \mathrm{D}^*_c w_a$$

Here, the left-hand side is a gauge tensor by definition of the operators D^*_e and D_i, as is $\mathrm{D}^*_c w_a$ on the right-hand side. Therefore, the coefficients of w_f and $\mathrm{D}^*_c w_a$ must be gauge tensors. Incidentally, this reasoning simultaneously provides another proof of the tensor nature of N^c_{ie} (Equation (5.63)).

5.3. We have

$$\partial N^a_i/\partial u^e = F t^k_e \, \partial N^a_i/\partial y^k = F t^k_e \, \partial u^a_{|i}/\partial y^k = F t^k_e (\partial u^a_k/\partial x^i - G_i{}^n \, \partial u^a_k/\partial y^n - G_i{}^n{}_k u^a_n)$$

and

$$A_e{}^a{}_i = -t^k_e U^a_{k|i} = -F t^k_e u^a_{k|i} = -F t^k_e (\partial u^a_k/\partial x^i - G_i{}^n \, \partial u^a_k/\partial y^n - \Gamma_i{}^n{}_k u^a_n)$$

so that

$$N^a_{ie} \overset{\text{def}}{=} \partial N^a_i/\partial u^e + A_e{}^a{}_i = t^k_e U^a_n (\Gamma_i{}^n{}_k - G_i{}^n{}_k)$$

or, because of (1.50),

$$N^a_{ie} = -t^k_e U^a_n C_i{}^n{}_{k|r} y^r. \tag{S.54}$$

This relation shows, in particular, that the gauge tensor N^a_{ie} vanishes in a Landsberg Finsler space, that is when $C_{ikj|r} y^r = 0$.

Similarly, with Equations (5.23′), (5.73), (5.77), and (S.54), we find after straightforward calculations that

$$A_a{}^f{}_{ie} = F t^k_a t^m_e U^f_n (\partial \Gamma_k{}^n{}_i/\partial y^m - C_k{}^n{}_{m|i} + C_k{}^l{}_{i|r} y^r C_l{}^n{}_m)$$

or, substituting Equation (1.48),

$$\begin{aligned} A_a{}^f{}_{ie} = F t^k_a t^m_e U^f_n (C_m{}^n{}_{i|k} - C_{mik}{}^{|n} - \\ - (C_i{}^n{}_l C_k{}^l{}_{m|r} - C_{kil} C^{ln}{}_{m|r}) y^r) \equiv t^k_a t^m_e U^f_n P_k{}^n{}_{im}. \end{aligned} \tag{S.55}$$

Equation (S.55) shows that $A_{afie} = -A_{faie}$.

5.4. $F_a{}^b{}_{in} = 0$ follows from the representation (S.31) and the identities $t^i_a l_i = 0$. As regards the tensor N^c_{in}, formulae (5.81) and (S.31) entail

$$N^c_{in} = N \sum_{A=1}^{N} C^c_A S^p_A (K_{pn} S^A_i - K_{pi} S^A_n)$$

which depends on x^k only.

5.5. Applying the operator $i t^k{}_\nu{}^\mu \mathrm{D}_k$ to Equation (5.154) and using the identity (5.162), we get

$$t^k{}_\nu{}^\mu t^j{}_\alpha{}^\nu \mathrm{D}_k \mathrm{D}_j v^\alpha + q^2 v^\mu = 0.$$

However, because of the specific form of the Pauli matrices (5.92),

$$\begin{aligned} & t_{(a)\nu}{}^\mu t_{(b)\alpha}{}^\nu + t_{(b)\nu}{}^\mu t_{(a)\alpha}{}^\nu = \\ & \quad = 2\delta^{\nu\beta'} \delta^{\mu\lambda'} (s_{(a)\alpha\beta'} s_{(b)\nu\lambda'} + s_{(b)\alpha\beta'} s_{(a)\nu\lambda'}) = \\ & \quad = 2\delta^{\mu\lambda'} \delta_{\alpha\lambda'} \delta_{ab} = 2\delta^\mu_\alpha \delta_{ab}. \end{aligned}$$

On the other hand, direct calculations can be performed to show that the definitions (5.126), (5.163) and (5.134′) taken with any symmetric connection coefficients $D_k{}^i{}_j$ entail that the following commutation formula is valid:

$$(D_k D_j - D_j D_k)v^\alpha = L^\alpha{}_{\beta kj} v^\beta + N^a_{kj} D_a v^\alpha$$

(cf. Equation (5.54)), where the gauge tensor N^a_{kj} is given by Equation (5.56), and

$$L^\alpha{}_{\beta kj} = \partial_k L^\alpha{}_{\beta j} - \partial_j L^\alpha{}_{\beta k} + N^b_k \partial_b L^\alpha{}_{\beta j} - N^b_j \partial_b L^\alpha{}_{\beta k} + \\ + L^\alpha{}_{\nu k} L^\nu{}_{\beta j} - L^\alpha{}_{\nu j} L^\nu{}_{\beta k} - N^a_{kj} L^\alpha{}_{\beta a}$$

may be called *the spin gauge curvature tensor*. Therefore, the sought-for equation will read

$$-h^{kj} D_k D_j v^\mu + q^2 v^\mu + t^k{}_\nu{}^\mu t^j{}_\alpha{}^\nu (L^\alpha{}_{\beta kj} v^\beta + N^a_{kj} D_a v^\alpha) = 0,$$

where in writing the first term we have used the relations (5.155) and (1.140) with $q^0 = -q^1 = -q^2 = -q^3 = 1$.

5.6. Using the explicit form (5.196) of $r_{ab}(u)$, we find

$$r^{ab}(u) = \delta^{ab} - u^a u^b / k^2$$

$(k = 1/K)$ and

$$q^{*c}_{a\ b}(u) = \frac{1}{k^2} \delta_{ab} u^c + \frac{1}{k^2(k^2 - u^2)} u^a u^b u^c$$

where $u^2 = \delta_{ab} u^a u^b$. The triads can be taken in the form

$$g^{*(c)}_{\ a} = \delta^c_a + \frac{k - (k^2 - u^2)^{1/2}}{u^2(k^2 - u^2)^{1/2}} u^a u^c,$$

$$g^{*\ a}_{(c)} = \delta^a_c + \frac{(k^2 - u^2)^{1/2} - k}{ku^2} u^a u^c,$$

which yields, after substituting in (5.166),

$$R^{(c)(d)}{}_a = -\frac{1}{k} \frac{1}{k + (k^2 - u^2)^{1/2}} (\delta^d_a u^c - \delta^c_a u^d).$$

The reader can readily verify that $D_a \Phi^\beta$ thus defined is identical to $\Phi^\beta{}_{||a}$.

5.7. The simplest Lagrangian density for deriving equations for the scalar-vector field $w_a(x, u)$ (the case of π-mesons) reads

$$L = J(x) I(x, u) g^{*ab}(x, u)(a^{ij}(x)(D_i w_a)(D_j w_b) - m^2 w_a w_b). \tag{S.56}$$

Here, $a_{ij}(x)$ is a Riemannian metric tensor, which can be taken as the osculating Riemannian metric tensor; g^{*ab} is the metric tensor of the isotopic space, that is, of the inidcatrix;

$$J(x) = |\det(a_{ij})|^{1/2}, \qquad I(x, u) = |\det(g^*_{ab})|^{1/2};$$

m denotes a real constant. The Euler–Lagrange derivative associated with L as given by (S.56) will be

$$E^a = \partial(\partial L/\partial w_{a,j})/\partial x^j + \partial(\partial L/\partial w_{ab})/\partial u^b - \partial L/\partial w_a \tag{S.57}$$

where $w_{a,j} = \partial w_a/\partial x^j$ and $w_{ab} = \partial w_a/\partial u^b$. The seven-fold action principle

$$\delta \int L d^4 x d^3 u = 0$$

will result in the field equations $E^a = 0$.

By construction, the Lagrangian L/JI is a scalar under the combined coordinate and gauge transformations $x^i = x^i(\bar{x}^j)$, $u^a = u^a(x, \bar{u}^b)$. Therefore, as is known from the calculus of variations,

$$\bar{E}^a = (\det(\partial x^i/\partial \bar{x}^j))(\det(\partial u^c/\partial \bar{u}^d)) \frac{\partial \bar{u}^a}{\partial u^b} E^b,$$

so that the equations $E^a = 0$ will be gauge invariant.

From (S.56) we get

$$\partial L/\partial w_{a,j} = 2JI\,\mathrm{D}^j w^a, \qquad \partial L/\partial w_{ab} = 2JIN_j^b\mathrm{D}^j w^a,$$
$$-\partial L/\partial w_a = 2JIA_b{}^a{}_j\mathrm{D}^j w^b + 2JIm^2 w^a.$$

Thus, according to the definition (S.57),

$$E^a/2JI = \frac{1}{J}\partial_j(J\mathrm{D}^j w^a) + A_b{}^a{}_j\mathrm{D}^j w^b + N_j^b\partial_b(\mathrm{D}^j w^a) + N_{jb}^b\mathrm{D}^j w^a + m^2 w^a. \tag{S.58}$$

Here, we used the fact that the contraction N_{jb}^b of the gauge tensor N_{ib}^c given by Equation (5.63) reads

$$N_{jb}^b \stackrel{\text{def}}{=} \partial N_j^b/\partial u^b + A_b{}^b{}_j = \partial N_j^b/\partial u^b + \partial \ln I/\partial x^j + N_j^b\,\partial \ln I/\partial u^b, \tag{S.59}$$

where in the second step the following identity has been taken into account:

$$0 = g^{*ab}\mathrm{D}_j g_{ab}^* \stackrel{\text{def}}{=} g^{*ab}(\partial_j g_{ab}^* - 2A_{abj} + N_j^c\partial_c g_{ab}^*) = 2(\partial_j \ln I - A_b{}^b{}_j + N_j^c\partial_c \ln I).$$

If we define the action of the operator D_j on $\mathrm{D}^j w^a$ in accordance with the definition (5.134′) where we assume that the connection coefficients $D_n{}^k{}_i$ are the Christoffel symbols deduced from $a_{ij}(x)$, we can rewrite (S.58) in the following explicitly gauge-invariant form:

$$E^a/2JI = \mathrm{D}_j\mathrm{D}^j w^a + N_{jb}^b\mathrm{D}^j w^a + m^2 w^a. \tag{S.60}$$

It will be noted that, by virtue of (S.54), the contraction N_{jb}^b is equal to $-C_i{}^n{}_{n|r}y^r$ and, therefore, vanishes in the case of the Berwald–Moór metric function.

Of course, it is also possible to use in the Lagrangian density (S.56) the parametrical metric tensor $b_{ij}(x,u)$ (Equation (5.2)) instead of the Riemannian $a_{ij}(x)$ and take L to be

$$L = J(x,u)I(x,u)g^{*ab}(x,u)(b^{ij}(x,u)(\mathrm{D}_i w_a)(\mathrm{D}_j w_b) - m^2 w_a w_b), \tag{S.61}$$

where

$$J(x,u) = |\det(b_{ij}(x,u))|^{1/2}.$$

In this case, the new term $N_j^b\partial_b \ln J$ should be added to the right-hand side of (S.60). Further,

$$\partial_j \ln J(x,u) + N_j^b\partial_b \ln J(x,u) = \tfrac{1}{2}g^{mn}(\partial g_{mn}(x,y)/\partial x^j +$$
$$+ 2C_{mnk}\,\partial t^k(x,u)/\partial x^j) + N_j^b g^{mn} t_b^k C_{mnk} =$$
$$= \tfrac{1}{2}g^{mn}(\partial g_{mn}(x,y)/\partial x^j + 2C_{mnk}(\partial t^k(x,u)/\partial x^j + N_j^b t_b^k))$$

which, because of $\mathrm{D}_j t^k = 0$ (Equation (5.90)), may finally be written as

$$\tfrac{1}{2}g^{mn}(\partial g_{mn}(x,y)/\partial x^j - 2C_{mnk}G_j{}^k).$$

Therefore, if we take into account the definition (1.32), we get

$$\partial_j \ln J(x,u) + N_j^b\partial_b \ln J(x,u) = \Gamma_k{}^k{}_j(x,t(x,u)) \tag{S.62}$$

where $\Gamma_k{}^n{}_j(x,y)$ are the Cartan connection coefficients. Thus, Equation (S.60) will retain its form, while the first term on the right-hand side of (S.60) will be written in terms of the Cartan connection coefficients.

Next, attempting to obtain equations of the type (5.154)–(5.157) for the isospinor field $v^\beta(x,u)$ (the case of K-mesons), we choose the Lagrangian density in the following form

$$L = J(x,u)I(x,u)\Big[-\frac{i}{2}\bar v_\beta(x,u)t^j{}_\alpha{}^\beta(x,u)\mathrm{D}_j v^\alpha(x,u) +$$
$$+\frac{i}{2}v^\alpha(x,u)t^j{}_\alpha{}^\beta(x;u)\mathrm{D}_j\bar v_\beta(x,u) + q\bar v_\beta(x,u)v^\beta(x,u)\Big]$$

which is real because of the relation (5.160). We get

$$\partial L/\partial v_{\alpha,j} = -\frac{i}{2} JI\bar{v}^{\beta} t^{j\alpha}{}_{\beta}, \qquad \partial L/\partial v_{\alpha,b} = -\frac{i}{2} JI\bar{v}^{\beta} t^{j\alpha}{}_{\beta} N^b_j,$$

$$\partial L/\partial v_{\alpha} = \frac{i}{2} JI(\bar{v}^{\beta} t^{j\nu}{}_{\beta} L^{\alpha}{}_{\nu j} + t^{j\alpha}{}_{\beta} D_j \bar{v}^{\beta}) + JIq\bar{v}^{\alpha},$$

so that the Euler-Lagrange derivative of L with respect to v_{α}:

$$E^{\alpha} \stackrel{\text{def}}{=} \partial_j(\partial L/\partial v_{\alpha,j}) + \partial_b(\partial L/\partial v_{\alpha,b}) - \partial L/\partial v_{\alpha}$$

will be given by a relation which, after taking into account the identities (5.162), (S.59), and (S.62), can be written as

$$-E^{\alpha}/JI = it^{j\alpha}{}_{\beta} D_j \bar{v}^{\beta} + q\bar{v}^{\alpha} + \frac{i}{2}\bar{v}^{\beta} t^{j\alpha}{}_{\beta} N^b_{jb}.$$

Finally, wishing to derive an equation of the Dirac type (5.141) from a variational principle (the case of nucleons), we can take the Lagrangian density in the conventional form

$$L = J(x)I(x,u)t_0{}^{\alpha\beta'}\Big[-\frac{i}{2}\bar{\Phi}_{\beta'}(x,u)\gamma^j(x)D_j\Phi_{\alpha}(x,u) + \\ + \frac{i}{2}(D_j\bar{\Phi}_{\beta'}(x,u))\gamma^j(x)\Phi_{\alpha}(x,u) + m\bar{\Phi}_{\beta'}(x,u)\Phi_{\alpha}(x,u)\Big] \tag{S.63}$$

which is real and involves the spinor Φ_{β} and its Dirac conjugate $\bar{\Phi}_{\beta'}$ in a symmetrical way. Taking into account the definition (5.135) and using the identities (5.139) and (5.148), we readily find that the Euler–Lagrange derivative E^{α} of (S.63) with respect to Φ_{α} is given by a relation of the form

$$t_{0\alpha\beta'}E^{\alpha}/JI = -i(D_j\bar{\Phi}_{\beta'}(x,u))\gamma^j(x) - m\bar{\Phi}_{\beta'}(x,u) - \frac{i}{2}N^b_{jb}(x,u)\bar{\Phi}_{\beta'}(x,u)\gamma^j(x).$$

Evidently, the invariant Lagrangians of interactions may be constructed in a similar way, for example,

$$L_{nn,\pi} = iJIt_a{}^{\alpha\beta'}\bar{\Phi}_{\beta'}\Phi_{\alpha}w^a,$$

$$L_{KK,\pi} = JIt_a{}^{\alpha\beta'}v^{\mathscr{C}'}_{\beta'}v_{\alpha}w^a.$$

5.8. Let us stipulate, as in Chapter 4, that an initial (x, y)-dependent scalar density is of the form (3.60), which, after substituting the representation (3.14), takes the form (4.5). Assuming the density (3.60) and, hence, (4.5) to be homogeneous of degree zero in y^i, we may substitute l^i for y^i in (4.5) and then use the parametrical representation $l^i = t^i(x, u)$ of the indicatrix, which leads to a scalar density of the form

$$L_{(t)} = L(t(x,u), S^A(x), d_m S^A(x), d^2_{mn} S^A(x)).$$

Finally, taking here the parametrical representation to be uniform:

$$t^i(x,u) = T^i(S^A(x), u)$$

(definition (5.41)), we obtain *the parametrical scalar density*

$$L^u = L^u(u^a, S^A(x), d_m S^A(x), d^2_{mn} S^A(x)). \tag{S.64}$$

This construction offers us the interesting possibility of treating $S^A(x)$ in (S.64) as gravitational field variables, as has been done in Chapter 4, and, at the same time, of regarding u^a merely as a set of independent parameters, according to which we may take u^a outside the operator $\partial/\partial x^m$ whenever this operator occurs. The Euler–Lagrange derivative $E^u{}_{A}{}^{q\dots}_{p\dots}$ of L^u with respect to the variables $S^{Ap\cdots}_{q\cdots}$ will be very similar to the Euler-Lagrange derivative $E_{Ap\ldots}^{q\cdots}$ associated with the osculating gravitational Lagrangian density (4.6) and given by (4.11). Indeed, comparing (S.64) with (4.6), we may immediately conclude that the replacement of $y^i(x)$ by $t^i(x, u)$ in $E_{Ap\ldots}^{q\cdots}$ results in the part of $E^{u}{}_{Ap\ldots}^{q\cdots}$ which does not involve the derivative of $T^i(S^A, y)$ with respect to S^A. So, the complete form of $E^{u}{}_{Ap\ldots}^{q\cdots}$ reads

$$E^{u}{}_{Ap\ldots}^{q\cdots} = E_{(1)Ap\ldots}^{q\cdots} - (\partial L/\partial y^n)\,\partial T^n/\partial S^{Ap\cdots}_{q\ldots}.$$

Here, $\partial L/\partial y^n$ means the partial derivative of the density (4.5) with respect to the argument y^n, and

$$E_{(1)Ap\ldots}^{q\cdots} = E_{Ap\ldots}^{q\cdots}\Big|_{y^i(x)\to t^i(x,u)} \tag{S.65}$$

where $E_{Ap\ldots}^{q\cdots}$ is given by Equation (4.12). According to (S.65), we are to replace $y^i(x)$ by $t^i(x, u)$ everywhere y^i occurs in Equations (4.13)–(4.18), for example

$$y^k(x) \to t^k(x, u), \qquad y^k_{,m} \to t^k_{,m} \stackrel{\text{def}}{=} \partial t^k(x, u)/\partial x^m,$$

$$y^k_{,mn} \to \partial t^k_{,m}(x, u)/\partial x^n, \qquad y^k_m \to t^k_m \stackrel{\text{def}}{=} t^k_{,m} + \Gamma_n{}^k{}_m(x, t)t^n,$$

$$y^k_{mn} \to t^k_{mn} \stackrel{\text{def}}{=} \partial t^k_m(x, u)/\partial x^n - \Gamma_m{}^i{}_n(x, t)t^k_i + \Gamma_i{}^k{}_n(x, t)t^i_m,$$

$$\mathrm{d}_m L^{ij,m} \stackrel{\text{def}}{=} \partial L^{ij,m}(x, y)/\partial x^m + y^n_{,m}\,\partial L^{ij,m}(x, y)/\partial y^n \to$$

$$\to L^{ij,m}{}_{,m}(x, t) + t^n_{,m}\, L^{ij,m}{}_n(x, t),$$

where $L^{ij,m}{}_{,m}(x, y) = \partial L^{ij,m}(x, y)/\partial x^m$ and $L^{ij,m}{}_n(x, y) = \partial L^{ij,m}(x, y)/\partial y^n$.

Bearing in mind these remarks, all the conclusions reached in Section 4.1–4.5 remain valid. The same is true of the possibility of separating out the ordinary divergence in the gravitational Lagrangian density (Section 4.6). In the particular case when the Finsler space is assumed to be of the 1-form type and the parametrical representation of the indicatrix is taken to be uniform (relation (5.45)), the parametrically represented reduced Lagrangian K^u_1 can be obtained merely by substituting $t^A(u^a)$ (see Equation (5.32)) for y^A in K_1 given by (4.114) (and putting $\mathrm{d}_B y^A$ to zero because we treat x^i and u^a as independent variables):

$$K^u_1(u, M^A{}_{BC}) = K_1(t^A(u^a), 0, M^A{}_{BC}).$$

This simplifies the expressions (4.117) and (4.118) for the Euler-Lagrange derivative, namely we obtain

$$E^u{}_A{}^i \stackrel{\text{def}}{=} \mathrm{d}_j(\partial(JK^u_1)/\partial \mathrm{d}_j S^A_i) - \partial(JK^u_1)/\partial S^A_i = \mathrm{d}_j(JK^u{}_A{}^{ji}) - JT^u{}_A{}^i,$$

where

$$T^u{}_A{}^i = (S^i_A + t^i C_A{}^B{}_B)K^u_1 - K^u{}_B{}^{ni} M^B{}_{nA},$$

$$K^u{}_B{}^{ni} = \partial K^u_1/\partial \mathrm{d}_j S^A_i,$$

and J is regarded as $J^u(u^a, S^A) = J(t^A(u), S^A)$. One can readily reformulate also the conservation laws stated in Section 4.7.

5.9. See Asanov (1984a).

5.10. Taking as an example an imaginary field $w_a(x, u)$, an immediate generalization of the Lagrangian density (S.61) will read

$$L = J(x, u)I(x, u)g^{*ab}(x, u)(b^{ij}(x, u)(\mathrm{D}_i w^{\mathscr{C}}_a)(\mathrm{D}_j w_b) - m^2 w^{\mathscr{C}}_a w_b) \tag{S.65'}$$

which is real; $\mathscr{C}$ denotes complex conjugation. Consider the change

$$w_a(x, u) \to \hat{w}_a(x, u) = S_a^b(x, u) w_b(x, u)$$

with imaginary coefficients S_a^b constrained by the condition

$$g^{*ab}(x, u) S^{\mathscr{C}\,e}_{\ a}(x, u) S_b^d(x, u) = g^{*ed}(x, u)$$

and then introduce gauge fields $E_a{}^b{}_i(x, u)$ subject to the transformation law

$$S_b^c \hat{E}_a{}^b{}_i = S_a^b E_b{}^c{}_i + D_i S_a^c$$

assuming at the same time that S_a^b and $E_a{}^b{}_i$ each have a tensor law of transformation under gauge transformations (5.36) and a scalar law of transformation under general coordinate transformations (5.37). Under these conditions, the sought-for Finslerian generalization of the SU (N-1)-covariant derivative reads

$$\mathscr{S}_i w_a = \partial w_a / \partial x^i + N_i^b \, \partial w_a / \partial u^b - A_a{}^b{}_i w_b - E_a{}^b{}_i w_b$$

implying

$$\mathscr{S}_i \hat{w}_a(x, u) = S_a^b(x, u) \mathscr{S}_i w_b(x, u).$$

Accordingly, the Lagrangian density (S.65′) with the operator D_i replaced by $\mathscr{S}_i$ will possess generalized SU(N-1)-invariance, in particular generalized SU(3)-invariance in the four-dimensional case.

5.11. Starting with a field $w(x, y)$ invariant under (1.141), that is $w'(x, y') \stackrel{\text{def}}{=} w(x, Z^*(x, y')) = w(x, y')$, we can define a derivative

$$G_i w = w_{|i} + M^j{}_i(x, y) \, \partial w / \partial y^j \equiv \partial w / \partial x^i + (M^j{}_i - G_i^j) \, \partial w / \partial y^j$$

which will possess the required property

$$G_i' w' \stackrel{\text{def}}{=} \partial w'(x, y') / \partial x^i + (M^j{}_i(x, y') - G_i^j(x, y')) \partial w'(x, y') / \partial y'^j =$$

$$= G_i w$$

if the transformation law of the gauge field $M^j{}_i$ reads

$$\partial Z^{*k}(x, y') / \partial x^i + (M^j{}_i(x, y') - G_i^j(x, y')) y_j^k = M^k{}_i(x, y) - G_i^k(x, y)$$

Because of the identity (1.161), this relation implies $y_j' M^j{}_i(x, y') = y_j M^j{}_i(x, y)$. The object $M^j{}_i$ will be a tensor under general coordinate transformations $x^n = x^n(\bar{x}^m)$. Considering the second derivative

$$G_n G_i w \stackrel{\text{def}}{=} (G_i w)_{|n} + M^k{}_n \, \partial(G_i w) / \partial y^k,$$

we find

$$(G_n G_i - G_i G_n) w = M^j{}_{ni} \, \partial w / \partial y^j,$$

where

$$M^j{}_{ni} = \partial(M^j{}_i - G_i^j) / \partial x^n - \partial(M^j{}_n - G_n^j) / \partial x^i +$$
$$+ (M^k{}_n - G_n^k) \, \partial(M^j{}_i - G_i^j) / \partial y^k - (M^k{}_i - G_i^k) \, \partial(M^j{}_n - G_n^j) / \partial y^k$$

is the associated gauge tensor: $M^j{}_{ni}(x, y) = y_k^j M^k{}_{ni}(x, y')$. The definition $M_m{}^j{}_{ni} = \partial M^j{}_{ni} / \partial y^m$ (cf. definitions (4.6.6) and (4.8.14) of Rund (1959)) may be regarded as the associated gauge curvature tensor; $y_m^h M_h{}^j{}_{ni}(x, y) = y_k^j M_m{}^k{}_{ni}(x, y')$. Notice that $G_i F = 0$ if and only if $y_j M^j{}_i = 0$. If we consider a vector field $w_k(x, y)$ invariant under (1.141), that is $w_k'(x, y') \stackrel{\text{def}}{=} w_k(x, Z^*(x, y')) = w_k(x, y')$, then the above definition may be retained: $G_i w_k = w_{k|i} + M^j{}_i \, \partial w_k / \partial y^j$, implying $G_i' w_k' = G_i w_k$.

Restricting the G-transformations to be metric (definitions (1.156)–(1.157)), the operator G_i can naturally be extended to act on any fields tensorial under metric G-transformations. In particular, given a vector field $w_n(x, y)$, $w'_n(x, y') \stackrel{\text{def}}{=} y^m_n w_m(x, Z^*(x, y')) = w_n(x, y')$, we put

$$\mathrm{G}_i w_n = w_{n|i} + w_m h^P_n(h^m_{P|i} + M^j{}_i \, \partial h^m_P/\partial y^j) + M^j{}_i \, \partial w_n/\partial y^j$$

which results in $\mathrm{G}'_i w'_n = y^m_n \mathrm{G}_i w_m$ and

$$\mathrm{G}_i h^P_n = \mathrm{G}_i l_n = \mathrm{G}_i y_n = \mathrm{G}_i F = \mathrm{G}_i g_{mn} = 0$$

together with $\mathrm{G}_i w_n = h^P_n \mathrm{G}_i(w_m h^m_P)$, where h^P_i are orthonormal n-tuples chosen to obey the G-metric property $h^P_i(x, y') = y^j_i h^P_j(x, y)$.

Chapter 6

6.1. The homogeneous Lagrangian (6.3) will read

$$L(x^n, \dot{x}^i) = (\tfrac{1}{2} m_{ab}(x^n) \dot{x}^a \dot{x}^b/(\dot{x}^N)^2 - V(x^n))\dot{x}^N$$

which entails that

$$p_a = L m_{ab} \dot{x}^b/\dot{x}^N,$$

$$p_N = -(\tfrac{1}{2} m^{ab} p_a p_b/H^2 + V)H,$$

where the tensor m^{ab} is reciprocal to m_{ab}. The last relation may be regarded as the equation for H:

$$\tfrac{1}{2} m^{ab} p_a p_b + p_N H(x, p) + V H^3(x, p) = 0.$$

In constructing the Hamilton–Jacobi equation $H(x, p) = 1$ (Equation (6.25)), we replace p_i by $\partial S/\partial x^i$, which converts the above equation to

$$\tfrac{1}{2} m^{ab} \frac{\partial S}{\partial x^a} \frac{\partial S}{\partial x^b} + \frac{\partial S}{\partial x^N} + V = 0.$$

This is identical with the classical Hamilton–Jacobi equation related to a holonomic dynamical system.

6.2. Suppose we are given two scalars of the form $v(x^i, p_j)$ and $w(x^i, p_j)$. The Poisson brackets may be defined in the conventional way:

$$\{v, w\} = \frac{\partial v}{\partial x^i} \frac{\partial w}{\partial p_i} - \frac{\partial w}{\partial x^i} \frac{\partial v}{\partial p_i}$$

The usual Jacobi identities will hold. The canonical equations (6.19) and (6.23) associated with the homogeneous case may be represented in terms of the Poisson brackets as follows:

$$\frac{\mathrm{d}x^i}{\mathrm{d}t} = \{x^i, \tfrac{1}{2}H^2\}, \qquad \mathrm{L}\frac{\mathrm{d}}{\mathrm{d}t}(p_i/L) = \{p_i, \tfrac{1}{2}H^2\}.$$

If we choose the parameter t to be the Finslerian arc-length s, the above equations take the same form as in nonhomogeneous classical mechanics:

$$\frac{\mathrm{d}x^i}{\mathrm{d}s} = \{x^i, H\}, \qquad \frac{\mathrm{d}p_i}{\mathrm{d}s} = \{p_i, H\}..$$

6.3. Let us consider a local congruence β of solutions of the Euler–Lagrange equations (6.6) associated with the homogeneous case. In view of the existence of the identity $\dot{x}^i \, \partial L/\partial \dot{x}^i = L$, there must, in general, exist no more than N-1 independent integrals for the congruence β. Denoting these integrals by u^A (A, B will be specified over $1, \ldots, N$-1), and choosing the Finslerian arc-length s to be the N-th coordinate, we may represent our congruence as follows:

$$x^A = x^A(u^B, s), \qquad x^N = s.$$

The Lagrange brackets of the congruence under study may be defined as usual:

$$[u^A, u^B] = \frac{\partial x^h}{\partial u^A}\frac{\partial p_h}{\partial u^B} - \frac{\partial x^h}{\partial u^B}\frac{\partial p_h}{\partial u^A}$$

and the reader can readily verify that the Lagrange brackets are constant along solutions of the Euler–Lagrange equations, i.e.,

$$\frac{\partial}{\partial s}[u^A, u^B] = 0.$$

Since $x^N = s$, we get $H = 1$, and the vorticity tensor (6.36) takes the form

$$\omega_{hj} = \partial p_h/\partial x^j - \partial p_j/\partial x^h$$

so that

$$[u^A, u^B] = \frac{\partial x^j}{\partial u^A}\frac{\partial x^h}{\partial u^B}\omega_{hj}$$

If, in particular, the coordinates x^i are chosen such that

$$x^i = u^1, \ldots, x^{N-1} = u^{N-1},$$

we get merely

$$[u^A, u^B] = -\omega_{AB}$$

so that, relative to such a coordinate system, the vorticity tensor ω_{AB} is constant along the streamlines of the considered congruence. (Rund, 1979, pp. 206–208; Baumeister, 1979, Section 2).

6.4. Introducing an auxiliary nonvanishing scalar density n(x), we may construct the Lagrangian density as follows

$$L(x^h, Q^a, Q^a_{,m}, P_a, n) = n(H(x^h, P_b Q^b_{,m}) - (Q^a, P_a)).$$

Then it can be directly verified that the associated Euler–Lagrange equations

$$E_{Q^a}(L) = 0, \qquad E_{P_a}(L) = 0, \qquad E_n(L) = 0$$

imply the validity of the associated canonical equations (6.33) and (6.34), together with the original canonical equations (6.23) and the continuity equation

$$\frac{\partial}{\partial x^j}\left(nH\frac{\partial H}{\partial p_j}\right) = 0.$$

(Rund, 1979, Section 3).

6.5. In view of the particular form of L, we have

$$p_i \overset{\text{def}}{=} L\,\partial L/\partial \dot{x}^i = L(a_{ij}\dot{x}^j S^{-1} + b_i)$$

where we have used the notation $S = (a_{ij}(x)\dot{x}^i\dot{x}^j)^{1/2}$. Therefore,

$$\dot{x}^j = SL^{-1}a^{ji}(p_i - Lb_i).$$

Further, it will be noted that the function $C(x, p) \overset{\text{def}}{=} S(x, \dot{x}(x, p))$ will be

$$C = (a^{ij}(p_i - Lb_i)(p_j - Lb_j)S^2L^{-2})^{1/2}$$

so that, because of $C = S$ and the definition $H(x, p) = L(x, \dot{x}(x, p))$, we obtain the following equation for the Hamiltonian function H:

$$H^2 = a^{ij}(p_i - Hb_i)(p_j - Hb_j).$$

From the last equation the Hamilton–Jacobi equation $H(x^m, \partial S/\partial x^n) = 1$ is obtained directly in the following form:

$$a^{ij}\left(\frac{\partial S}{\partial x^i} - b_i\right)\left(\frac{\partial S}{\partial x^j} - b_j\right) = 1.$$

(Rund, 1966b, Section 3.6; the above derivation of the Hamilton–Jacobi equation is very similar to a method proposed by Vanstone, 1963).

6.6. In terms of the unit vectors $l_i(x) = p_i(x)/H(x, p(x))$ the definition (6.36) reads

$$\omega_{hj} = \partial_j l_h - \partial_h l_j.$$

Since the field $l_i(x)$ represents a congruence of Finslerian geodesics, the identity $l^h \omega_{hj} = 0$ will hold in accordance with the solution of Problem 1.2.

This identity shows that the bivector ω_{hj} (the skew-symmetric tensor of valence two is called a bivector) cannot be of the highest rank. In particular, in the four-dimensional case $N = 4$, this means that the Clebsch representation of $l_i(x)$ tangent to a congruence of Finslerian geodesics is at most of the form

$$l_i = \partial_i Z + P\,\partial_i Q$$

with three independent scalars Z, P, Q, so that

$$\omega_{hj} = \partial_j P\,\partial_h Q - \partial_h P\,\partial_j Q,$$

that is, the bivector ω_{hj} is simple.

Because of the identity $l^h \omega_{hj} = 0$, the equation (6.37) which describes the motion of ω_{hj} will be invariant under the replacement of p^l by Mp^l for any nonvanishing scalar $M(x)$. Therefore, choosing the coordinate system x^i such that $p^l = M(x)\delta^l_N$, $M(x)$ being a scalar [to get such a representation of $p^l(x)$, it is sufficient to use the Clebsch representation of $p^l(x)$ (see Equation (4.68)) and take a special coordinate system of the type (4.73)], and replacing further p^l by $p^l/M = \delta^l_N$, we find that equation (6.37) says that, relative to such a coordinate system, the vorticity tensor ω_{hj} is constant along geodesics of the congruence (which is another proof of the statement formulated at the end of paragraph 6.3).

For the electromagnetic case treated in the preceding paragraph 6.5, we have $l_i = u_i + b_i$, where $u_i = a_{ij}\dot{x}^j/S$ is the Riemannian four-dimensional velocity, so that

$$\omega_{hj} = \omega^*_{hj} + F_{jh}.$$

Here, $\omega^*_{hj} = \partial_j u_h - \partial_h u_j$ is the Riemannian vorticity tensor of the congruence of trajectories of test electric charges, that is, the vorticity tensor attached to the congruence from the viewpoint of the Riemannian background; $F_{jh} = \partial_j b_h - \partial_h b_j$ is the electromagnetic field tensor. Thus, from the above statements it can be inferred that the sum of the vorticity tensor of a congruence of trajectories of test electric charges and the electromagnetic field tensor is a simple bivector, and that this sum is constant along the trajectories relative to a special coordinate system.

6.7. From Equation (2.168) which gives the corresponding Finslerian metric tensor it is clear that the Finslerian triad $h^a_i(x, \dot{x})$ is related to the Riemannian triad $r^a_i(x)$ (related to the Riemannian metric tensor $a_{ij}(x)$ entering Equation (2.168)) as follows

$$r^a_i = h^a_i(S/L)^{1/2}$$

where we have used the notation $S(x, \dot{x})$ instead of $a(x, \dot{x})$ adopted in Equation (2.168); $a = 1, 2, 3$. A direct calculation of $G_i{}^k(x, \dot{x})$ on the basis of formulae (2.168) and (2.169) yields the following result

$$2G_i{}^k = SF_i{}^k + u_i a^k + u^k I_i + LS^{-1}\delta^k_i\,\mathrm{d}^2 s/\mathrm{d}q^2.$$

Here, $F_i{}^k = a^{kn}F_{in}$, $a^k = \dot{x}^n F_n{}^k$ is the Lorentz force, $u^k = \dot{x}^k/S$ is the Riemannian four-dimensional velocity, and s and q denote the Finslerian and Riemannian arc-lengths respectively, that is

$ds = L(x, dx)$ and $dq = S(x, dx)$. The explicit form of the vector I_i is not of interest to us here because $u^k h_k^a = 0$, and therefore I_i will not appear in the final result. With the above formulae, the sought-for result is

$$\frac{dr_i^a}{dq} - \tfrac{1}{2}(F_i^{\ j} + u_i u^k F_k^{\ j}) r_i^a = 0,$$

where $u_i = a_{ik} u^k$.

This equation should be compared with the general law

$$\frac{dS_i}{dq} = \left(\frac{g}{2} F_i^{\ j} + \left(\frac{g}{2} - 1\right) u_i F^j{}_k u^k\right) S_j$$

which describes the motion of the spin four-vector S_i of a charged particle with gyromagnetic ratio g (Bargmann *et al.*, 1959). We observe that the Finslerian δ-parallel transportation under consideration corresponds to $g = 1$, that is to the transportation of a rigidly rotating charged particle (classical magnetic moment).

Chapter 7

7.1. The matrix of a special Lorentz transformation reads

$$\|L_P^Q\| = \left\| \begin{matrix} f & fV/c & 0 & 0 \\ fV/c & f & 0 & 0 \\ 0 & 0 & 1 & 0 \\ 0 & 0 & 0 & 1 \end{matrix} \right\|$$

where $f = (1 - V^2/c^2)^{-1/2}$. The problem of finding the eigenvalues and eigenvectors of such a matrix is readily solved, and the solution can be written in the form

$$S_1^P = \begin{pmatrix} 1/4 \\ 1/4 \\ 0 \\ 0 \end{pmatrix}, \quad S_2^P = \begin{pmatrix} 1/4 \\ -1/4 \\ 0 \\ 0 \end{pmatrix}, \quad S_3^P = \begin{pmatrix} 0 \\ 0 \\ 1 \\ 0 \end{pmatrix}, \quad S_4^P = \begin{pmatrix} 0 \\ 0 \\ 0 \\ 1 \end{pmatrix},$$

$$k_1 = ((1 + V/c)(1 - V/c))^{1/2} = 1/k_2, \qquad k_3 = k_4 = 1.$$

This solution belongs to the type (c) (p. 215). The eigenvalues k_1 and k_2 coincide with the familiar expressions for the Doppler effect for light signals propagating, respectively, along and against the direction of relative motion. The components C_3^3 and C_4^4 will be infinitely large which corresponds to the circumstance that, in conventional special relativity theory, motion does not deform the scales perpendicular to the direction of motion.

If we limit our considerations to the plane $x^0 \times x^1$, then all formulae derived in Chapter 7 will be applicable to this plane after restricting them to the two-dimensional case. For instance, instead of (7.30) we shall write

$$u_M = \tfrac{1}{2}\left(\frac{S_M^1}{S_N^1 u^N} + \frac{S_M^2}{S_N^2 u^N}\right)$$

where the indices M and N take on the range 0, 1. Substituting the values

$$S_M^1 = 4S_1^M, \qquad S_M^2 = 4S_2^M$$

in the above formula yields the usual Lorentzian relations

$$u_0 = u^0, \qquad u_1 = -u^1.$$

Similarly, the fundamental kinematic relation (7.33) will reduce to the relation $U_1 = -U^1$ known as the reciprocity principle (Equation (7.57)).

7.2. In order to derive the desired law of addition one should resort to the transformation law (7.13) which describes the transition between the inertial frames of references β_1 and β_2. We have:

$$R^a(1) = R^b(2)K^a_b + R^0(2)K^a_0, \qquad R^0(1) = R^b(2)K^0_b + R^0(2)K^0_0 \tag{S.66}$$

where K^P_Q are functions of the proper relative velocity U^a of β_2 relative to β_1. Let a material point M move relative to β_2 with velocity U_2, i.e. relative to β_2 we have a set of events $(R^0(2), R^a(2))$ such that $R^a(2) = U^a_2 R^0(2)$ where U^a_2 are fixed, and $R^0(2)$ are arbitrary. By definition, the velocity of motion of the point M with respect to β_1 will be equal to $U^a_1 = R^a(1)/R^0(1)$, where $R^a(1)$ and $R^0(1)$ are given by relation (S.66). Thus the desired law of addition of the proper relative velocities will read

$$U^a_1 = \frac{K^a_b(U)U^b_2 + K^a_0(U)}{K^0_b(U)U^b_2 + K^0_0(U)}.$$

Substituting here (7.45) and (7.46), we explicitly find

$$U^a_1 = \frac{U^a + U^a_2 + (1/c)r_b{}^a{}_d U^b_2 U^d}{1 + \dfrac{1}{c^2}\sum_{b=1}^3 U^b U^b_2} \tag{S.67}$$

where the anisotropy tensor $r_b{}^a{}_d$ is given by (7.47). This law of addition is in many ways similar to the ordinary law of addition of the special relative velocities

$$V_1 = \frac{V + V_2}{1 + \dfrac{1}{c^2}VV_2}. \tag{S.68}$$

At low velocities the laws of addition (S.67) and (S.68) have the same form.

7.3. The close intrinsic relationship between conventional Lorentzian special relativity theory and the Maxwell equations for the electromagnetic field arises from the Lorentz-invariance rooted in the Maxwell equations. This invariance can be displayed as follows. Suppose we are given an inertial frame of reference with proper coordinates x^P related to the pseudo-Euclidean metric tensor $q^{PQ} = \delta^{PQ} q^P$, where $q^0 = -q^1 = -q^2 = -q^3 = 1$. Then the Maxwell equations claim that the electromagnetic field can be represented by a tensor $F_{PQ}(x^R)$ subject to the following two requirements. First, the tensor is a curl, i.e. there exists a covariant vector field $A_P(x)$ such that

$$F_{PQ} = \partial A_Q/\partial x^P - \partial A_P/\partial x^Q. \tag{S.69}$$

And second,

$$\partial F^P{}_Q/\partial x^P = \frac{4\pi}{c} J_Q, \tag{S.70}$$

where J_Q is the current density; $F^P{}_Q = q^{PM} F_{MQ}$; the indices take on the range 0, 1, 2, 3.

Let us make a non-degenerate linear transformation of the four-dimensional system of reference under consideration. Denoting the coefficients of such a transformation by L^P_Q, we get: $A_P \to \tilde{A}_P = L^Q_P A_Q$, $F_{PQ} \to \tilde{F}_{PQ} = L^M_P L^N_Q F_{MN}$, $\partial/\partial x^P \to \partial/\partial\tilde{x}^P = L^P_Q \partial/\partial x^Q$, $J_Q \to \tilde{J}_Q = L^P_Q J_P$. Hence, Equation (S.69) has the same form under any non-degenerate linear transformation of the system of reference, whereas Equation (S.70) turns out to be transformed into the form

$$L^P_N q^{NR} L^{*M}_R \, \partial\tilde{F}_{MQ}/\partial\tilde{x}^P = \frac{4\pi}{c}\tilde{J}_Q \tag{S.71}$$

where L^{*M}_R are the entries of the matrix inverse to $||L^M_R||$. From Equation (S.71) it follows that equation (S.70) remains invariant under the examined transformations if and only if the coefficients satisfy the relation $L^P_N q^{NR} L^{*M}_R = q^{PM}$, which is merely the definition of the Lorentz transformations.

So, we have verified that the Maxwell equations are Lorentz-invariant. Simultaneously, this proof has demonstrated the purely passive sense of the Lorentz-invariance of the Maxwell equations, namely this invariance manifests itself as the form-invariance of the Maxwell equations with respect to a certain class of transformations of the proper systems of reference of a fixed inertial frame of reference. Essentially, the proof has never dealt with the problem of how the electromagnetic field transforms under real, that is active, transitions to other frames of reference. The Lorentz-invariance proved above leaves us ignorant as to whether active kinematic transformations are described by Lorentz transformations or by some other ones. Repeating the arguments of Section 7.2, we may contend that, whilst observations are made from a laboratory on the Earth and instruments are not really carried to other material bodies moving (at a sufficiently high velocity) relative to the Earth, electromagnetic phenomena cannot yield any information on how the electromagnetic field will behave under active kinematic transformations.

It is not difficult to distinguish clearly between the passive and active kinematic transformations of the electromagnetic field. Indeed, suppose that the basic metric function is Finslerian, so that the active K-transformations differ from the passive Lorentzian ones, and denote by $A_i(x^n)$ the electromagnetic vector potential relative to an arbitrary admissible coordinate system x^m of the background manifold. If, as in Section 7.6, we consider the terestrial laboratory to be referred to the coordinates x^P and denote by h^P_i the proper Lorentzian orthonormal system of reference, then the vector potential will have the components $A_P(x^Q) = h^i_P A_i(h^n_Q x^Q)$ giving rise to the components $F_{PQ}(x^R) = \partial A_Q/\partial x^P - \partial A_P/\partial x^Q \equiv h^i_P h^j_Q(\partial A_j/\partial x^i - \partial A_i/\partial x^j)$ of the electromagnetic field tensor relative to the terestrial laboratory under consideration. The Maxwell equations will be of the form (S.69)–(S.70). Relative to another inertial frame of reference β, these fields will be represented by the components $A_{\beta P}(R^Q) = h^i_P(U)A_i(h^n_Q(U)R^Q)$ and $F_{\beta PQ}(R^M) = \partial A_{\beta Q}/\partial R^P - \partial A_{\beta P}/\partial R^Q$. Here, U^P is the velocity of motion of β relative to the Earth, $h^n_Q(U)$ is the proper orthonormal system of reference of β, and $R^P = K^{*P}_{Q} x^Q$ (see Equation (7.13)). Accordingly, the active kinematic transformations of the electromagnetic field will read

$$A_{\beta P} = K^Q_P A_Q, \qquad F_{\beta PQ} = K^M_P K^N_Q F_{MN}, \tag{S.72}$$

and the equation (S.70) will take the form $\partial F_{\beta}{}^{P}{}_{Q}/\partial R^P = 4\pi/c\, J_Q(R)$, where $F_{\beta}{}^{P}{}_{Q} = g^{PM}(U)F_{\beta MQ}$ and $g^{PM}(U)$ is the reciprocal of the tensor (7.59). It will be noted that, because $\det(K^P_Q) = 1$, the function $F_{PQ}F_{MN}$, $I = e^{PQMN} F_{PQ}F_{MN}$ where e^{PQMN} denotes the permutation symbol, will be an invariant of the active kinematic transformations as well as of the passive Lorentzian ones.

7.4. Under active kinematic transformations, the wave vector k_P of the photon transforms similarly to (S.72), namely, $k_{\beta P} = K^Q_P k_Q$, where we have to substitute the decompositions (7.45)–(7.48) and (7.54). Under passive kinematic Lorentzian transformations, $k_P \to \tilde{k}_P = L^Q_P k_Q$. Thus, the Finslerian corrections to the Doppler effect will be given by $\Delta k_P = (K^Q_P - L^Q_P)k_Q$. Once again, these corrections may only be observed when a light signal receiver moves at a sufficiently high velocity relative to the Earth.

Accordingly, we shall have two Doppler effects, passive and active, of which the first comprises all the phenomena concerning the changes of the directions of the light rays and of the photons' frequencies that are describable by exploiting the Lorentz-rotated tetrads $L^P_Q h^Q_i(u_1)$ in the fixed laboratory β_1 (say, on the Earth). At the same time, the active Doppler effect would include the totality of similar phenomena appearing as a result of real transitions from β_1 to other laboratories β_2, these kinds of phenomena being described by K-transformations. The distinction between the two different effects evaporates only when one postulates that the tensor g_{ij} is Riemannian. The active kinematic Doppler effect may generally be expected to be highly anisotropic and to involve terms of the first order in q.

Also, we may suspect that the real change of clock functioning under the transition from β_1 in β_2 will not be describable exactly by the Lorentzian relations. This means, in particular, that the lifetime of an unstable particle in its proper reference frame will depend on the velocity of the particle relative to the Earth. In fact, since the apparent dilatation of the lifetime of a relativistic unstable particle relative to the Earth (the reference frame β_1) is correctly described by the

Lorentzian law $t_1 = tL_0^0$, then, because at the same time we have $t_1 = t_2 K_0^0$ by definition, we find that the actual lifetime t_2 of the particle in its proper reference frame β_2 used as a laboratory will be equal to $t_2 = tL_0^0/K_0^0$. From relations (7.48), (7.49), and (7.54) it follows that t_2 is of the third order of magnitude in q.

7.5. This can be done by using the tetrads of the tensor g_{ij} and even without assuming g_{ij} to be Finslerian. Indeed, given any direction-dependent nondegenerate symmetric tensor $g_{ij}(x, y)$ of signature $(+ - - -)$, the kinematic transformations will be $K_Q^P = h_n^P(u_1)h_Q^n(u_2)$ similarly to Equation (7.14). As in Chapter 7, the proper relative velocities will be defined as follows: $u^P = h_n^P(u_1)h_0^n(u_2)$, $u_P = h_P^n(u_1)h_n^0(u_2)$, $u^{*P} = h_n^P(u_2)h_0^n(u_1)$ and $u_P^* = h_P^n(u_2)h_n^0(u_1)$. It is true that $g_{PQ}(u^R)u^P u^Q = g^{PQ}(u_R)u_P u_Q = 1$ together with $g_{PQ}(u^R)u^P = u_P$ (and the same for u_P^*), so that u_P may be regarded as the momentum related to the four-dimensional velocity u^P.

In the Finslerian case, at low velocities we obtain the following relation between the direct and reverse proper relative velocities: $u^{*0} = 1 + \frac{1}{2}q^2 + O(q^3)$, $u^{*a} = -u^a - Fh_{ij}^a h_b^i h_c^j u^b u^c + O(q^3)$, where $q^2 = \delta_{ab}u^a u^b$ and $h_{ij}^a = \partial h_i^a(u)/\partial u^j$. Also, $u^0 = 1 + \frac{1}{2}q^2 + O(q^3)$ and $u_0 = u^0 + O(q^3)$, $u_a = -u^a + C_{abc}u^b u^c + O(q^3)$.

In the case when the Finslerian metric function is taken to be (2.9) with $N = 4$ and $r^A = (1, -1, -1, -1)$, the tetrads were found explicitly in Solution 2.3. In the case of this metric function we also know the metric G-transformations (see Solution 1.12), which makes it possible to represent the kinematic transformations in terms of the metric G-transformations as follows: $K_Q^P = \Sigma_{A,B} L_B^A \cdot (u_2^B/u_1^A)_2^{f-1} h_Q^B(u_1)h_A^P(u_1)$. They may be interpreted as generalized Lorentz transformations.

Appendix A

A.1. The substitution of (A.10) in (A.16) yields

$$DW^j = -W_l^{*j} \wedge dx^l - \frac{\partial \Gamma_h{}^j{}_k}{\partial y^m} Dy^m \wedge dx^k \wedge dx^h.$$

On the other hand, from Equation (A.14) and (A.18) it follows that

$$DW^j = \tfrac{1}{2}S_h{}^j{}_{k|l}\, dx^h \wedge dx^k \wedge dx^l + \tfrac{1}{2}\frac{\partial S_h{}^j{}_k}{\partial y^m} dx^h \wedge dx^k \wedge Dy^m -$$

$$- S_m{}^j{}_l W^m \wedge dx^l.$$

Comparing these two relations yields the sought-for result. It will be noted that the cyclic identity thus obtained has a structure which is identical with that of the standard theory of linear connections dependent solely on x^i, and reduces to (1.84) in the proper Finslerian case.

A.2. Taking the exterior derivative of (A.12), we get

$$DW_l^{*j} - W_l^{*j}{}_m \wedge Dy^m = -\tfrac{1}{2}(K_l{}^j{}_{hk|m} - K_l{}^j{}_{pm} S_h{}^p{}_k) dx^h \wedge dx^k \wedge dx^m.$$

When this relation is compared with the Bianchi identities (A.22), we obtain the required result.

A.3. It can be readily verified that the curvature 2-forms associated with $p_h{}^j$ vanish identically:

$$dp_h{}^j + p_l{}^j \wedge p_h{}^l = 0.$$

From this fact it follows in turn that the decomposition of the curvature 2-forms $W_h{}^j$ associated with $\beta_h{}^j$ results in

$$W_h{}^j = DL_h{}^j - L_m{}^j \wedge L_h{}^m.$$

Appendix B

B.1. As is well known (see, e.g., Kramer *et al.* (1980)), the metric tensor $r_{ij}(x)$ of a Riemannian space of constant curvature K takes (locally) the conformal form

$r_{ij}(x) = (1 + (K/4)\delta_{mn}x^m x^n)^{-2} e_i \delta_{ij}$ with respect to some coordinate system $\{x^i\}$, where $e_i = \pm 1$ represents the signature of r_{ij}. In the same vein, we may say that we deal with a positive-definite internal space of constant curvature $K(x)$ if a system $\{z^P\}$ exists such that $g_{PQ}(x, z) = b(x, z)\delta_{PQ}$ with $b = (1 + (K(x)/4)\delta_{RS} z^R z^S)^{-2}$. According to this definition, the curvature K of the internal space may have different values at different points x^i. Clearly, the simplest way of specifying the gauge field $N_i^P(x, z)$ will be to stipulate that $\mathrm{D}_i b \overset{\text{def}}{=} \mathrm{d}_i b \equiv \partial_i b + N_i^P \mathrm{d}_p b = 0$, that is, $\delta_{RS} x^R z^S \partial_i K + 2KN_i^P z^Q \delta_{QP} = 0$, which obviously admits the solution

$$N_i^P(x, z) = z^P C_i(x),$$

where $C_i = -(\partial_i K)/2K$. Substituting this solution in the representation (B.26) yields $M_j{}^P{}_i = 0$. At the same time, we have $\mathrm{d}_i g_{PQ} = 0$, so that postulating the metric condition (B.125) results in $D_{PQi} = -D_{QPi}$. So, Equations (B.202), (B.209), and $D_P{}^P{}_m = 0$ established in the linear case prove to be applicable to the present space of constant curvature. However, the vanishing of the gauge field $Q_S{}^P{}_T$, which was postulated in Section B.12 (Equation (B.186)), does not have an invariant meaning under the gauge transformations leaving the metric tensor $g_{PQ} = b\delta_{PQ}$ invariant. The substitution of this metric tensor in the metric expression (B.135) yields

$$Q_P{}^T{}_R = \tfrac{1}{2}(\delta_P^T a_R + \delta_R^T a_P - \delta_{PR}\delta^{TS} a_S),$$

where $a_P = (d_P b)/b \equiv -Kz^P b^{1/2}$ and the torsion tensor $Y_P{}^T{}_R$ has been put to zero: $Y_P{}^T{}_R = 0$.

Under these conditions simple calculations yield the following results in the three-dimensional case $M = 3$:

$$F^P{}_{ij} \overset{\text{def}}{=} -\tfrac{1}{2} e^{PQR} E_{QRij} = \mathrm{d}_i A_j^P - \mathrm{d}_j A_i^P - e^P{}_{QR} A_i^Q A_j^R,$$

$$Q^P{}_{Rj} \overset{\text{def}}{=} \tfrac{1}{2} e^{PTS} Q_{TSRj} = -[\mathrm{d}_R A_j^P + \tfrac{1}{2}(\delta^{TP}\delta_{QR} - \delta_Q^T \delta_R^P) a_T A_j^Q] \overset{\text{def}}{=} -\mathrm{S}_R A_j^P,$$

$$Q_{PTRj} = -Q_{TPRj},$$

and $Z_{PQMN} = K(g_{PM}g_{QN} - g_{PN}g_{QM})$. Here, we put $D_{PQi} = e_{PQR}A_i^R$, where $e_{PQR} = b^{3/2}\ \varepsilon_{PQR}$ is the totally skew tensor with $\varepsilon_{123} = 1$. Because of the Equations $\mathrm{d}_i g_{PQ} = 0$ and $D_{PQi} = -D_{QPi}$, we have

$$\mathrm{D}_i e_{PQR} \overset{\text{def}}{=} \mathrm{d}_i e_{PQR} - D_P{}^S{}_i e_{SQR} - D_Q{}^S{}_i e_{PSR} - D_R{}^S{}_i e_{PQS} = 0,$$

and hence $\mathrm{D}_m^* E^P{}_Q{}^{jm} = -e^P{}_{QR} \mathrm{D}_m^* F^{Rim}$ (such a term enters the Euler-Lagrange derivative (B.70″)). Notice also that $\mathrm{S}_N Z^P{}_Q{}^{RN} = 0$, which annuls the last term in Equation (B.70‴).

B.2. If we put

$$\mathrm{D}z^P \overset{\text{def}}{=} \mathrm{d}z^P - N_i^P \,\mathrm{d}x^i,$$

it follows directly from the second member of Equations (B.8) that $\mathrm{D}z^P$ is *the covariant differential* of z^P, that is, $\mathrm{D}z^P = z_Q^P \mathrm{D}z'^Q$, whereupon the covariant differential of a tensor given by the transformation laws (B.5)–(B.6) will read

$$\mathrm{D}w_n{}^m{}_P{}^Q = (\mathrm{D}_i w_n{}^m{}_P{}^Q)\mathrm{d}x^i + (\mathrm{S}_R w_n{}^m{}_P{}^Q)\mathrm{D}z^R.$$

Following this observation, we introduce two (x, z)-*dependent connection 1-forms*

$$b_h{}^j = L_h{}^j{}_k \mathrm{d}x^k$$

(cf. the definition (A.1)) and

$$b_Q{}^P = D_Q{}^P{}_i \mathrm{d}x^i + Q_Q{}^P{}_R \mathrm{D}z^R$$

to obtain

$$\mathrm{D}w_n{}^m{}_P{}^Q = \mathrm{d}w_n{}^m{}_P{}^Q + b_k{}^m w_n{}^k{}_P{}^Q - b_n{}^k w_k{}^m{}_P{}^Q + b_R{}^Q w_n{}^m{}_P{}^R - \\ - b_P{}^R w_n{}^m{}_R{}^Q$$

so that

$$\mathrm{D}(\mathrm{D}w^{jP}) \overset{\text{def}}{=} \mathrm{d}(\mathrm{D}w^{jP}) + b_m{}^j \wedge \mathrm{D}w^{mP} + b_Q{}^P \wedge \mathrm{D}w^{jQ} = W_i{}^j w^{iP} + \\ + W_Q{}^P w^{jQ},$$

where

$$W_i{}^j = \mathrm{d}b_i{}^j + b_n{}^j \wedge b_i{}^n$$

(cf. Equation (A.4)) and

$$W_Q{}^P = \mathrm{d}b_Q{}^P + b_R{}^P \wedge b_Q{}^R$$

are the associated (x, z)-*dependent curvature 2-forms. The equations of structure* will read

$$W_h{}^j = -\tfrac{1}{2}L_h{}^j{}_{mn}\,\mathrm{d}x^m \wedge \mathrm{d}x^n + B_h{}^j{}_{kP}\,\mathrm{D}z^P \wedge \mathrm{d}x^k, \\ W_Q{}^P = -\tfrac{1}{2}E_Q{}^P{}_{mn}\,\mathrm{d}x^m \wedge \mathrm{d}x^n + Q_P{}^T{}_{Rj}\,\mathrm{d}x^j \wedge \mathrm{D}z^R - \tfrac{1}{2}Z_Q{}^P{}_{MN}\,\mathrm{D}z^M \wedge \mathrm{D}z^N. \tag{S.73}$$

The (x, z)-*dependent torsion 2-forms* are

$$W^j \overset{\text{def}}{=} -\mathrm{D}(\mathrm{d}x^j) = \tfrac{1}{2}S_h{}^j{}_k\,\mathrm{d}x^h \wedge \mathrm{d}x^k, \\ W^P \overset{\text{def}}{=} -\mathrm{D}(\mathrm{D}z^P) = \tfrac{1}{2}M_j{}^P{}_i\,\mathrm{d}x^j \wedge \mathrm{d}x^i + A_i{}^P{}_Q\,\mathrm{d}x^i \wedge \mathrm{D}z^Q + \\ + \tfrac{1}{2}Y_Q{}^P{}_R\,\mathrm{D}z^Q \wedge \mathrm{D}z^R \tag{S.74}$$

(cf. Equation (A.14)). The cyclic and Bianchi identities found in Section B.4 will be comprised by the following elegant *exterior identities*:

$$\mathrm{D}W_h{}^j \overset{\text{def}}{=} \mathrm{d}W_h{}^j + b_k{}^j \wedge W_h{}^k - b_h{}^k \wedge W_k{}^j = 0,$$

$$\mathrm{D}W_P{}^Q \overset{\text{def}}{=} \mathrm{d}W_P{}^Q + b_R{}^Q \wedge W_P{}^R - b_P{}^R \wedge W_R{}^Q = 0,$$

$$\mathrm{D}W^j = -W_h{}^j \wedge \mathrm{d}x^h, \ \mathrm{D}W^P = -W_Q{}^P \wedge \mathrm{D}z^Q.$$

It will be noted that, treating the background manifold of x's as the base space, and the manifold of z's supported by a point x^i as the fibre supported by this x^i, it is appropriate, in view of Equations (S.73)–(S.74), to adopt the following terminology indicating the geometric meaning of studied objects in the obvious and brief way: h-connection coefficients $L_m{}^i{}_n$, v-connection coefficients $Q_P{}^R{}_T$; h-, hv-, vh-, $v(hv)$-, and v-curvature tensors $L_h{}^j{}_{mn}$, $B_h{}^j{}_{kP}$, $E_Q{}^P{}_{mn}$, $Q_P{}^T{}_{Rj}$, and $Z_Q{}^P{}_{MN}$, respectively; h-, vh-, $v(hv)$-, and v-torsion tensors $S_h{}^j{}_k$, $M_j{}^P{}_i$, $A_i{}^P{}_Q$, and $Y_Q{}^P{}_R$, respectively, where the abbreviations h and v mean *horizontal* and *vertical*, respectively. The tensor $A_i{}^P{}_Q = -D_i{}^P{}_Q - \mathrm{d}_Q N_i^P$ (Equation (B.26)) may also be called *the deflection tensor* in the sense that the tensor represents the deflection of the gauge field $D_Q{}^P{}_i$ from the derivative $-\mathrm{d}_Q N_i^P$. The terminology of such a type was frequently used in recent papers devoted to Finsler geometry or its generalizations. Clearly, the Yang–Mills-type gauge field $D_P{}^Q{}_i$ and gauge tensor $E_P{}^Q{}_{mn}$ (or their (x, y)-dependent counterparts $D_p{}^q{}_i(x, y)$ and $E_p{}^q{}_{mn}(x, y)$) have no analogs in proper Finslerian approaches.

B.3. Using the gauge-covariant differential D introduced in the preceding Problem B.2, the covariant derivative of a tensor $w^{jP}(x, z)$ along a curve $C = (x^i(t), z^P(t))$ can be defined as follows:

$$\mathrm{D}w^{jP}/\mathrm{d}t = (\mathrm{D}_i w^{jP})\dot{x}^i + (\mathrm{S}_R w^{jP})a^R \equiv \\ \equiv \mathrm{d}w^{jP}/\mathrm{d}t + (L_k{}^j{}_i w^{kP} + (D_Q{}^P{}_i - N_i^R Q_Q{}^P{}_R)w^{jQ})\dot{x}^i + \dot{z}^R Q_Q{}^P{}_R w^{jQ}, \tag{S.75}$$

where

$$\dot{x}^i \overset{\text{def}}{=} \mathrm{d}x^i/\mathrm{d}t = \mathrm{D}x^i/\mathrm{d}t \tag{S.76}$$

and

$$a^R \overset{\text{def}}{=} \dot{z}^R - N_i^R \dot{x}^i = \mathrm{D}z^R/\mathrm{d}t \tag{S.77}$$

($\dot{z}^R = \mathrm{d}z^R/\mathrm{d}t$) are the covariant derivatives of x^i and z^P along C. The external and internal metric tensors $a_{ij}(x,z)$ and $g_{PQ}(x,z)$ give rise to the following two gauge-invariant Lagrangians:

$$L_{(h)} = (a_{ij}\dot{x}^i\dot{x}^j)^{1/2} \tag{S.78}$$

and

$$L_{(v)} = (g_{PQ}a^P a^Q)^{1/2}, \tag{S.79}$$

which are of the form $L(x, z, \dot{x}, \dot{z})$. Since

$$\delta\int_{(x_1,z_1)}^{(x_2,z_2)} L(x,z,\dot{x},\dot{z})\,\mathrm{d}t = -\int_{(x_1,z_1)}^{(x_2,z_2)} (E_i\delta x^i + E_P\delta^* z^P)\,\mathrm{d}t$$

with

$$E_P = \mathrm{d}(\partial L/\partial\dot{z}^P)/\mathrm{d}t - \partial L/\partial z^P, \tag{S.80}$$

$$E_i = \mathrm{d}(\partial L/\partial\dot{x}^i)/\mathrm{d}t - \partial L/\partial x^i + N_i^P E_P, \tag{S.81}$$

and $\delta^* z^P = \delta z^P - N_i^P\delta x^i$ being the gauge-covariant differential of z^P, we may contend that thus defined Euler–Lagrange derivatives (S.80) and (S.81) are respectively gauge-covariant and gauge-invariant ($E'_P = z_P^Q E_Q$ and $E'_i = E_i$) and hence are of a gauge-invariant meaning. These E_P and E_i may be called respectively *the vertical and horizontal projections of the Euler–Lagrange derivative associated with L*. In our particular cases (S.78) and (S.79), simple direct calculations yield the representations

$$\begin{aligned} L_{(h)}E_{(h)}{}^i = {} & \mathrm{D}\dot{x}^i/\mathrm{d}t + \dot{x}^n a^P a^{im}\mathrm{S}_P a_{mn} + \\ & + a^{ik}\dot{x}^m\dot{x}^n(\mathrm{D}_n a_{km} - \tfrac{1}{2}\mathrm{D}_k a_{mn} + S_{kmn}) - \dot{x}^i(\mathrm{d}L_{(h)}/\mathrm{d}t)/L_{(h)}, \end{aligned} \tag{S.82}$$

$$L_{(h)}E_{(h)P} = -\tfrac{1}{2}\dot{x}^i\dot{x}^j\mathrm{S}_P a_{ij} \tag{S.83}$$

and

$$\begin{aligned} L_{(v)}E_{(v)}{}^P = {} & \mathrm{D}a^P/\mathrm{d}t + \dot{x}^i a^R g^{PT}\mathrm{D}_i g_{TR} - a^Q\dot{x}^i A_{iQ}{}^P + \\ & + g^{PT}a^Q a^R(\mathrm{S}_R g_{TQ} - \tfrac{1}{2}\mathrm{S}_T g_{QR} + Y_{TQR}) - a^P(\mathrm{d}L_{(v)}/\mathrm{d}t)/L_{(v)}, \end{aligned} \tag{S.84}$$

$$L_{(v)}E_{(v)i} = a^P(\dot{x}^n M_{ipn} + A_{iPQ}a^Q) - \tfrac{1}{2}a^P a^Q\mathrm{D}_i g_{PQ} \tag{S.85}$$

which are explicitly of a gauge-invariant meaning. It is seen that the metric conditions (Section B.8) and the torsionless conditions ($S_m{}^n{}_i = 0$ or $Y_P{}^Q{}_R = 0$) will essentially simplify the representations (S.82)–(S.85). The equations $E_{(h)i} = E_{(h)P} = 0$ represent *the horizontal geodesics* (they may also be called *the external geodesics*) and the equations $E_{(v)P} = E_{(v)i} = 0$ *the vertical geodesics* (*the internal geodesics*). If the parameter t is chosen to be *the horizontal arc-length*, that is, $\mathrm{d}t = \sqrt{a_{ij}\mathrm{d}x^i\mathrm{d}x^j}$, we shall have $L_{(h)} = 1$ and hence $\mathrm{d}L_{(h)}/\mathrm{d}t = 0$; similarly, in the case when t is taken to be *the vertical arc-length*, that is $\mathrm{d}t = \sqrt{g_{PQ}\mathrm{D}z^P\mathrm{D}z^Q}$, we obtain $L_{(v)} = 1$ and then $\mathrm{d}L_{(v)}/\mathrm{d}t = 0$ which drops the last term in Equation (S.84). By analogy with the definition given after Equation (1.56), a curve C may be called *autoparallel* with respect to the Lagrangian $L(x, z, \dot{x}, \dot{z})$ if the geodesic equations $E_P = E_i = 0$, when written in terms of the arc-length parameter t defined by $\mathrm{d}t = L(x, z, \mathrm{d}x, \mathrm{d}z)$, tell us that $\mathrm{D}\dot{x}^i/\mathrm{d}t = \mathrm{D}a^P/\mathrm{d}t = 0$. In general, the horizontal or vertical geodesics are not autoparallel, as is seen from Equations (S.82)–(S.85).

B.4. Starting with a Lagrangian $L(x, z, \dot{x}, a)$ possessing the property of the positive homogeneity of degree one in $(\dot{x}^n, a^P)$, that is, $L(x, z, k\dot{x}, ka) = kL(x, z, \dot{x}, a)$, $k > 0$ (the Lagrangians (S.78) and (S.79) obviously belong to such a type), we put $b_P = L\partial L/\partial a^P$ and $c_i = L\partial L/\partial\dot{x}^i$ and assume that these equations can be resolved for a^P and $\dot{x}^i$. Under these conditions, we obtain the generalized Hamiltonian function $H(x, z, c, b) \overset{\mathrm{def}}{=} L(x, z, \dot{x}(x, z, c, b), a(x, z, c, b))$ (cf. the proper Finslerian definition (6.18)), which enables us to introduce *the generalized Hamilton–Jacobi equation*

$$H(x, z, S_i, S_P) = 1 \tag{S.86}$$

by analogy with Equation (6.26); $S_i = D_i S \equiv d_i S$ and $S_P = S_P S \equiv d_P S$ with $S(x, z)$ being a scalar function. Thus we define the generalized momenta $p_i = S_i$ and $p_Q = S_Q$ to postulate the first canonical equations

$$\dot{x}^i = \partial H/\partial p_i, \qquad a^Q \equiv Dz^Q/dt = \partial H/\partial p_Q, \tag{S.87}$$

whereupon we calculate the derivatives dp_i/dt and dp_Q/dt paying attention to the identities $\partial H/\partial x^i + \dot{x}^n \partial S_n/\partial x^i + a^P \partial S_P/\partial x^i = 0$ and $\partial H/\partial z^P + \dot{x}^n d_P S_n + a^Q d_P S_Q = 0$ ensuing from Equation (S.86) (cf. the proof of the Proposition 3 in Section 6.2)). The result will be the following: *the characteristic curves associated with the generalized Hamilton–Jacobi equation* (S.86) *are presented by the generalized canonical equations* (S.87) *and*

$$dp_i/dt = -(\partial H/\partial x^i + N_i^P \partial H/\partial z^P + a^Q p_P D_Q{}^P{}_i) - \dot{x}^n p_Q M_i{}^Q{}_n - a^Q p_P A_i{}^P{}_Q \tag{S.88}$$

$$dp_Q/dt = -\partial H/\partial z^Q - \dot{x}^n p_R d_Q N_n^R \tag{S.89}$$

whose gauge-covariance can readily be verified directly. In general these characteristics are distinguishable from the geodesics as well as from the autoparallel curves treated in the preceding Solution B.3, in contrast to proper Finslerian case where the three types of curves are always identical. It will also be noted that the characteristics given by the equations (S.87)–(S.89) are the curves which are obtainable in the geometric optic approximation (see Equation (S.53)) of the generalized Klein–Gordon equation derivable from the Lagrangian $H^2(x, z, D_i w, S_P w) - M^2 w^2$ (cf. Equation (S.44)), where $w(x, z)$ is a scalar field.

B.5. In the usual Riemannian approach the Lagrangian responsible for the interaction of the electric charge q with the electromagnetic field is of the form $L_{(q)} = qB_i(x)\dot{x}^i$, where $B_i(x)$ is the electromagnetic vector potential. The Riemannian gauge transformation $B_i(x) = B_i'(x) + \partial a(x)/\partial x^i$ gives rise to adding but an exact differential to the integrand of the variational problem $\delta \int L_{(q)}\, dt = 0$, so that the gauge-invariance holds. In the (x, z)-dependent approach, however, the choice $L_{(q)} = qB_i(x, z)\dot{x}^i$ would be inconsistent with the requirement of the gauge-invariance of the variational problem, for the transformation law of $B_i(x, z)$ given by the fifth member of Equations (B.17) will entail $B_i dx^i - B_i' dx^i = d \ln|Z| - (\partial \ln|Z|/\partial z'^P) Dz'^P$, where only the first term in the right-hand side is an exact differential. This circumstance makes us use the vector $D_i(x, z) = B_i(x, z) + D_P{}^P{}_i(x, z)$ (the second member of Equation (B.54)) instead of the gauge field $B_i(x, z)$ to define $L_{(q)} = q\dot{x}^i D_i$, obtaining the following gauge-covariant result: $E_{(q)P} = -q\dot{x}^n S_P D_n$ and then $E_{(q)i} = q(\dot{x}^n T_{ni} + a^P S_P D_i)$, where the Euler–Lagrange derivatives designated by the symbol E are defined in accordance with the rules (S.80) and (S.81) and $T_{ni} \stackrel{\text{def}}{=} d_n D_i - d_i D_n \equiv F_{ni} + Q_R M_n{}^R{}_i$ in terms of the tensors (B.47) and (B.54). Thus, *the gauge-covariant generalization of the usual electromagnetic Lorentz force* $f_i(x, \dot{x}) = q\dot{x}^n(\partial B_n(x)/\partial x^i - \partial B_i(x)/\partial x^n)$ *reads*

$$f_i(x, z, \dot{x}, \dot{z}) = q\dot{x}^n(F_{in} + Q_R M_i{}^R{}_n + a^P S_P D_i), \tag{S.90}$$

so that $f_i'(x, z', \dot{x}, \dot{z}') = f_i(x, z, \dot{x}, \dot{z})$. This "horizontal Lorentz force" may be supplemented by the concept of "the vertical Lorentz force" (the internal Lorentz force)

$$\begin{aligned} E_{(p)P} &= p(D_n Q_P + (S_R Q_P - S_P Q_R)a^P - Q_R \dot{x}^i A_i{}^R{}_P) \equiv \\ &\equiv p(\dot{x}^n(b_{nP} + S_P D_j - Q_R{}^R{}_{Pj}) + (H_{RP} + S_Q{}^Q{}_{PR})a^R), \end{aligned} \tag{S.91}$$

$$\begin{aligned} E_{(p)i} &= p(Q_R \dot{x}^n M_i{}^R{}_n + (Q_Q A_i{}^R{}_P - D_i Q_P)a^P) \equiv \\ &\equiv p(-(b_{iP} + S_P D_i + Q_R{}^R{}_{Pi})a^P + Q_R \dot{x}^n M_i{}^R{}_n) \end{aligned} \tag{S.92}$$

ensuing from the Lagrangian $L_{(p)} = pQ_R a^R$; the notation (B.47) has been used in the second steps.

By analogy with Equations (S.90)–(S.92) we may also define *the gauge-Lorentz force* ($E_{(g)i}$, $E_{(g)P}$) as given by the gauge-invariant Lagrangian

$$L_{(g)} = g A_{iPQ} \dot{x}^i a^P a^Q \tag{S.93}$$

representing the interaction of the internal charge g with the Yang–Mills–type gauge field $D_P{}^Q{}_i$ through the gauge tensor $A_i{}^Q{}_P = -D_P{}^Q{}_i - d_P N_i^Q$ (Equation (B.26)). In case of the Lagrangian

(S.93), simple straightforward calculations yield the following result:

$$\tfrac{1}{g}E_{(g)i} = \dot{x}^n a^P a^Q [\mathrm{D}_n A_{iPQ} - \mathrm{D}_i A_{nPQ} + A_i{}^R{}_P(A_{nRQ} + A_{nQR})] -$$
$$- \dot{x}^j \dot{x}^n a^Q (A_{jRQ} + A_{jQR}) M_n{}^R{}_i + a^P a^Q a^R \mathrm{S}_P A_{iQR} +$$
$$+ (A_{iPQ} + A_{iQP}) a^Q \mathrm{D}a^P/\mathrm{d}t$$

and

$$\tfrac{1}{g}E_{(g)P} = \dot{x}^i a^Q a^R [\mathrm{S}_R(A_{iPQ} + A_{iQP}) - \mathrm{S}_P A_{iRQ}] + \dot{x}^i \dot{x}^n a^Q$$
$$[\mathrm{D}_n(A_{iPQ} + A_{iQP}) - A_n{}^R{}_P(A_{iQR} + A_{iRQ})] +$$
$$+ (A_{iPQ} + A_{iQP})(a^Q \mathrm{D}\dot{x}^i/\mathrm{d}t + \dot{x}^i \mathrm{D}z^Q/\mathrm{d}t),$$

where the expression in the first brackets can be expressed through the Yang–Mills-type gauge tensor $E_P{}^Q{}_{in}$ with the help of the identity (B.57). $E_{(g)i}$ and $E_{(g)P}$ represent respectively the horizontal and vertical projections of the gauge-Lorentz force.

B.6. For the Yang–Mills-type gauge field $D_Q{}^P{}_i$, this can be done by using the gauge tensor $A_i{}^P{}_Q = -D_Q{}^P{}_i - \mathrm{d}_Q N_i^P$ (Equation (B.26)), the coefficient $n_2/2d_3$ in the Euler–Lagrange derivative (B.70″) being in fact the negative of squared mass of the gauge field $D_P{}^Q{}_i$. It should be noted in this respect that the ideas of the gauge field theory currently used in the present-day models of physical field interactions can be traced back to the work by Utiyama (1956) in which the action of gauge transformations on the physical field wave function was treated as the action on only the form of the function, and not on its argument (this method was used in the solution of the Problem 5.10 in case of SU(3)-invariance). Such a treatment, however, is faced with the well-known difficulty (see, e.g., Chaichian and Nelipa (1983); Bilenky *et al.* (1982)) that the mesons associated with the non-Abelian Yang-Mills gauge fields are massive, while the squared gauge field is evidently not a gauge scalar and hence cannot, without violation the gauge-invariance, be introduced in the Lagrangian as the term representing the mass of the gauge field. As a remedy, the interaction of the scalar field with hypothetic Higgs particles is frequently invoked (*op. cit.*). However, in the gauge-geometric approach developed in the Appendix B, where the Utiyama-type approach was generalized in fact by allowing the fields to depend on the coordinates z^P of points of the internal space, and the gauge transformations to affect not only the form of the wave functions but also its argument z^P (in passing, it should be noted that the introduction of the dependence on the internal points is characteristic of the modern Kaluza–Klein theories (Ferrara *et al.* (1983)) we need not hypothesize such an interaction with the purpose of giving the gauge field $D_P{}^Q{}_i$ the mass. Instead, the gauge-invariant mass term $n_2 A_{iPQ} A^{iPQ}$ involves the interaction of $D_P{}^Q{}_i$ with the derivative $\mathrm{d}_Q N_i^P$ of the gauge field N_i^P which is characteristic of the present gauge-geometric approach. Similarly, the electromagnetic gauge fields B_k and H_p as well as the gauge field E_k and the contractions $L_m{}^m{}_k$, $D_P{}^P{}_k$, and $Q_P{}^P{}_R$ all acquire the mass through the gauge-scalar mass term $m_1 L_k L^k + m_2 D_k D^k + m_3 Q_R Q^R$, where L_k, D_k, and Q_R are the gauge tensors given by Equation (B.54). As regards the remaining gauge fields $L_m{}^i{}_n$, N_i^P, and $Q_P{}^Q{}_R$, we are unable in general to make them massive, for there does not exist a gauge tensor containing the term $L_m{}^i{}_n, N_i^P$, or $Q_P{}^Q{}_R$ as an addend (in other words, there are no deflection tensors other than $A_P{}^Q{}_i$, L_k, D_k, and Q_R). This notwithstanding, new possibilities can be offered in particular cases of gauge transformations because specifying the gauge transformations gives rise, as a rule, to new gauge tensors. For example, in the particular case when the homogeneity condition (B.172) is imposed on the gauge transformations, we shall have the identity $z'^R z_R^P = z^P$ which tells us that the z^P becomes the gauge vector. Therefore, the objects $A_i{}^P \overset{\text{def}}{=} -N_i^P - D_Q{}^P{}_i z^Q \equiv -\mathrm{D}_i z^P$ and $Q_Q{}^P \overset{\text{def}}{=} Q_R{}^P{}_Q z^R \equiv \mathrm{S}_Q z^P - \delta_Q^P$ will become gauge tensors (such implications can be continued; for instance, putting $W^P = z^P$ in the commutators of the type (B.34), (B.36), and (B.41) will yield new gauge-covariant identities). Thus, we obtain new deflection gauge tensor $A_i{}^P$, so that the gauge-scalar term $mA_{iP}A^{iP}$ will assign the mass $-(m/2n_1)^{1/2}$ to the gauge field N_i^P. Moreover, in the homogeneous case under consideration we may make the z^P itself the massive gauge vector field by the help of the gauge-scalar mass term $mg_{PQ} z^P z^Q$. Finally, in case of linear gauge transformations (Section B.12) the gauge field $Q_P{}^Q{}_R$ becomes the gauge tensor (Equation (B.185′)) and hence is made massive by introducing the term $mQ^{PRS}Q_{PRS}$ in the Lagrangian.

Bibliography

Adler, R., Bazin, M., and Schiffer, M.: 1975, *Introduction to Relativity*, McGraw-Hill, New York.

Aringazin, A. K. and Asanov, G. S.: 1985, 'Finslerian Post-Riemannian Corrections to the Equations of Geodesics', *General Relativity and Gravitation* **17**, 900–905.

Asanov, G. S.: 1975, 'Electromagnetic Field as Finsler Manifold', *Izvestiya Vysshikh Uchebnykh Zavedenii, Fizika* 1975 (**1**), 86–90.

Asanov, G. S.: 1976, 'Gravitational Field in Finsler Space Based on Notion of Volume', *Vestnick Moskovskogo Universiteta, Seriya Fiziki i Astronomii* **17**, 288–296.

Asanov, G. S.: 1977a, 'The Finslerian Structure of Space-Time Defined by Its Absolute Parallelism', *Annalen der Physik* **34**, 169–174.

Asanov, G. S.: 1977b, 'Motion of the Rest Frame of the Electric Charge Defined by the Finslerian Structure of the Electromagnetic Field', *Reports on Mathematical Physics* **11**, 221–226.

Asanov, G. S.: 1978, 'Scalar and Electromagnetic Fields in Finslerian Space-Time with the Absolute Parallelism', *Reports on Mathematical Physics* **14**, 239–248.

Asanov, G. S.: 1979a, 'New Examples of S3-Like Finsler Spaces', *Reports on Mathematical Physics* **16**, 329–333.

Asanov, G. S.: 1979b, 'On Finslerian Relativity', *Nuovo Cimento* **B49**, 221–246.

Asanov, G. S.: 1979c, 'On 1-Form Finsler spaces', *Preprint No.* 195, *Institute of Mathematics, Polish Ac. Sci.*, 1–34.

Asanov, G. S.: 1980, 'C-reducible Finsler Spaces. Finsler Spaces with Randers and Kropina Metrics', *Problems of geometry*, VINITI Akademii Nauk USSR **11**, 65–88 (in Russian).

Asanov, G. S.: 1981, 'A Finslerian Extension of General Relativity', *Foundations of Physics* **11**, 137–154.

Asanov, G. S.: 1982, 'Variational Principle for the Finslerian Extension of General Relativity', *Aequationes Mathematicae* **24**, 207–229.

Asanov, G. S.: 1983a, 'Gravitational Field Equations Based on Finsler Geometry', *Foundations of Physics* **13**, 501–527.

Asanov, G. S.: 1983b, 'Finslerian Geometrization of Internal Symmetries of Physical Fields', in *Proceedings of* 10*th International Conference on General Relativity and Gravitation, Contributed Papers* **1**, Padova 4–9 July 1983, Consiglio Nazionale Delle Ricerche, Roma, pp. 463–465.

Asanov, G. S.: 1984a, Derivation of the Finslerian Gauge Field Equations', *Annalen der Physik* **41**, 222–227.

Asanov, G. S.: 1984b, 'Examples of Finslerian Metric Tensors of the Space-Time Signature' *Annalen der Physik* **41**, 263–266.

Asanov, G. S.: 1985, 'Gauge Fields in General Parametrical Approach and Their Finslerian Reduction', *Annalen der Physik* **42**, 400–410.

Asanov, G. S. and Kirnasov, E. G.: 1982, 'On Finsler Spaces Satisfying the T-Condition', *Aequationes Mathematicae* **24**, 66–73.

Bargmann, V., Michel, L., and Telegdi, V. L.: 1959, 'Precession of the Polarization of Particles Moving in a Homogeneous Electromagnetic Field', *Phys. Rev. Lett.* **3**, 435–436.

Baumeister, R.: 1978, 'Generalized Hamilton-Jacobi Theories', *J. Math. Phys.* **19**, 2377–2387.

Baumeister, R.: 1979, 'Clebsch Representations and Variational Principles in the Theory of Relativistic Dynamical Systems', *Utilitas Mathematica* **16**, 43–72.

Beem, J. K.: 1970, 'Indefinite Finsler Spaces and Timelike Spaces', *Canad. J. Math.* **22**, 1035–1039.

Beem, J. K.: 1973, 'On the Indicatrix and Isotropy Group in Finsler Spaces with Lorentz Signature', *Lincei Rend.* (8) **54**, 385–392.

Beem, J. K.: 1976, 'Characterizing Finsler Spaces which are Pseudo-Riemannian of Constant Curvature', *Pacific J. Math.* **64**, 67–77.

Beem, J. K. and Kishta, M. A.: 1974, 'On Generalized Indefinite Finsler Spaces', *Indiana Univ. Math. J.* **23**, 845–853.

Bergmann, P. G.: 1962, 'The General Theory of Relativity', in *Handbuch der Physik*, Bank IV, Springer, Berlin, pp. 203–272.

Berwald, L.: 1939, 'II. Invarianten bei der Variation vielfacher Integrale und Parallelhyperflächen in Cartanschen Räumen', *Composito mat.* **7**, 141–176.

Berwald, L.: 1941, 'III. Two-Dimensional Finsler Spaces with Rectilinear Extremals', *Ann. Math.* (2), **42**, 84–112.

Berwald, L: 1946, 'Über Beziehungen zwischen den Theorien der Parallelübertragung in Finslerschen Räumen', *Nederl. Akad. Wetensch. Proc., Ser.* **A49**, 642–647.

Berzi, V. and Gorini, V.: 1969, 'Reciprocity Principle and Lorentz Transformations', *J. Math. Phys.* **10**, 1518–1524.

Bilenky, S. M. and Hosek, J.: 1982, 'Glashow-Weinberg-Salam Theory of Electroweak Interactions and the Neutral Currents', *Physics Reports* **90C**, No. 2, 73-157.

Bogoljubov, N. N. and Shirkov, D. V.: 1959, *Introduction to the Theory of Quantized Fields*, Interscience, New York-London.

Bondi, H.: 1960, Cosmology, Cambridge University Press.

Brickell, F.: 1967, 'A Theorem on Homogeneous Functions', *London Math. Soc.* **42**, 325–329.

Busemann, H.: 1967, *Timelike spaces*, PWN, Warszawa.

Cavalleri, G. and Spinelli, G.: 1980, 'Field-Theoretic Approach to Gravity in the Flat Space-Time', *La Revista del Nuovo Cimento* **3**, N 8, 1–92.

Chaichian, M. and Nelipa, N. F.: 1983, *Introduction to the Gauge Theory*, Springer, Berlin.

Chang, P. and Gürsey, F.: 1967, 'Nonlinear Pion-Nucleon Lagrangians', *Phys. Rev.* **164**, 1752–1761.

Coley, A. A.: 1982, 'Clocks and Gravity', *General Relativity and Gravitation* **14**, 1107–1114.

Deicke, A.: 1953, 'Über die Finsler-Räume mit $A_i = 0$', *Arch. Math.* **4**, 45–51.

Drechsler, W. and Mayer, M.E.: 1977, *Fiber Bundle Techniques in Gauge Theories*, Springer, Berlin.

Einstein, A.: 1928, 'Riemann-Geometrie mit Aufrechterhaltung des Begriffes des Fernparallelismus', *Sitzungsber. preuss. Akad. Wiss., phys.-math. Kl.*, 217–221.

Eisenhart, L. P.: 1927, *Non-Riemannian Geometry*, Amer. math. Soc. Coll. Publ., New York.

Eisenhart, L. P.: 1950, *Riemannian Geometry*, Princeton.

Eliopoulos, H. A.: 1965, 'A Generalized Metric Space for Electromagnetic Theory', *Acad. Roy. Belg., Bull. Cl. Sci.* **51**, 986–995.

Ferrara, S., Taylor, J. G. and van Nieuwenhuizen, P. (ed.): 1983, *Supersymmetry and Supergravity* '82, World Scientific, Singapore.

Finsler, P.: 1918, *Über Kurven und Flächen in allgemeinen Räumen: Dissertation*, Göttingen.

Fock, V. A.: 1959, *Theory of Space, Time and Gravitation*, Pergamon Press, London.

Gasiorowicz, S.: 1966, *Elementary Particle Physics*, John Wiley & Sons, Inc., New York, London, Sydney.

Gold, T. (ed.): 1967, *The Nature of Time*, Cornell University Press, New York.

Hehl, F. W.: *On the Gauge Field Theory of Gravitation*. 3 lectures given at the Dublin Institute for Advanced studies from 18–21 November 1981.

Hlavatý, V.: *Geometry of Einstein's Unified Field Theory*, Noorhoff, Groningen 1957.

Holland, P. R.: 1982, 'Electromagnetism, Particles and Anholonomy', *Phys. Lett.* **91A**, 275–278.

Holland, P. R. and Philippidis, C.: 1984, 'Anholonomic Deformations in the Ether: A Significance for the Potentials in a New Physical Conception of Electrodynamics' (to be published).
Horváth, J. I.: 1958, 'New Geometrical Methods of the Theory of Physical Fields', *Nuovo Cimento* **9**, 444–496.
Horváth, J. I.: 1963, 'Internal Structure of Physical Fields', *Acta physica et chemica*, Szeged 9, 3–24.
Horváth, J. I. and Moór, A.: 1952, 'Entwicklung einer einheitlichen Feldtheorie begründet auf die Finslersche Geometrie', *Zeitsch. Phys.* **131**, 544–570.
Ikeda, S.: 1981a, 'On the Conservation Laws in the Theory of Fields in Finsler Spaces', *J. Math. Phys.* **22**, 1211–1214.
Ikeda, S.: 1981b, 'On the Theory of Fields in Finsler Spaces', *J. Math. Phys.* **22**, 1215–1218.
Infeld, L. and van der Waerden, B. L.: 1933, 'Die Wellengleichung des Elektrons in der allgemeinen Relätivitätstheorie', *Sitzungsber. Preuss. Akad. Wiss. Phys.-Math. Kl.*, 380–390.
Ingarden, R. S.: 1957, 'On the Geometrically Absolute Optical Representation in the Electron Microscope', *Prace Wroclawckiego Tow. Nauk.*, *Wroclaw* **B45**, 1–60.
Ingarden, R. S.: 1976, 'Differential Geometry and Physics', *Tensor* **30**, 201–206.
Ishikawa, H.: 1981, 'Note on Finslerian Relativity', *J. Math. Phys.* **22**, 995–1004.
Izumi, H. and Sakaguchi, T.: 1982, 'Identities in Finsler Space', *Memoirs of the National Defence Academy* **22**, 7–15.
Källen, G.: 1964, *Elementary Particle Physics*, Addison-Wesley, Massachusetts, Palo Alto, London.
Kawaguchi, H.: 1972, 'On Finsler Spaces with the Vanishing Second Curvature Tensor', *Tensor* **26**, 250–254.
Kikkawa, K. (ed.) *Gauge Theory and Gravitation*. Springer-Verlag, Berlin 1983.
Konopleva, N. P. and Popov, V. N.: 1981, *Gauge Fields*, Harwood Academic Publisher, Chur, London, New York.
Kramer, D., Stephani, H., Maccallum, M. and Herlt, E.: 1980, *Exact Solutions of Einstein's Field Equations*, VEB, Berlin.
Kropina, V. K.: 1961, 'Projective Two-Dimensional Finsler Spaces with the Special Metric', *Trudy Sem. Vektor. Tenzor. Anal.* (in Russian) **11**, 277–291.
Landau, L. D. and Lifshitz, E. M.: 1962, *The Classical Theory of Fields*, Addison-Wesley, Reading, Mass.
Lovelock, D. and Rund, H.: 1975, *Tensors, Differential Forms and Variational Principles*, Wiley-Interscience, New York.
Madore, J.: 1981, 'Geometric Methods in Classical Field Theory', *Physics Reports* **75**, No. 3, 125–204.
Mandelshtam, L. I.: 1972, *Lectures in Optics, Relativity Theory and Quantum Mechanics*, Nauka, Moscow (in Russian).
Matsumoto, M.: 1971, 'On Finsler Spaces with Curvature Tensors of some Special Forms', *Tensor* **22**, 201–204.
Matsumoto, M.: 1972a, 'V-Transformations of Finsler Spaces. I.', *J. Math. Kyoto Univ.* **12**, 479–512.
Matsumoto, M.: 1972b, 'On C-reducible Finsler Spaces', *Tensor* **24**, 29–37.
Matsumoto, M.: 1974, 'On Finsler Spaces with Randers Metric and Special Forms of Important Tensors', *J. Math. Kyoto Univ.* **14**, 447–498.
Matsumoto, M.: 1975, 'On Einstein's Gravitational Field Equation in a Tangent Riemannian Space of a Finsler Space', *Reports on Math. Phys.* **8**, 103–108.
Matsumoto, M. and Hōjō, S., 1978, 'A Conclusive Theorem on C-reducible Finsler Spaces', *Tensor* **32**, 225–230.
Matsumoto, M. and Shimada, H. 1978a, 'On Finsler Spaces with 1-form Metric', *Tensor* **32**, 161–169.
Matsumoto, M.: 1978b, 'On Finsler Spaces with 1-form Metric, II: Berwald-Moór's Metric $L=(y^1 y^2 \dots y^n)^{1/n}$', *Tensor* **32**, 275–278.
McKiernan, M. A.: 1966, 'Fatique Spaces in Electromagnetic-Gravitational Theory', *Canad. Math. Bull.* **9**, 489–507.

Meetz, K.: 1969, 'Realization of Chiral Symmetry in a Curved Isospin Space', *J. Math. Phys.* **10**, 589–593.

Mercier, A., Treder, H.-J., and Yourgrau, W.: 1979, *On General Relativity: An Analysis of the Fundamentals of the Theory of General Relativity and Gravitation*, Akademie Verlag, Berlin.

Miron, R.: 1983, 'Metrical Finsler Structures and Metrical Finsler Connections', *J. Math. Kyoto Univ.* **23**, 219–224.

Moór, A.: 1954, 'Ergänzung zu meiner Arbeit: "Über die Dualität von Finslerschen und Cartanschen Räumen"', *Acta Math.* **91**, 187–188.

Okubo, K.: 1979, 'Some Theorems of $S_3(K)$ Metric Spaces', *Reports on Math. Phys.* **16**, 401–408.

Palatini, A.: *Rend. Circ. Mat. Palermo* **43** (1919), 203.

Penrose, R. and Rindler, W.: 1984, *Spinors and Space-Time*, Cambridge University Press, Cambridge.

Pirani, F. A. E.: 1962, 'Gravitational Radiation', in L. Witten (ed.), *Gravitation: Introduction to Current Research*, John Wiley & Sons, Inc., New York, Lond, pp. 199–226.

Randers, G.: 1941, 'On an Assymetrical Metric in the Four-Space of General Relativity', *Phys. Rev.* **59**, 195–199.

Riemann, B.: 1892, '*Über die Hypothesen welche der Geometrie zugrunde liegen. Habilitationsvortrag.*', Ges. math. Werke, Leipzig, pp. 272–287.

Rund, H.: 1959, *The Differential Geometry of Finsler Spaces*, Springer, Berlin.

Rund, H.: 1962, 'Über Finslersche Räume mit speziellen Krümmungseigenschaften', *Monatsh. Math.* **66**, 241–251.

Rund, H. 1966a, 'Hamiltonian Formalism in Relativistic Field Theories', in *Perspectives in geometry and relativity, Halavatý Festschrift*, University of Indiana Press, pp. 328–339.

Rund, H.: 1966b, *The Hamilton-Jacobi Theory in the Calculus of Variations*, D. Van Nostrand, London and New York (revised and augmented edition, Krieger Publications, New York, 1973).

Rund, H.: 1967, 'Generalized Metrics on Complex Manifolds', *Math. Nach.* **34**, 55–77.

Rund, H.: 1972, 'Invariant Equations of Motion in General Relativity', *Tensor* **23**, 365–368.

Rund, H.: 1976, 'Generalized Clebsch Representations on Manifolds', in *Topics in Differential Geometry*, Academic Press, New York, pp. 111–133.

Rund, H.: 1977, 'Clebsch Potentials and Variational Principles in the Theory of Dynamical Systems', *Archive for Rational Mechanics and Analysis* **65**, 305–344.

Rund, H.: 1979, 'Clebsch Representations and Relativistic Dynamical Systems', *Archive for Rational Mechanics and Analysis* **67**, 111–122.

Rund, H.: 1980, Connection and Curvature Forms Generated by Invariant Variational Principles', *Resultate der Mathematik* **3**, 74–120.

Rund, H.: 1981a, *The Differential Geometry of Finsler Spaces*, Russian translation, Nauka, Moscow.

Rund, H.: 1981b, 'An Extension of Noether's Theorem to Transformations Involving Position-Dependent Parameters and Their Derivatives', *Foundations of Physics* **11**, 809–838.

Rund, H.: 1982, 'Differential-Geometric and Variational Background of Classical Gauge field Theories', *Aequationes Mathematicae* **24**, 121–174.

Rund, H.: 1983, 'Invariance Identities Associated with Finite Gauge Transformations and the Uniqueness of the Equations of Motion of a Particle in a Classical Gauge Field', *Foundations of Physics* **13**, 93–114.

Rund, H. and Beare, J. H.: 1972, *Variational Properties of Direction-Dependent Metric Fields*, University of South Africa, Pretoria.

Sachs, M., 1982, *General Relativity and Matter, D.* Reidel, Dordrecht.

Shibata, C.: 1978a, 'On Finsler Spaces with Kropina Metric', *Reports on Math. Phys.* **13**, 117–128.

Shibata, C.: 1978b, 'On the Curvature Tensor R_{hijk} of Finsler Spaces of Scalar Curvature', *Tensor* **32**, 311–317.

Slebodziński, W.: 1970, *Exterior Forms and Their Applications*, PWN, Warszawa.

Stephenson, G.: 1953, 'Affine Field Structure of Gravitation and Electromagnetism', *Nuovo Cimento* **10**, 354–355.

Stephenson, G.: 1957, 'La géometrie de Finsler at les théories du champ unifie', *Ann. Inst. H. Poincaré* **15**, 205–215.

Stephenson, G., and Kilmister, C. W.: 1953, 'A Unified Field Theory of Gravitation and Electromagnetism', *Nuovo Cimento* **10**, 230–235.

Sternberg, S.: 1964, *Lectures on Differential Geometry*, Prentice Hall. Inc. Englewood Cliffs, N.J.

Synge, J. L.: 1936, *Tensorial Methods in Dynamics*, University of Toronto.

Synge, J. L.: 1960, *Relativity: the General Theory*, North-Holland, Amsterdam.

Szabó, Z. I.: 1981, 'Positive Definite Finsler Spaces Satisfying the T-condition are Riemannian', *Tensor* **35**, 247–248.

Takano, Y.: 1964, 'Theory of Fields in Finsler Spaces', *Progress of Theoretical Physics* **32**, 365–366.

Takano, Y.: 1982, 'Gauge Fields in Finsler Spaces', *Lett. Nuovo Cimento* **35**, 213–217.

Törnebohm, H.: 1970, 'Two Studies Concerning the Michelson-Morley Experiment', *Foundations of Physics* **1**, 47–56.

Trautman, A.: 1962, 'Conservation Laws in General Relativity', in L. Witten (ed.), *Gravitation: Introduction to Current Research*, John Wiley & Sons, Inc., New York, London.

Treder, H. J.: 1971, *Gravitationstheorie und Äquivalenzprinzip*, Akademie Verlag, Berlin.

Treder, H.-J., Borzeskowski, H., v. der Merwe, Al. and Yourgrau, W.: 1980, *Fundamental Principles of General Relativity Theories, Local and Global Aspects of Gravitation and Cosmology*, Plenum Press, New York, London.

Utiyama, R.: 1956, 'Invariant Theoretical Interpretation of Interaction', *Phys. Rev.* **101**, 1597–1607.

Vagner, V. V.: 1949, 'The Geometry of Finsler as a Theory of the Field of Local Hypersurfaces in X_n', *Trudy Sem. Vektor. Tenzor. Anal.* (In Russian) **7**, 65–166.

Vanstone, J. R.: 1963, 'The Hamilton-Jacobi Equation for a Relativistic Charged Particle', *Canad. Math. Bull.* **6**, 341–349.

Watanabe, S. and Ikeda, F.: 1981, 'On some Properties of Finsler Spaces Based on the Indicatrices', *Publ. Math. Debrecen* **28**, 129–136.

Weinberg, S.: 1968, 'Nonlinear Realizations of Chiral Symmetry', *Phys. Rev.* **166**, 1568–1577.

Whittaker, E. T.: 1960, *A History of the Theories of Aether and Electricity*, Harper, New York.

LIST OF PUBLICATIONS ON FINSLER GEOMETRY*

Agnihotri, A. K.: 1975, Cs-curves in the hypersurface of a Finsler space, Ranchi Univ. Math. J. 6, 6-11.
Agnihotri, A. K.: 1976, Curvature of Cs-curves, Ranchi Univ. Math. J. 7, 46-49.
Agnihotri, A. K.: 1977, Torsion of Cs-curves, Ranchi Univ. Math. J. 8, 49-56.
Agrawal, P.: 1970, On special curves of a subspace of a Finsler space, Indian J. Pure Appl. Math. 1, 354-359.
Agrawal, P.: 1971, On K_λ-curvature of a vector field in Finsler space, Indian J. Pure Appl. Math. 2, 80-86.
Agrawal, P.: 1972, On the concircular geometry in Finsler spaces, Tensor, N. S. 23, 333-336.
Agrawal, P.: 1973a, On K_λ-curves and hyperasymptotic curves of a subspace in a Finsler space, Tensor, N. S. 27, 177-182.
Agrawal, P.: 1973b, On hyperasymptotic and hypernormal curves, Tensor, N. S. 27, 16-20.
Agrawal, P. and Behari, R.: 1970, Relative curvature tensors in a subspace of a Finsler space, J. Math. Sci. 5, 23-34.
Agrawal, P. and Das, A.: 1974, Indefinite Finsler manifold F_n and orthonormal ennuples, Tensor, N. S. 28, 53-58.
Aikou, T.: 1981, On Generalized Berwald Spaces: Thesis, Kagoshima, 49 p.
Aikou, T. and Hashiguchi, M.: 1981, On the paths in generalized Berwald spaces, Rep. Fac. Sci. Kagoshima Univ. (Math. Phys. Chem.) 14, 1-8.
Akbar-Zadeh, H.: 1954, Sur la réductibilité d'une variété finslérienne, C. R. Acad. Sci. Paris 239, 945-947.
Akbar-Zadeh, H.: 1956, Sur les isométries infinitésimales d'une variété finslérienne, C. R. Acad. Sci. Paris 242, 608-610.
Akbar-Zadeh, H.: 1957, Sur une connexion euclidienne d'espace d'éléments linéaires, C. R. Acad. Sci. Paris 245, 26-28.
Akbar-Zadeh, H.: 1958, Sur une connexion coaffine d'espace d'éléments linéaires, C. R. Acad. Sci. Paris 247, 1707-1710.
Akbar-Zadeh, H.: 1961a, Sur les espaces de Finsler isotropes, C. R. Acad. Sci. Paris 252, 2061-2063.
Akbar-Zadeh, H.: 1961b, Transformations infinitésimales conformes des variété finslériennes compactes, C. R. Acad. Sci. Paris 252, 2807-2809.
Akbar-Zadeh, H.: 1963, Les espaces de Finsler et certaines de leurs généralisations, Ann. Sci. Ecole Norm. Sup. (3) 80, 1-79.
Akbar-Zadeh, H.: 1964, Une généralisation de la géométrie finslérienne, in Topologie et et Geometrie Differentielle, Inst. H. Poincare, Paris, pp. 1-9.
Akbar-Zadeh, H.: 1965, Sur les automorphismes de certaines structures presque cosymplectiques, Canad. Math. Bull. 8, 39-57.
Akbar-Zadeh, H.: 1966, Sur les homothéties infinitésimales des variétés finslériennes, C. R. Acad. Sci. Paris 262, 1058-1060.
Akbar-Zadeh, H.: 1967, Sur quelques théorémes issus du calcul des variations, C. R. Acad. Sci. Paris A 264, 517-519.
Akbar-Zadeh, H.: 1968, Sur les sous-variétés des variétés finslériennes, C. R. Acad. Sci. Paris A 266, 146-148.
Akbar-Zadeh, H.: 1969, Sur les invariants conformes des variétés finslériennes, C. R. Acad. Sci. Paris A 268, 402-404.
Akbar-Zadeh, H.: 1971, Sur le noyau de l'opérateur de courbure d'un variété finslérienne, C. R. Acad. Sci. Paris A 272, 807-810.
Akbar-Zadeh, H.: 1972a, Espaces de nullité de certains opérateurs en géométrie des sous-variétés, C. R. Acad. Sci. Paris A 274, 490-493.
Akbar-Zadeh, H.: 1972b, Espaces de nullité en géométrie finslérienne, Tensor, N. S. 26, 89-101.
Akbar-Zadeh, H.: 1974, Sur les isométries infinitésimales d'une variété finslérienne compacte, C. R. Acad. Sci. Paris A 278, 871-874.
Akbar-Zadeh, H.: 1975a, Sur les transformations infinitésimales projectives des variétés finslériennes compactes, C. R. Acad. Sci Paris A 280, 591-593.
Akbar-Zadeh, H.: 1975b, Transformations infinitésimales projectives des variétés finslériennes compactes, C. R. Acad. Sci. Paris A 280, 661-663.
Akbar-Zadeh, H.: 1975c, Transformations infinitésimales conformes des variétés finslériennes compactes, C. R. Acad. Sci. Paris A 281, 655-657.

* This comprehensive bibliography was provided unexpectedly by the author at a late proof stage. It was decided to include it in typewritten form so as not to delay publication.

Akber-Zadeh, H.: 1977, Remarques sur les isométries infinitésimales d'une variété finslérienne compacte, C. R. Acad. Sci. Paris A 284, 451-453.

Akbar-Zadeh, H.: 1978, Sur les transformations infinitésimales conformes de certaines variétés finslériennes compactes, C. R. Acad. Sci. Paris A 286, 177-179.

Akbar-Zadeh, H.: 1979a, Transformations infinitésimales conformes des variétés finslériennes compactes, Ann. Polon. Math. 36, 213-229.

Akbar-Zadeh, H.: 1979b, Sur les espaces de Finsler isotropes, C. R. Acad. Sci. Paris A 288, 53-56.

Akbar-Zadeh, H.: 1979c, Opérateurs elliptiques sur le fibré unitaire tangent à une variété finslérienne, C. R. Acad. Sci. Paris A 289, 405-408.

Akbar-Zadeh, H.: 1980a, Opérateurs elliptiques sur le fibré unitaire finslérien, Sem. Math. Pure, Univ. Catholique de Louvain, No. 102, 1-10.

Akbar-Zadeh, H.: 1980b, Fibré unitaire finslérien et groupes de transformations, C. R. Acad. Sci. Paris A 291, 595-598.

Akbar-Zadeh, H.: 1981, Champ de vecteurs de Liouville sur le fibré unitaire finslérien, C. R. Acad. Sci. Paris A 293, 593-595.

Akbar-Zadeh, H.: 1982, Remarques sur les espaces de Finsler isotropes, C. R. Acad. Sci. Paris A 294, 239-242.

Akbar-Zadeh, H.: 1983, Sur les espaces de Finsler isotropes à courbures non positives, C. R. Acad. Sci. Paris A 296, 73-76.

Akbar-Zadeh, H. and Bonan, E.: 1964, Structure presque kählérienne naturelle sur le fibré tangent à une variété finslérienne, C. R. Acad. Sci. Paris 258, 5581-5582.

Akbar-Zadeh, H. and Wegrzynowska, A.: 1976, Sur la géométrie du fibré tangent à une variété finslérienne, C. R. Acad. Sci. Paris A 282, 325-328.

Akbar-Zadeh, H. and Wegrzynowska, A.: 1979, Sur la géométrie du fibré tangent à une variété finslérienne compactes, Ann. Polon. Math. 36, 231-244.

Albu, I. P. and Opris, D.: 1981, Géométrie des spineurs de Finsler, Ann. Univ. Timisoara, Sti. Mat. 19, 5-22.

Albu, I. P. and Opris, D.: 1982a, On the differential geometry of Finsler tangent bundle. I, Proc. Nat. Sem. Finsler Spaces, Brasov, pp. 17-24.

Albu, A. C. and Opris, D.: 1982b, Holonomy groups of continuous connections. Finsler continuous connections, Tensor, N. S. 39, 105-112.

Alt, F.: 1937, Dreiecksungleichung und Eichkörper in verallgemeinerten Minkowskischen Räumen, Ergebn. Math. Kolloq. 8, 32-33.

Amari, S.: 1962, A theory of deformations and stresses of ferromagnetic substances by Finsler geometry, RAAG Mem. No. 3D, 193-214.

Amari, S.: 1964, On the analysis of the plastic structures of ferromagnetic substances by Finsler geometry, Japan Material Sci. 1, 27-34.

Anastasiei, M.: 1978, Transformations of Finsler connections, Proc. Colloq. Geometry and Topology, Cluj-Napoca (Rom.), pp. 82-89.

Anastasiei, M.: 1980, Finsler geometric objects and their Lie derivatives, Proc. Nat. Sem. Finsler Spaces, Brasov, pp.11-25.

Anastasiei, M.: 1981, Some tensorial Finsler structures on the tangent bundle, Anal. Sti. Univ. Iasi Ia 27, Suppl. 9-16.

Anastasiei, M.: 1982, Some existence theorems in Finsler geometry, Proc. Nat. Sem. Finsler Spaces, Brasov, pp. 25-33.

Anastasiei, M. and Popovici, I.: 1980, An intrinsic characterization of Finsler connections, Proc. Nat. Sem. Finsler Spaces, Brasov, pp. 27-39.

Andricioaei, G.: 1976, Sur les espaces Finsler bi-récurrents, Bul. Inst. Politeh. Iasi I 22 (26) 61-64.

Andricioaei, G.: 1977, Sur les tenseurs projectifs dans les espaces Finsler (u, a)-récurrents et (v,b)-récurrents, Bul. Inst. Politeh. Iasi I, 23 (27), 59-61.

Andricioaei, G. and Chinea, I.: 1978, Invariants des groups de transformation de connexion finslérienne conforme presque symplectique, Bul. Inst. Politeh. Iasi I 24 (28), 59-64 (in Roumanian).

Anh, K. Q.: 1980, The configuration Myller M(C, ζ^i_j, T^m, y^i). Applications to the study of the submanifolds F_m in F_m in F_n, Rev. Roum. Math. Pures Appl. 25, 67-75.

Anosov, D. V.: 1975, Geodesic in Finslerian geometry, Proc. Intern. Congr. Math. Vancouver, II, 293-297 (in Russian). English translation: 1977, Amer. Math. Soc. Transl. (2) 109, 81-85.

Arazoza, R. H.: 1982, Structural equations and tensors in fibred manifolds of Finsler spaces, Ciens. Mat. Habana 3, 3-18.

Aronszajn, N.: 1937, Sur quelques problèmes concernant les espaces de Minkowski et les espaces vectoriels généraux, Lincei Rend. (6) 26, 374-376.

Asanov, G. S.: 1975a, Finslerian method of investigation of the noninertial reference frame of the electric charge, Izv. Vyss. Uchebn. Zaved., Fizika, No. 5, 127-129 (in Russian).

Asanov, G. S.: 1975b, Projective invariance anf fifth coordinate, Izv. Vyss. Uchebn. Zaved., Fizika, No. 7, 71-75 (in Russian).

Asanov, G. S.: 1975c, Finslerian geometry of locally anisotropic space-time, VI All-Union Geometrical Congr., Vil'nyus, pp. 19-21 (in Russian).

Asanov, G. S.: 1976a, Observables in the general theory of relativity. II Finslerian approach, Izv. Vyss. Uchebn. Zaved., Fizika, No. 7, 84-88 (in Russian).

Asanov, G. S.: 1976b, A Finslerian substructure of the Riemannian structure of space-time, in Modern Theoretical and Experimental Problems of the Theory of Relativity and Gravitation, Minsk University, pp. 251-253.

Asanov, G. S.: 1977a, Finsler spaces with the algebraic metric defined by the field of the ennuples, Problems of Geometry 8, 67-87, 279 (in Russian). English translation: 1980, J. Sov. Math. 13, 588-600.

Asanov, G. S.: 1977b, Local anisotropic effects of the space-time which are defined by the Finslerian structure, in Problems of the Theory of Gravitation and Elementary Particles, Atomizdat, Moscow 8, 37-43 (in Russian).

Asanov, G. S.: 1977c, Structure of the scalar and electromagnetic fields on the tangent bundle of the space-time with the absolute parallelism, Vestnik Moskows. Univ., Fizika, Astronimija 18, No. 6, 19-24 (in Russian).

Asanov, G. S.: 1978a, Some consequences of the Finslerian structure of the space-time with the absolute parallelism, Reports on Math. Phys. 13, 13-23.

Asanov, G. S.: 1978b, On invariant meaning of the concept of the world time reference frame, Izv. Vyss. Uchebn. Zaved., Fizika, No. 2, 49-52 (in Russian).

Asanov, G. S. 1978c, Finslerian transportation of the frame of reference of the electric charge, in Problems of the Theory of Gravitation and Elementary Particles, Atomizdat, Moscow 9, 140-144 (in Russian).

Asanov, G. S.: 1979a, Basic principles of Finslerian relativity. I, II, Izv. Vyss. Uchebn. Zaved., Fizika, No. 7, 58-62, 104-107 (in Russian).

Asanov, G. S.: 1980a, On Finslerian generalization of the hypothesis of Riemannian geodesics, in Abstracts of Contributed Papers for the Discussion Groups, v. 1. IX Intern. Conf. General Relativity and Gravitation, Fridrich Schiller Univ., Jena, pp. 152-153.

Asanov, G. S.: 1980b, On Finslerian relativity theory. I, Izv. Vyss. Uchebn. Zaved., Fizika, No. 4, 49-51 (in Russian).

Asanov, G. S.: 1980c, On Finslerian relativity theory. II, Izv. Vyss. Uchebn. Zaved., Fizika, No. 5, 25-27 (in Russian).

Asanov, G. S.: 1981a, 'C-reducible Finsler spaces. Finsler spaces with Randers and Kropina metrics, J. Sov. Math. 17, 1610-1624.

Asanov, G. S.: 1981b, On special Finsler spaces. In Finslerian extension of general relativity, in Rund, H.: The Differential Geometry of Finsler Spaces, Russian translation, Nauka, Moscow, pp.398-471.

Asanov, G. S.: 1981c, A relationship between Finslerian geodesics and field equations. I, Izv. Vyss. Uchebn. Zaved., Fizika, No. 2, 70-74.

Asanov, G. S.: 1981d, A relationship between Finslerian geodesics and field equations. II', Izv. Vyss. Uchebn. Zaved., Fizika, No. 10, 19-22 (in Russian).

Asanov, G. S.: 1981e, On Finslerian generalization of the hypothesis of Finslerian geodesics for the notion of test bodies. I, Izv. Vyss. Uchebn. Zaved., Fizika, No. 9, 87-91 (in Russian).

Asanov, G. S.: 1981f, On Finslerian generalization of the hypothesis of Finslerian geodesics for the notion of test bodies. II, Izv. Vyss. Uchebn. Zaved., Fizika, No. 10, 19-22 (in Russian).

Asanov, G. S.: 1981g, Covariant law of conservation of the energy-momentum of the classical fields from the standpoint of the Clebsch representation and invariance identities, in Modern Theoretical and Experimental Problems of Relativity Theory and Gravitation, Moscow University, p. 53 (in Russian).

Asanov, G. S.: 1981h, On Finslerian generalization of relativity theory, in Modern Theoretical and Experimental Problems of Relativity Theory and Gravitation, Moscow University, p. 139 (in Russian).

Asanov, G. S.: 1983a, A Finslerian geometrization of isotopic invariance, Space Res. in the Ukraine 17, 34-35 (in Russian).

Asanov, G. S.: 1983b, A possible unification of gravitational and strong interactions on the basis of Finslerian geometrization of the space-time, Space Res. in the Ukraine 17, 35-36 (in Russian).

Asanov, G. S.: 1984a, Finslerian formulation of the equations of physical fields with internal symmetries, Izv. Vyss. Uchebn. Zaved., Fizika, No. 8, 37-41 (in Russian).

Asanov, G. S.: 1984b, Finslerian kinematic relativistic effects, Izv. Vyss. Uchebn. Zaved., Fizika, No. 8, 41-44 (in Russian).

Asanov, G. S.: 1984c, Finslerian kinematic relativistic effects, in Problems of the Theory of Gravitation and Elementary Particles, Atomizdat, Moscow 15, pp. 11-15 (in Russian).

Asanov, G. S.: 1985a, Finsler spaces, in Mathematical Encyclopedia, v. 5, Sovetskaja Encyklopedija, Moscow, pp. 622-623 (in Russian).
Asanov, G. S.: 1985b, On Finslerian generalization of gravitational field equations, in Problems of the Theory of Gravitation and Elementary Particles, Energoatomizdat, Moscow 16, pp. 22-33 (in Russian).
Asanov, G. S. and Kirnasov, E. G.: 1984, On 1-form Finsler spaces, Reports on Math. Phys. 19, 303-319.
Atanasiu, G.: 1979, Sur les transformations de connexions Finsler conformes métriques, Bul. Univ. Brasov C21, 9-16.
Atanasiu, G.: 1980a, Structures et connexions Finsler presque complexes et presque Hermitiénnes, Proc. Nat. Sem. Finsler Spaces, Brasov, pp. 41-53.
Atanasiu, G.: 1980b, Connexions de Finsler générales presque cosymplectiques,C. R. Acad. Sci. Paris A 290, 969-972.
Atanasiu, G.: 1981, Über die Finslerschen Räume mit konform verallgemeinerter Struktur, Manuscripta Math. 35, 165-171.
Atanasiu, G.: 1982a, Variétés différentiables douées de couples de structures Finsler. Proc. Nat. Sem. Finsler Spaces, Brasov, pp. 35-67.
Atanasiu, G.: 1982b, Über die Finslerschen Räume mit verallgemeinerter Metrik, J. Geom. 19, 1-17.
Atanasiu, G., Banu, O., Cazacu, M. and Coca, M.: 1979a, Couples de connexions Finsler conformes presque symplectiques, Bul. Univ. Brasov, C21, 17-27 (in Roumanian).
Atanasiu, G., Banu, O., Cazacu, M. and Coca, M.: 1979b, Tenseurs invariants aux transformations de connexions Finsler conformes presque symplectiques, Bul. Univ. Brasov, C21, 29-34 (in Roumanian).
Atanasiu, G. and Ghinea, I.: 1979, Connexions Finslériennes générales presque symplectiques, Anal. Sti. Univ. Iasi 25, Supl. 11-15.
Atanasiu, G. and Klepp, F. C.: 1982a, Invariant tensors for transformations of almost complex Finsler connections, Bul. Sti. Teh. Inst. Politeh. Timisoare, Mat.-Fiz. 26(40) (1981), 37-40.
Atanasiu, G. and Klepp, F. C.: 1982b, Almost contact Finsler connections, Bul. Sti. Teh. Inst. Politeh. Timisoare, Mat.-Fiz. 26(40) (1981), 45-51.
Atanasiu, G. and Stoica, E.: 1982, Almost Hermitian hyperbolic Finsler structures and connections, Proc. Nat. Sem. Finsler Spaces, Brasov, pp. 69-73.
Atkin, C.J. 1978, Bounded complete Finsler structures I., Studia Math. 62, 219-228.
Auslander, L.: 1955a, On curvature in Finsler geometry, Trans. Amer. Math. Soc. 79, 378-388.
Auslander, L.: 1955b, The use of forms in variational calculations, Pacific J. Math. 5, 5, 853-859.
Auslander, L.: 1956, Remark on the use of forms in variational calculations, Pacific J. Math. 6, 209-210.
Awasthi, G. D.: 1974, A study of certain special Finsler spaces, Univ. Nac. Tucumán, Rev. A24, 163-166.
Awasthi, G. D.: 1980, On some curve in a subspace of a Finsler space, C. R. Acad. Bulgare Sci. 33, 893-896.
Awasthi, G. D. and Shukla, A. K.: 1982a, On some curve in a subspace of a Finsler space, Demonstratio Math. 15, 311-318.
Awasthi, G. D. and Shukla, A. K.: 1982b, On subspaces of subspaces of a Finsler space, Demonstratio Math. 15, 571-578.
Barbilian, D.: 1960, Les J-géométries naturelles finslériennes, Acad. R. P. Romine. Stud. Cerc. Mat. 11, 7-47 (in Roumanian).
Barbilian, D. and Radu, N.: 1962, Les J-métriques finslériennes naturelles et la fonction de représentation de Riemann, Acad. R. P. Romine. Stud. Cerc. Mat. 13, 21-36 (in Roumanian).
Barthel, W.: 1953a, Zum Inhaltsbegriff in der Minkowskischen Geometrie, Math. Z. 58, 358-375.
Barthel, W.: 1953b, Über eine Parallelverschiebung mit Längeninvarianz in lokal-Minkowskischen Räumen, I, II, Arch. Math. 4, 346-365.
Barthel, W.: 1953c, Über Minkowskische und Finslersche Geometrie, Conveg. Geom. Diff., Roma, 71-76.
Barthel, W.: 1954a, Zur Flächentheorie in Finslerschen Räumen, Proc. Intern. Math. Congr. Amsterdam, 194-195.
Barthel, W.: 1954b, Über die Minimalflächen in gefaserten Finslerraümen, Ann. Mat. Pura Appl. (4) 36, 159-190.
Barthel, W.: 1955a, Variationsprobleme der Oberflächenfunktion in der Finslerschen Geometrie, Math. Z. 62, 23-36.
Barthel, W.: 1955b, Extremalprobleme in der Finslerschen Inhaltsgeometrie, Ann. Univ. Sarav. Naturwiss. Sci. 3-IV, 171-183.
Barthel, W.: 1959a, Zur isodiametrischen und isoperimetrischen Ungleichung in der Relativgeometrie, Comm. Math. Helv. 33, 241-257.

Barthel, W.: 1959b, Zur Minkowski-Geometrie, begründet auf dem Flächeninhaltsbegriff Monatsh. Math. 63, 317-343.
Barthel, W.: 1959c, Natürliche Gleichungen einer Kurve in der metrischen Differentialgeometrie, Arch. Math. 10, 392-400.
Barthel, W.: 1963, Nichtlineare Zusammenhänge und deren Holonomiegruppen, J. Reine Angew. Math. 212, 120-149.
Barthel, W.: 1978, Nichtlineare Differentialgeometrie, insbesondere Minkowski-Geometrie, Contributions to Geometry, Proc. Symp., Stegen, 301-312.
Barthel, W.: 1981, Die Verallgemeinerung Christoffelscher Zusammenhänge in der nichtlinearen Differentialgeometrie, in E. B. Christoffel, Birkhäuser, Basel-Boston, Massachusetts, pp. 611-623.
Barthel, W. and Pabel, H.: 1978, Zur Kurventheorie in der vierdimensionalen metrischen Differentialgeometrie, Result. Math. 1, 1-41.
Báscó, S.: 1983, Eine Charakterisierung der Finslerschen Räume von konstanter Krümmung, Acta Math. Hungar. 42, 233-236.
Basilova, V. I.: 1974, An example of a Minkowski geometry with a non-symmetric indicatrix, in Differential Geometry, No. 1, Saratov University, pp. 93-99 (in Russian).
Beem, J. K.: 1970, Indefinite Minkowski spaces, Pacific. J. Math. 33, 29-41.
Beem, J. K.: 1971, Motions in two-dimensional indefinite Finsler spaces, Indiana Univ. Math. J. 21, 551-555.
Beem, J. K.: 1975, Symmetric perpendicularity for indefinite Finsler metrics on Hilbert manifolds, Geom. Dedicata 4, 45-49.
Beem, J. K.: 1976, Characterizing Finsler spaces which are pseudo-Riemannian of constant curvature, Pacific J. Math. 64, 67-77.
Beem, J. K. and Woo, P. Y.: 1969, Doubly timelike surfaces, Mem. Amer. Math. Soc. 92, 1-115.
Behari, R. and Prakash, N.: 1960, A study of normal curvature of a vector field in Minkowskian Finsler space, J. Indian Math. Soc., N. S. 24, 443-456.
Bernstein, I. N. and Gerver, M. L.: 1978, A problem on integral geometry for a family of geodesics and an inverse kinematic seismics problem, Dokl. Akad. Nauk SSSR 243, 302-305 (in Russian).
Berwald, L.: 1919/1920, Über die erste Krümmimg der Kurven bei allgemeiner Massbestimmung, Lotos Prag 67/68, 52-56.
Berwald, L.: 1926a, Über Parallelübertragung in Räumen met allgemeiner Massbestimmung, Jber. Deutsch. Math.-Verein. 34, 213-220.
Berwald, L.: 1926b, Untersuchung der Krümmung allgemeiner metrischer Räume auf Grund des in ihnen herrschenden Parallelismus, Math. Z. 25, 40-73.
Berwald, L.: 1926b, Zur Geometrie ebener Variationsprobleme, Lotos Prag 74, 43-52.
Berwald, L.: 1927a, Untersuchung der Krümmung allgemeiner metrischer Räume auf Grund des in ihnen herrschenden Parallelismus, Math. Z. 26, 176.
Berwald, L.: 1927b, Über zweidimensionale allgemeine metrische Räume. I, II, J. Reine Angew. Math. 156, 191-222.
Berwald, L.: 1927c, Sui differenziali secondi covarianti, Lincei Rend. (6) 5, 763-768.
Berwald, L.: 1928a, Parallelübertragung in allgemeinen Räumen, Atti Congr. Intern. Mat. Bologna 4, 263-270.
Berwald L.: 1928b, Una forma normale invariante della seconda variazione, Lincei Rend. (6) 7, 301-306.
Berwald, L.: 1929a, Über eine charakteristische Eigenschaft der allgemeinen Räume konstanter Krümmung mit geradlinigen Extremalen, Monatsh. Math. Phys. 36, 315-330.
Berwald, L.: 1929b, Über die n-dimensionalen Geometrien konstanter Krümmung, in denen die Geraden die kürzesten sind, Math. Z. 30, 449-469.
Berwald, L.: 1935, Über Finslersche und verwandte Räume, Cas. Mat. Phys. 64, 1-16.
Berwald, L.: 1936, Über die Hauptkrümmungen einer Fläche in dreidimensionalen Finslerschen Raum, Monatsh. Math. Phys. 43, 1-14.
Berwald, L.: 1941a, Über Finslersche und Cartansche Geometrie I. Geometrische Erklärungen der Krümmung und des Hauptskalars eines zweidimensionalen Finslerschen Räumes, Math. Timisoara 17, 34-58.
Berwald, L.: 1941b, On Finsler and Cartan geometries. III. Two-dimensional Finsler spaces with rectilinear extremals, Ann. of Math. (2) 42, 84-112.
Berwald, L.: 1946a, Über die Beziehungen zwischen den Theorien der Parallelübertragung in Finslerschen Räumen, Nederl. Akad. Wetensch. Proc. 49, 642-647.
Berwald, L.: 1946b, Über die Beziehungen zwischen den Theorien der Parallelübertragung in Finslerschen Räumen, Indag. Math. 8, 401-406.
Berwald, L.: 1947a, Über Systeme von gewöhnlichen Differentialgleichungen zweiter Ordnung deren Integralkürven mit dem System der geraden Linien topologisch aequivalent sind, Ann. of Math. (2), 48, 193-215.
Berwald, L.: 1947b, Über Finslersche und Cartansche Geometrie IV. Projektivkrümmung allgemeiner affiner Räume und Finslersche Räume skalarer Krümmung, Ann. of Math. (2) 48, 755-781.

Bielecki, A. and Golab, S.: 1945, Sur un problème de la métrique angulaire dans les espaces de Finsler, Ann. Soc. Polon. Math. 18, 134-144.
Blaschke, W.: 1912, Über die Figuratrix in der Variationsrechnung, Arch. Math. Phys. 20, 28-44.
Blaschke, W.: 1920, Geometrische Untersuchungen der Variationsrechnung, I. Über Symmetralen, Math. Z. 6, 281-285.
Blaschke, W.: 1936a, Integralgeometrie 11. Zur Variationsrechnung, Abh. Math. Sem. Hamburg Univ. 11, 359-366.
Blaschke, W.: 1936b, Integralgeometrie 12. Über vollkomene optische Instrumente, Abh. Math. Sem. Hamburg Univ. 11, 409-412.
Blaschke, W.: 1954, Zur Variationsrechnung, Rev. Univ. Istanbul 19, 106-107.
Bliss, G. A.: 1906, A generalization of the notion of angle, Trans. Amer. Math. Soc. 7, 184-196.
Bliss, G. A.: 1915, Generalizations of geodesic curvature and a theorem of Gauss concerning geodesic triangles, Amer. J. Math. 37, 1-18.
Bliznikas, V. I.: 1959, On theory of curves in a metric space of line-elements, Dokl. Akad. Nauk SSSR, N. S. 127, 9-12 (in Russian).
Bliznikas, V. I.: 1960a, On the differential geometry of bilinear-metric spaces of linear elements, Vilniaus Valst. Univ. Mokslo Darbai Mat. Fiz. 9, 97-106 (in Russian).
Bliznikas, V. I: 1960b, On the differential geometry of metric line-element spaces, Uch. Zap. Vil'nyussk. Gos. Ped. Inst. 10, 11-29 (in Russian).
Bliznikas, V. I.: 1960c, Congruence of cemtroidal geodesic curves of a metric space of line elements, Dokl. Akad. Nauk SSSR 132, 735-738 (in Russian).
Bliznikas, V. I.: 1961a, Über einige geometrischen Objekten des metrischen Linienelementraumes, Liet. Mat. Rinkinys 1, 15-23 (in Russian).
Bliznikas, V. I.: 1961b, Certain aspects of the differential geometry of bilinear metric line-element spaces, Liet. Mat. Rinkinys 1, 372-373 (in Russian).
Bliznikas, V. I.: 1962, Zur Theorie der Hyperflächen in metrischen Linienelementraum mit euklidischen Zusammenhang, Liet. Mat. Rinkinys 2, 9-16 (in Russian).
Bliznikas, V. I.: 1963a, Euklidischer Zusammenhang der Art von Cartan des metrischen Linienelementraumes, Liet. Mat. Rinkinys 2, 33-37 (in Russian).
Bliznikas, V. I.: 1963b, Certain varieties of support elements', Liet. Mat. Rinkinys 3, 221-222 (in Russian).
Bliznikas, V. I.: 1964, Affine connection in a support element space, Rep. 3rd Siberian Conf. Math. Mech., 181-183 (in Russian).
Bliznikas, V. I.: 1965, Zur Krümmungstheorie im Stützelelementraum Liet. Mat. Rinkinys 5, 9-24 (in Russian).
Bliznikas, V. I.: 1966a, Nicht-holonome Liesche Differenzierung und lineare Zusammenhänge im Stützelementraum, Liet. Mat. Rinkinys 6, 141-209 (in Russian).
Bliznikas, V. I.: 1966b, Linear differential-geometric connections of higher order in the space of supporting elements, Izv. Vyss. Uchebn. Zaved. Mat. No. 5(54), 13-24 (in Russian).
Bliznikas, V. I.: 1967, Finsler spaces and their generalizations, Algebra, Topology, Geometry pp. 73-125 (in Russian)
Bogoslovskii, G. Y.: 1973, The special relativity theory of anisotropic space-time, Dokl. Akad. Nauk SSSR 213, 1055-1058 (in Russian).
Bogoslovskii, G. Y.: 1977, A special-relativistic theory of the locally anisotropic space-time. I. The metric and group of motions of the anisotropic space of events. II. Mechanics and electrodynamics in the anisotropic space, Nuovo Cimento 40B, 99-115, 116-134.
Bohnenblust, F.: 1938, Convex regions and projections in Minkowski spaces, Ann. of Math. (2) 39, 301-308.
Bompiani, E.: 1952a, Sulle connessioni affini non-posizionali, Arch. Math. 3, 183-186.
Bompiani, E.: 1952b, Sulla curvatura pangeodetica di una curva di una superficie dello spazio proiettivo, Boll. Un. Mat. Ital. (3) 7, 103-106.
Bompiani, E.: 1963, Alcuni tipi di spazi metrici, Archimede 15, 281-288.
Borisovic, J. G. and Gliklih, J. E.: 1980, Fixed points of mappings of Banach manifolds and some applications, Nonlinear Anal., Theory Methods Appl. 4, 165-192.
Bosquet, J. P.: 1929a, Quelques formules fondamentales de la théorie invariante du calcul des variations, Bull. Acad. Bruxelles (5) 15, 270-277.
Bosquet, J. P.: 1929b, Contribution à la théorie invariantive du calcul des variations. Bull. Acad. Bruxelles (5) 15, 1002-1017.
Bouligand, G.: 1956, Types d'intégrales généralisées pour une classe d'équations aux dérivées partielles du premier ordre, C. R. Acad. Sci. Paris 242, 2423-2426.
Bouligand, G.: 1957, Sur quelques modèles spatiaux, Acad. Roy. Belg. Bull. Cl. Sc. (5) 43, 133-138.
Bouligand, G.: 1958, Observations au sujet de la Note de M. Marcel Coz., C. R. Acad. Sci. Paris 246, 880.

Brickell, F.: 1965a, A new proof of Deicke's theorem on homogeneous functions, Proc. Amer. Math. Soc. 16, 190-191.
Brickell, F.: 1965b, On the differentiability of affine and projective transformations, Proc. Amer. Math. Soc. 16, 567-574.
Brickell, F.: 1967, Differentiable manifolds with an area measure, Canad, J. Math. 19, 540-549.
Brickell, F.: 1972, A relation between Finsler and Cartan structures, Tensor, N. S. 25, 360-364.
Brickell, F.: 1974, Area measures om a real vector space, in Global Analysis & its Applications, Intern. Atomic Energy Agency, Vienna, II, pp. 23-32.
Brickell, F., Clark, R. S. and Al-Borney, M. S.: 1976, (G, E) Structures, in Topics in Differential Geometry, Academic Press, New York, pp. 29-43.
Brickell, F. and Yano, K.: 1974, Concurrent vector fields and Minkowski structures, Kōdai Math. Sem. Rep. 26, 22-28.
Brickell, F. and Al-Borney, M. S.: 1982, The local equivalence problem for (G. H.)-structures and conformal Finsler geometry, Tensor, N. S. 39, 151-161.
Browder, F. E.: 1965a, Infinite dimensional manifolds and non-linear elliptic eigenvalue problems, Ann. of Math. (2) 82, 459-477.
Browder, F. E.: 1965b, Lusternik-Schnirelman category and nonlinear elliptic eigenvalue problems, Bull. Amer. Math. Soc. 71, 644-648.
Browder, F. E.: 1970, Existence theorems for nonlinear partial differential equations, Proc. Symp. Pure Math. 16, 1-60.
Brown, G. M.: 1968a, A study of tensors which characterize a hypersurface of a Finsler space, Canad. J. Math. 20, 1025-1036.
Brown, G. M.: 1968b, Gaussian curvature of a subspace in a Finsler space, Tensor, N. S. 19, 195-202.
Bucur, I.: 1955, Sur une propriété globale des ligues géodésiques d'un espace de Finsler, Com. Acad. R. P. Romania 5, 965-968 (in Roumanian).
Busemann, H.: 1932, Über die Geometrien, in denen die 'Kreise mit unendlichem Radius' die kürzesten Linien sind, Math. Ann. 106, 140-160.
Busemann, H.: 1933, Über Räume mit konvexen Kügeln und Parallelenaxiom, Nachr. Ges. Wiss. Göttingen I 38, 116-140.
Busemann, H.: 1939, Lokale Eigenschaften der zu Variationsproblemen gehörigen metrischen Räume, Fundam. Math. 32, 265-287.
Busemann, H.: 1941, Metric conditions for symmetric Finsler spaces, Proc. Nat. Acad. Sci. U. S. A. 27, 533-535.
Busemann. H.: 1942, Metric methods in Finsler spaces and in the foundations of geometry, Annals of Math. Studies 8, Princeton, 243 p.
Busemann, H.: 1947, Intrinsic area, Ann. of Math. (2) 48, 234-267.
Busemann, H.: 1949, The isoperimetric problem or Minkowski area, Amer. J. Math. 71, 743-762.
Busemann, H.: 1950a, The geometry of Finsler spaces, Bull. Amer. Math. Soc. 56, 5-16.
Busemann, H.: 1950b, The foundations of Minkowskian geometry, Comment. Math. Helv. 24, 156-187.
Busemann, H.: 1950c, On geodesic curvature in two-dimensional Finsler spaces, Ann. Mat. Pura Appl. (4) 31, 281-295.
Busemann, H.: 1953, Metrics on the torus without conjugate points, Bol. Soc. Mat. Mexicana 10, 1-2, 12-29 (in Spanish). English translation: ibid. 10, 3-4, 1-18.
Busemann, H.: 1955a, Quasihyperbolic geometry, Rend. Circ. Mat. Palermo (2) 4, 256-269.
Busemann, H.: 1955b, The geometry of geodesics, Pure and Appl. Math. 6, Acad. Press Inc., New York, 422 p.
Busemann, H.: 1955c, On normal coordinates in Finsler spaces, Math. Ann. 129, 417-423.
Busemann, H.: 1961, The synthetic approach to Finsler space in the large, C. I. M. E. Geom. Calcolo Varaz 1-2, 70 p.
Busemann, H.: 1966, On Hilbert's fourth problem, Uspechi Mat. Nauk 21, 155-164 (in Russian).
Busemann, H.: 1969/1971, Transitive geodesics in Minkowski planes, Math. Chronicle 1, 27-29.
Busemann, H.: 1974, Spaces with homothetic spheres, J. Geom. 4, 175-186.
Busemann, H.: 1975, Planes with analogous to Euclidean angular bisectors, Math. Scand. 36, 5-11.
Busemann, H.: 1976a, Remark on 'planes with analogues to Euclidean angular bisectors, Math. Scand. 38, 81-82.
Busemann, H.: 1976b, Problem IV: Desarguesian spaces, Math. Dev. Hilbert Problem., Proc. Symp. Pure Math. 28, 131-141.
Busemann, H. and Mayer, W.: 1941, On the foundations of calculus of variations, Trans. Amer. Math. Soc. 49, 173-198.
Busemann, H. and Phadke, B. B.: 1979, Minkowskian geometry, convexity conditions and the parallel axiom, J. Geom. 12, 17-33.

Busemann, H. and Straus, E. G. 1960, Area and normality, Pacific J. Math. 10, 35-72.
Cairns, S. S.: 1938, Normal coordinates for extremals transversal to a manifold, Amer. J. Math. 60, 423-435.
Carathéodory, C.: 1906, Über die starken Maxima und Minima bei einfachen Integralen, Math. Ann. 62, 449-503.
Cartan, E.: 1930, Sur un problème d'équivalence et la théorie des espaces métriques généralisés, Mathematica Cluj 4, 114-136. Oeuvres Comp. III 2, 1131-1153.
Cartan, E.: 1933a, Observations de M. Elie Cartan sur la Communication précédente, C. R. Acad. Sci. Paris 196, 27-28.
Cartan, E. 1933b, Sur les espaces de Finsler, C. R. Acad. Sci. Paris 196, 582-586. Oeuvres Compl. III 2, 1245-1248.
Cartan, E.: 1934, Les espaces de Finsler, Actualités 79, Hermann, Paris, 41 pp. (2nd edit. 1971).
Cartan, E.: 1937, Les espaces de Finsler, Trudy Sem. Vektor. Tenzor. Anal., Moskau 4, 70-81. Russian, 82-94. Oeuvres Compl. III 2, 1385-1396.
Castilio, L. D.: 1976, Tenseurs de Weyl d'une gerbe de directions, C. R. Acad. Sci. Paris A 282, 595-598.
Cavalleri, G. and Spinelli, G.: 1977a, Relativistic Lagrangian equations of motion with constraints: Check on the continuum, Nuovo Comento 39B, 87-92.
Cavalleri, G. and Spinelli, G.: 1977b, Gravity theory allowing for point particles and Zitterbewegung, Nuovo Cimento 39B, 93-104.
Chakerian, G. D.: 1960, The isoperimetric problem in the Minkowski plane, Amer. Math. Monthly 67, 1002-1004.
Chakerian, G. D.: 1962, Integral geometry in the Minkowski plane, Duke Math. J. 29 375-381.
Chakerian, G. D. and Klamkin, M. S.: 1973, Inequalities for sums of distances, Amer. Math. Monthly 80, 1009-1017.
Chandra, A.: 1972, Neo-covariant derivative and its application, Ganita 23, 33-39.
Chatterji, L. D. and Kumar, A.: 1977, Some theorems on recurrence vector in Finsler space, J. Math. Phys. Sci. 11, 461-469.
Chawla, M. and Behari, R.: 1971, Almost complex space with Finsler metric, Indian J. Pure Appl. Math. 2, 401-408.
Chegodaev, J. M.: 1973, Geodesic fields of directions in a Finsler space with a Sasakian bundle, and complex structures, Trudy Moskov. Inst. Radiotechn., Elektron. i Avtomat. Mat. 67, 15-31, 150 (in Russian).
Chegodaev, J. M.: 1975, A geodesic field of q-dimensional directions in a Finsler space, and an included system of paths, Izv. Vyss. Uchebn.Zaved. Mat. No. 4(155), 74-86 (in Russian).
Ch'en, I.: 1965, A special deformation of a curve in Finsler space, Acta Sci. Nat. Univ. Amoiensis 12 (2), 123-127.
Chern, S.: 1943, On the Euclidean connections in a Finsler space, Proc. Nat. Acad. Sci. U.S.A. 29, 33-37.
Chern, S.: 1948, Local equivalence and Euclidean connections in Finsler spaces, Sci. Rep. Nat. Tsing Hua Univ. A5, 95-121.
Chetyrkina, Z. M.: 1966, Homotheties and motions in two-dimensional Finsler spaces, Volz. Mat. Sb. 5, 366-373 (in Russian).
Chetyrkina, Z. M.: 1971, Conformal transformations in Finsler spaces, Ukrain. Geom. Sb. 11, 95-98 (in Russian).
Chetyrkina, Z. M.: 1981, Homothetically moving three-dimensional Randers spaces, Vestsi Akad. Nauk BSSR, Fiz.-Mat. No. 1, 43-51, 138 (in Russian).
Choquet, G.: 1944a, Etude différentielle des minimisantes dans les problèmes réguliers du calcul des variations, C. R. Acad. Sci. Paris 218, 540-542.
Choquet, G.: 1944b, Etude métrique des espaces de Finsler. Nouvelles méthodes pour les théorèmes d'existence en calcul des variations, C. R. Acad. Sci. Paris 219, 476-478.
Chowdhury, V. S. P.: 1980, On some special Finsler spaces, Ann. Soc. Sci. Bruxelles I 94, 179-184.
Chowdhury, V. S. P.: 1981, On a concept of parallelism in a Finsler space, Bull. Calcutta Math. Soc. 73, 11-16.
Chu, Y. H.: 1970, Studies on Finsler spaces, Hak-sul Chi 11, 645-655 (in Korean).
Clark, R. S.: 1951, The conformal geometry of a general differential metric spaces, Proc. London Math. Soc. (2) 53, 294-309.
Comić, I.: 1970, Hyperplanes of Finsler spaces of the constant intrinsic curvature, Mat. Vesnik 7 (22), 257-267 (in Serbo-Croatian).
Comić, I.: 1971, Relations between induced curvatur tensors of Finsler hypersurface F_{n-1} and curvature tensors of imbedding space F_n, Publ. Inst. Math. Beograd, N. S. 11 (25), 43-52.
Comić, I.: 1972a, The induced curvature tensors of a subspace in a Finsler space, Tensor N. S. 23, 21-34.

Comić, I.: 1972b, The intrinsic curvature tensors of a subspace in a Finsler space, Tensor N. S. 24, 19-28.
Comić I.: 1977, Relations between induced and intrinsic curvature tensors of a subspace in the Finsler space, Mat. Vesnik 1 (14) (29), 65-72.
Comić I.: 1978, Induced and intrinsic curvature tensors of a subspace in the Finsler space, Publ. Inst. Math. Beograd, N. S. 23(37), 67-74.
Comić, I.: 1982a, Family of subspaces in a Finsler space, Tensor, N. S. 37, 187-197.
Comić, I.: 1982b, Bianchi identities for the induced and intrinsic curvature tensors of a subspace in Finsler space, in Differential Geometry, North-Holland, Amsterdam-New York, 1982.
Comić, I.: 1981, Connections between the double alternated absolute differential of curvature tensors of the Finsler space and induced curvature tensors of its subspace, Univ. u Novom Sadu Zb. Rad. R. Prirod.-Mat. Fak. Ser. Mat. 11, 177-187 (in Serbo-Croatian).
Coz, M.: 1957, Sur les cas riemanniens dans la classe (C) de métriques variationelles du type ds = f(u, v; du, dv), Acad. Roy. Belg. Bull. Cl. Sci. (5) 43, 139-145.
Coz, M.: 1958, Une classe d'équations intégrales liées à des métriques variationelles à indicatrices centrées, C. R. Acad. Sci. Paris 246, 877-880.
Coz, M.: 1959, Une classe de modèles euclidiens pour métriques variationnelles, Acad. Roy. Belg. Cl. Sci. Mém. Coll. in-8° (2)31-4, 42 p.
Coz. M.: 1961, Métriques variationnelles régulières et équations intégrales, Acad. Roy. Belg. Cl. Sci. Mém. Coll. in-8° (2) 33-2, 123 p.
Damköhler, W.: 1940, Zur Frage der Äquivalenz indefiniter Variationsprobleme mit definiten, S.-B. Bayer Akad. Wiss., pp. 1-14.
Damköhler, W. and Hopf E.: 1947, Über einige Eigenschaften von Kurvenintegralen und über die Äquivalenz von indefiniten mit definiten Variationsproblemen, Math. Ann. 120, 12-20.
Davies, E. T.: 1939, Lie derivation in generalized metric spaces, Ann. Mat. Pura Appl. (4) 18, 261-274.
Davies, E. T.: 1945, Subspaces of a Finsler space, Proc. London. Math. Soc. (2) 49, 19-39.
Davies, E. T.: 1947, On metric spaces based on a vector density, Proc. London. Math. Soc. (2) 49, 241-259.
Davies, E. T.: 1949, On the second variation of a simple integral with movable end points, J. London Math. Soc. 24, 241-247.
Davies, E. T.: 1959, Applicazioni del Calcolo delle variazioni alla geometria differenziale, Conf. Sem. Mat. Univ. Bari 47, 11 p.
Davies, E. T.: 1961, Vedute generali sugli spazi variazionali, C. I. M. E. Geom. Calcolo Variaz. 1-2, 93 p.
Davies, E. T.: 1963, On the use of osculating spaces, Tensor, N. S. 14, 86-98.
Davies, E. T.: 1966, Some applications of the theory of parallel distributions, in Perspectives in Geometry and Relativity, Indiana, Bloomingten-London, pp. 80-95.
Davies, D. R.: 1928, The inverse problem of the calculus of variations in higher space, Trans. Amer. Math. Soc. 30, 710-736.
Davies, D. R.: 1929, The inverse problem of the calculus of variations in a space of (n+1) dimensions, Bull. Amer. Math. Soc. 35, 371-380.
Davies, D. R.: 1931, Integral whose extremals are a given 2n-parameter family of curves, Trans. Amer. Math. Soc. 33, 244-251.
Dazord, P.: 1964, Tenseur de structure d'une G-structure dérvée, C. R. Acad. Sci. Paris 258, 2730-2733.
Dazord, P.: 1966a, Sur une généralisation de la notion de 'spray', C. R. Acad. Sci. Paris A 263, 543-546.
Dazord, P.: 1966b, Connexion de direction symétrique associée à un 'spray' généralisé, C. R. Acad. Sci. Paris A 263, 576-578.
Dazord, P.: 1968a, Variétés finslériennes à géodésiques fermées, C. R. Acad. Sci. Paris A 266, 348-350.
Dazord, P.: 1968b, Variétés finslériennes de dimension paire δ-pincées, C. R. Acad. Sci. Paris A 266, 496-498.
Dazord, P.: 1968c, Variétés finslériennes en forme de sphères, C. R. Acad. Sci. Paris A 267, 353-355.
Dazord, P.: 1969, Propriétés glabales des géodésiques des espaces de Finsler: Thèse, Lyon, 193 p.
Dazord, P.: 1970, Sur la formule de Gauss-Bonnet en géométrie finslérienne, C. R. Acad. Sci. Paris A 270, 1241-1243.
Dazord, P.: 1971, Tores finslériennes sans points conjugués, Bull. Soc. Math. France 99, 171-192, 397.
Debever, R.: 1942, Sur quelques problèmes de géométries dérivées du calcul des variations. I.', Bull. Acad. Roy. Belg. Cl. Sci. (5) 28, 794-808.

Debever, R.: 1943, Sur quelques problèmes de géométries dérivées du calcul des variations. II., Bull. Acad. Roy. Belg. Cl. Sci. (5) 29, 194-203.
Deicke, A.: 1953, Über die Darstellung von Finsler-Räumen durch nichtholonome Mannifaltigkeiten in Riemannschen Räumen, Arch. Math. 4, 234-238.
Deicke, A.: 1955, Finsler spaces as non-holonomic subspaces of Riemannian spaces, J. London Math. Soc. 30, 53-58.
Delens, P.: 1933, Sur certains problèmes relatifs aux espaces de Finsler, C. R. Acad. Sci. Paris 196, 1356-1358.
Delens P.: 1934, La métrique angulaire des espaces de Finsler et la géometrie différentielle projective, Actualités 80, Paris, 39 p.
Delens, P. and Devisme, J.: 1933, Sur certaines formes différentielles et les métriques associées, C. R. Acad. Sci. Paris 196, 518-521.
Devisme, J.: 1939, Sur l'espace dont l'élément linéaire est défini par $ds^3 = dx^3 + dy^3 + dz^3 - 3dx\,dy\,dz$, C. R. Acad. Sci. Paris 208, 1773-1775.
Devisme, J.: 1940, Sur un espace dont l'élément linéaire est défini par $ds^3 = dx^3 + dy^3 + dz^3 - 3dx\,dy\,dz$, J. Math. Pures Appl. (9) 19, 359-393.
Devisme, J.: 1941, Sur quelques propriétés des trièdres d'Appell, C. R. Acad. Sci. Paris 212, 43-45.
Dhawan, M.: 1965, Curvature properties of a subspace embedded in a Finsler space, Ganita 16, 25-36.
Dhawan, M.: 1969, A study of complex structures in Riemannian and Finsler manifolds and their prolongation to tangent and cotangent bundles: Thesis, Delhi. 267 p.
Dhawan, M. and Prakash, N.: 1964, Generalizations of Gauss-Codazzi equations in a subspace imbedded in a Finsler manifold, Tensor, N. S. 15, 159-167.
Dhawan, M. and Prakash, N.: 1970, Adjoint complex Finsler manifolds, Indian J. Pure Appl. Math. 1, 8-16.
Diaz, J. G.: 1972, Quelques propriétés des tenseurs de courbure en géométrie finslérienne, C. R. Acad. Sci. Paris A 274, 569-572.
Diaz, J. G. and Grangier, G.: 1976, Courbure et holonomie des variétés finslériennes, Tensor, N. S. 30, 95-109.
Douglas, J.: 1926, The transversality relative to a surface of $\int F(x, y, z, y', z',)dx$ = minimum, Bull. Amer. Math. Soc. 32, 669-674.
Douglas, J.: 1927/1928, The general geometry of paths, Ann. of Math. (2) 29, 143-168.
Dowling, J.: 1970, Finsler geometry on Sobolev manifolds, Proc. Symp. Pure Math. 15, 1-10.
Dragomir, S.: 1981, The theorem of K. Nomizu on Finsler manifolds, An. Univ. Timisoara, Sti. Mat. 19, 117-127.
Dragomir, S.: 1982a, On the holonomy groups of a connection in the induced Finsler bundle, Proc. Nat. Sem. Finsler Spaces, Brasov, pp. 83-97.
Dragomir, S.: 1982b, p-distributions on differentiable manifolds, An. Sti. Univ. Iasi Ia 28, Suppl. 55-58.
Dragomir, S.: 1983, The theorem of É, Cartan on Finsler manifolds, Proc. Nat. Colloquium on Geometry and Topology, Busteni 1981, Univ. Bucuresti, Bucharest, pp. 103-112.
Dragomir, S. and Ianus, S.: 1982a, On the holomorphic sectional curvature of the Kaehlerian Finsler spaces, Proc. Nat. Sem. Finsler Spaces, Brasov, pp. 99-108.
Dragomir, S. and Ianus, S.: 1982b, On the holomorphic sectional curvature of Kaelerian Finsler spaces, Tensor, N. S. 39, 95-98.
Dragunov, V. K.: 1974, A certain way of representing the geometry of the hyperbolic plane locally as a Minkowski geometry, in Differential Geometry, No. 1, Saratov University, pp. 100-106 (in Russian).
Dubey, R. S. D.: 1981, A note on recurrent Finsler space, Progr. Math. (Allahabad) 15, 69-74.
Dubey, R. S. D.: 1982, (a,u,v,)-recurrent Finsler spaces, in Differential Geometry (Budapest, 1979), Colloq. Math. Soc. János Bolyai, 31, North-Holland, Amsterdam-New York, pp. 769-782.
Dubey, R. S. D. and Singh, H.: 1979, Finsler spaces with recurrent pseudocurvature tensor, Proc. Indian Acad. Sci. Sect. A Math. Sci. 88, 363-367.
Dubey, R. S. D. and Singh, H.: 1980a, Hypersurfaces of recurrent Finsler spaces, Indian J. Pure Appl. Math. 11, 1125-1129.
Dubey, R. S. D. and Singh, H.: 1980b, Commutation formulae in conformal Finsler space. II., J. Math. Phys. Sci. 14, 501-513.
Dubey, R. S. D. and Singh, H.: 1981, Study of some motions and curvature collineations in subspaces of Finsler spaces, J. Math. Phys. Sci. 15, 581-592.
Dubey, R. S. D. and Srivastava, A. K.: 1980, CA-collineation in a generalized birecurrent Finsler manifold, Bul. Inst. Politeh. Iasi I 26(30), 63-65.
Dubey, R. S. D. and Srivastava, A. K.: 1981, On recurrent Finsler spaces, Bull. Soc. Math. Belg. B 33, 283-288.

Dubey, S. K. D.: 1978, S-recurrent Finsler space, J. Math. Phys. Sci. 12, 491-496.
Duc, T.: 1973a, Connexions non linéaires et équations de structure, Rev. Roumaine Math. Pures Appl. 18, 1269-1274.
Duc, T.: 1973b, Fibrés tangents d'ordre supérieur, Rev. Roumaine Math. Pures Appl. 18, 1275-1281.
Duc, T.: 1975, Sur la géométrie différentielle des fibrés vectoriels, Kōdai Math. Sem. Rep. 26, 349-408.
Duschek, A.: 1937, Über geometrische Variationsrechnung, Trudy Sem. Vektor. Tenzor. Anal. 4, 95-99 (in Russian).
Duschek, A. and Mayer, W.: 1933, Zur geometrischen Variationsrechnung. II: Über die zweite Variation des eindimensionalen Problems, Monatsh. Math. Phys. 40, 294-308.
Earle, C. J. and Eelis J.: 1967, Foliations and fibrations, J. Diff. Geom. 1, 33-41.
Earle, C. J. and Hamilton, R. S.: 1970, A fixed point theorem for holomorphic mappings, Proc. Symp. Pure Math. 16, 61-65.
Eelis, J.: 1958, On the geometry of function spaces, Symp. Intern. Topol. Alg. Mexico City, 303-308.
Eelis, J.: 1966, A setting for global analysis, Bull. Amer. Math. Soc. 72, 751-807.
Eggleston, H. G.: 1958, Notes on Minkowski geometry (I): Relations between the circumradius, diameter, inradius and minimal width of a convex set, J. London. Math. Soc. 33, 76-81.
Egorov, I. P.: 1967, Motions in generalized differential-geometric spaces, Algebra, Topology, Geometry 1965, pp. 375-428 (in Russian). English translation: 1970, Progr. Math. 6, 171-227.
Egorov, A. I.: 1971, Some maximally mobile spaces, Penz. Ped. Inst. Uchen Zap. 124, 47-50 (in Russian).
Egorov, A. I.: 1974, Maximally mobile Finsler spaces, in Motions in Generalized Spaces, Ryazanskii Gos. Ped. Inst., pp. 17-21 (in Russian).
Egorov, A. I. 1977, Maximally mobile spaces of linear and contravariant vector elements with affine connection, Penz. Ped. Inst. Uchen. Zap. 124, 54-60 (in Russian).
Egorov, A. I.: 1979, Maximally movable spaces of linear elements of affine connection. I, Ukrain. Geom. Sb. 22, 47-59 (in Russian).
Egorov, A. I.: 1981, Lacunary Finsler spaces, Mat. Sb. 116(158), 310-314, 463 (in Russian).
Egorov, I. P. and Egorov, A. I.: 1971, Spaces with generalized affine connection, Penz. Ped. Inst. Uchen, Zap. 124, 10-12 (in Russian).
Egorov, I. P. and Egorov, A. I.: 1982, Some problems of automorphisms in generalized spaces, in Motions in generalized spaces, Ryazanskii Gos. Ped. Inst., Ryazan, pp. 41-52 (in Russian).
Egorova, L. I.: 1978, Singular trajectories in Finsler spaces, Problems of geometry 10, 25-27 (in Russian).
Egorova, L. I.: 1982, Homotheties in generalized metric spaces, Ukrain. Geom. Sb. No. 25, 51-56, 141 (in Russian).
Egorov, I. P.: 1978, Automorphisms in generalized spaces, Problems of geometry 10, 147-191, 224 (in Russian).
Eguchi, K. and Ichijyō, Y.: 1969, General projective connections and Finsler metric, J. Math. Tokushima Univ. 3, 1-20.
Eisenhart, L. P.: 1948, Finsler spaces derived from Riemann spaces by contact transformations, Ann. of Math. (2) 49, 227-254.
Eliasson, H. I.: 1970, Variation integrals in fibre bundles, Proc. Symp. Pure Math. 16, 67-89.
Eliasson, H. I.: 1974, Introduction to global calculus of variations, Global Anal. Appl. Intern. Sem. Course Trieste 1972, II, pp. 113-135.
Eliopoulos, H. A.: 1956, Methods of generalized metric geometry with applications to mathematical physics: Thesis, Toronto.
Eliopoulos, H. A.: 1959a, Sur la définition de la courbure totale d'une hypersurface plongée dans un espace de Finsler localement minkowskien, Bull. Acad. Roy. Belg. Cl. Sci. (5) 45, 203-214.
Eliopoulos, H. A.: 1959b, Subspaces of a generalized metric space, Canad. J. Math. 11, 235-255.
Eliopoulos, H. A.: 1965, A generalized metric space for electromagnetic theory, Bull. Acad. Roy. Belg. Cl. Sci. (5) 51, 986-995.
Eliopoulos, H. A.: 1966, Multi-particle theory derived from the geometry of a locally Minkowskian Finsler space, Bull. Acad. Roy. Belg. Cl. Sci. (5) 52, 69-75.
Enghis, P. and Tarinà, M.: 1982, E-connexions dans un espace de Finsler, Proc. Nat. Sem. Finsler Spaces, Brasov, pp. 109-114.
Ermakov, Y. I.: 1958, Three-dimensional space with a cubic semimetric, Dokl. Akad. Nauk SSSR 118, 1070-1073 (in Russian).
Ermakov, Y. I.: 1959, Spaces X_n with an algebraic metric and semimetric, Dokl. Akad.

Nauk SSSR 128, 460-463 (in Russian).
Ermakov, Y. I.: 1974, Hypersurfaces of a space with a cubic metric, in Differential Geometry, No. 1, Saratov University, pp. 20-25, 135 (in Russian).
Ermakov, Y. I.: 1975, Differentiable manifolds with a biquadratic metric, in Differential Geometry, No. 2, Saratov University, pp. 21-40, 117 (in Russian).
Evtushik, L. E.: 1973, Holonomy of nonlinear connections, Sibir. Math. Z. 14, 536-548 (in Russian). English translation: 1973, Siber. Math. J. 14, 370-379.
Fava, F.: 1979, Derivations of graded algebras and nonlinear connections of arbitrary order, Conf. Sem. Mat. Univ. Bari 170, 29 p.
Fava, F. and Misra, R. B.: 1976-1977, Eulerian curvature tensors and the conformal mappings, Univ. Politec. Torino, Rend. Sem. Mat. 35, 311-326.
Fernandes, A.: 1935, Derivazione tensoriale composta negli spazii non puntuali, Lincei Rend. (6) 21, 555-562.
Fernandes, A.: 1941, Axiomatics of spaces of linear elements, Portugaliae Math. 2, 7-12 (in Portuguese).
Ferrand, J.: 1980, Le groupe des automorphismes conformes d'une variété de Finsler compacte, C. R. Acad. Sci. Paris A 291, 209-210.
Ferzaliev, A. S.: 1974a, Equiaffine spaces of linear elements, Izv. Vyss. Uchebn. Zaved. Mat., No. 9(148), 81-85 (in Russian).
Ferzaliev, A. S.: 1974b, Weyl spaces of linear elements with affine connection, Izv. Vyss. Uchebn. Zaved. Mat., No. 10(149), 75-83 (in Russian).
Ferzaliev, A. S.: 1974c, Equiaffine spaces of tensor support elements, in Motions in Generalized Spaces, Ryazanskii Gos. Ped. Inst., pp. 38-47 (in Russian).
Ferzaliev, A. S.: 1975a, Projective mapping and special type of line-element spaces, Izv. Vyss. Uchebn. Zaved. Mat., No. 1(152), 126-130 (in Russian).
Ferzaliev, A. S.: 1975b, Projective mappings of a space of linear elements with support vector tangent to the affine paths, Trudy Geom. Sem. Kazan. Univ. 8, 95-108 (in Russian).
Ferżaliev, A. S.: 1975c, Projective mappings and special types of spaces of linear elements with general connections. I, Izv. Vyss. Uchebn. Zaved. Mat., No. 12(163), 87-91 (in Russian).
Ferzaliev, A. S.: 1976, Projective mappings and special types of spaces of linear elements with general connections. II, Izv. Vyss. Uchebn. Zaved. Mat., No. 2(165), 114-118 (in Russian).
Finsler, P.: 1939, Über die Krümmungen der Kurven und Flächen, Reale Accad. Italia, Fond. A. Volta, Atti dei Convegni 9, 463-478.
Finsler, P.: 1940, Über eine Verallgemeinerung des Satzes von Meusnier, Vierteljschr. Naturforsch. Ges. Zürich 85, Beibl. 32, 155-164.
Finsler, P.: 1951, Über Kurven und Flächen in allgemeinen Räumen: Dissertation, Birkhäuser Verlag, Basel 1951.
Fischer, H. R.: 1972, On a certain class of C(X)-modules, in Topics in Topology, Amsterdam-London, North-Holland, pp. 309-315.
Freeman, J. G.: 1944, First and second variations of the length integral in a generalized metric space, Quart. J. Math. Oxford 15, 70-83.
Freeman, J. G.: 1946, Theory of a ruled two-space in a generalized metric space, Quart. J. Math. Oxford 17, 119-128.
Freeman, J. G.: 1948, A generalizaton of minimal varieties, Edinburgh Math. Proc. (2) 8, 66-72.
Freeman, J. G.: 1956, Finsler-Riemann systems, Quart. J. Math. Oxford (2) 7, 100-109.
Freeman, J. G.: 1957, Complete Finsler-Riemann systems, Quart. J. Math. Oxford (2) 8, 161-171.
Freeman, J. G.: 1970, Parallel transport in a Finsler space defined by a field of oriented vectors, Math. Nachr. 45, 161-166 (in Russian).
Freidina, M. G.: 1947, Dual systems admitting a group of motions, Dokl. Akad. Nauk SSSR 57, 547-550.
Friesecke, H.: 1925, Verktorübertragung, Richtungübertragung, Metrik, Math. Ann. 93, 101-118.
Fujinaka, M.: 1953, On Finsler spaces and dynamics with special refence to equations of hunting, Proc. 3rd Japan Nat. Congr. Appl. Mech. 433-436.
Fukui, M. and Yamada, T.: 1979, On the indicatrized tensor of $S_{ijkh|1}$ in Finsler spaces, Tensor, N. S. 33, 373-379.
Fukui, M. and Yamada, T.: 1981, On projective mappings in Finsler geometry, Tensor, N. S. 35, 216-222.
Fukui, M. and Yamada, T.: 1982, On Finsler spaces of constant curvature, Tensor, N. S. 38, 129-134.
Funk, P.: 1919, Über den Begriff 'extremale Krümmung' und eine kennzeichnende Eigenschaft der Ellipse, Math. Z. 3, 87-92.

Funk, P.: 1929, Über Geometrien, bei denen die Geraden die Kürzesten sind, Math. Ann. 101, 226-237.
Funk, P.: 1930, Über Geometrien, bei denen die Geraden die kürzesten Linien sind und die Äquidistanten zu einer Geraden wieder Gerade sind, Monatsh. Math. Phys. 37, 153-158.
Funk, P.: 1935, Über zweidimensionale Finslersche Räume, insbesondere über solche mit geradlinigen Extemalen und positiver konstanter Krümmung, Math. Z. 40, 86-93.
Funk, P.: 1948, Beiträge zur zweidimensionalen Finslerschen Geometrie, Monatsh. Math. Phys. 52, 194-216.
Funk, P.: 1963, Eine Kennzeichung der zweidimensionalen elliptischen Geometrie, Österreich. Akad. Wiss. Math. Natur. Kl. S.-B. II 172, 251-269.
Funk, P. and Berwald, L.: 1919/1920, Flächeninhalt und Winkel in der Variationsrechnung, Lotos Prag 67/68, 45-49.
Galvani, O.: 1946a, Sur la réalisation des espaces de Finsler, C. R. Acad. Sci. Paris 222, 1067-1069.
Galvani, O.: 1946b, Les connexions finslériennes de congruences de droites, C. R. Acad. Sci. Paris 222, 1200-1202.
Galvani, O.: 1946c, Sur l'immersion du plan de Finsler dans certains espaces de Riemann à trois dimensions, C. R. Acad. Sci. Paris 223, 1088-1090.
Galvani, O.: 1951, La réalisation des connextions euclidiennes d'éléments linéaires et des espaces de Finsler, Ann. Inst. Fourier 2, 123-146.
Galvani, O.: 1952, La rélisation des espaces de Finsler, C. R. Congr. Soc. Savantes Paris et Départ. Grenoble, Sect. Sci., pp. 57-60.
Galvani, O.: 1955, Réalisations euclidiennes des plans de Finsler, Ann. Inst. Fourier 5, 421-454.
Garbiero, S.: 1980, Symmetry of differentiable manifolds equipped with induced connections and Finsler connections, Rend. Sem. Mat. Univ. Politec. Torino 38, No. 3, 125-145 (In Italian).
Gariepy, R. and Pepe, W. D.: 1972, On the level sets of a distance function in a Minkowski space, Proc. Amer. Math. Soc. 31, 255-259.
Gheorghiu, G. T.: 1977, Sur les espaces de Finsler, Math.-Rev. Anal. Numér. Théo. Appro. Math. 19(42), 45-33.
Ghinea, I.: 1976, Connexions Finslériennes conformes métriques, Anal. Univ. Timisoara, Sti. Mat. 14, 101-115.
Ghinea, I.: 1977a, Connexions Finslériennes presque symplectiques, Anal. Univ. Timisoara, Sti. Mat. 15, 93-102.
Ghinea, I.: 1977b, Connexions Finslériennes presque complexes, Math.-Rev. Anal. Numér. Théo. Appro. Math. 19(42), 55-61.
Ghinea, I.: 1977c, Connexions Finslériennes presque complexes, Math.-Rev. Anal. Numér. Théo. Appro. Math. 19(42), 153-161.
Ghinea, I.: 1977d, Almost symplectic Finslerian conformal connections, Bul. Inst. Politeh. Iasi I 23(27), No. 3-4, 37-42 (in Roumanian).
Ghinea, I.: 1978a, Invariants des groupes de transformations des connexions finslériennes conformes métriques, Bul. Inst. Politeh. Iasi I 24(28), No. 1-2, 43-46 (in Roumanian).
Ghinea, I.: 1978b, Courbures invariantes aux transformations de connexion finslériennes métrique. I. Bul. Inst. Politeh. Iasi I 24(28), No. 3-4, 47-51 (in Roumanian).
Ghinea, I.: 1979a, Invariant curvatures to transformations of conformal metric Finsler connections. II. Bul. Inst. Politeh Iasi I 25(29), No. 1-2, 55-59.
Ghinea, I.: 1979b, Kawaguchi connections associated to a Finsler almost symplectical structure, Bul. Inst. Politeh. Iasi I 25(29), No. 3-4, 57-61.
Ghinea, I.: 1979c, Kawaguchi connections associated to a Finsler conformal structure, An. Univ. Timisoara, Sti. Mat. 17, 119-124.
Ghinea, I.: 1980a, Kawaguchi connections associated to a Finsler conformal almost symplectical structure, Bul. Inst. Politeh. Iasi I 26(30), 43-47.
Ghinea, I.: 1980b, Conformal Finsler connections, Proc. Nat. Sem. Finsler Spaces, Brasov, pp. 55-68.
Ghinea, I.: 1980c, On a class of Finsler connections, Bul. Inst. Politeh. Iasi I 26(30), No. 3-4, 59-62.
Ghinea, I.: 1982, Generalized Finsler connections compatible with a couple of metrical or almost symplectical structure, Proc. Nat. Sem. Finsler Spaces, Brasov, pp. 115-124.
Glogovskii, V. V.: 1968, Bisectors in non-generalized Minkowski spaces. Priklad. Geom. Inz. Grafika, Kiev 7, 118-126 (in Russian).
Glogovskii, V. V.: 1970, Bisectors on the Minkowski plane with norm $(|x|^p + |y|^p)^{1/p}$, Visnik L'viv. Politeh. Inst. 44, 192-198, 218 (in Ukrainian).
Gohman, A. V.: 1969, Differential-Geometric Foundations of the Classical Dynamics of Systems, Saratov. Univ., 93 p. (in Russian).
Golab, S.: 1932a, Quelques problèmes métriques de la géométries de Minkowski, Trav. Acad. Mines Cracovie Fasc. 6, 1-79 (in Polish).

Golab, S.: 1932b, Einige Bemerkungen über Winkelmetrik in Finslerschen Räumen, Verh. Intern. Math. Kongr. Zürich II, pp. 178-179.
Golab, S.: 1933a, Sur la représentation conforme de l'espace de Finsler sur l'espace euclidien, C. R. Acad. Sci. Paris 196, 25-27.
Golab, S.: 1933b, Sur un invariant intégral relatif aux espaces métriques généralisés, Lincei Rend. (6) 17, 515-518.
Golab S.: 1933c, Sur la représentation conforme de deux espaces de Finsler, C. R. Acad. Sci. Paris 196, 986-988.
Golab, S.: 1933d, Contribution à un théorème de M. M. S. Knebelman, Prace Mat. Fiz. 41, 97-100.
Golab, S.: 1935a, Sur une condition nésessaire et suffisante afin qu'un espace de Finsler soit un espace riemannien, Lincei Rend. (6) 21, 133-137.
Golab, S.: 1935b, Sur la mesure des aires dans les espaces de Finsler, C. R. Acad. Sci. Paris 200, 197-199.
Golab, S.: 1935c, Les transformations par polaires réciproques dans la géométrie de Finsler, C. R. Acad. Sci. Paris 200, 1462-1464.
Golab, S.: 1935d, Sur le rapport entre les notions des mesures des angles et des aires dans les espaces de Finsler, C. R. Acad. Sci. Paris 201, 250-251.
Golab, S.: 1951, On Finsler's measurement of an angle, Ann. Soc. Polon. Math. 24, 78-84.
Golab, S.: 1957, Zur Theorie der Ubertragungen, Schr. Forschungsinst. Math. 1, 162-177.
Golab, S.: 1966, Sur la longueur de l'indicatrice dans la géométrie plane de Minkowski, Colloq. Math. 15, 141-144.
Golab, S.: 1970, On a certain problem of Minkowski geometry, Prace Nauk. Inst. Mat. Fiz. Teor. Politech. Wroclaw Ser. Studia Materiely 2, 3-5 (in Polish).
Golab, S.: 1971, Sur un problème de la métrique angulaire dans la géométrie de Minkowski, Aequationes Math. 6, 121-129.
Golab, S.: 1972, Sur quelques conditions suffisantes pour que l'espace de Minkowski à deux dimensions soit euclidien I. Le role de l'axiome de L. Dubikajtes, Tensor, N. S. 24, 383-388.
Golab, S. and Härlen, H.: 1931, Minkowskische Geometrie, I, II, Monatsh. Math. Phys. 38, 387-398.
Golab, S. and Tamássy, L.: 1960, Eine Kennzeichnung der euklidischen Ebene unter den Minkowskischen Ebenen, Publ. Math. Debrecen 7, 187-193.
Gorshkova, L. S.: 1971a, Finsler spaces F_4 $(x,\dot{x})$ that admit groups of motions G_r, $r \geqslant 3$, Penz. Ped. Inst. Uchen. Zap. 124, 31-35 (in Russian).
Gorshkova, L. S.: 1971b, Four-dimensional Finsler spaces with groups of motions G_4, Penz. Ped. Inst. Uchen, Zap. 124, 36-41 (in Russian).
Gorshkova, L. S.: 1971c, Finsler spaces F_4 that admit groups of motions with a nonabelian three-parameter subgroup, Penz. Ped. Inst. Uchen. Zap. 124, 42-46 (in Russian).
Gorshkova, L. S.: 1973a, Finsler spaces F_4that admit five-parameter groups of motions with an abelian subgroup G_3, Volz. Mat. Sb. 23, 6-25 (in Russian).
Gorshkova, L. S.: 1973b, Motions in Finsler spaces, Volz. Mat. Sb. 23, 3-6 (in Russian).
Gorshkova, L. S.: 1974, Solvable five-parameter groups of motions in Finsler spaces F_4, in Motions in Generalized Spaces, Ryazanskii Gos. Ped. Inst., pp. 3-17 (in Russian).
Graham, R. L., Witsenhausen, H. S. an Zassenhaus, H. J.: 1972, On tightest packings in the Minkowski plane, Pacific J. Math. 41, 699-715.
Grangier, G.: 1973, Sur l'holonomie des variétés finslériennes, C. R. Acad. Sci. Paris A 276, 289-292.
Grifone, J.: 1969a, Connexions non linéaires conservatives, C. R. Acad. Sci. Paris. A 268, 43-45.
Grifone, J.: 1969b, Prolongement linéaire d'une connexion de directions, C. R. Acad. Sci. Paris A 269, 90-93.
Grifone, J.: 1971, Sur les connexions d'une variété finslérienne et d'un système mécanique, C. R. Acad. Sci. Paris A 272, 1510-1513.
Grifone, J.: 1972a, Structure presque tangente et connexions, I., Ann. Inst. Fourier 22, 287-334.
Grifone, J.: 1972b, Structure presque tangente et connexions, II, Ann. Inst. Fourier 22, 291-338.
Grifone, J.: 1971c, Structure presque tangente et connexions non homogenes: Thèse, Grenoble, 107 p.
Grifone, J.: 1975a, Transformations infinitésimales conformes d'une variété finslérienne, C. R. Acad. Sci. Paris A 280, 519-522.
Grifone, J.: 1975b, Sur les transformations infinitésimales conformes d'une variété finslérienne, C. R. Acad. Sci. Paris A 280, 583-585.
Grifone, J.: 1976, Sur les connexions induite et intrinsèque d'une sous-variété d'une variété finslérienne, C. R. Acad. Sci. Paris A 282, 599-602.

Grossman, N.: 1967, On real projective spaces as Finsler manifolds, Proc. Amer. Math. Soc. 18, 325-326.

Grove, K., Karcher, H. and Ruh, E. A.: 1974, Jacobi fields and Finsler metrics on compact Lie groups with an application to differentiable pinching problems, Math. Ann. 211, 7-21.

Grove, K., Karcher, H. and Ruh, E. A.: 1975, Group actions and curvature, Bull. Amer. Math. Soc. 81, 89-92.

Grünbaum, B.: 1957/1958a, Borsak's partition conjecture in Minkowski planes, Bull. Res. Council. Israel. F 7, 25-30.

Grünbaum, B.: 1957/1958b, On a problem of S. Mazur, Bull. Res. Council. Israel F 7, 133-135.

Grünbaum, B.: 1959, On some covering and intersection properties in Minkowski spaces, Pacific J. Math. 9, 487-494.

Grünbaum, B.: 1966, The perimeter of Minkowski unit discs, Colloq. Math. 15, 135-139.

Grüss, G.: 1928, Über Gewebe auf Flächen in dreidimensionalen allgemeinen metrischen Räumen, Math. Ann. 100, 1-31.

Grüss, G.: 1929, Eine Bemerkung über geodätische Kegelschnitte auf Flächen allgemeiner Metrik, Jber. Deutsch. Math.-Verein. 38, 83-91.

Grüss, G.: 1930, Beiträge zur Differentialgeometrie zweidimensionaler allgemein-metrischer Räume, Math. Ann. 103, 162-184.

Guggenheimer, H. W.: 1965, Pseudo-Minkowski differential geometry, Ann. Mat. Pura Appl. (4) 70, 305-370.

Guggenheimer, H. W.: 1972, Approximation of curves, Pacific J. Math. 40, 301-303.

Haimovici, A.: 1965, Sur les espaces à connexion non linéaire à parallélisme absolu, Rev. Roum. Math. Pures et Appl. 10, 1121-1128.

Haimovici, M.: 1934a, Formules fondamentales dans la théorie des hypersurfaces d'un espace de Finsler, C. R. Acad. Sci. Paris 198, 426-427.

Haimovici, M.: 1934b, Sur les espaces généraux qui se correspondent point par point avec conservation du parallélisme de M. Cartan, C. R. Acad. Sci. Paris 198, 1105-1108.

Haimovici, M.: 1934c, Sur quelques types de métriques de Finsler, C. R. Acad. Sci. Paris 199, 1091-1093.

Haimovici, M.: 1935, Les formules fondamentales dans la théorie des hypersurfaces d'un espace général, Ann. Sci. Univ. Jassy I 20, 39-58.

Haimovici, M.: 1937, Sur les espaces de Finsler à connexion affine, C. R. Acad. Sci. Paris 204, 837-839.

Haimovici, M.: 1938a, Le parallélisme dans les espaces de Finsler et la différentiation invariante de M. Levi-Civita, Ann. Sci. Univ. Jassy I 24, 214-218.

Haimovici, M.: 1938b, Sulle superficie totalmente geodetiche negli spazi di Finsler, Lincei Rend. (6) 27, 633-641.

Haimovici, M.: 1939, Variétés totalement extrémales et variétés totalement géodésiques dans les espaces de Finsler, Ann. Sci. Univ. Jassy I 25, 559-644.

Hamel, G.: 1901, Über die Geometrien, in denen die Geraden die Kürzesten sind: Dissertation, Göttingen.

Hamel, G.: 1903, Über die Geometrien, in denen die Geraden die Kürzesten sind, Math. Ann. 57, 231-264.

Hammer, P. C.: 1963, Convex curves of constant Minkowski breadth, Proc. Symp. Pure Math. 7, 291-304.

Hashiguchi, M.: 1958, On parallel displacements in Finsler spaces, J. Math. Soc. Japan 10, 365-379.

Hashiguchi, M.: 1969, On determinations of Finsler connections by deflection tensor fields, Rep. Fac. Sci. Kagoshima Univ. (Math. Phys. Chem.) 2, 29-39.

Hashiguchi, M.: 1971, On the hv-curvature tensors of Finsler spaces, Rep. Fac. Sci. Kagoshima Univ. (Math. Phys. Chem.) 4, 1-5.

Hashiguchi, M.: 1975, On Wagner's generalized Berwald space, J. Korean Math. Soc. 12, 51-61.

Hashiguchi, M.: 1976, On conformal transformations of Finsler metrics, J. Math. Kyoto Univ. 16, 25-50.

Hashiguchi, M.: 1980, Wagner connections and Miron connections of Finsler spaces, Rev. Roum. Math. Pures et Appl. 25, 1387-1390.

Hashiguchi, M.: 1983, Some problems on generalized Berwald spaces, Rep. Fac. Sci. Kagoshima Univ. (Math. Phys. Chem.) 16, 59-63.

Hashiguchi, M., Hōjō, S. and Matsumoto, M.: 1973, On Landsberg spaces of two dimensions with (α, β)-metric, J. Korean Math. Soc. 10, 17-26.

Hashiguchi, M. and Ichijyō, Y.: 1975, On some special (α, β)-metrics, Rep. Fac. Sci. Kagoshima Univ. (Math. Phys. Chem.) 8, 39-46.

Hashiguchi, M. and Ichijyō, Y.: 1977, On conformal transformations of Wagner spaces, Rep. Fac. Sci. Kagoshima Univ. (Math. Phys. Chem.) 10, 19-25.

Hashiguchi, M. and Ichijyō, Y.: 1980, Randers spaces with rectilinear geodesics, Rep. Fac. Sci. Kagoshima Univ. (Math. Phys. Chem.) 13, 33-40.
Hashiguchi, M. and Ichijyō, Y.: 1982, On generalized Berwald spaces, Rep. Fac. Sci. Kagoshima Univ. (Math. Phys. Chem.) 15, 19-32.
Hashiguchi, M. and Varga, T.: 1979, On Wagner spaces of W-scalar curvature, Studia Sci. Math. Hungar. 14, 11-14.
Hassan, B. T. M.: 1967, The theory of geodesics in Finsler spaces: Thesis, Southampton, 108 p.
Hassan, B. T. M.: 1973, The cut locus of a Finsler manifold, Lincei Rend. (8) 54, 739-744.
Hassan, B. T. M.: 1975a, A theorem on compact Finsler surfaces, Proc. Math. Phys. Soc. Egypt 40, 1-6.
Hassan, B. T. M.: 1975b, Isometric immersions of a compact Finsler space into a Finsler space, Proc. Math. Phys. Soc. Egypt. 40, 15-18.
Hassan. B. T. M.: 1979, Connections associated with linear maps on the induced bundle of a Finsler space. I, Proc. Math. Phys. Soc. Egypt 47, 1-6.
Hassan, B. T. M.: 1981, Sprays and Jacobi fields in Finsler geometry, Ann. Univ. Timisoara, Sti. Mat. 19, 129-139.
Hassan, B. T. M. and Tamin, A. A.: 1982, On a Finsler space with generalized Randers metric in Differential Geometry, North-Holland, Amsterdam-New York, pp. 259-266.
Heil, E.: 1965a, A relation between Finslerian and Hermitian metrics, Tensor, N. S. 16, 1-3.
Heil, E.: 1965b, Zur affinen Differentialgeometrie der Eilinien, Dissertation, Darmstadt, 51 p.
Heil, E.: 1966, Eine Charakterisierung lokal-Minkowskischer Räume Math. Ann. 167, 64-70.
Heil, E.: 1967a, Abschätzungen für einige Affininvarianten konvexer Kurven, Monatsh. Math. 71, 405-423.
Heil, E.: 1967b, Scheitelsätze in der euklidischen, affinen und Minkowskischen Geometrie, Darmstadt, 73 p.
Heil, E.: 1970, Der Vierscheitelsatz in Relativ- und Minkowski-Geometrie, Monatsh. Math. 74, 97-107.
Heil, E. and Laugwitz, D.: 1974, Finsler spaces with similarity are Minkowski spaces, Tensor, N. S. 28, 59-62.
Hermes, H.: 1972, The geometry of time-optimal control, SIAM J. Control 10, 221-229.
Heskia, S.: 1971, A new viewpoint on the space-time model of elementary particles. I. II, Progr. Theor. Phys. 45, 277-294, 640-648.
Hiramatu, H.: 1954a, Groups of homothetic transformations in a Finsler space, Tensor, N. S. 3, 131-143.
Hiramatu, H.: 1954b, On some properties of groups of homothetic transformations in Riemannian and Finslerian spaces, Tensor, N. S. 4, 28-39.
Hiramatu, H.: 1959, On n-dimensional Finslerian manifolds admitting homothetic transformation groups of dimension > n(n-1)/2 + 1, Kumamoto J. Sci. A 4, 4-10.
Hit, R.: 1975, Decomposition of Berwald's curvature tensor fields, Ann. Fac. Sci. Kinshasa, Zaire, Math. Phys. 1, 220-226.
Hōjō, S.: 1980, On geodesics of certain Finsler metrics, Tensor, N. S. 34, 211-217.
Hōjō, S.: 1981a, Structure of fundamental functions of S3-like Finsler spaces, J. Math. Kyoto Univ. 21, 787-807.
Hōjō, S.: 1981b, On the determination of generalized Cartan connections and fundamental functions of Finsler spaces, Tensor, N. S. 35, 333-344.
Hōjō, S.: 1982, On genralizations of Akbar-Zadeh's theorem in Finsler geometry, Tensor, N. S. 37, 285-290.
Hokari, S.: 1936, Winkeltreue Transformationen und Bewegungen im Finslerschen Raume, J. Fac. Sci. Hokkaido Univ. I 5, 1-8.
Hölder, E.: 1957, Über die auf Extremalintegrale gegründeten metrischen Raume, Schr. Forschungsinst. Math. 1, 178-193.
Hölder, E.: 1970, Navigationsformel zu A. Busemanns Variationsproblem der Raumfahrt, Celestial Mech. 2, 435-447.
Hombu, H.: 1934, Konforme Invarianten im Finslerschen Raume, J. Fac. Sci. Hokkaido Univ. I 2, 157-168.
Hombu, H.: 1935, Zur Theorie der unitären Geometrie, Fac. Sci. Hokkaido Univ. I 3, 27-42.
Hombu, H.: 1935-1936, Konforme Invarianten im Finslerschen Raume, II, Fac. Sci. Hokkaido Univ. I 4, 51-66.
Hombu, H.: 1936-1937, Die Krümmungstheorie im Finslerschen Raume, Fac. Sci. Hokkaido Univ. I 5, 67-94.
Horváth, J. I.: 1950, A geometrical model for the unified theory of physical fields, Phys. Rev. (2), 80, 901.
Horváth, J. I.: 1956a, On the theory of the electromagnetic field in moving dielectrics, Bull. Acad. Polon. Sci. III 4, 447-452.

Horváth, J. I.: 1956b, Contribution to Stephenson-Kilmister's unified theory of gravitation and electromagnetism, Nuovo Cimento (10) 4, 571-576.

Horváth, J. I.: 1956c, Contributions to the unified theory of physical fields, Nuovo Cimento (10) 4, 577-581.

Horváth, J. I.: 1958a, Classical theory of physical fields of second kind in general spaces, Acta. Phys. Chem. Szeged 4, 3-17.

Horváth, J. I.: 1958b, New geometrical methods of the theory of physical fields, Nuovo Cimento (10) 9, Suppl. 444-496.

Horváth, J. I.: 1961, A possible geometrical interpretation of the isospace and of its transformations, Acta Phys. Chem. Szeged 7, 3-16.

Horváth, J. I.: 1963, Internal structure of physical fields, Acta Phys. Chem. Szeged 9, 3-24.

Horváth, J. I.: 1967, Zero-point kinetic energy of relativistic Fermion gases, Acta Phys. Chem. Szeged 13, 3-19.

Horváth, J. I.: 1968a, On the hyper-geometrization of relativistic phase-space formalism, I, II, Acta Phys. Acad. Sci. Hungar. 24, 205-223, 347-371.

Horváth, J. I.: 1968b, On the hyper-geometrization of relativistic phase-space formalism, III, Acta Phys. Acad. Sci. Hungar. 25, 1-15.

Horváth, J. I. and Gyulai, J.: 1956, Über die Erhaltingssätze des elektromagnetischen Felden in bewegten Dielektriken, Acta. Phys. Chem. Szeged 2, 39-48.

Horváth, J. I. and Moór, A.: 1952, Entwicklung einer einheitlichen Feldtheorie begründet auf die Finslersche Geometrie, Z. Physik 131, 544-570.

Horváth, J. I. and Moór, A.: 1955, Entwicklung einer Feldtheorie begründet auf einen allgemeinen metrischen Linienelementraum. I, II, Nederl. Akad. Wetensch. Proc. A 58 = Indag. Math. 17, 421-429, 581-587.

Hosokawa, T.: 1930, On the various linear displacements in the Berwald-Finsler's manifold, Sci. Rep. Tokyo 19, 37-51.

Hosokawa, T.: 1932, Conformal property of a manifold B_n, Jap. Math. 9, 59-62.

Hosokawa, T.: 1938, Finslerian wave geometry and Milne's world-structure, J. Sci. Horoshima Univ. A 8, 249-270.

Householder, A. S.: 1933-1937, The dependence of a focal point upon curvature in the calculus of variations, Contrib. Calcul. Varia., pp. 485-526, Thesis, Chicago.

Hozyo, S.: 1964, Pair connections on homogeneous spaces which are invariant under tangential transformations, Mem. Fac. Eng. Yamaguchi Univ. 14, 135-139.

Hozyo, S.: 1965, On the connection mapping of the induced (C1)-pair connections on homogeneous spaces, Mem. Fac. Eng. Yamaguchi Univ. 16, 33-37.

Hu, H.: 1959, A Finslerian product of two Riemannian spaces, Sci. Record, N. S. 3, 446-448.

Humbert, P.: 1939a, Sur l'espace attaché à la forme Δ_3, C. R. Acad. Sci. Paris 208, 1965-1966.

Humbert, P.: 1939b, Sur les courbes planes de l'espace attaché à l'opérateur Δ_3, C. R. Acad. Sci. Paris 209, 590-591.

Humbert, P.: 1940, Sur certaines figures planes de l'espace attaché à l'operateur Δ_3, C. R. Acad. Sci. Paris 211, 530-531.

Humbert, P.: 1941a, Sur une extension de la notion d'angle: angles d'un faisceau de trois droites, C. R. Acad. Sci. Paris 213, 970-971.

Humbert, P.: 1941b, Sur la géométrie plane dans l'espace attaché à l'operateur Δ_3, Ann. Univ. Lyon (3) A 4, 93.

Humbert, P.: 1942a, Géométrie plane dans l'espace attaché à l'operateur Δ_3, J. Math. Pures Appl. (9) 21, 141-153.

Humbert, P.: 1942b, Sur certaines figures planes de l'espace attaché à l'operateur Δ_3, Bull. Sci. Math. (2) 66, 145-154.

Humbert, P.: 1944, Bitétraèdres de l'espace attaché à l'opérateur Δ_3, Bull. Sci. Math. (2) 68, 50-59.

Humbert, P.: 1946, Formules trigonométriques dans le plan et l'espace attaché à l'operateur Δ_3, Ann. Soc. Sci. Bruxelles I 60, 196-199.

Hurt, N. E.: 1972a, Topology of quantizable dynamical systems and the algebra of observables, Ann. Inst. Poincaré, N. S. A 16, 203-217.

Hurt, N. E.: 1972b, Homogeneous fibered and quantizable dynamical systems, Ann. Inst. Poincaré, N. S. 16, 219-222.

Hyland, G. J.: 1979, A nonlocaö spinor field theory of matter, General Relativity and Gravitation 10, 231-252.

Ichijyō, Y.: 1967, Almost complex structures of tangent bundles and Finsler metrics, J. Math. Kyoto Univ. 6, 419-452.

Ichijyō, Y.: 1976a, Finsler manifolds modeled on a Minkowski space, J. Math. Kyoto Univ. 16, 639-652.

Ichijyō, Y.: 1976b, Finsler manifolds with a linear connection, J. Math. Tokushima Univ. 10, 1-11.

Ichijyō, Y.: 1978a, On special Finsler connections with the vanishing hv-curvature tensor, Tensor, N. S. 32, 149-155.

Ichijyō, Y.: 1978b, On the Finsler connection associated with a linear connection satisfying $P^h_{ikj} = 0$, J. Math. Tokushima Univ. 12, 1-7.

Ichijyō, Y.: 1979, On the conditions for a {V, H}- manifold to be locally Minkowskian or conformally flat, J. Math. Tokushima Univ. 13, 13-21.

Ichijyō, Y.: 1980, On the G-connections and motions on a {V, G}- manifold, J. Math. Tokushima Univ. 14, 11-23.

Ichijyō, Y.: 1982, Almost Hermitian Finsler manifolds, Tensor, N. S. 37, 279-284.

Ikeda, F.: 1979a, On two-dimensional Landsberg spaces, Tensor, N. S. 33, 43-48.

Ikeda, F.: 1979b, On the tensor T_{ijkl} of Finsler spaces, Tensor, N. S. 33, 203-209.

Ikeda, F.: 1980, On the tensor T_{ijkl} of Finsler spaces, II, Tensor, N. S. 34, 85-93.

Ikeda, F.: 1981, On S3- and S4-like Finsler spaces with the T-tensor of a special form, Tensor, N. S. 35, 345-351.

Ikeda, S.: 1975a, Prolegomena to Applied Geometry, Mahā Shobō, Koshigaya-city, 215 p.

Ikeda, S.: 1975b, A structurological consideration on a unified field, Lett. Nuovo Cimento 13, 497-500.

Ikeda, S.: 1976a, Some structurological remarks on a nonlocal field, Inter. J. Theor. Phys. 15, 377-387.

Ikeda, S.: 1976b, Some Finslerian features underlying the theory of physical fields, Lett. Nuovo Cimento 15, 623-626.

Ikeda, S.: 1977, Some physico-geometrical remarks on the theory of fields in Finsler spaces, Lett. Nuovo Cimento 18, 29-32.

Ikeda, S.: 1978a, On the geometrical theory of 'nonlocal' fields characterized by internal variables, Lett. Nuovo Cimento 21, 165-168.

Ikeda, S.: 1978b, Some constructive comments on the theory of fields in Finsler spaces, Lett. Nuovo Cimento 21, 567-571.

Ikeda, S.: 1978c, Some structurological remarks on the theory of fields in Finsler spaces, Lett. Nuovo Cimento 23, 449-454.

Ikeda, S.: 1979a, Some physico-geometrical remarks on the relationship between the micro- an macro-gravitational fields, Lett. Nuovo Cimento 25, 21-25.

Ikeda, S.: 1979b, Some physico-geometrical features underlying the theory of fields in Finsler spaces, Lett. Nuovo Cimento 25, 26-31.

Ikeda, S.: 1979c, On the theory of gravitational field in Finsler spaces, Lett. Nuovo Cimento 26, 277-281.

Ikeda, S.: 1979d, Some constructive remarks on the theory of fields in Finsler spaces, Lett. Nuovo Cimento 26, 313-316.

Ikeda, S.: 1980a, On the theory of fields in Finsler spaces, Post-RAAG Reports, No.130, 27 p.

Ikeda, S.: 1980b, On the theory of gravitational field in Finsler spaces, Proc. Einstein Centenary Symp., Napur, India, pp. 155-164.

Ikeda, S.: 1980c, Some physical aspects underlying the theory of fields in Finsler spaces, Lett. Nuovo Cimento 28, 541-544.

Ikeda, S.: 1980d, On the theory of fields in Finsler spaces, Acta Phys. Polon. B13, 321-324.

Ikeda, S.: 1980e, A differential geometrical consideration on a 'nonlocal' field, Rep. Math. Phys. 18, 103-110.

Ikeda, S.: 1981a, Some physical aspects induced by the internal variables in the theory of fields in Finsler spaces, Nuovo Cimento (11) 61B, 220-228.

Ikeda, S.: 1981b, Some structurological features induced by the internal variable in the theory of fields in Finsler spaces, Progr. Theor. Phys. 65, 2075-2078.

Ikeda, S.: 1981c, A structural consideration on the Brans-Dicke Scalar $\phi(x)$, Progr. Theor. Phys. 66, 2284-2286.

Ingarden, R. S.: 1948, Problèmes P 58, Colloq. Math. 1, 334.

Ingarden, R. S.: 1954, Über die Einbettung eines Finslerschen Raumes in einem Minkowskischen Raum, Bull. Acad. Polon. Sci. III 2, 305-308.

Ionusauskas, A.: 1965, Über die Existenz von invarianten Finslerschen Metriken in homogenen Räumen, Liet. Mat. Rinkinys 5, 45-55 (in Russian).

Ionusauskas, A.: 1966a, Existenz von invarianten Finslerschen Metriken in homogenen Räumen mit linearer Isotropiegruppe tensorischen Typus, Liet. Mat. Rinkinys 6, 51-57 (in Russian).

Ionusauskas, A.: 1966b, The existence of invariant Finsler metrics in certain uniform spaces, Liet. Mat. Rinkinys 6, 621-622 (in Russian).
Ionusauskas, A.: 1967, Existenz von invarianten Finslerschen Metriken in homogenen Räumen mit linearer Isotropiegruppe tensorschen Typus. II, Liet. Mat. Rinkinys 7, 619-631 (in Russian).
Ispas, C. I.: 1952a, Identités de type Ricci dans l'espace de Finsler, Com Acad. R. P. Romane 2, 13-18 (in Roumanian).
Ispas, C. I.: 1952b, Les identités de Veblen dans les espaces généralisés, Acad. R. P. Romane, Bul. Sti. Sect. Sti. Mat. Fiz. 4, 533-539 (in Roumanian).
Ispas, C. I.: 1955, Les déformations itérées. Sur les déformations des géodésiques dans les espaces généralises, Acad. R. P. R. Baza Timisoara, Lucr. Consf. Geom. Dif., pp. 253-261 (in Roumanian).
Ispas, C. I.: 1967, Die Gleichwertigkeit der Bianchischen und Veblenschen Identitäten in den Finslerschen Räumen, Rev. Roum. Math. Pures Appl 12, 1467-1478.
Ispas, C. I.: 1972, On the equivalence of Bianchi and Veblen identities in Finsler-Cartan-Kawaguchi spaces, Tensor, N. S. 26, 468-476.
Ispas, C. I.: 1974, On the equivalence of Bianchi and Veblen identities in Finsler-Cartan-Kawaguchi spaces (II), Proc. Inst. Math. Iasi, pp. 51-55.
Ispas C. I.: 1980, Finsler-Cartan-Kawaguchi spaces, Proc. Nat. Sem. Finsler Spaces, Brasov, pp. 77-106.
Ispas, C. I.: 1982, Infinitesimal deformations theory and Lie derivatives, Tensor, N. S. 39, 249-258.
Ispas, C. I.: 1983, Finsler recurrent spaces of 'p' order; Bianchi and Veblen identities: their equivalence, Proc. Nat. Colloq. Geometry and Topology, Busteni, 1981, Univ. Bucharest, pp. 171-178.
Ispas, C. I. and Ispas, M. I.: 1973, Les espaces Finsler-Berwald, Bul. Stil Inst. Const. Bucuresti 16, 23-28 (in Roumanian).
Ispas, C. I. and Ispas, M. I.: 1980, About some problems of plasticity theory, Proc. Nat. Sem. Finsler Spaces, Brasov, pp. 107-119.
Ispas, C. I. and Oncescu, G. F.: 1967, A new formulation of the problem of continuous deformations in Finsler spaces, Bull. Inst. Petrol. Gaze si Geologie 16, 257-267 (in Roumanian).
Ispas, M. I.: 1969a, Systèmes de coordonées convectives dans la théorie de la plasticité, dans le cadre des espaces de Finsler, Bul. Sti. Inst. Const. Bucuresti 12, 23-25 (in Roumanian).
Ispas, M. I.: 1969b, Plasticity equations in Finsler's three-dimensional spaces with convective coordinates, Bul. Sti. Inst. Const. Bucuresti 12, 27-30.
Izumi, H.: 1976, On *P-Finsler spaces, I, Memo. Defense Academy Japan 16, 133-138.
Izumi, H.: 1977a, Conformal transformations of Finsler spaces. I. Concircular transformations of a curve with Finsler metric, Tensor, N. S. 31, 33-41.
Izumi, H.: 1977b, On *P-Finsler spaces, II. Memo. Defense Academy Japan 17, 1-9.
Izumi, H.: 1980, Conformal transformations of Finsler spaces. II. An h-conformally flat Finsler space, Tensor, N. S. 34, 337-359.
Izumi, H.: 1982, On *P-Finsler space of scalar curvature, Tensor, N. S. 38, 220-221.
Izumi, H.: 1983, Non-holonomic frames in a Finsler space with a 1-form metric, Tensor, N. S. 40, 189-192.
Izumi, H. and Srivastava, T. N.: 1978, On R3-like Finsler spaces, Tensor, N. S. 32, 339-349.
Izumi, H. and Yoshida, M.: 1978, On Finsler spaces of perpendicular scalar curvature, Tensor, N. S. 32, 219-224.
Izumi, H. and Yoshida, M.: 1983, Remarks on Finsler spaces of perpendicular scalar curvature and the property H, Tensor, N. S. 40, 215-220.
Jaglom, I. M.: 1971, A certain extremal property of the number π, Moskov, Gos. Ped. Inst. Uchen. Zap. 401, 135-138 (in Russian).
John, V. N.: 1980, Investigations in Randers and Finsler spaces: Thesis, Gorakhpur.
Johnson, M. M.: 1931, Tensors of the calculus of variations, Amer. J. Math. 53, 103-116.
Kabanov, N. I.: 1961, A singular Finsler space defined to within Carathéodory mappings, Sibirsk. Mat. Z. 2, 655-671 (in Russian).
Kabanov, N. I.: 1968, Differential-geometric methods in the calculus of variations, Algebra, Topology, Geometry, pp. 191-224 (in Russian).
Kagan, F. I.: 1964, On two-dimensional Finsler spaces admitting a singular embedding in a three-dimensional affine space with a vector metric, Izv. Vyss. Uchebn. Zaved. Mat., No. 1(38), 46-55 (in Russian).
Kagan, F. I.: 1965, On groups of pseudo-motions in spaces of Finsler and Riemann, Volz. Mat. Sb. 3, 190-196 (in Russian).
Kagan, F. I.: 1966a, On the operation of an infinitesimal S-extension with respect to a one-dimensional S-distribution on X_n, Volz. Mat. Sb. 4, 103-108 (in Russian).

Kagan, F. I.: 1966b, The operation of an infinitesimal S-extension, Izv. Vyss. Uchebn. Zaved. Math., No. 1(50), 66-78 (in Russian).
Kaljuznyi, V. N.: 1974, Commutative groups of isometries of Minkowski spaces, Sibirsk. Mat. Z. 15, 1138.1142, 1182 (in Russian). English translation: 1974, Siber. Math. J. 15, 801-803.
Kashiwabara, S.: 1958, On Euclidean connections in a Finsler manifolds, Tohoku Math. J. (2) 10, 69-80.
Katok, A. B.: 1973, Ergodic perturbations of degenerate integrale Hamiltonian systems, Izv. Akad. Nauk SSSR Mat. 37, 539-576 (in Russian). English translation: 1973, Math. USSR, Izv. 7, 535-571.
Katsurada, Y.: 1950, On the connection parameters in a non-holonomic space of line elements, J. Fac. Sci. Hokkaido Univ. I, 11, 129-149.
Katsurada, Y.: 1951, On the theory of non-holonomic systems in the Finsler space, Tohoku Math. J. (2) 3, 140-148.
Kaul, R. N.: Curvatures in Finsler space, Bull. Calcutta Math. Soc. 50, 189-192.
Kawaguchi, A.: 1937, Beziehung zwischen einer metrischen linearen Uebertragung und einer nichtmetrischen in einem allgemeinen metrischen Raume, Akad. Wetensch. Amsterdam Proc. 40, 596-601.
Kawaguchi, A.: 1952, On the theory of non-linear connections I. Introduction to the theory of general non-linear connections, Tensor, N. S. 2, 123-142.
Kawaguchi, A.: 1953, On the theory of non-linear connections, Conveg. Intern. Geom. Diff. Italia, pp. 27-32.
Kawaguchi, A.: 1956, On the theory of non-linear connections II. Theory of Minkowski spaces and of non-linear connections in a Finsler space, Tensor. N. S. 6, 165-199.
Kawaguchi, A. and Laugwitz. D.: 1957, Remarks on the theory of Minkowski spaces, Tensor, N. S. 7, 190-199.
Kawaguchi, H.: 1977, Geometry of Finsler spaces as a mathematical information model, Sem. Note, Fac. Eng., Hokkaido, pp. 1-30 (in Japanese).
Kawaguchi, H.: 1978, A Minkowski space closely related to a certain special Finsler space, Tensor, N. S. 32, 114-118.
Kawaguchi, H.: 1980, Some characterization of the vector field A_i proper to Finsler spaces, Tensor, N. S. 34, 367-372.
Kawaguchi, T.: On the application of Finsler geometry to engineering dynamical systems, Period. Math. Hungar. 8, 281-289.
Kawaguchi-Kanai, T.: 1969a, On Finsler spaces with special line elements, Gakuen Rev. 15, 39-44.
Kawaguchi-Kanai, T.: 1969b, On the covariant differentiation of rheonimic Finsler spaces, Gakuen Rev. 15, 45-51.
Kawaguchi-Kanai, T.: 1970, On the connection parameters and the stretch tensor of rheonomic Finsler spaces, Gakuen Rev. 16, 21-28.
Kawaguchi-Kanai, T.: 1971, On the curvature strong tensors and the identities of Ricci and Bianchi in rheonomic Finsler spaces, Gakuen Rev. 18, 63-75.
Kelly, L. M.: 1975, On the equilateral feeble four-point property, Geom. Metric Lin. Spaces, Proc. Conf. East Lansing 1974, Lect. Notes Math. 490, pp. 14-16.
Kern, J.: 1971a, Fastriemannsche Finslersche Metriken, Manuscripta Math. 4, 285-303.
Kern, J.: 1971b, Das Pinchingproblem in Fastriemannschen Finslerschen Mannigfaltigkeiten, Manuscripta Math. 4, 341-350.
Kern, J.: 1974, Lagrange geometry, Arch. Math. 25, 438-443.
Kern, J.: 1977, Finsler manifolds, Proc. 9th Brazil Math. Coll., I, 227-231 (in Portuguese).
Kerner, E. H.: 1976, Extended inertial frames and Lorentz transformations. II., J. Math. Phys. 17, 1797-1807.
Kikuchi, S.: 1952, On the theory of subspace in a Finsler space, Tensor, N. S. 2, 67-79.
Kikuchi, S.: 1962, Theory of Minkowski space and of non-linear connections in a Finsler space, Tensor, N. S. 12, 47-60.
Kikuchi, S.: 1968, On some special Finsler spaces, Tensor, N. S. 19, 238-240.
Kikuchi, S.: 1979, On the condition that a space with (α, β)-metric be locally Minkowskian, Tensor, N. S. 33, 242-246.
Kilmister, C. W. and Stephenson, G.: 1954, An axiomatic criticism of unified field theories, I, II, Nuovo Cimento 11, Suppl. 91-105, 118-140.
Kirznits, D. A. and Chechin, V. A.: 1972, Ultra-high-energy cosmic rays and a possible generalization of relativistic theory, Soviet J. Nuclear Phys. 15, 585-589.
Kishta, M. A.: 1975, Einstein-Finsler spaces, Dirāsāt Res. J. Natur. Sci. 2, 108-116.
Kitamura, S.: 1960, On Finsler spaces with the fundamental function $L = g_{ij}(x)\dot{x}^i\dot{x}^j + a_i(x)\dot{x}^i$: Thesis, Kyoto, 36 p.
Kitayama, M. and Kikuchi, S.: 1984, On Finsler spaces with $R^{\ i}_{0hk} = 0$, Tensor, N. S. 41, 90-92.

Klein, J.: 1954, Sur les trajectories d'un système dynamique dans un espace finslérien ou variationnel généralisé, C. R. Acad. Sci. Paris 238, 2144-2146.
Klein, J.: 1962, Espaces variationnels et mécanique, Ann. Inst. Fourier 12, 1-124.
Klein, J. and Voutier, A.: Formes extérieures génératrices de sprays, Ann. Inst. Fourier 18, 241-260.
Klepp, F. C.: 1978, On some technical applications of Finsler spaces, Bul. Sti. Teh. Inst. Politeh. Timisoara, Mat.-Fiz. 23(37), 106-108 (in Roumanian).
Klepp, F. C.: 1980, Applications of Finsler spaces in the study of the physical fields, Proc. Nat. Sem. Finsler Spaces, Brasov, pp. 121-129.
Klepp, F. C.: 1981, Finsler geometry on vector bundles, An. Sti. Univ. Iasi, Ia 27, Suppl. 37-42.
Klepp, F. C.: 1982, Remarkable Finsler structures and connections on Finsler vector bundles, Proc. Nat. Sem. Finsler Spaces, Brasov, pp. 125-140.
Klepp, F. C.: 1983a, Metrical almost product Finsler structures, An. Sti. Univ. Iasi Ia 29, No. 2, Suppl. 21-25.
Klepp, F. C.: 1983b, Almost product Finsler structures and connections, Proc. Nat. Colloq. Geometry and Topology, Busteni, 1981, pp. 187-193.
Knebelman, M. S. 1927a, Groups of collineations in a space of paths, Proc. Nat. Acad. Sci. U. S. A. 13, 396-400.
Knebelman, M. S. 1927b, Motion and collineations in general space, Proc. Nat. Acad. Sci. U. S. A. 13, 607-611.
Knebelman, M. S.: 1929a, Conformal geometry of generalized metric spaces, Proc. Nat. Acad. Sci. U. S. A. 15, 376-379.
Knebelman, M. S.: 1929b, Collineations and motions in generalized spaces, Amer. J. Math. 51, 527-564.
Kondo, K.: 1953, On the theoretical investigation based on abstract geometry of dynamical systems appearing in engineering, Proc. 3rd. Japan Nat. Congr. Appl. Mech., pp. 425-432.
Kondo, K.: 1962, A Finslerian approach to space-time and some microscopic as well as macroscopic criteria with references to quantization, mass spectrum and plasticity, RAAG Memo., No. 3-E-VIII, 307-318.
Kondo, K.: 1963, Non-Riemannian and Finslerian approaches to the theory of yielding, Intern. J. Engineering Sci. 1, 71-88, 422.
Kondo, K.: 1973, A pseudo-Finslerian picture of energy conservation, Post RAAG Memo., No. 4, 24 p.
Kondo, K.: 1974, On the osculatory character of a geometrical picture of the Voigt transformation and its modification, Post RAAG Memo. No. 14, 30 p.
Kondo, K.: 1980, Reorganization of the higher order space picture of the world of elementary particles. I: Imbedding the quantum-mechanical formalism in the geometry of higher order spaces, Post RAAG Rep., No. 109, 37 p.
Kondo, K. and Amari, S.: 1968, A constructive approach to the non-Riemannian features of dislocation and spin distributions in terms of Finsler's geometry and a possible extension of the space-time formalism, RAAG Memo. No. 4-D, 225-238.
Kondo, K. and Fujinaka, M.: 1955, Geometrization of dissipation and general non-linearity in dynamical equations with a preliminary reference to hysteresis, RAAG Memo., No. I-B, 335-355.
Kondo, K. and Kawaguchi, T.: 1966, On the basic principle of the analysis of magnetic hysteresis by a Finslerian approach, RAAG Res. Notes, No. 3-111, 20 p.
Kondo, K. and Kawaguchi, T.: 1968, On the origin of the hysteresis in the Finslerian magnetic dynamical system, RAAG Memo., No. 4-B, 65-72.
Kosambi, D.: 1938, Les espaces des paths généralisés qu'on peut associer avec un espace de Finsler, C. R. Acad. Sci. Paris 206, 1538-1541.
Koschmieder, L.: 1931, Die neuere formale Variationsrechnung, Jber. Deutsch. Math.-Verein. 40, 109-132.
Kovanzov, N. I. and Nagy, P.: 1971, The moving frame method in a space of line elements with a general transformation law, Publ. Math. Debrecen 18, 77-87 (in Russian).
Kowolik, J.: 1975, A remark on geodesics, Zeszyty Nauk 18, Zastos. Math., 17-19 (in Polish).
Kramer, H. an Németh, A. B.: 1973, Equally spaced points for families of compact convex sets in Minkowski spaces, Math. Cluj 15 (38), 71-78.
Kreter, R.: 1956/1957, Zusammenhänge in Finslerschen Räumen, Wiss. Z. Humbolt Univ. Math. Nat. R. 6, 353-365 (in Russian).
Kropina, V. K.: 1959a, On projective Finsler spaces with a metric of some special form, Nauchn. Dokl. Vyss. Skoly, Fiz-Mat. Nauki No. 2, 38-42 (in Russian).
Kropina, V. K.: 1959b, On projective Finsler spaces, Uch. Zap. Arkhang. Gos. Ped. Inst. 4, 111-118 (in Russian).
Kropina, V. K.: 1960, On introduction of absolute differentiation in Finsler space, Uch. Zap. Yaroslavsk. Gos. Ped. Inst. 34, 113-123 (in Russian).

Kropina, V. K.: 1961, On projective two-dimensional Finsler spaces with a special metric, Trudy Sem. Vektor. Tenzor. Anal. 11, 277-292 (in Russian).
Ku, C.: 1956, Imbedding of a Finsler space in a Minkowski space, Acta Math. Sinica 6, 215-232 (in Chinese, Russian summary).
Ku, C.: 1957, On Finsler spaces admitting a group of motions on the greatest order, Sci. Record., N. S. 1, 215-218.
Ku, C.: 1958, Embedding of Finsler manifolds in a Minkowski space, Acta Math. Sinica 8, 272-275 (in Chinese, Russian summary). English tranlations: 1964, Amer. Math. Soc. Transl. II 37, 253-257, 1966, Chinese Math. 8, 878-882.
Kumar, A.: 1975a, On special projective tensor fields, Lincei Rend. (8) 58, 184-189.
Kumar, A.: 1975b, Decomposition of pseudo-projective tensor fields in a second order recurrent Finsler space, Lincei Rend. (8) 59, 77-82.
Kumar, A.: 1976a, On a generalised 2-projective recurrent Finsler space, Bull. Acad. Serbe Sci. Arts Cl. Sci. Math. Natur. 55, No. 9, 49-54.
Kumar, A.: 1976b, Some theorems in projective recurrent Finsler space of second order, Acta Cienc. Indica 2, 275-277.
Kumar, A.: 1976c, On W-generalised 2-recurrent Finsler space, Acta Cienc. Indica 2, 393-396.
Kumar, A.: 1976d, On a special recurrent Finsler space of second order, Lincei Rend. (8) 61, 432-437.
Kumar, A.: 1977a, On the existence of affine motion in a recurrent Finsler space, Indian J. Pure Appl. Math. 8, 791-800.
Kumar, A.: 1977b, On the existence of projective affine motion in a projective recurrent Finsler space, Acta Cienc. Indica 3, 243-249.
Kumar, A.: 1977c, Decomposition of projective recurrent tensor field of second order, Acta Cienc. Indica 3, 89-92.
Kumar, A.: 1977d, On a projective recurrent Finsler space, Acta Cienc. Indica 3, 177-180.
Kumar, A.: 1977e, On the existence of special projective affine motion in a recurrent Finsler space. II, Acta Cienc. Indica 3, 372-377.
Kumar, A.: 1977f, On recurrence vectors in a F_n^*-space, Lincei Rend (8) 62, 463-470.
Kumar, A.: 1977g, On some type of affine motion in bi-recurrent Finsler spaces. II, Indian J. Pure Appl. Math. 8, 505-513.
Kumar, A.: 1977h, Some theorems of affine motion in a recurrent Finsler space. IV, Indian J. Pure Appl. Math. 8, 672-684.
Kumar, A.: 1977i, Some theorems on affine motion in a recurrent Finsler space, Indian J. Pure Appl. Math. 8, 1176-1181.
Kumar, A.: 1978a, On some type of affine motion in birecurrent Finsler spaces, J. Math. Phys. Sci. 12, 529-545.
Kumar, A.: 1978b, Decomposition of projective tensor fields in a recurrent Finsler space of second order, Acta Cienc. Indica 4, 52-55.
Kumar, A.: 1978c, On a W-recurrent Finsler space, Acta Cienc. Indica 4, 208-211.
Kumar, A.: 1978d, A remark on S-PRF-space of the first kind, J. Math. Phys. Sci. 12, 287-295.
Kumar, A.: 1978e, Curvature collineation in a projective symmetric Finsler space, Acta Cienc. Indica 4, 298-301.
Kumar, A.: 1978f, Some theorems on special projective recurrence vector in Finsler space, Acta Cienc. Indica 4, 429-432.
Kumar, A.: 1978g, Projective motion and projective curvature collineation in a Finsler space, Acta Cienc. Indica 4, 433-435.
Kumar, A.: 1978h, On a special B-R-F_n-space, J. Math. Phys. Sci. 12, 483-490.
Kumar, A.: 1978i, On projective motion in a recurrent Finsler space, J. Math. Phys. Sci. 12, 497-505.
Kumar, A.: 1978j, On some type of affine motion in bi-recurrent Finsler space, J. Math. Phys. Sci. 12, 529-545.
Kumar, A.: 1978k, On projective motion in an F_n^*-space, J. Math. Phys. Sci. 12, 559-567.
Kumar, A.: 1978l, On a special bi-recurrent Finsler space. I, II, Indian J. Pure Appl. Math. 9, 1241-1253.
Kumar, A.: 1979a, On a Q-recurrent Finsler space of second order, Publ. Math. Debrecen 26, 13-15.
Kumar, A.: 1979b, Some theorems on a special bi-SPRF$_n$-space, J. Math. Phys. Sci. 13, 21-28.
Kumar, A.: 1979c, On special quasisymmetric Finsler spaces and BR-F_n-space, Acta Cienc. Indica 5, 198-200.
Kumar, A.: 1979d, On a remained affine motion in F_n^*-space, J. Math. Phys. Sci. 13, 257-264.
Kumar, A.: 1980, Some theorems on affine motion in a recurrent Finsler space II, Indian J. Pure Appl. Math. 11, 40-53.

Kumar, A.: 1981a, On affine motion in a recurrent Finsler space, Istanbul Univ. Fen. Fak. Mecm. Ser. A 41 (1976), 13-20.
Kumar, A.: 1981b, Affine motion in a recurrent Finsler space, Indian J. Math. 23, 193-199.
Kumar, A.: 1981c, Some theorem on special conformal motion in a F^*_n-space, Acta Cienc. Indica 7, 87-91.
Kumar, A.: 1981d, On a special quasiprojective symmetric space and BSPR-F_n-space, Acta Cienc. Indica 7, 92-95.
Kumar, A.: 1981e, On special BRP-F_n spaces. II, Acta Cienc. Indica 7, 141-145.
Kumar, A.: 1982a, On affine motion in a recurrent Finsler space, III, Acta Cienc. Indica 8, 79-88.
Kumar, A.: 1982b, Some theorems on special conformal motion in a symmetric Finsler space, Acta Cienc. Indica 8, 195-198.
Kumar, A. and Dubey, G. C.: 1975, General decomposition of pseudoprojective curvature tensor field in recurrent Finsler space, Lincei Rend. (8) 58, 708-712.
Kumar, A. and Pandey, J. P.: 1976, On the existence of special projective motion in a recurrent Finsler space, Rev. Fac. Sci. Univ. Istanbul Ser. A 41, 107-118.
Kumar, A. and Pandey, T. N.: 1980, Projective motion in a projective symmetric Finsler space, J. Math. Phys. Sci. 14, 515-522.
Künzle, H. P.: 1969, Degenerate Lagrangean systems, Ann. Inst. Poincaré, N. S. A 11, 393-414.
Kurita, M.: 1963, On the dilatation in Finsler spaces, Osaka Math. J. 15, 87-98.
Kurita, M.: 1966, Theory of Finsler spaces based on the contact structure, J. Math. Soc. Japan 18, 119-134.
Kurita, M.: 1970, Formal foundation of analytical dynamics based on the contact structure, Nagoya Math. J. 37, 107-119.
Laget, B.: 1971, Sur une méthode de déformation en géométrie Banachique, Publ. Dépt. Math. Lyon 8, 71-85.
Lal, K. B. and Agrawal, P.: 1971, Relative curvature tensors of subspace of a Finsler space, Indian J. Pure Appl. Math. 2, 769-777.
Lal, K. B. and Prasad, C. M.: 1966, The generalised curvatures of a congruence of curves in the subspace of a Finsler space, Rev. Fac. Sci. Univ. Istanbul 31, 49-56.
Lal, K. B. and Singh, S. S.: 1970, The commutation formulae involving partial derivatives and projective covariant derivatives in a Finsler space, Lincei Rend. (8) 49, 250-257.
Lal, K. B. and Singh, S. S.: 1971, Commutation formulae in conformal Finsler space. II, Lincei Rend. (8) 51, 346-351.
Lal, K. B. and Singh, S. S.: 1972, Commutation formulae in conformal Finsler space. I, Ann. Mat. Pura Appl. (4) 91, 119-127.
Landsberg, G.: 1907a, Über die Totalkrümmung, Jber. Deutsch. Math.-Verein. 16, 35-46.
Landsberg, G.: 1907b, Krümmungstheorie und Variationsrechnung, Jber Deutsch. Math.-Verein. 16, 547-551.
Landsberg, G.: 1908, Über die Krümmung in der Variationsrechnung, Math. Ann. 65, 313-349.
Laptev, B. L.: 1937, Covariant integration in a Finsler space of two and three dimensions, Izv. Fiz.-Mat. Obsch., Kazan 3(9), 61-76.
Laptev, B. L.: 1938, Lie derivative for the objects which are function of a point and direction, Iz. Fiz.-Mat. Obsch. Kazan. 3(10), 3-38 (in Russian).
Laptev, B. L.: 1940, The invariant form of the second variation obtainable by means of Lie differentiation in Finsler space, Izv. Fiz.-Mat. Obsch. Kazan. 3(12), 3-8.
Laptev, B. L.: 1956a, Lie derivative of geometric objects in a support element space, Proc. 3rd All-Union Math. Congr. 1, 157.
Laptev, B. L.: 1956b, The Lie derivation in the space of supporting elements, Trudy Sem. Vektor. Tenzor. Anal. 10, 227-248.
Laptev, B. L.: 1957, Lie derivative of geometric objects in a support element space, Uch. Zap. Kazansk. Univ. 112 (2), 16-18.
Laptev, B. L.: 1958a, The covariant differential and the theory of differential invariants in the space of tensor support elements, Kazan. Gos. Univ. Uchen. Zap. 118 (4), 75-147 (in Russian).
Laptev, B. L.: 1958b, Application of Lie differentiation to investigation of geodesics displacement in a space of linear elements, Izv. Vyss. Uchebn. Zaved. Mat. No. 2(3), 173-181 (in Russian).
Laptev, B. L.: 1961, The space of support elements. Proc. 4th All-Union Math. Congr. Leningrad, II, 221-226 (in Russian).
Laptev, B. L.: 1967, Lie differentiation, Algebra, Topology, Geometry, 1965, pp. 429-465 (in Russian). English translation: 1970, Progr. Math. 6, 229-269.
Laugwitz, D.: 1954, Konvexe Mittelpunktsbereiche und normierte Räume, Math. Z. 61, 235-244.
Laugwitz, D.: 1955, Zur geometrischen Begründung der Parallelverschiebung in Finslerschen Räumen, Arch. Math. 6, 448-453.

Laugwitz, D.: 1956a, Die Vektoübertragungen in der Finslerschen Geometrie und der Wegegeometrie, Nederl. Akad. Wetensch. Proc. A 59 = Indag. Math. 18, 21-28.
Laugwitz, D.: 1956b, Grundlagen für die Geometrie der unendlichdimensionalen Finslerräume, Ann. Mat. Pura Appl. (4) 41, 21-41.
Laugwitz, D.: 1956c, Zur projektiven und konformen Geometrie der Finsler-Räume, Arch. Math. 7, 74-77.
Laugwitz, D.: 1956d, Geometrische Behandlung eines inversen Problems der Variationsrechnung, Ann. Univ. Sarav. 5, 235-244.
Laugwitz, D.: 1957, Zur Differentialgeometrie der Hyperflächen in Vektorräumen und zur affingeometrischen Deutung der Theorie der Finsler-Räume, Math. Z. 67, 63-74.
Laugwitz, D.: 1957/58, Eine Beziehung zwischen affiner und Minkowskischer Differentialgeometrie, Publ. Math. Debrecen 5, 72-76.
Laugwitz, D.: 1958, Die Geometrie von H. Minkowski, Mathematikunterricht Heft 4, 27-42.
Laugwitz, D.: 1959, Beiträge zur affinen Flächentheorie mit Anwendungen auf die allgemein-metrische Differentialgeometrie, Bayer. Akad. Wiss. Math.-Natur. Kl. Abh. 93, 1-59.
Laugwitz, D.: 1960, Differentialgeometrie, Teubner, Stuttgart, 183 p.
Laugwitz, D.: 1961, Geometrical methods in the differential geometry of Finsler spaces, C. I. M. E., Geom. Calcolo Varia. 1-3, 49 p.
Laugwitz, D.: 1963, Über die Erweiterung der Tensoranalysis auf Mannigfaltigkeiten unendlicher Dimension, Tensor, N. S. 13, 295-304.
Laugwitz, D.: 1965, Differentialgeometrie in Vektorräumen, unter besonderer Berücksichtigung der unendlichdimensionalen Räumen, Friedr. Vieweg & Sohn, Braunschweig; VEB Deutsch. Verlag der Wiss., Berlin, 89 p.
Laugwitz, D.: 1975, A characterization of Minkowski spaces, Boll. Un. Mat. Ital. (4) 12, Suppl. 3, 267-270.
Laugwitz, D. and Lorch, E. R.: 1956, Riemann metrics associated with convex bodies and normed spaces, Amer. J. Math. 78, 889-894.
Lehmann, D.: 1964, Théorie de Morse en géométrie finslérienne, in Topologie et Géométrie Differentielle, Inst. H. Poincaré, Paris, pp. 1-9.
Levashov, A. E.: 1938, Generalized Finsler geometry and classical mechanics, Bull. Univ. Asie Centr. Tashkent 22, 109-118 (in Russian).
Levine, J.: 1951, Collineations in generalized spaces, Proc. Amer. Math. Soc. 2, 447-455.
Lewis, D. C.: 1949, Metric properties of differential equations, Amer. J. Math. 71, 294-312.
Liber, A. E.: 1952, On two-dimensional spaces with an algebraic metric, Trudy Sem. Vektor. Tenzor. Anal. 9, 319-350 (in Russian).
Lichnerowicz, A.: 1942, Sur une généralisation des espaces de Finsler, C. R. Acad. Sci. Paris 214, 599-601.
Lichnerowicz, A.: 1943a, Sur une extension du calcul des variations, C. R. Acad. Sci. Paris 216, 25-28.
Lichnerowitz, A.: 1943b, Les espaces variationnels généralisés, C. R. Acad. Sci. Paris 217, 415-418.
Lichnerowicz, A.: 1945, Les espaces variationnels généralisés, An. Sci. Ecole Norm. Sup. (3) 62, 339-384.
Lichnerowicz, A.: 1946, Sur une extension de la formule d'Allendoerfer-Weil à certaines variétés finslériennes, C. R. Acad. Sci. Paris 223, 12-14.
Lichnerowicz, A.: 1949, Quelques théorèmes de géométrie différentielle globale, Comment. Math. Helv. 22, 271-301.
Lichnerowicz, A. and Thiry, Y.: 1947, Problèmes de calcul des variations liés à la dynamique classique et à la théorie unitaire du champ, C. R. Acad. Sci. Paris 224, 529-531.
Lippmann, H.: 1957, Zur Winkeltheorie in zweidimensionalen Minkowski- und Finsler-Räumen, Nederl. Akad. Wetensch. Proc. A 60 = Indag. Math. 19, 162-170.
Lippmann, H.: 1958, Metrische Eigenschaften verschiedener Winkelmase im Minkowski- und Finslerraum. I, II, Nederl. Akad. Wetensch. Proc. A 61 = Indag. Math. 20, 223-238.
Losik, M. V.: 1964, On infinitesimal connections in tengent fiber spaces, Izv. Vyss. Uchebn. Zaved. Mat., No. 5(42), 54-60 (in Russian).
Lovelock, D.: 1969a, Complex spaces with locally product metrics: Special spaces, Ann. Mat. Pura Appl. (4) 83, 43-51.
Lovelock, D.: 1969b, Complex spaces with locally product metrics: General theory, Ann. Mat. Pura Appl. (4) 83, 53-72.
Lumiste, J. G.: 1969, The theory of connections in fibered spaces, Algebra, Topology, Geometry, pp. 123-168 (in Russian).
Maerbashi, T.: 1959, A weakly osculating Riemann space of the Finsler space and its application to a theory of subspaces in the Finsler space, Tensor, N. S. 9, 62-72.
Maerbashi, T.: 1960-1961, Vector fields and space forms, J. Fac. Sci. Hokkaido Univ. I 15, 62-92.
Maerbashi, T.: 1961, A certain type of vector field. I, II, III, Proc. Japan Acad. 37, 23-26.

Maerbashi, T.: 1977, On line geometry, Kumamoto J. Sci. Math. 12, 56-61.
Makai, I.: 1972, Differential concomitants in spaces of linear elements, Reports on Math. Phys. 3, 77-89.
Mangione, V.: 1971, Proprietà geometriche negli spazi di Finsler quasi hermitiani, Rend. Mat. (6) 4, 867-876.
Marei, L.: 1980, Sous-variétés finslériennes et principle de moindre courbure: Thése, Grenoble, 45 p.
Matsumoto, M.: 1960/1961, A global foundation of Finsler geometry, Memo. Coll. Sci. Univ. Kyoto A 33, 171-208.
Matsumoto, M.: 1963, Affine transformations of Finsler spaces, J. Math. Kyoto Univ. 3, 1-35.
Matsumoto, M.: 1963/1964a, Linear transformations of Finsler connections, J. Math. Kyoto Univ. 3, 145-167.
Matsumoto, M.: 1963/1964b, Paths in a Finsler space, J. Math. Kyoto Univ. 3, 305-318.
Matsumoto, M.: 1965, On R. Sulanke's method deriving H. Rund's connection in a Finsler space, J. Math. Kyoto Univ. 4, 355-368.
Matsumoto, M.: 1966a, A Finsler connection with many torsions, Tensor, N. S. 17, 217-226.
Matsumoto, M.: 1966b, Connections, Metrics and almost complex structures of tangent bundles, J. Math. Kyoto Univ. 5, 251-278.
Matsumoto, M.: 1967, Theory of Finsler spaces and differential geometry of tangent bundles, J. Math. Kyoto Univ. 7, 169-204.
Matsumoto, M.: 1968, Intrinsic transformations of Finsler metrics and connections, Tensor, N. S. 19, 303-313.
Matsumoto, M.: 1969a, On F-connections and associated nonlinear connections, J. Math. Kyoto Univ. 9, 25-40.
Matsumoto, M.: 1969b, A geometrical meanings of a concept of isotropic Finsler spaces, J. Math. Kyoto Univ. 9, 405-411.
Matsumoto, M.: 1970, The theory of Finsler connections, Publ. of the Study Group of Geometry 5, Dept. Math. Okayama Univ., 220 p.
Matsumoto, M.: 1971a, On h-isotropic and C^h-recurrent Finsler spaces, J. Math. Kyoto Univ. 11, 1-9.
Matsumoto, M.: 1971b, On some transformations of locally Minkowskian spaces, Tensor, N. S. 22, 103-111.
Matsumoto, M.: 1973, A theory of three-dimensional Finsler spaces in terms of scalars, Demonstratio Math. 6, 223-251.
Matsumoto, M.: 1974, Theory of non-linear connections from the standpoint of Finsler connections, Tensor, N. S. 28, 69-77.
Matsumoto, M.: 1975a, On three-dimensional Finsler spaces satisfying the T- and B^p-conditions, Tensor, N. S.: 29, 13-20.
Matsumoto, M.: 1975a, Metrical differential geometry, Kiso Sūgaku Sensho 14, Shokabō, Tokyo, 229 p.
Matsumoto, M.: 1976, What is the Finsler geometry?, Bull. Korean Math. Soc. 13, 121-126.
Matsumoto, M.: 1977a, On the indicatrices of a Finsler space, Period. Math. Hungar. 8, 185-191.
Matsumoto, M.: 1977b, Strongly non-Riemannian Finsler spaces, Anal. Sti. Univ. Iasi 23, 141-149.
Matsumoto, M.: 1977c, Foundation of Finsler Geometry and Special Finsler Spaces (unpublised), 300 pp.
Matsumoto, M.: 1978a, Finsler spaces with the hv-curvature tensor P_{hijk} of a special form, Reports on Math. Phys. 14, 1-13.
Matsumoto, M.: 1978b, On geodesics in the tangent space of a Finsler space, J. Korean Math. Soc. 14, 167-183.
Matsumoto, M.: 1979, The length and the relative length of tangent vectors of Finsler spaces, Reports on Math. Phys. 15, 375-386.
Matsumoto, M.: 1980a, Fundamental functions of S3-like Finsler spaces, Tensor, N. S. 34, 141-146.
Matsumoto, M.: 1980b, Projective changes of Finsler metrics and projectively flat Finsler spaces, Tensor, N. S. 34, 303-315.
Matsumoto, M.: 1981a, Differential-geometric properties of indicatrix bundle over Finsler space, Publ. Math. Debrecen 28, 281-291.
Matsumoto, M.: 1981b, Berwald connections with (h)h-torsion and generalized Berwald spaces, Tensor, N. S. 35, 223-229.
Matsumoto, M.: 1982a, Theory of curves in tangent planes of two-dimensional Finsler spaces, Tensor, N. S. 37, 35-42.
Matsumoto, M.: 1982b, On Wagner's generalized Berwald spaces of dimension two, Tensor, N. S. 36, 303-311.
Matsumoto, M.: 1983, A relative theory of Finsler spaces, J. Math. Kyoto Univ. 23, 25-37.

Matsumoto, M. and Eguchi, K.: 1974, Finsler spaces admitting a concurrent vector field, Tensor, N. S. 28, 239-249.
Matsumoto, M. and Miron, R.: 1977, On an invariant theory of the Finsler spaces, Period. Math. Hungar. 8, 73-82.
Matsumoto, M. and Numaţa, S.: 1979, On Finsler spaces with a cubic metric, Tensor, N. S. 33, 153-162.
Matsumoto, M. and Numata, S.: 1980, On semi-C-reducible Finsler spaces with constant coefficients and C2-like Finsler spaces, Tensor, N. S. 34, 218-222.
Matsumoto, M. and Okada, T.: Connections in Finsler spaces, Sem. Diff. Geom 4, Kyoto Univ., 146 pp.
Matsumoto, M. and Shibata, C.: 1976, On the curvature tensor R_{ijkh} of C-reducible Finsler spaces, J. Korean Math. Soc. 13, 141-144.
Matsumoto, M. and Shibata, C.: 1979, On semi-C-reducibility, T-tensor = 0 and S4-likeness of Finsler spaces, J. Math. Kyoto Univ. 19, 301-314.
Matsumoto, M. and Shimada, H.: 1977, On Finsler spaces with the curvature tensors P_{hijk} and S_{hijk} satisfying special conditions, Reports on Math. Phys. 12, 77-87.
Matsumoto, M. and Tamássy, L.: 1980, Scalar and gradient vector fields of Finsler spaces and holonomy groups of nonlinear connections, Demonstratio Math. 13, 551-564.
Matsumoto, M. and Tamássy, L.: 1981, Scalar and gradient vector fields of Finsler spaces and holonomy groups of nonlinear connections, Demonstratio Math. 14, 529.
Matthias, H. H.: 1978, Eine Finslermetrik auf S^2 mit nur zwei geschlossenen Geodätischen, Bonn. Math. Schrift. 102, 27-50
Matthias, H. H.: 1980, Zwei Verallgemeinerungen eines Satzes von Gromoll und Meyer, Bonn. Math. Schr. 126, 90 p.
Maugin, G.: 1971, Etude des déformations d'un milieu continu magnétiquement saturé, employant la notion d'espace de Finsler, C. R. Acad. Sci. Paris A 273, 474-476.
Maurin, K.: 1955, Eingliedrige Gruppen der homogenen kanonischen Transformationen und Finslersche Räume, Ann. Polon. Math. 2, 97-102.
Maurin, K.: 1976, Calculus of variations and classical field theory. Part I, Lecture Notes 34. Aarhus: Mat. Inst. Aarhus Univ., 100 pp.
Mayer, W.: 1929, Beitrag zur geometrischen Variationsrechnung, Jber. Deutsch. Math.-Verein. 38, 260-281.
Meher, F. M.: 1972a, Some commutation formulae arising from the generalised Lie differentiation in Finsler space, Rev. Univ. Istanbul 37, 121-125.
Meher, F. M.: 1972b, On the existence of affine motion in a HR-F_n, Indian J. Pure Appl. Math. 3, 219-225.
Meher, F. M.: 1972c, Projective motion in a symmetric space, Tensor, N. S. 23, 275-278.
Meher, F. M.: 1973, An SHR-F_n admitting an affine motion. II, Tensor, N. S. 27, 208-210.
Meher, F. M.: 1978, Projectively flat manifolds with recurrent curvature, Boll. Un. Mat. Ital. (5) B 15, 828-834.
Meher, F. M. and Patel, L. M.: 1973/1974, Semi-symmetric connections in a Finsler space, Rend. Accad. Naz. XL (4) 24/25, 281-286.
Meher, F. M. and Patel, L. M.: 1979, Projective motion in a PS-F_n, Boll. Un. Mat. Ital. A (5) 16, 554-559.
Menger, K.: 1930, Untersuchungen über allgemeine Metrik. Vierte Untersuchung, Math. Ann. 103, 466-501.
Mercier, A. and Treder, H. J.: 1972, Future of GRG (general relativity and gravitation), Math. Student 40, 49-64.
Mercuri, F.: 1976, Closed geodesics on Finsler manifolds, Lincei Rend. (8) 60, 111-118.
Mercuri, F.: 1977, The critical points theory for the closed geodesic problem, Math. Z. 156, 231-245.
Merza, J.: 1973, On a special Finsler geometry, Demonstratio Math. 6, 761-769.
Minkowski, H.: 1910, Geometrie der Zahlen, Leipzig & Berlin, 256 pp. Reproduct. by Chelsea Publ., New York, 1953.
Miron, R.: 1961, Sur les connexions pseudo-euclidiennes des espaces de Finsler à métrique indéfinite, Acad. R. P. Romine Fil. Iasi, Sti. Mat. 12, 125-134 (in Roumanian).
Miron, R.: 1968, Espace de Finsler ayant un groupe de Lie comme groupe d'invariance, Rev. Roum. Math. Pures Appl. 13, 1409-1412.
Miron R.: 1977, Research of Romanian mathematicians in Finsler spaces, Mat. Coll. on Geometry and Topology, Soc. Sti. Mat. Univ. Timisoara, pp. 9-13 (in Roumanian).
Miron, R.: 1979, On a class of Finsler connections, Bul. Inst. Politeh. Iasi 25, 53-55.
Miron, R.: 1980, Introduction to the theory of Finsler spaces, Proc. Nat. Sem. Finsler spaces. Brasov, pp. 131-183.
Miron, R.: 1981a, On transformation groups of Finsler connections, Tensor, N. S. 35, 235-240.
Miron, R.: 1981b, Structures de Finsler presque symplectiques, Actualités Math., Actes 6, Congr. Group. Math. Expr. Latine, Luxembourg, pp. 289-291.

Miron, R.: 1982a, Vector bundles Finsler geometry, Proc. Nat. Sem. Finsler Spaces, Brasov, pp. 147-188.
Miron, R.: 1982b, On almost complex Finsler structures, An. Sti. Univ. Iasi I a, 28, 13-17.
Miron, R.: 1983, Metrical Finsler structures and metrical Finsler connections, J. Math. Kyoto Univ. 23, 219-224.
Miron, R.: 1984, A nonstandard theory of hypersurfaces in Finsler spaces, An. Sti. Univ. Iasi, I a, 30, 35-53.
Miron, R.: 1985, A Lagrangian theory of relativity, Sem. Geometrie si Topologie, Univ. Timisoara, Fac. Sti. Ale Naturii, 53 pp.
Miron, R. and Anastasiei, M.: 1981, On the notion of Finsler geometric object, Mem. Sect. Sti. Acad. Republ. Soc. Romania. IV 4, 25-31.
Miron, R. and Anastasiei, M.: 1983, Existence et arbitrariété des connexions compatibles à une structure Riemann généralisée du type presque k-horsymplectique métrique, Kodai Math. J. 6, 228-237.
Miron, R. and Hashiguchi, M.: 1983, Conformal almost symplectic Finsler stucture, Rep. Fac. Sci. Kagoshima Univ. (Math. Phys. Chem.) 16, 49-57.
Miron, R. and Bejancu, A.: 1984, A new method in geometry of Finsler subspaces, An. Sti. Univ. Iasi I a, 30, 55-59.
Miron, R. and Hashiguchi, M.: 1979, Metrical Finsler connections, Rep. Fac. Sci. Kagoshima Univ. (Math. Phys. Chem.) 12, 21-35.
Miron, R. and Hashiguchi, M.: 1981a, Conformal Finsler connections, Rev. Roum. Math. Pures Appl. 26, 861-878.
Miron, R. and Hashiguchi, M.: 1981b, Almost symplectic Finsler structures, Rep. Fac. Sci. Kagoshima Univ. (Math. Phys. Chem.) 14, 9-19.
Mishra, R. S., Misra, R. B. and Kishore, N.: 1977, On bisymmetric Finsler manifolds, Boll. Unione Mat. Ital (5) A 14, 157-164.
Mishra, R. S., Misra, R. B. and Kishore, N.: 1978, On a symmetric Finsler manifold admitting an affine motion, Bull. Soc. Math. Belg. A 30, 39-44.
Mishra, R. S. and Pande, H. D.: 1966, Conformal identities, Rev. Univ. Istanbul 31, 39-48.
Mishra, R. S. and Pande, H. D.: 1967, The Ricci identity, Ann. Mat. Pura Appl. (4) 75, 355-361.
Mishra, R. S. and Pande, H. D.: 1968a, Recurrent Finsler space, J. Indian Math. Soc., N. S. 32, 17-22.
Mishra, R. S. and Pande, H. D.: 1968b, Certain projective changes in Finsler space, Rev. Mat. Hisp.-Amer (4) 28, 49-55.
Mishra, R. S. and Singh, U. P.: 1966, On the union curvature of a curve of a Finsler space, Tensor, N. S. 17, 205-211.
Mishra, R. S. and Singh, U. P.: 1967, The generalisation of the first curvature of a curve of a subspace of a Finsler space, Rend. Circ. Mat. Palermo (2) 16, 33-38.
Mishra, R. S. and Sinha, R. S.: 1964, Congruences of curves through points of a subspace and hypersurface of a Finsler space, Rev. Univ. Istanbul 29, 71-87.
Mishra, R. S. and Sinha, R. S.: 1965a, Bianchi identities satisfied by the first and second of Cartan's curvature tensors, Bull. Calcutta Math. Soc. 57, 99-101.
Mishra, R. S. and Sinha, R. S.: 1965b, Relative Frenet formulae for curves in a subspace and a hypersurface of a Finsler space, Tensor, N. S. 16, 114-132.
Mishra, R. S. and Sinha, R. S.: 1965c, Union and hyperasymptotic curves of a Finsler subspace and hypersurface, Rend. Circ. Mat. Palermo (2) 14, 119-128.
Mishra, R. S. and Sinha, R. S.: 1965d, Union curvature of a curve in a Finsler space, Tensor, N. S. 16, 160-168.
Mishra, R. S. and Sinha, R. S.: 1966, Some identities satisfied by Cartan's curvature tensors, Proc. Nat. Acad. Sci. India A 36, 534-538.
Misra, R. B.: 1966a, The projective transformation in a Finsler space, Ann. Soc. Sci. Bruxelles I 80, 227-239.
Misra, R. B.: 1966b, Projective tensors in a conformal Finsler space, Acad. Roy. Belg. Bull. Cl. Sci. (5) 52, 1275-1279.
Misra, R. B.: 1967a, The commutation formulae in a Finsler space, I, II, Ann. Mat. Pura Appl. (4) 75, 363-383.
Misra, R. B.: 1967b, The Bianchi identities satisfied by curvature tensors in a conformal Finsler space, Tensor. N. S. 18, 187-190.
Misra, R. B.: 1967c, Some Problems in Finsler Spaces: Thesis, Allahabad.
Misra, R. B.: 1968, On the deformed Finsler space, Tensor, N. S. 19, 241-250.
Misra, R. B.: 1969a, Hyper-asymptotic curves of a subspace of a Finsler space, Bull. Polon. Math. Astro. Phys. 17, 65-69.
Misra, R. B.: 1969b, The generalized Killing equation in Finsler space, Rend. Circ. Mat. Palermo (2) 18, 99-102.
Misra, R. B.: 1970a, Projective invariants in a conformal Finsler space, Tensor, N. S, 21, 186-188.

Misra, R. B.: 1970b, On the generalized Lie differentiation arising from Su's infinitesimal transformation, Rev. Univ. Istanbul 35, 5-15.
Misra, R. B.: 1972a, A symmetric Finsler space, Tensor, N. S. 24, 346-350.
Misra, R. B.: 1972b, A projectively symmetric Finsler space, Math. Z. 126, 143-153.
Misra, R. B.: 1973, On a recurrent Finsler space, Rev. Roum. Math. Pures Appl 18, 701-712.
Misra, R. B.: 1975, A bi-recurrent Finsler manifold with affine motion, Indian J. Pure Appl. Math. 6, 1441-1448.
Misra, R. B. 1977, A turning point in the theory of recurrent Finsler manifolds, J. South Gujarat Univ. 6, 72-96.
Misra, R. B.: 1979, A turning point in the theory of recurrent Finsler manifolds. II: certain types of projective motions, Boll. Un. Mat. Ital. (5) B 16, 32-53.
Misra, R. B. and Kishore, N.: 1978, On curvature collineations in Finsler manifolds, Atti Accad. Sci. Lett. Atti Palermo I(4) 36 (1976/1977), 521-534.
Misra, R. B., Kishore, N. and Pandey, P. N.: 1977, Projective motion in an SNP-F_n, Boll. Un. Mat. Ital. A (5) 14, 513-519.
Misra, R. B. and Kumar, A.: 1982, The special projective tensor fields (with 2-recurrent and special quasisymmetric properties) in nonsymmetric Finsler space F^*_n, Indian J. Pure Appl. Math. 13, 1011-1017.
Misra, R. B. and Meher, F. M.: 1971a, Projective motion in an RNP-Finsler space, Tensor, N. S. 22, 117-120.
Misra, R. B. and Meher, F. M.: 1971b, Some commutation formulae arising from Lie differentiation in a Finsler space, Lincei Rend. (8) 50, 18-23.
Misra, R. B. and Meher, F. M.: 1971c, A SHR-F_n admitting an affine motion, Acta Math. Acad. Sci. Hungar. 22, 423-429.
Misra, R. B. and Meher, F. M.: 1972a, On the existence of affine motion in a HR-F_n, Indian J. Pure Appl. Math. 3, 219-225.
Misra, R. B. and Meher, F. M.: 1972b, Lie differentiation and projective motion in the projective Finsler space, Tensor, N. S. 23, 57-65.
Misra, R. B. and Meher, F. M.: 1972c, A Finsler space with special concircular projective motion, Tensor, N. S. 24, 288-292.
Misra, R. B. and Meher, F. M.: 1973, A recurrent Finsler space of second order, Rev. Roum. Math. Pures Appl. 18, 563-569.
Misra, R. B. and Meher, F. M.: 1975, CA-motion in a PS-F_n, Indian J. Pure Appl. Math. 6, 522-526.
Misra, R. B., Meher, F. M. and Kishore, N.: 1978, On a recurrent Finsler manifold with a concircular vector field, Acta Math. Acad. Sci. Hungar. 32, 287-292.
Misra, R. B. and Mishra, R. S.: 1965, Lie-derivatives of various geometric entities in Finsler space, Rev. Univ. Istanbul 30, 77-82.
Misra, R. B. and Mishra, R. S.: 1966, The Killing vector and generalised Killing equation in Finsler space, Rend. Circ. Mat. Palermo (2) 15, 216-222.
Misra, R. B. and Mishra, R. S.: 1969, Curvature tensor arising from non-linear connection in a Finsler space, Bull. Polon. Math. Astro. Phys. 17, 755-760.
Misra, R. B. and Pande, K. S.: 1970a, On the Finsler space admitting a holonomy group, Ann. Mat. Pura Appl. (4) 85, 327-346.
Misra, R. B. and Pande, K. S.: 1970b, On Misra's covariant differentiation in a Finsler space, Lincei Rend. (8) 48, 199-204.
Misra, S. B. and Dubey, V. J.: 1980, Some theorems on infinitesimal conformal transformation in projective symmetric Finsler space, Indian J. Pure Appl. Math. 11, 54-59.
Misra, S. B. and Misra, A. K.: 1983a, F^*_n-recurrent Finsler space of first order, Acta Cienc. Indica 9, 91-95.
Misra, S. B. and Misra, A. K.: 1983b, Q^+-recurrent Finsler space of first order, Acta Cienc. Indica 9, 187-194.
Misra, S. B. and Misra, A. K.: 1983c, On the decomposition of pseudo curvature tensor W^{*i}_{jkh} in a recurrent Finsler space, Acta Cienc. Indica 9, 232-238.
Misra, S. B. and Misra, A. K.: 1984, Affine motion in RNP-Finsler space, Indian J. Pure Appl. Math. 15, 463-472.
Mizoguchi, N. and Katō, S.: 1968, Umbilical points on subspaces of Finslerian and Minkowskian spaces, Memo. Kitami Inst. Tech. 2, 319-325.
Moalla, M. F.: 1964a, Espaces de Finsler complets, C. R. Acad. Sci. Paris 258, 2251-2254.
Moalla, M. F.: 1964b, Espaces de Finsler complets à courbure de Ricci positive, C. R. Acad. Sci. Paris 258, 2734-2737.
Moalla, M. F.: 1965, Espaces de Finsler sans points conjugués, C. R. Acad. Sci. Paris 260, 6510-6512.
Moalla, M. F.: 1966, Sur quelques théorèmes globaux en géométrie finslérienne, Ann. Mat. Pura Appl. (4) 73, 319-365.
Montesinos, A.: 1979a, Geometry of spacetime founded on spacelike metric, J. Math. Phys. 20(5), 953-965.

Montesinos, A.: 1979b, On Finsler connections, Rev. Mat. Hisp. Amer. (4) 39, 99-110.
Moór, A.: 1950a, Espaces métriques dont le scalaire de courbure est constant, Bull. Sci. Math. (2) 74, 13-32.
Moór, A.: 1950b, Finslersche Räume mit der Grundfunktion L = f/g, Comment. Math. Helv. 24, 188-195.
Moór, A.: 1950c, Généralisation du scalaire de courbure et du scalaire principal d'un espace finslérien à n dimensions, Canad. J. Math. 2, 307-313.
Moór, A.: 1951, Einfürung des invarianten Differentials und Integrals in allgemeinen metrischen Räumen, Acta Math. 86, 71-83.
Moór, A.: 1952a, Finslersche Räume mit algebraischen Grundfunktionen, Publ. Math. Debrecen 2, 178-190.
Moór, A.: 1952b, Quelques remarques sur la généralisation du scalaire de courbure et du scalaire principal, Canad. J. Math. 4, 189-197.
Moór, A.: 1952c, Über die Dualität von Finslerschen und Cartanschen Räumen, Acta Math. 88, 347-370.
Moór, A.: 1952d, Über oskulierende Punkträume von affinzusammenhängenden Linienelement-mannigfaltigkeiten, Ann. of Math. (2) 56, 397-403.
Moór, A.: 1955, Metrische Dualität der allgemeinen Räume, Acta Sci. Math. Szeged 16, 171-196.
Moór, A.: 1956a, On the extremals of the generalized metric spaces, Bul. Inst. Politeh. Iasi I 2, 19-26.
Moór, A.: 1956b, Entwicklung einer Geometrie der allgemeinen metrischen Linienelementräu-me, Acta Sci. Math. Szeged 17, 85-120.
Moór, A.: 1956c, Allgemeine metrische Räume von skalarer Krümmung, Publ. Math. Debrecen 4, 207-228.
Moór, A.: 1957a, Über die Torsions- und Krümmungsinvarianten der dreidimensionalen Finslerschen Räumen, Math. Nachr. 16, 85-99.
Moór, A.: 1957b, Über die autoparallele Abweichung in allgemeinen metrischen Linienele-menträumen, Publ. Math. Debrecen 5, 102-118.
Moór, A.: 1957c, Über den Schurschen Satz in allgemeinen metrischen Linienelementräumen, Nederl. Akad. Wetensch. Proc. A 60 = Indag. Math. 19, 290-301.
Moór, A.: 1959, Über nicht-holonome allgemeine metrische Linienelementräume, Acta Math. 101, 201-233.
Moór, A.: 1960, Untersuchungen über die kovariante Ableitung in Linienelementräumen, Publ. Math. Debrecen 7, 41-53.
Moór, A.: 1961, Über affine Finslerräume von skalarer Krümmung, Acta Sci. Math. Szeged 22, 157-189.
Moór, A.: 1963a, Untersuchungen über Finslerräume von rekurrenter Krümmung, Tensor, N. S. 13, 1-18.
Moór, A.: 1963b, Über projektive Veränderung der Übertragung in Linienelementmannigfaltig-keiten, Acta Sci. Math. Szeged 24, 119-128.
Moór, A.: 1963c, Eine Verallgemeinerung der metrischen Übertragung in allgemeinen metri-schen Räumen, Publ. Math. Debrecen 10, 145-150.
Moór, A.: 1963d, Untersuchungen über die oskulierenden Punkträume der metrischen Linien-elementräume, Liet. Mat. Rinkinys. 3(2), 212-213.
Moór, A.: 1964a, Linienelementräume mit nicht-symmetrischen Fundamentaltensor, Publ. Math. Debrecen 11, 245-256.
Moór, A.: 1964b, Gleichung der autoparallelen Abweichung in n-dimensionalen Linienele-menträumen, Acta. Sci. Math. Szeged 25, 266-282.
Moór, A.: 1965a, Untersuchungen über die oskulierenden Punkträume der metrischen Linienelementräume, Acta Math. Acad. Sci. Hungar. 16, 57-74.
Moór, A.: 1965b, Über eine skalare Form de Gleichung der autoparallelen Abweichung im affinen Raum, Publ. Math. Debrecen 12, 281-291.
Moór, A.: 1966a, Übertragungstheorie bezüglich der allgemeinen Linienelementtransforma-tionen, Publ. Math. Debrecen 13, 263-287.
Moór, A.: 1966b, Äquivalenztheorie der allgemeinen metrischen Linienelementräume, in Perspectives in Geometry and Relativity, Indiana, Bloomingten-London, pp. 238-244.
Moór, A.: 1967a, Über eine Charakterisierung von differentialgeometrischen Räumen skalarer Krümmung, Rev. Roum. Math. Pures Appl. 12, 93-100.
Moór, A.: 1967b, Einige Bemerkungen über Kurven und Grundelementfolgen in allgemeinen Räumen, Publ. Math. Debrecen 14, 337-347.
Moór, A.: 1968, Objektentheoretische Untersuchungen über die kovarianten Ableitungen in allgemeinen Linienelementräumen, Acta. Sci. Math. Szeged 29, 177-186.
Moór, A.: 1969, Grundzüge der möglichen Hyperflächentheories der M_n-Räume, Publ. Math. Debrecen 16, 161-174.
Moór, A.: 1970, Über die kovariante Ableitung der Vektoren in verallgemeinerten Linien-elementräumen, Acta Sci. Math. Szeged 31, 129-139.
Moór, A.: 1971, Über Finslerräume von zweifach rekurrenter Krümmung, Acta Math. Acad. Sci. Hungar 22, 453-465.

Moór, A.: 1972a, Unterräume von rekurrenter Krümmung in Finslerräumen, Tensor, N. S. 24, 261-265.
Moór, A.: 1972b, Über die Nichtexistenz gewisser Type von Linienelementräumen von rekurrenter Krümmung, Publ. Math. Debrecen 19, 169-175.
Moór, A.: 1973a, Untersuchungen in vierdimensionalen verallgemeinerten Finslerräumen mit Hilfe eines natürlichen Vierbeins, Publ. Math. Debrecen 20, 241-258.
Moór, A.: 1973b, Finslerräume von identischer Torsion Acta Sci. Math. Szeged 34, 279-288.
Moór, A.: 1974, Finslerräume von rekurrenter Torsion, Publ. Math. Debrecen 21, 255-265.
Moór, A.: 1975a, Über eine Übertragungstheorie der metrischen Linienelementräume mit rekurrentem Grundtensor, Tensor, N. S. 29, 47-63.
Moór, A.: 1975b, Über die Veränderung der Krümmung bei einer Torsion der Übertragung, Acta Math. Acad. Sci. Hungar. 26, 97-111.
Moór, A.: 1975b, Übertragungstheorien in Finslerschen und verwandten Räumen, Berichte Math.-Stat. Sekt. Forschungszentrum Graz 42, 1-14.
Moór, A.: 1975d, Über die Charakterisierung von speziellen 3- and 4-dimensionalen Finslerräumen durch Invarianten, Boll. Un. Mat. Ital. (4) 12, Suppl. 3, 189-199.
Moór, A.: 1976a, Über spezielle Type von Hyperflächen in verallgemeinerten Finslerräumen, Publ. Math. Debrecen 23, 27-39.
Moór, A.: 1976b, Über allgemeine Übertragungstheorien in metrischen Linienelementräumen, Acta Math. Acad. Sci. Hungar. 28, 321-334.
Moór, A.: 1977, Über Unteräume von Linienelementräumen mit rekurrentem Masstensor, Tensor, N. S. 31, 1-7.
Moór, A.: 1979, Über die durch Kurven bestimmten Sektionalkrümmungen der Finslerschen und Weylschen Räume, Publ. Math. Debrecen 26, 205-214.
Moór, A.: 1981, Über Linienelementräume von skalarer Krümmung und mit rekurrenten Grundtensor, Tensor, N. S. 35, 125-135.
Moór, A.: 1982a, Über die Begründung von Finsler-Otsukischen Räumen und ihre Dualität, Tensor, N. S. 37, 121-129.
Moór, A.: 1982b, On eigencurves and eigenvectors in spaces of line elements, in Differential Geometry, North-Holland, Amsterdam-New-York, pp. 465-480.
Mosharrafa, A. M.: 1945, On the metric of space and the equations of motion of a charged particle, Proc. Math. Phys. Soc. Egypt 3, 19-24.
Mosharrafa, A. M.: 1948, The metric of space and mass deficiency, Philos. Mag. (7) 39, 728-738.
Murgescu, V. C.: 1962a, Sur l'introduction d'une métrique dans le plan affin, Gaz. Mat. Fiz. A 14 (67), 581-587 (in Roumanian).
Murgescu, V. C.: 1962b, Sur l'introduction d'une métrique dans l'espace affin, Bull. Inst. Politeh. Iasi I 8(12), No. 3-4, 37-46.
Muto, Y. and Yano, K.: 1939, Sur les transformations de contact et les espaces de Finsler, Tohoku Math. J. 45, 295-307.
Myers, S. B.: 1938, Arc length in metric and Finser manifolds, Ann. of Math. (2) 39, 463-471.
Myers, S. B.: 1945, Arcs and geodesics in metric spaces, Trans. Amer. Math. Soc. 57, 217-227.
Nagai, T.: 1960, On groups of rotations in Minkowski space. I, II, J. Fac. Sci. Hokkaido Univ. I 15, 29-61.
Nagano, T.: 1959, Isometries on complex-product spaces, Tensor, N. S. 9, 47-61.
Nagata, Y.: 1955, Normal curvature of a vector field in a hypersurface in a Finsler space, Tensor, N. S. 5, 17-22.
Nagata, Y.: 1958, Remarks on the normal curvature of a vector field, Tensor, N. S. 8, 177-183.
Nagata, Y.: 1960, Formulas of Frenet for a vector field in a Finsler space, Mem. Muroran Inst. of Tech. 3, 157-160.
Nagata, Y.: 1978, On some geometrical properties in three dimensional Finsler space, Mem. Ehime Univ. Mat. Sci. A 8-3, 1-2.
Nagata, Y. and Anh, K. O.: 1978, On the curvatures of a vector field on a subspace of a Finsler space, Tensor, N. S. 32, 303-310.
Nagy, P. T.: 1969, Über die Äquivalenztheorie der verallgemeinerten Linienelementräume, Publ. Math. Debrecen 16, 79-89.
Nagy, P. T.: 1975, On the theory of Finsler connections especially their equivalence, Acta Sci. Math. Szeged 37, 331-338.
Nakajima, S.: 1926, Geometrische Untersuchungen zur Variationsrechnung, Japanese J. Math. 3, 61-64.
Nasu, Y.: 1949a, On the structure of a Finsler space whose holonomy group preserves invariant more than one point, Holonomy-gun Kenkyu 12, 40-47 (in Japanese).
Nasu, Y.: 1949b, On the structure of a Finsler space admitting a vector field with torse-forming direction, Holonomy-gun Kenkyo 13, 47-53 (in Japanese).
Nasu, Y.: 1950, Non Euclidean geometry in the space with general projective connection. I, II, Holonomy-gun Kenkyu 15, 33-45 (in Japanese).

Nasu, Y.: 1952a, Non Euclidean geometry in Finsler spaces, Kumamoto J. Sci. A I, 8-12.

Nasu, Y.: 1952b, On the torse-forming directions in Finsler spaces, Tohoku Math. J. (2) 4, 99-102.

Nasu, Y.: 1954, On the normality in Minkowskian space, Kumamoto J. Sci. A 2, 11-17.

Nasu, Y.: 1959a, On similarities and transitive abelian groups of motions in Finsler spaces, Kumamoto J. Sci. A 4, 103-110.

Nasu, Y.: 1959b, On similarities in a Finsler space, Tensor, N. S. 9, 175-189.

Nasu, Y.: 1960, On local properties in a metric space, Tensor, N. S. 10, 81-89.

Nasu, Y.: 1961, On characterization of spaces of constant curvature, Tensor, N. S, 11, 6-15.

Nasu, Y.: 1962, On Desarquesian spaces, Math. J. Okayama Univ. 11, 19-26.

Nazim Terzioglu, A.: 1936, Über Finslersche Räume: Dissertation, München.

Nazi, Terzioglu, A.: 1948, Über den Satz von Gauss-Bonnet im Finslerschen Raum, Univ. Istanbul, Fac. Sci., pp. 26-32.

Newhouse, S. and Palis, J.: 1971, Bifurcations of Morse-Smale dynamical systems, Proc. Symp. Univ. Bahia, Salvador, pp. 303-366.

Nitka, W. and Wiatrowska, L.: 1969, Linearity in the Minkowski space with non-strictly convex spheres, Colloq. Math. 20, 113-115.

Nobuhary, T. and Nagai, T.: 1952, On the special Finsler space of three dimensions, Tensor, N. S. 2, 175-180.

Noether, A. A.: 1918, Invarianten beliebiger Differentialausdrücke, Nachr. Göttingen 25, 37-44.

Noether, A. A.: 1923, Algebraische und Differentialinvarianten, Jber. Deutsch. Math.-Verein. 32, 177-184.

Norden, A. P.: 1976, Affinely connected spaces, Nauka, Moskau, 432 pp. (in Russian).

Numata, S.: 1975a, On the curvature tensor S_{hijk} and the tensor T_{hijk} of generalized Randers spaces, Tensor, N. S. 19, 35-39.

Numata, S.: 1975b, On Landsberg spaces of scalar curvature, J. Korean Math. Soc. 12, 97-100.

Numata, S.: 1978, On the torsion tensors R_{hjk} and P_{hjk} of Finsler spaces with a metric $ds = (g_{ij}(dx)dx^i dx^j)^{\frac{1}{2}} + b_i(x)\, dx^i$, Tensor, N. S. 32, 27-31.

Numata, S.: 1980, On C3-like Finsler spaces, Reports Math. Phys. 18, 1-10.

Nutku, Y.: 1974, Geometry of dynamics in general relativity, Ann. Inst. Poincaré, N. S. A 21, 175-183.

O'Byrne, B.: 1969, On Finsler geometry and applications to Teichmüller spaces, Proc. Conf. Stony Brook, N. Y., pp. 317-328.

Ohkubo, T.: 1940, Geometry in a space with a generalized metric, Tensor, O. S. 3, 48-55 (in Japanese).

Ohkubo, T.: 1941a, On a symmetric displacement in a Finsler space, Tensor, O. S. 4, 53-55 (in Japanese).

Ohkubo, T.: 1941b, Geometry in a space with generalized metrics. II, J. Fac. Sci. Hokkaido Imp. Univ. I 10, 157-178.

Ohkubo, T.: 1954, On relations among various connections in Finslerian space, Kumamoto J. Sci. A 1, No. 3, 1-6.

Okada, T.: 1962, Connections in the frame bundle on the tangent vector bundle and Finsler connections: Thesis, Kyoto, 21 p. (in Japanese).

Okada, T.: 1964, Theory of pair-connections, Sci. Engin. Rev. Doshisha Univ. 5, 113-152 (in Japanese).

Okada, T.: 1975, The holonomy group of Finsler connection. I. The holonomy group of the V-connection, Tensor, N. S. 27, 229-239.

Okada, T.: 1982, Minkowskian product of Finsler spaces and Berwald connection, J. Math. Kyoto Univ. 22, 323-332.

Okada, T. and Numata, S.: 1981, On generalized C-reducible Finsler spaces, Tensor, N. S. 35, 313-318.

Okubo, K.: 1967, On Finsler metrics, Memo. Fac. Education, Shiga Univ. 17, 1-5 (in Japanese).

Okubo, K.: 1975, On Finsler spaces with the curvature S_{ijkl} of a special form, Memo. Fac. Education, Shiga Univ. 25, 1-3.

Okubo, T.: 1964, On the connections of the tangent bundle of a Finsler space, Yokohama Math. J. 12, 23-37.

Okubo, T. and Houh, C.: 1972, Some cross-section theorems on the tangent bundle over a Finslerian manifold, Ann. Mat. Pura Appl. (4) 92, 129-138.

Omori, H.: 1967, Manifolds of Infinite Dimensions and the Problem of variations, Rep. Sem. Toritsu Univ., 36 p (in Japanese).

Opris, D.: 1978, Sur les couples de connexions de Finsler compatibles avec un (0, 2)-tenseur de Finsler, An. Univ. Timisoara, Sti. Mat. 15, 113-132.

Opris, D.: 1980, Fibrés vectoriels de Finsler et connexions associés, Proc. Nat. Sem. Finsler Spaces, Brasov, pp. 185-193.

Oproiu, V.: 1980, Some properties of the tangent bundle related to the Finsler geometry, Proc. Nat. Sem. Finsler Spaces, Brasov, pp. 195-207.
Otsuki, T.: 1957a, On geodesic coordinates in Finsler spaces, Math. J. Okayama Univ. 6, 135-145.
Otsuki, T.: 1957b, Theory of affine connections of the space of tangent directions of a differentiable manifold. I, II, III, Math. J. Okayama Univ. 7, 1-74.
Otsuki, T.: 1957c, Note on homotopies of some curves in tangent bundles, Math. J. Okayama Univ. 7, 191-194.
Otsuki, T.: 1958, Note on curvature of Finsler manifolds, Math. J. Okayama Univ. 8, 107-116.
Owens, O. G.: 1952, The integral geometry definition of arc length for two-dimensional Finsler spaces, Trans. Amer. Math. Soc. 73, 198-210.
Palais, R. S.: 1966, Lusternik-Schnirelman theory on Banach manifolds, Topology 5, 115-132.
Palais, R. S.: 1870, Critical point theory and the minimax principle, Proc. Symp. Pure Math. 15, 185-212.
Pande, H. D.: 1967, The projective transformation in a Finsler space, Lincei Rend. (8) 43, 480-484.
Pande, H. D.: 1968a, Projective invariants of an orthogonal ennuple in a Finsler space. I, Nanta Math. 2, 26-30.
Pande, H. D.: 1968b, Projective invariants in a conformal Finsler space, Nanta Math. 2, 31-35.
Pande, H. D.: 1968c, Various commutation formulae in conformal Finsler space, Progr. Math. Allahabad 2, 55-60.
Pande, H. D.: 1968d, On some commutation formulae in conformal Finsler spaces, Torino Fis. Mat. 103, 71-77.
Pande, H. D.: 1968e, Projective invariants in a Finsler space, Bull. Math. Soc. Sci. Math. R. S. Roumanie, N. S. 12 (60), 163-168.
Pande, H. D.: 1968f, Some identities in conformal Finsler space, Lincei Rend. (8) 45, 278-282.
Pande, H. D.: 1968g, Projective invariants of an orthogonal ennuple in a Finsler space, Ann. Inst. Fourier 18, 337-342.
Pande, H. D.: 1968h, Various projective invariants in a conformal Finsler space, Lincei Rend. (8) 44, 349-353.
Pande, H. D.: 1968i, The commutation formulae in conformal Finsler space, Ann. Mat. Pura Appl. (4) 79, 381-390.
Pande, H. D.: 1968j, Projective invariants of an orthogonal ennuple in a conformal Finsler space, Inst. Lombardo Accad. Rend. A 102, 421-426.
Pande, H. D.: 1969a, On some projective invariants of an orthogonal ennuple in a conformal Finsler space. I. Bull. Polon. Math. Astro. Phys. 17, 71-75.
Pande, H. D.: 1969b, Various Ricci identities in Finsler space, J. Austral. Math. Soc. 9, 228-232.
Pande, H. D.: 1969c, The conformal transformation in a Finsler space, Math. Nachr. 41, 247-252.
Pande, H. D.: 1969d, Some theorems on the projective derivative of certain entities in conformal Finsler spaces, Czechosl. Math. J. 19, 343-348.
Pande, H. D.: 1969e, On some projective invariants of an orthogonal ennuple in a conformal Finsler space. II, Monatsh. Math. 73, 406-410.
Pande, H. D.: 1969f, Certain projective entities for the unit tangent vector in conformal Finsler space, Rev. Roum. Math. Pures Appl. 14, 1303-1309.
Pande, H. D.: 1970, On some orthogonal ennuples in Finsler geometry, C. R. Acad. Bulgare Sci. 23, 623-626.
Pande, H. D.: 1971, Lie derivation in a Minkowskian Finsler space, Lincei Rend. (8) 50, 699-702.
Pande, H. D.: 1972, Orthogonal ennuples in Finsler space, Acad. Serbe Sci. Arts. Bull. 48, Cl. Sci. Math. Natur. Sci. Math., N. S. 8, 9-10.
Pande, H. D.: 1974, Bianchi identities in Finsler space, Univ. Nac. Tucumán Rev. A 24, 157-161.
Pande, H. D.: 1976a, Projective entities in Finsler space, Rev. Mat. Hisp.-Amer. (4) 36, 133-137.
Pande, H. D.: 1976b, Recurrent Finsler spaces of third order, Indian J. Pure Appl. Math. 7, 1333-1336.
Pande, H. D., Chandra, K. and Misra, S. B.: 1981, The commutation formulae involving Berwald's and projective covariant derivatives, J. Math. Phys. Sci. 15, 31-38.
Pande, H. D. and Dubey, V. J.: 1978, Generalised Gauss-Codazzi equations in a hypersurface of a Finsler space, J. Math. Phys. Sci. 12, 325-333.
Pande, H. D. and Dubey, V. J.: 1979, Decomposition of Berwald's curvature tensor field in generalized 2-recurrent Finsler space, Indian J. Pure Appl. Math. 10, 33-38.

Pande, H. D. and Dubey, V. J.: 1980, Some symmetric properties in projective Finsler space, J. Math. Phys. Sci. 14, 477-483.
Pande, H. D. and Dubey, V. J.: 1982a, Decomposition of $H^i_{jkh}(x,\dot{x})$ in generalised recurrent Finsler space of second order, Mat. Vesnik 6(19)(34), No. 4, 401-408.
Pande, H. D. and Dubey, V. J.: 1982b, Decomposition of projective recurrent tensor field of third order, Acta Cienc. Indica 8, 89-93.
Pande, H. D. and Dubey, V. J.: 1982c, On recurrent conformal motion and Q-collineation in symmetric Finsler space, Acta Cienc. Indica 8, 189-194.
Pande, H. D. and Dubey, V. J.: 1982d, Some theorems on conformal motion in special projective symmetric Finsler space, Acta Cienc. Indica 8, 209-213.
Panda, H. D. and Gupta, K. A.: 1979a, Bianchi's identities in a Finsler space with non-symmetric connections, Bull. Cl. Sci. Math. Nat., Sci. Math., N. S. 10, 41-46.
Pande, H. D. and Gupta, K. A.: 1979b, Projective entities, J. Math. Phys. Sci. 13, 425-435.
Pande, H. K. and Gupta, K. A.: 1979c, Some special properties of R^h_{ijk} and Q^h_{ijk} in recurrent Finsler spaces, Indian J. Pure Appl. Math. 10, 1020-1030.
Panda, H. K. and Gupta, K. A.: 1980, An R-recurrent Finsler space of third order, Acta Cienc. Indica 6, 177-183.
Pande H. D. and Khan, T. A.: 1973a, General decomposition of Berwald's curvature tensor fields in recurrent Finsler space, Lincei Rend. (8) 55, 680-685.
Pande, H. D. and Khan, T. A.: 1973b, Recurrent Finsler space with Cartan's first curvature tensor field, Lincei Rend. (8) 55, 224-227.
Pande, H. D. and Khan, T. A.: 1974a, General decomposition of relative curvature tensor field in recurrent Finsler space, Progr. Math. Allahabad 8-2, 9-13.
Pande, H. D. and Khan, T. A.: 1974b, Decomposition of Berwald's curvature tensor field in second order recurrent Finsler space, Lincei Rend. (8) 57, 565-569.
Pande, H. D. and Khan, T. A.: 1978, On generalised 2-recurrent Berwald's curvature tensor field in Finsler space, Acta Cienc. Indica 4, 56-59.
Pande, H. D. and Kumar, A.: 1972a, Bianchi and Veblen identities for the special pseudo-projective curvature tensor fields, Lincei Rend. (8) 53, 384-388.
Pande, H. D. and Kumar, A.: 1972b, The effect of conformal change over some entities in Finsler space, Lincei Rend. (8) 53, 60-70.
Pande, H. D. and Kumar, A.: 1972c, Cartan's covariant derivative with projective connection parameter, Progr. Math. Allahabad 6-2. 51-57.
Pande, H. D. and Kumar, A.: 1973, Pseudo-projective curvature identities in Finsler space, Lincei Rend. (8) 55, 676-679.
Pande, H. D. and Kumar, A.: 1974a, Special curvature collineation and projective symmetry in Finsler space, Lincei Rend. (8) 57, 75-79.
Pande, H. D. and Kumar, A.: 1974b, Special infinitesimal projective transformation in a Finsler space, Lincei Rend. (8) 57, 190-193.
Pande, H. D. and Kumar, A.: 1974c, Conformal motion in a recurrent Finsler space, Rev. Univ. Istanbul A 39, 63-67.
Pande, H. D. and Kumar, A.: 1975a, Generalisation of Gauss-Codazzi equations for the curvature tensor $M_h{}^i{}_{jk}(x, \dot{x})$ in a hypersurface of a Finsler space, Progr. Math. Allahabad 9, No. 2, 1-8.
Pande, H. D. and Kumar, A.: 1975b, Infinitesimal special projective transformation in Finsler space, Pure Appl. Math. Sci. 2, 11-15.
Pande, H. D. and Kumar, A.: 1975c, Generalisation of Gauss-Codazzi equations for Berwald's curvature tensor in a hypersurface of a Finsler space, Publ. Math. Debrecen 22, 263-267.
Pande, H. D. and Kumar, A.: 1975d, Special conformal motion in a special projective symmetric Finsler space, Lincei Rend. (8) 58, 713-717.
Pande, H. D. and Kumar, A.: 1975e, General decomposition of Weyl's curvature tensor in a recurrent Finsler space, Rev. Fac. Sci. Univ. Istanbul A 40, 171-176.
Pande, H. D. and Kumar, A.: 1976a, On infinitesimal special projective transformation in a Finsler space, Acta Cienc. Indica 2, No. 2, 169-172.
Pande, H. D. and Kumar, A.: 1976b, Generalized Gauss-Codazzi equations for the curvature tensor $R^i_{jhk}(x,\dot{x})$ in a hypersurface of a Finsler space, Jnanabha Sect. A 6, 53-58.
Pande, H. D. and Kumar, A.: 1976c, Projective curvature collineation in a Finsler space, Jnanabha Sect. A 6, 103-107.
Pande, H. D. and Kumar, A.: 1977, General decomposition of projective entities in a recurrent Finsler space, Progr. Math. Allahabad 11, 63-70.
Pande, H. D., Kumar, A. and Khan, T. A.: 1975, Curvature collineations in a Finsler space, Acta Cienc. Indica 1, 357-360.
Pande, H. D. and Mishra, R. S.: 1967, General projective geometry of geodesics in a Finsler space, Rend. Circ. Mat. Palermo (2) 16, 239-246.

Pande, H. D. and Mishra, R. S.: 1968, Projective transformation in the theory of non-linear connections, Proc. Math. Phys. Soc. U. A. R. 32, 21-27.
Pande, H. D. and Misra, S. B.: 1975, Some special Ricci identities, Lincei Rend. (8) 58, 190-194.
Pande, H. D. and Misra, S. B.: 1976a, Projective Ricci identities in a Finsler space, Lincei Rend. (8) 59, 251-257.
Pande, H. D. and Misra, S. B.: 1976b, Bianchi and Veblen identities in a generalised 2-recurrent Finsler space, J. Math. Phys. Sci. 10, 464.
Pande, H. D. and Misra, S. B.: 1976c, An W^{*i}_{hj} generalised 2-recurrent Finsler space, Rev. Fac. Sci. Univ. Istanbul, A 41, 7-12.
Pande, H. D. and Misra, S. B.: 1976d, On 2-generalized projective recurrent Finsler space, Acta Cienc. Indica 2, 272-274.
Pande, H. D. and Pandey, J. P.: 1976 A W-recurrent Finsler space of the second order, Bull. Acad. Serbe Sci. Arts. Cl. Sci. Math. Natur. 55, 66-71.
Pande, H. D. and Pandey, J. P.: 1978, On generalised $\tilde{K}$-recurrent Finsler space, Pure Appl. Math. Sci. 8, 1-4.
Pande, H. D. and Pandey, J. P.: 1980, Decomposition of Weyl's curvature tensor fields in a projective recurrent Finsler space, Ann. Soc. Sci. Bruxelles I 94, 185-190.
Pande, H. D. and Pandey, J. P.: 1982, Decomposition of $H^i_{hjk}(x, \dot{x})$ in a recurrent Finsler space, Acta Cienc. Indica 8, 1-15, 214-218.
Pande, H. D. and Pandey, J. P.: 1983a, Some theorems on special curvature collineation and projective motion in a projective symmetric Finsler space, Acta Cienc. Indica 9, 155-159.
Pande, H. D. and Pandey, J. P.: 1983b, On the equality of Cartan's curvature tensor and Berwald's curvature tensor in special Finsler spaces, Tamkang J. Math. 14, 195-201.
Pande, H. D. and Pandey, J. P.: 1984, On the three-dimensional Finsler spaces with T-tensor of a special form, Acta Math. Hung. 43, 47-52.
Pande, H. D. and Shukla, H. S.: 1977a, On infinitesimal conformal transformation in a Finsler space, Indian J. Pure Appl. Math. 8, 288-294.
Pande, H. D. and Shukla, H. S.: 1977b, On the decomposition of curvature tensor fields K^i_{jhk} and H^i_{jhk} in recurrent Finsler spaces, Indian J. Pure Appl. Math. 8, 418-424.
Pande, H. D. and Singh, B.: 1973/1974, On projective recurrent Finsler spaces of the first order, Rend. Accad. Naz. XL(4) 24/25, 275-280.
Pande, H. D. and Singh, B.: 1974a, An W^{*i}_{hj} recurrent Finsler space, Progr. Math. Allahabad 8, 41-46.
Pande, H. D. and Singh, B.: 1974b, Recurrent Finsler spaces with pseudo projective tensor field, Lincei Rend. (8) 57, 70-74.
Pande, H. D. and Singh, B.: 1977, On existence of the affinely connected Finsler space with recurrent tensor fields, Indian J. Pure Appl. Math. 8, 295-301.
Pande, H. D. and Singh, B.: 1980, Berwald's recurrent Finsler spaces of the first order, Bull. Calcutta Math. Soc. 71, 94-102.
Pande, H. D. and Tripathi, S. D.: 1977, On generalized K-recurrent Finsler spaces with a symmetric non-zero recurrence tensor field, Indian J. Pure Appl. Math. 8, 282-287.
Pande, H. D. and Tripathi, S. D.: 1981, On H-recurrent Finsler space, Acta Cienc. Indica 7, 120-126.
Pande, H. D. and Tripathi, S. D.: 1983a, On generalised $3RF_n$, Acta Cienc. Indica 9, 145-149.
Pande, H. D. and Tripathi, S. D.: 1983b, Some theorems on projective Ricci identities in Finsler space, Acta Cienc. Indica 9, 160-166.
Pandey, J. P. and Pandey, J. D.: 1984, On the three-dimensional Finsler spaces with T-tensor of a special form, Acta Math. Hungar. 43, 47-52.
Pandey, P. N. 1977a, Contra projective motion in a Finsler manifold, Math. Educ. A 11, 25-29, 72.
Pandey, P. N.: 1977b, On bisymmetric Finsler manifolds, Math. Educ. A 11, 77-80.
Pandey, P. N.: 1978, CA-collineation in a bi-recurrent Finsler manifold, Tamkang J. Math. 9, 79-81.
Pandey, P. N.: 1979a, Groups of conformal transformations in conformally related Finsler manifolds, Lincei Rend. (8) 65 (1978), 269-274.
Pandey, P. N.: 1979b, Decomposition of curvature tensor in a recurrent Finsler manifold, Tamking J. Math. 10, 31-34.
Pandey, P. N.: 1979c, A recurrent Finsler manifold admitting special recurrent transformations, Progr. Math. 13, 85-98.
Pandey, P. N.: 1980a, A recurrent Finsler manifold with a concircular vector field, Acta Math. Acad. Hungar. 15, 465-466.
Pandey, P. N.: 1980b, Conformal covariant derivative in a Finsler manifold, Atti. Accad. Sci. Lett. Arti Palermo, Parte I (4) 37 (1977/1978), 341-350.
Pandey, P. N.: 1980c, Affine motion in recurrent Finsler manifold, Ann. Fac. Sci. Univ. Nat. Zaire (Kinshasa) Sect. Math.-Phys. 6, 51-63.

Pandey, P. N.: 1980d, On NPR-Finsler manifolds, Ann. Fac. Sci. Nat. Zaire (Kinshasa) Sect. Math.-Phys. 6, 65-67.
Pandey, P. N.: 1981a, A note on recurrence vector, Proc. Nat. Acad. Sci. India A 51, 6-8.
Pandey, P. N.: 1981b, Some identities in an NPR-Finsler manifold, Proc. Nat. Acad. Sci. India A 51, 185-189.
Pandey, P. N.: 1981c, On decomposability of curvature tensor of a Finsler manifold, Acta Math. Acad. Sci. Hungar. 38, 109-116.
Pandey, P. N.: 1982a, On Lie recurrent Finsler manifolds, Indian J. Math. 24, 135-143.
Pandey, P. N.: 1982b, On a Finsler space of zero projective curvature, Acta Math. Sci. Hungar. 39, 387-388.
Pandey, P. N. and Dwivedi, V. J.: 1977, Normal projective curvature tensor in a conformal Finsler manifold, Proc. Nat. Acad. Sci. India A 47, 115-118.
Pandey, P. N. and Misra, R. B.: 1981, Projective recurrent Finsler manifolds. I, Publ. Math. Debrecen 28, 191-198.
Pauc, C.: 1951, Les théorèmes fort et faible de Vitali et les conditions d'évanescence de halos, C. R. Acad. Sci. Paris 232, 1727-1729.
Penot, J. P.: 1972, Topologie faible sur des variétés de Banach, C. R. Acad. Sci. Paris A 274, 405-408.
Penot, J. P.: 1974, Topologie faible sur des variétés de Banach, Application aus géodésiques des variétés de Sobolev, J. Diff. Geom. 9, 141-168.
Perscy, M.: 1974, 'A characterization of classical Minkowski planes over a perfect field of characteristic two, J. Geom. 5, 191-204.
Petty, C. M.: 1971, Equilaterial sets in Minkowski spaces, Proc. Amer. Math. Soc. 29, 369-374.
Phadke, B. B.: 1972, Equidistant loci and the Minkowskian geometries, Canad. J. Math. 24, 312-327.
Phadke, B. B.: 1975, A triangular world with hexagonal circles, Geom. Dedicata 3, 511-520.
Phadke, B. B.: 1976, The theorem of Desargues in planes with analogues to Euclidean angular bisectors, Math. Scand. 39, 191-194.
Pihl. M.: 1955, Classical mechanics in a geometrical description, Danske Vid. Selsk. Mat.-Pys. Medd. 30, No. 12, 1-26 (in Danish).
Pimenov, R. I.: 1981, Finsler kinematics, Sibirsk. Mat. Z. 22, No. 3, 136-146, 237 (in Russian).
Pinl, M.: 1932, Quasimetrik auf totalisotropen Flächen. I, Akad. Wetensch. Proc. 35, 1181-1188.
Pinl, M.: 1933, Quasimetrik auf totalisotropen Flächen. II, Akad. Wetensch. Proc. 36, 550-557.
Pinl, M.: 1935, Quasimetrik auf totalisotropen Flächen. III, Akad. Wetensch. Proc. 38, 171-180.
Popescu, I. P.: 1980, Remarques sur les connexions de Finsler, Proc. Nat. Sem. Finsler Spaces, Brasov, 209-218.
Popovici, I. and Anastasiei, M.: 1980, Sur les bases de la géométrie finslérienne, C. R. Acad. Sci. Paris 290, 807-810.
Prakash, N.: 1961, Generalised normal curvature of a curve and generalised principal directions in Finsler space, Tensor, N. S. 11, 51-56.
Prakash, N.: 1962, Kaehlerian Finsler manifolds, Math. Student 30, 1-12.
Prakash, N. and Behari, R.: 1960a, Union curves and union curvature in Finsler space, Proc. Nat. Inst. Sci. India A 26, Suppl. II, 21-30.
Prakash, N. and Behari, R.: 1960b, Derivations from parallelism and equidistance in Finsler space, Proc. Indian Acad. Sci. A 52, 209-227.
Prakash, N. and Behari, R.: 1960c, Generalizations of Codazzi's equations in a subspace imbedded in a Finsler manifold, Proc. Nat. Inst. Sci. India A 26, 532-540.
Prakash, N. and Dhawan, M.: 1969, A study of contravariant and covariant almost analytic vectors in a Finsler space, J. Math. Sci. 4, 69-76.
Prasad, B. N.: 1970, Curvature collineations in Finsler space, Lincei Rend. (8) 49, 194-197.
Prasad, B. N.: 1971a, Certain investigations in Finsler and complex spaces: Thesis, Gorakhpur, 186 p.
Prasad, B. N.: 1971b, On union curves and union curvature of a vector field. (From the standpoint of Cartan's Euclidean connection.), Lincei Rend. (8) 51, 199-205.
Prasad, B. N.: 1979a, A transformation associated with the set of n fundamental forms of Minkowskian hypersurface, J. Math. Phys. Sci. 13, 251-256.
Prasad, B. N.: 1979b, On a transformation associated with sets of n fundamental forms of Minkowskian hypersurfaces, Indian J. Pure Appl. Math. 10, 1068-1075.
Prasad, B. N. 1980, Finsler spaces with the torsion Tensor P_{ijk} or a special form, Indian J. Pure Appl. Math. 11, 1572-1579.
Prasad, B. N. and Bose, V. V.: 1981, The Lie derivatives in complex areal space, Publ. Math. Debrecen 28, 79-88.

Prasad, B. N. and Singh, V. P.: 1980, Finsler spaces with the Berwald's curvature tensor of a special form, Indian J. Pure Appl. Math. 11, 1270-1277.
Prasad, C. M.: 1969, Special curves of a hypersurface of a Finsler space, Rev. Univ. Istanbul 34, 97-101.
Prasad, C. M.: 1970, Δ-curvatures and Δ-geodesic principal directions of a congruence of curves in the subspace of a Finsler space, Rev. Univ. Istanbul 34, 97-101.
Prasad, C. M.: 1971a, Union congruence in a subspace of a Finsler space, Ann. Mat. Pura Appl. (4) 88, 143-154.
Prasad, C. M.: 1971b, On the hypersymptotic and hypergeodesic curvatures of a curve of a Finsler hypersurface, Acad. Roy. Belg. Bull. Cl. Sci. (5) 57, 1196-1201.
Prasad, C. M.: 1971c, Hyperasymptotoc and hypernormal congruences in a subspace of a Finsler space, Lincei Rend. (8) 51, 206-212.
Prasad, C. M.: 1972a, Hyperasymptotic and hypernormal congruences in a subspace of a Finsler space, Tensor, N. S. 24, 206-210.
Prasad, C. M.: 1972b, Relative associate curvature of a congruence and λ-pseudogeodesics of a Finsler subspace, Lincei Rend. (8) 52, 702-707.
Prasad, C. M.: 1972c, Special congruence in a subspace of a Finsler space, Ann. Mat. Pura Appl. (4) 92, 29-36.
Prasad, C. M.: 1973a, On Δ-principal directions of a congruence of curves in a Finsler hypersurface, Math. Nachr. 57, 39-50.
Prasad, C. M.: 1973b, Generalization of Meusnier's theorem in a Finsler space, Lincei Rend. (8) 55, 219-223.
Prasad, C. M. and Srivastava, R. K.: 1979a, On special congruence of a Finsler space, Ann. Math. Pura Appl. (4) 119, 195-203.
Prasad, C. M. and Srivastava, R. K.: 1979b, Decomposition of neo-recurrent curvature tensor field of the second order, Indian J. Pure Appl. Math. 10, 420-427.
Prasad, C. M. and Srivastava, R. K.: 1981, Super geodesic congruence in a subspace of a Finsler space, Publ. Math. Debrecen 28, 145-151.
Prvanovic, M.: 1958, Les transformations conformes et projectives d'espace riemannian généralisé au sens de T. Takasu, Ann. Fac. Lett. Sci. Novi Sad. 3, 265-272 (in Serbo-Croatian).
Rachevsky, P. K.: 1935a, Dualité métrique dans la géométrie à deux dimensions de Finsler, en particulier, sur une surface arbitraire, Dokl. Akad. Nauk SSSR 3, 147-150.
Rachevsky, P. K.: 1935b, Une géométrie métrique duale, fondée sur les espaces de Cartan généralisés, C. R. Acad. Sci. Paris 201, 921-923.
Rachevsky, P. K.: 1935c, Système bimétrique dual, C. R. Acad. Sci. Paris 201, 1088-1090.
Rachevsky, P. K.: 1936, Systèmes trimétrique et la métrique de Finsler généralisée, C. R. Acad. Sci. Paris 202, 1237-1239.
Rachevsky, P. K.: 1941, Polymetric geometry, Trudy Sem. Vektor. Tenzor. Anal. 5, 21-147 (in Russian).
Radivoivici, M. S.: 1983, Holonomy groups and Finsler metric connections, Proc. Nat. Colloq. Geometry and Topology, Buşteni, 1981, Univ. Bucharest, pp. 312-315.
Radon, H.: 1916, Über eine besondere Art ebener konvexer Kurven, Leipzig Ber. 68, 123-128.
Radu, C. F.: 1973, On extensions of the Lie derivative to tensors that depend on point and direction, A. Univ. Bucuresti Mat.-Mec. 22-1, 95-101 (in Roumanian).
Ramanujan, P. B.: 1972, A note on parallel vector fields in Finsler spaces, Indian J. Math. 14, 105-106.
Rani, N.: 1968, Deviations from parallelism and equidistance of the congruence of curves, Progr. Math. Allahabad 2, No. 2, 15-20.
Rani, N.: 1969a, A note on a complex Finsler manifold, Tensor, N. S. 20, 91-94.
Rani, N.: 1969b, Theory of Lie derivatives in a Minkowskian space, Tensor, N. S. 20, 100-102
Rapcsák, A.: 1949, Kurven auf Hyperflächen im Finslerschen Räume, Hungar. Acta Math. 1, 21-27.
Rapcsák, A.: 1954, Eine neue Definition der Normalkoordinaten im Finslerschen Raum, Acta Univ. Debrecen 1, 109-116. ad. 1. (1955), 17 (in Hungarian).
Rapcsák, A.: 1955a, Invariante Taylorsche Reihe in einem Finslerschen Raum, Publ. Math. Debrecen 4, 49-60.
Rapcsák, A.: 1955b, Theorie der Bahnen in Linienelementmannigfaltigkeiten und eine Verallgemeinerung ihrer affinen Theorie, Acta Sci. Math. Szeged 16, 251-265.
Rapcsák, A.: 1957a, Eine neue Charakterisierung Finslerscher Räume skalarer und konstanter Krümmung und projektiv-ebene Räume, Acta Math. Acad. Sci. Hungar. 8, 1-18.
Rapcsák, A.: 1957b, Metrische Charakterisierung der Finslerschen Räume mit verschwindender projektiver Krümmung, Acta Sci. Math. Szeged 18, 192-204.
Rapcsák, A.: 1959, Hyperebenen in Finslerschen Räumen, Acta Univ. Debrecen 4, 85-87 (in Hungarian).
Rapcsák, A.: 1960, Über die begründung der Lokalen metrischen Differentialgeometrie, Publ. Math. Debrecen 7, 382-393.
Rapcsák, A.: 1961a, Über die bahntreuen Abbildungen affinzusammenhängender Räume, Publ. Math. Debrecen 8, 225-230.

Rapcsák, A.: 1961b, Über die bahntreuen Abbildungen metrischer Räume, Publ. Math. Debrecen 8, 285-290.
Rapcsák, A.: 1961c, Bahntreue Abbildungen von metrische und affinzusammenhängenden Bahnräumen, Magyar Tud. Akad. III, Mat. Fiz. 11, 339-369 (in Hungarian).
Rapcsák, A. 1962a, Die Bestimmung der Grundfunktionen projektiv-ebener metrischer Räume, Publ. Math. Debrecen 9, 164-167.
Rapcsák, A.: 1962b, Über die Metrisierbarkeit affinzusammenhängender Bahnräume, Ann. Mat. Pura Appl. (4) 57, 233-238.
Rastogi, S. C.: 1970, On the Bianchi and Veblen identities in a Finsler space, An. Univ. Timisoara, Sti. Mat. 8, 95-100.
Rastogi, S. C.: 1971a, Generalised curves in a Finsler space, Univ. Nac, Tucumán Rev. A 21, 15-23.
Rastogi, S. C.: 1971b, On a curve in a subspace of a Finsler space, Univ. Nac. Tucumán Rev. A 21, 271-276.
Rastogi, S. C.: 1971-1972, On relative δ-derivative in a Finsler space, Univ. Lisboa, Rev. Fac. Sci. (2) A 14, 9-18.
Rastogi, S. C.: 1976, Submanifolds of a Finsler manifold, Tensor, N. S. 30, 140-144.
Rastogi, S. C.: 1980, On Finsler space of recurrent curvature tensors, Kyungpook Mat. J. 20, 37-54.
Rastogi, S. C.: 1981, On projective invariants based on nonlinear connections in a Finsler space, Publ. Math. Debrecen 28, 121-128.
Rastogi, S. C.: 1982, On recurrent Finsler spaces, J. Nigerian Math. Soc. 1, 68-81.
Rastogi, S. C. and Agrawal, R. C.: 1970, On generalisations of the first curvature vector in a hypersurface of Finsler space, Ganita 21, 105-116.
Rastogi, S. C. and Sharma, I. D.: 1973, A study of some curves in a subspace of a Finsler space, Ganita 24, 49-58.
Rastogi, S. C. and Trivedi, H. K. N.: 1972, On generalized curves in a Finsler space, Indian J. Pure Appl. Math. 3, 1058-1063.
Rastogi, S. C. and Trivedi, H. K. N.: 1974, A study of congruences of curves in a Finsler space, Ganita 25, 19-28.
Reeb, G.: 1951a, Quelques propriétés globales des géodésiques d'un espace de Finsler et des Variétés minima d'un espace de Cartan, Colloq. Topol. Strasbourg, No. 2, pp. 1-9.
Reeb, G.: 1951b, Sur certaines propriétés globales des trajectoires de la dynamique, dues à l'existence de l'invariant intégral de M. Elie Cartan, Colloq. Topol. Strasbourg, No. 3, pp. 1-7.
Reeb, G.: 1952a, Sur les éléments de contact linéaires de second ordre attachés à un système différentiel, J. Reine Angew. Math. 189, 186-189.
Reeb, G.: 1952b, Sur certaines propriétés topologiques des trajectoires des systèmes dynamiques, Acad. Roy. Belg., Cl. Sci. Mem. Coll. 8°, 27-9, 1-64.
Reeb, G.: 1953, Sur les espaces de Finsler et les espaces de Cartan, Colloq. Intern. Géom. Diff. Strasbourg, pp. 35-40.
Reiffen, H. J.: 1963, Die differentialgeometrischen Eigenschaften der invarianten Distanzfunktion von Carathéodory, Schr. Math. Inst. Münster 26, 1-66.
Reiffen, H. J.: 1977/1978, Metrische Grössen in C^q-Räumen, Rend. Accad. Naz. XL(5) 3, 1-29.
Rider, P. R.: 1926, The figuratrix in the calculus of variations, Trans. Amer. Math. Soc. 28, 640-653.
Riedler, K.: 1975, Globale Finslersche Geometrie, Ber. Math.-Stat. Sekt. Forschungszentrum Graz 43, 3-18.
Rinow, W.: 1936, Über vollständige differentialgeometrische Räume, Deutsch. Mat. 1, 46-63.
Rizza, G. B.: 1962, Strutture di Finsler sulle varietà quasi complesse, Lincei Rend. (8) 33, 271-275.
Rizza, G. B.: 1963, Strutture di Finsler di tipo quasi Hermitiano, Riv. Mat. Univ. Parma (2) 4, 83-106.
Rizza, G. B.: 1964, F-forme quadratiche ed hermitiane, Rend. Mat. Appl. (5) 23, 221-249.
Rizza, G. B.: 1975, Monogenic functions on the real algebras and conformal mappings, Boll. U. M. I. (4) 12, Suppl. 3, 437-450.
Roxburgh, I. W. and Tavakol, R.: 1975, The gravitational theories of Poincaré and Milne and the non-Riemannian kinematic models of the universe, Mon. Not. R. Astr. Soc. 170, 599-610.
Ruiz, Monroy, O. R.: 1977, Existence of brake orbits in Finsler mechanical systems, in Geometry and Topology. Lect. Notes in Math. 597. Springer, Berlin, pp. 542-567.
Rund, H.: 1950, Finsler spaces considered as generalised Minkowskian spaces: Thesis, Cape Town.
Rund, H.: 1951a, Über die Parallelverschiebung in Finslerschen Räumen, Math. Z. 54, 115-128.
Rund, H.: 1951b, A theory of curvature in Finsler spaces, Colloq. Topol. Strasbourg, No. 4, 1-12.
Rund, H.: 1952a, Zur Begründung der Differentialgeometrie der Minkowskischen Räume, Arch. Math. 3, 60-69.

Rund, H.: 1952b, Eine Krümmungstheorie der Finslerschen Räume, Math. Ann. 125, 1-18.
Rund, H.: 1952c, The theory of subspaces of a finsler space. I, Math. Z. 56, 363-375.
Rund, H.: 1952d, Die Hamiltonische Funktion bei allgemeinen dynamischen Systems, Arch. Math. 3, 207-215.
Rund, H.: 1953a, The theory of subspaces of a Finsler space. II, Math. Z. 57, 193-210.
Rund, H.: 1953b, The scalar form of Jacobi's equations in the calculus of variations, Ann. Mat. Pura Appl. (4) 35, 183-202.
Rund, H.: 1953c, Application des méthodes de la géométrie métrique généralisée à la dynamique théorique, Colloq. Intern. Géom. Diff. Strasbourgh, pp. 41-51.
Rund, H.: 1953d, On the geometry of generalised metric spaces, Convegno Intern. Geom. Diff. Italia, pp. 114-121.
Rund. H.: 1954a, On the analytical properties of curvature tensors in Finsler spaces, Math. Ann. 127, 82-104.
Rund, H.: 1954b, Über nicht-holonome allgemeine metrische Geometrie, Math. Nachr. 11, 61-80.
Rund, H.: 1956, Hypersurfaces of a Finsler space, Canad. J. Math. 8, 487-503.
Rund, H.: 1957, Über allgemeine nicht-holonome und dissipative dynamische Systeme, Schr. Forschungsinst. Math. 1, 269-279.
Rund, H.: 1958, Some remarks concerning the theory of non-linear connections, Nederl. Akad. Wetensch. Proc. A 61, = Indag. Math. 20, 341-347.
Rund, H.: 1959, Some remarks concerning Carathéodory's method of 'equivalent' integrals in the calculus of variations, Nederl. Akad. Wetensch. Proc. A 62 = Indag. Math. 21, 135-141.
Rund, H.: 1961a, The theory of problems in the calculus of variations whose Langrangian function involves second order derivatives: A new approach, Ann. Mat. Pura Appl. (4) 55, 77-104.
Rund, H.: 1961b, Dynamics of particles with internal 'spin', Ann. Physik (7) 7, 17-27.
Rund, H.: 1962, Note on the Lagrangian formalism in relativistic mechanics, Nuovo Cimento (10) 23, 227-232.
Rund, H.: 1963, Curvature properties of hypersurfaces of Finsler and Minkowskian space, Tensor, N. S. 14, 226-224.
Rund, H.: 1965a, The intrinsic and induced curvature theories of subspaces of a Finsler space, Tensor, N. S. 16, 294-312.
Rund, H.: 1965b, Calculus of variations on complex manifolds, 3rd Colloq. Calculus Varia. Univ. South Africa, pp. 10-85.
Rund, H.: 1966, Finsler spaces of scalar and constant curvature in geodesic correspondence, Tydskr. Natuurwet. 6, 243-254.
Rund, h.: 1967, Invariant theory of variational problems for geometric objects, Tensor, N. S. 18, 239-258.
Rund, H.: 1968, A geometrical theory of multiple integral problems in the calculus of variations, Canad. J. Math. 20, 639-657.
Rund, H.: 1972, The curvature theory of direction-dependent connections on complex manifolds, Tensor, N. S. 24, 189-205.
Rund, H.: 1975, A divergence theorem for Finsler metrics, Monatsh. Math. 79, 233-252.
Rund, H.: 1980, Direction-dependent connection and curvature forms, Abh. Math. Sem. Hamb. 50, 188-209.
Rund, H.: 1981, Subspaces of manifolds with direction-dependent connections, Tensor, N. S. 35, 139-154.
Rund, H.: 1982, Invariance theory of distributions associated with multiple integral problems in the calculus of variations, Tensor, N. S. 39, 45-66.
Sakaguchi, T.: 1980, On Finsler spaces with $F_{n\ jk}^{\ i} = 0$, Tensor, N. S. 34, 326-337.
Sakaguchi, T.: 1982, On Finsler spaces of scalar curvature', Tensor, N. S. 38, 211-219.
Sakaguchi, T.: 1983, Remarks on Finsler spaces with 1-form metric, Tensor, N. S. 40, 173-183.
Sakata, S.: 1970, A constructive approach to non-teleparallelism and non-metric representations of plastic material manifold by generalized diakoptical tearing. II. Finslerian tearing, RAAG Res. Notes 3-149, pp. 1-27.
Sándor, I.: 1961, Die Darbouxschen Formeln der Eichfläche eines Minkowskischen Raumes, Publ. Math. Debrecen 8, 187-192.
Sanini, A.: 1971-1973, Derivazioni su distribuzioni e connessioni di Finsler, Univ. Politec. Torino, Rend. Sem. Mat. 31, 157-184.
Sanini, A.: 1971-1972, Alcuni tipi di connessioni su varietà quasi prodotto, Atti. Accad. Torino, Cl. Sci. Fis. 106, 317-332.
Sanini, A.: 1973-1974, Su un tipo di struttura quasi Hermintiana del fibrato tangente as uno spazio di Finsler, Univ. Politec. Torino, Rend. Sem. Mat. 32, 303-316.
Sanini, A.: 1974, Connessioni lineari del tipo di Finsler e strutture quasi hermitiane, Riv. Mat. Univ. Parma (3), 3, 239-252.
Sapukov, N. N.: 1961, Extremal displacement of a minimal hypersurface in Riemannian and Finsler spaces, Izv. Vyss. Uchebn. Zaved. Mat. No. 5(24), 112-116 (in Russian).
Sasaki, S.: 1937, Non-euclidean geometry in general space, Sci. Rep. Tohoku Univ. I 26, 313-322.

Sattarov, A. E.: 1978, Invariant form of the second variation in the metric space of support vector densities, Trudy Geom. Sem. Kazan. Univ. 10, 78-85 (in Russian).
Sattarov, A. E.: 1983, Some problems of differential geometry of one class of spaces of supporting elements, Arch. Math. (Brno) 19, 219-224.
Savage, L. J.: 1943, On the crossing of extremals at focal points, Bull. Amer. Math. Soc. 49, 467-469.
Schaer, J.: 1960, De la possibilité d'une théorie unitaire finslérienne de l'électromagnétisme et de la gravitation, Arch. Sci. Genève 13, 542-549.
Schneider R.: 1968, Über die Finslerräume mit $S_{ijkl} = 0$, Arch. Math. 19, 656-658.
Schöne, W.: 1969, Über eine Verallgemeinerung der Minkowskischen Distanzfunktion und ihre Anwendung in der Finslerschen Geometrie, Math. Nachr. 40, 305-325.
Schöne, W.: 1973, Ein Eigenwertproblem des metrischen Tensors und seine geometrische Deutung, Demonst. Math. 6, 343-365.
Schouten, J. A. and Haantjes, J.: 1936, Über die Festlegung von allgemeinen Massbestimmungen und Übertragungen in bezug auf ko- und kontravariante Vektordichten, Monatsh. Math. Phys. 43, 161-176.
Schröder, E. M.: 1970, Darstellung der Gruppenräume Minkowskischer Ebenen, Arch. Math. 21, 308-316.
Segre, B.: 1949, Geometria non euclidea ed ottica geometrica. I, II, Lincei Rend. (8) 7, 16-26.
Sen, R. N.: 1967a, Application of an algebraic system in Finsler geometry, Tensor, N. S. 18, 191-195.
Sen, R. N.: 1967b, On curvature tensors in Finsler geometry, Tensor, N. S. 18, 217-299.
Sen, R. N.: 1967c, Some generalised formulae for curvature tensors in Finsler geometry, Indian J. Math. 9, 211-221.
Sen, R. N.: 1968, Finsler spaces of recurrent curvature, Tensor, N. S. 19, 291-229.
Sevrjuk, V. P.: 1973, Covariant derivatives of a spinor in Finsler geometry, Isz. Vyss. Uchebn. Zaved. Fizika, No. 2, 43-49.
Shamihoke, A. C.: 1961a, On the subspaces of a Finsler space, J. Indian Math. Soc., N. S. 25, 215-220.
Shamihoke, A. C.: 1961b, Frenet's formulae and curvatures in a generalised Finsler space, Rev. Univ. Istanbul 26, 69-75.
Shamihoke, A. C.: 1962a, Contributions to the differential geometry of Riemannian, Finsler and generalised Finsler spaces: Thesis, Delhi.
Shamihoke, A. C.: 1962b, Some properties of curvature tensors in a generalised Finsler space, Tensor, N. S. 12, 97-109.
Shamihoke, A. C.: 1962c, Normal curvature of a vector field in a hypersurface of a generalised Finsler space, Rev. Univ. Istanbul 27, 9-14.
Shamihoke, A. C.: 1963a, Hypersurfaces of a generalised Finsler space, Tensor, N. S. 13, 129-144.
Shamihoke, A. C.: 1963b, Subspaces of a generalised Finsler space. I, Ganita 14, 43-59.
Shamihoke, A. C.: 1964a, Parallelism and covariant differentiation in a generalized Finsler space of n dimensions, Riv. Mat. Univ. Parma (2) 5, 189-200.
Shamihoke, A. C.: 1964b, A note on a curvature tensor in a generalized Finsler space, Tensor, N. S. 15, 20-22.
Sharma, I. D. and Trivedi, K. N.: 1972, On the Einstein's law of gravitation in Finsler space, Ganita 23, 13-18.
Sharma, N. K. and Srivastava, A.: 1983a, Some problems on a conformal Randers-space, J. Math. Phys. Sci. 17, 37-40.
Sharma, N. K. and Srivastava, A.: 1983b, On a special Finsler space, Demonstratio Math. 16, 619-627.
Shibata, C.: 1984, On invariant tensors of β-changes of Finsler metrics, J. Math. Kyoto Univ. 24, 163-188.
Shibata, C. and Shimada, H.: 1981, The g-hypercone of a Minkowski space, Tensor, N. S. 35, 73-85.
Shibata, C., Shimada, H., Azuma, M., and Yasuda, H.: 1977, On Finsler spaces with Randers' metric', Tensor, N. S. 31, 219-226.
Shibata, C., Singh, U. P., and Singh, A. K.: 1983, On induced and intrinsic theories of hypersurfaces of Kropina spaces, J. Hokkaido Univ. II A 34, 1-11.
Shimada, H.: 1977, On the Ricci tensors of particular Finsler spaces, J. Korean Math. Soc. 14, 41-63.
Shimada, H.: 1979a, Finsler spaces of recurrent torsion and constant curvature, Research Rep. Kushiro Tech. College 13, 173-180.
Shimada, H.: 1979b On Finsler spaces with the metric $L = \sqrt[m]{a_{i_1 i_2 \ldots i_m}(x) y^{i_1} y^{i_2} \ldots y^{i_n}}$, Tensor, N. S. 33, 365-372.

Shing, D. K.: 1959, On the symmetric properties in some Finsler spaces, Acta Math. Sinica 9, 191-198 (in Chinese). English translation: 1967, Chinese Math. 9, 498-506.
Singh, A. K.: 1980, Decomposition of neo-curvature tensor field, J. Math. Phys. Sci. 14, 529-536.
Singh, A. K.: 1981, A study of the differential geometry of Finsler and Kropina spaces: Thesis, Gorakhpur, 191 p.
Singh, A. K.: 1981, Decomposition of neo-curvature tensor field in a recurrent Finsler space of second order, Acta Cienc. Indica 7, 56-61.
Singh, O. P.: 1973, On the projective motion in a projective Finsler space of recurrent curvature, Mat. Vesnik N. S. 10 (25), 105-110.
Singh, O. P.: 1975, A remark 'On projective motion in a projective Finsler space of recurrent curvature', Tokohama Math. J. 23, 1-4.
Singh, S. P.: 1973, Generalisation of Gauss-Codazzi equations in a generalised Finsler space, Yokohama Math. J. 21, 103-114.
Singh, S. P.: 1975, On decomposition of recurrent curvature tensor fields in generalised Finsler spaces, Kyungpook Math. J. 15, 201-212.
Singh, S. P.: 1976, Veblen identity and some tensors in Finsler space, Ann. Fac. Sci., Sect. Math.-Phys. Kinshasa 2, 285-294.
Singh, S. P. 1978, Some tensors in Finsler space, Ann. Fac. Sci. Sect. Math.-Phys. Kinshasa 4, 205-216.
Singh, S. P.: 1980, Finsler space of generalised recurrent curvature tensor field of second order, Ann. Fac. Sci. Univ. Nat. Zaire (Kinshasa) Sec. Math.-Phys. 6, 31-49.
Singh, S. P.: 1982a, Recurrent generalised Finsler spaces, Kyungpook Math. J. 22, 257-264.
Singh, S. P.: 1982b, Decomposition of recurrent curvature tensor field in 2-R-generalised Finsler spaces, Kyungpook Math. J. 22, 265-277.
Singh, U. P.: 1967a, Hypernormal curves of a Finsler subspaces, Progr. Math. Allahabad 1, 47-50.
Singh, U. P.: 1967b, On Union curve and union curvature of a vector field at points of the curve of a Finsler space, Bull. Calcutta Math. Soc. 59, 31-35.
Singh, U. P.: 1967c, On pseudogeodesic and relative parallelism in Finsler spaces, Bull. Calcutta Math. Soc. 59, 119-125.
Singh, U. P.: 1967d, On Δ-principal directions and minimal hypersurfaces in a Finsler space, Proc. Math. Phys. Soc. U. A. R. 31, 27-34.
Singh, U. P.: 1968a, On relative associate curvature of a vector-field and relative parallelism in Finsler spaces, J. Indian Math. Soc., N. S. 32, 229-242.
Singh, U. P.: 1968b, On union curves and pseudogeodesics in Finsler spaces, Univ Nac. Tucumán Rev. A 18, 35-41.
Singh, U. P.: 1968c, On union curves and U-parallelism in Finsler spaces, Gorakhpur Univ. Res. J. 2, 23-28.
Singh, U. P.: 1969, On principal directions in Finsler subspaces, Proc. Nat. Inst. Sci. India A 35, Suppl. 2, 130-137.
Singh, U. P.: 1970a, On union curves and pseudo-geodesics in Finsler spaces (from the standpoint of Cartan's Euclidean connection), Bull. Calcutta Math. Soc. 62, 165-172.
Singh, U. P.: 1970b, The curvature of a congruence relative to a vector-field of a Finsler hypersurface, Indian J. Pure Appl. Math. 1, 341-346.
Singh, U. P.: 1971a, The curvature of a congruence of a Finsler hypersurface, Progr. Math. Allahabad 5, 20-29.
Singh, U. P.: 1971b, The intrinsic and induced theories of a Finsler subspace, Univ. Nac. Tucumán Rev. A 21, 119-121.
Singh, U. P.: 1972a, Curvature invariants associated with fundamental forms of a Finsler hypersurface, Lincei Rend. (8) 52, 41-47.
Singh, U. P.: 1972b, On the induced theory of Finsler hypersurfaces from the standpoint of nonlinear connections, Lincei Rend. (8) 53, 541-548.
Singh, U. P.: 1973, On Finsler spaces admitting parallel vectorfields, Tensor, N. S. 27, 170-172.
Singh, U. P.: 1974, On hyper asymptotic curves in generalised spaces, Math. Student 42, 201-204.
Singh, U. P.: 1978, A note on the theory of non-linear connections in Finsler spaces, Progr. Math. 12, 13-19.
Singh, U. P.: 1980, Hypersurface of C-reducible Finsler spaces, Indian J. Pure Appl. Math. 11, 1278-1285.
Singh, U. P. and Agrawal, P. N.: 1981, On a transformation associated with some special Finsler spaces, Ganita 32, 89-98.
Singh, U. P., Agrawal, P. N., and Prasad, B. N.: 1982a, On Finsler spaces with T-tensor of some special forms, Indian J. Pure Appl. Math. 13,172-182.
Singh, U. P., Agrawal, P. N., and Prasad, B. N.: 1982b, On a special form of T-tensor and T2-like Finsler space, Indian J. Pure Appl. Math. 13, 997-1005.

Singh, U. P. and Chaubey, G. C.: 1977, On hypersurfaces of recurrent Finsler spaces, Indian J. Pure Appl. Math. 8, 497-504.
Singh, U. P. and Chaubey, G. C.: 1979, Curvature theories of Finsler subspaces with the standpoint of semi-symmetric connections, Indian J. Pure Appl. Math. 10, 112-123.
Singh, U. P. and John, V. N.: 1978, On induced and intrinsic connections of Randers' hypersurface, Indian J. Pure Appl. Math. 9, 1188-1196.
Singh, U. P., John, V. N. and Prasad, B. N.: 1979, Finsler spaces preserving killing vector fields, J. Math. Phys. Sci. 13, 265-271.
Singh, U. P. and Khan, K. A.: 1980a, Projective motion in a Finsler space with concircular and concurrent transformations, Ann. Soc. Sci. Bruxelles I 94, 191-198.
Singh, U. P. and Khan, K. A.: 1980b, On supergeodesic curvature of a curve, J. Math. Phys. Sci. 14, 421-428.
Singh, U. P. and Khan, K. A.: 1981a, General union and general special curves of a Finsler hypersurface from the standpoint of nonlinear connections, Acta Cienc. Indica 7, 51-56.
Singh, U. P. and Khan, K. A.: 1981b, On W^i_{jkh} transformation of projective recurrent Finsler spaces, Acta Cienc. Indica 7, 135-140.
Singh, U. P. and Mishra, R. D.: 1983, On bi-recurrent neo-pseudo projective tensor field of a Finsler space, Acta Cienc. Indica 9, 96-100.
Singh, U. P. and Prasad, B. N.: 1969, On hyper Darboux lines in Finsler subspace, Rev. Univ. Istanbul 34, 83-90.
Singh, U. P. and Prasad, B. M.: 1970a, The Bianchi and Veblen identities in Finsler space, Acad. Roy. Belg. Bull. Cl. Sci. (5) 56, 1100-1109.
Singh, U. P. and Prasad, B. N.: 1970b, On intrinsic normal curvature of a curve of a Finsler hypersurface, Proc. Math. Phys. Soc. A. R. E. 34, 81-86.
Singh, U. P. and Prasad, B. N.: 1971a, Special curvature collineations in Finsler space, Lincei Rend. (8) 50, 122-127.
Singh, U. P. and Prasad, B. N.: 1971b, On relative associate intrinsic curvature tensor, Univ. Nac. Tucumán Rev. A 21, 7-13.
Singh, U. P. and Prasad, B. N.: 1983, Modification of a Finsler space by a normalized semiparallel vector field, Period. Math. Hungar 14, 31-41.
Singh, U. P. and Singh, A. K.: 1979, On neo-pseudo projective tensor fields, Indian J. Pure Appl. Math. 10, 1196-1201.
Singh, U. P. and Singh, A. K.: 1981a, On the N-curvature collineation in Finsler spaces, Ann. Soc. Sci. Bruxelles I 95, 69-77.
Singh, U. P. and Singh, A. K.: 1981b, On the N-curvature collineation in Finsler spaces. II, Indian J. Pure Appl. Math. 12, 1208-1212.
Singh, U. P. and Singh, A. K.: 1981c, Finsler spaces admitting recurrent neo-pseudoprojective tensor fields, Indian J. Pure Appl. Math. 12, 1201-1207.
Singh, U. P. and Singh, A. K.: 1983, Hypersurfaces of semi-C-reducible and S4-like Finsler spaces, Period. Math. Hungar 14, 81-91.
Singh, U. P. and Singh, U. B.: 1973, Geodesic torsion of a curve of a Finsler hypersurface, Lincei Rend, (8), 55, 210-213.
Singh, U. P. and Singh, U. B.: 1976, Geodesics torsion of a curve in a Finsler space, Tensor, N. S. 30, 44-46.
Singh, U. P. and Singh, U. B.: 1978a, G-union and G-special curves of a Finsler subspace, J. Math. Phys. Sci. 12, 85-97.
Singh, U. P. and Singh, U. B.: 1978b, On semi-symmetric connection parameters in Finsler subspaces, Indian J. Pure Appl. Math. 9, 1254-1262.
Singh, U. P. and Singh, V. P.: 1974, On union curves and pseudogeodesics in a Finsler subspace from the standpoint of non-linear connections, Lincei Rend. (8) 56, 530-536.
Singh, U. P. and Srivastava, R. J.: 1979, On induced theory of semi-symmetric and quater-symmetric linear connections, Indian J. Pure Appl. Math. 10, 479-485.
Singh, U. P. and Srivastava R. J.: 1981, Special curvature collineation in Finsler space, Ganita 32, 16-22.
Singh, U. P. and Yadav, P. C.: 1973, On hyper Darboux lines of a Finsler hypersurface from the standpoint of the non-linear connections, Lincei Rend. (8) 54, 730-738.
Singh, U. P. and Yadev, P. C.: 1973/1974, On subspaces of subspaces of a Finsler space, Rend. Accad. Naz. XL (4), 24/25, 63-70.
Singh, U. P. and Yadav, P. C.: 1974, Subspaces of subspaces of a Finsler space, Univ. Nac. Tucumán Rev. A. 24, 213-221.
Singh, U. P. and Yadev, P. C.: 1975, On decomposition of curvature tensor in recurrent Finsler spaces, Univ. Nac. Tucumán Rev. A 25, 49-55.
Singh, V. P.: 1974, On union curve and union curvature of a Finsler hypersurface from the standpoint of non-linear connections, Progr. Math. Allahabad 8, 23-30.
Sinha, B. B.: 1965, Projective invariants, Math. Student 33, 121-127.
Sinha, B. B.: 1969, Union curve in Finsler space, Rend. Circ. Mat. Palermo (2) 18, 288-292.

Sinha, B. B.: 1971a, A generalization of Gauss and Codazzi's equations for intrinsic curvature tensor, Indian J. Pure Appl. Math. 2, 270-274.
Sinha, B. B.: 1971b, On projective mapping in a Finsler space, Tensor, N. S. 22, 326-328.
Sinha, B. B.: 1971c, On projectively flat Finsler space and pseudo deviation tensor, Progr. Math. Allahabad 5, 88-92.
Sinha, B. B.: 1972, Decomposition of recurrent curvature tensor fields of second order, Progr. Math. Allahabad 6, 7-14.
Sinha, B. B. and Singh, B.: 1981, Finsler space with recurrent neo-pseudo projective curvature tensor, Indian J. Pure Appl. Math. 12, 1063-1068.
Sinha, B. B. and Singh, G.: 1983a, Decomposition of recurrent curvature tensor fields of rth order in Finsler manifolds, Publ. Inst. Math. Beograd, N. S. 33 (47) 217-220.
Sinha, B. B. and Singh, G.: 1983b, On decomposibility of neo-pseudo projective curvature tensor fields of second order in recurrent Finsler space, Tamkang J. Math. 14, 115-122.
Sinha, B. B. and Singh, G.: 1983c, On Finsler spaces with neo-pseudo-projective curvature tensor of scalar curvature, An. Sti. Univ. Iasi Ia 29, No. 2, Suppl. 39-40.
Sinha, B. B. and Singh, S. P.: 1968, On the recurrent tensor fields of Finsler spaces, Rev. Univ. Istanbul 33, 117-124.
Sinha, B. B. and Singh, S. P.: 1969, Generalised Finsler spaces of recurrent curvature, Progr. Math. Allahabad 3, 85-92.
Sinha, B. B. and Singh, S. P.: 1969/1970, On recurrent spaces of second order in Finsler spaces, J. Sci. Res. Banaras Hindu Univ. 20, 37-42.
Sinha, B. B. and Singh, S. P.: 1970a, On decomposition of recurrent curvature tensor fields in Finsler spaces, Bull. Calcutta Math. Soc. 62, 91-96.
Sinha, B. B. and Singh, S. P.: 1970b, On recurrent spaces of second order in Finsler spaces, Yokohama Math. J. 18, 27-32.
Sinha, B. B. and Singh, S. P.: 1970c, On decomposition of recurrent curvature tensor fields, Proc. Nat. Acad. Sci. India A 40, 105-110.
Sinha, B. B. and Singh, S. P.: 1970d, K*-curves of a subspace of a Finsler space, Proc. Nat. Acad. Sci. India A 40, 157-161.
Sinha, B. B. and Singh, S. P.: 1971a, On recurrent Finsler spaces, Rev. Roum. Math. Pures Appl. 16, 977-986.
Sinha, B. B. and Singh, S. P.: 1971b, On pseudo-projective tensor fields, Tensor, N. S. 22, 317-320.
Sinha, B. B. and Singh, S. P.: 1971c, Recurrent Finsler space of second order, Tokohama Math. J. 19, 79-85.
Sinha, B. B. and Singh, S. P.: 1971d, A generalization of Gauss and Codazzi equation for Berwald's curvature tensor, Tensor, N. S. 22, 112-116.
Sinha, B. B. and Singh, S. P.: 1972, A generalization of Gauss and Codazzi equations for the first curvature of Cartan, Progr. Math. Allahabad 6, 1972, 1-6.
Sinha, B. B. and Singh, S. P.: 1973, Recurrent Finsler spaces of second order. II, Indian J. Purc Appl. Math. 4, 45-50.
Sinha, B. B. and Tripathi, R. S.: 1974, On decomposition of curvature tensor fields in recurrent Finsler spaces of second order, Indian J. Pure Appl. Math. 5, 426-429.
Sinha, B. B. and Yadava, S. L.: 1981, Almost product structures on tangent bundles, Indian J. Pure Appl. Math. 12, 61-70.
Sinha, R. S.: 1968a, Projective curvature tensors of a Finsler space, Acad. Roy. Belg. Bull. Cl. Sci. (5) 54, 272-279.
Sinha, R. S.: 1968b, Generalization of Gauss-Codazzi equations for the first of Cartan's curvature tensors in a hypersurface of a Finsler space, Tensor, N. S. 19, 275-278.
Sinha, R. S.: 1969a, Generalisation of Gauss-Codazzi equations for Berwald's curvature tensors in a hypersurface of a Finsler space, Tensor, N. S. 20, 13-19.
Sinha, R. S.: 1969b, Generalisations of Gauss-Codazzi equations for second of Cartan's curvature tensors in a hypersurface of a Finsler space, J. Indian Math. Soc., N. S. 33, 133-140.
Sinha, R. S.: 1969c, Affine motion in recurrent Finsler spaces, Tensor, N. S. 20, 261-264.
Sinha, R. S.: 1969d, Conformal transformation of curvature tensors of a Finsler space, Univ. Nac. Tucumán Rev. A 19, 35-43.
Sinha, R. S.: 1970, On projective motions in a Finsler space with recurrent curvature, Tensor, N. S. 21, 124-126.
Sinha, R. S.: 1971a, Infinitesimal conformal transformation in Finsler spaces, Univ. Nac. Tucumán Rev. A 21, 283-287.
Sinha, R. S.: 1971b, Infinitesimal projective transformation in Finsler spaces, Progr. Math. Allahabad 5, 30-34.
Sinha, R. S.: 1972, Some formulae in the theory of Lie derivatives in Finsler spaces, Tensor, N. S. 25, 332-336.
Sinha, R. S. and Chowdhury, V. S. P.: 1979, Affine motion in recurrent Finsler space. II, J. Math. Phys. Sci. 13, 437-443.

Sinha, R. S. and Farudi, S. A.: 1979a, Induced and intrinsic nonlinear connections of a hypersurface of a Finsler space, Tensor, N. S. 33, 394-399.
Sinha, R. S. and Farudi, S. A.: 1979b, On projectively related recurrent and projective recurrent Finsler space, Indian J. Pure Appl. Math. 10, 394-399.
Sinkunas, J.: 1966a, The space of support linearis, Liet. Mat. Rinkinys. 6, 449-455.
Sinkunas, J.: 1966b, Connections in spaces of special support elements, Liet. Mat. Rinkinys. 6, 622.
Sinkunas, J.: 1972, The space of support linears of a Finsler structure, Liet. Mat. Rinkinys 12, 221-227 (in Russian).
Sirokov, A. P.: 1950, A gonometric system in the geometry of Finsler, Trudy Sem. Vektor. Tenzor. Anal. 8, 414-424 (in Russian).
Slobodjan, J. S.: 1971, Certain properties of the totally geodesic hypersurfaces of a Finsler space, Ukrain. Geom. Sb. 10, 65-71 (in Russian).
Slobodjan, J. S.: 1972, The Gauss-Codazzi conditions for totally geodesic surfaces in a Finsler space, Ukrain. Geom. Sb. 12, 124-131, 170 (in Russian).
Sobczyk, A.: 1967, Minkowski planes, Math. Ann. 173, 181-190.
Soós, G.: 1954, Über Gruppen von Affinitäten und Bewegungen in Finslerschen Räumen, Acta Math. Acad. Sci. Hungar. 5, 73-83.
Soós, G.: 1956, Über Gruppen von Automorphismen in affinzusammenhängenden Räumen von Linienelementen, Publ. Math. Debrecen 4, 294-302.
Soós, G.: 1957, Über eine spezielle Klasse von Finslerschen Räumen, Publ. Math. Debrecen 5, 150-153.
Soós, G. 1959, Über die homothetische Gruppe von Finslerschen Räumen, Acta Math. Acad. Sci. Hungar 10, 391-394.
Soós, G.: 1960, Über einfache Finslersche Räume, Publ. Math. Debrecen 7, 364-373.
Soós, G.: 1963, On the theory of fiber spaces of Finsler type, Magyer Tud. Akad. Mat. Fiz. Oszt. Kötl. 13, 17-64.
Sorokin, V. A.: 1964, Certain questions of a Minkowski geometry with a non-symmetric indicatrix, Orehovo-Zuev. Ped. Inst. Uchen. Zap. Kaf. Mat. 22, No. 3, 138-147 (in Russian).
Sorokin, V. A.: 1967, Jung's and Blaschke's theorems for a Minkowski plane, Moskov. Gos, Ped. Inst. Uchen. Zap. 271, 145-153 (in Russian).
Spivak, M.: 1970, A Comprehensive Introduction to Differential Geometry. II, Brandies Univ. Waltham, Mass., 425 p.
Srivastava, S. C.: 1971, Generalized geodesics in a Finsler space, Acad. Roy. Belg. Bull. Cl. Sci. (5), 1224-1232.
Srivastava, S. C. and Sinha, R. S.: 1969, Generalised Gauss-Codazzi equation for a subspace of a Finsler space, Univ. Nac. Tucumán Rev. A 19, 123-129.
Srivastava, S. C. and Sinha, R. S.: 1971, Asymptotic lines in a hypersurface of a Finsler space, Lincei Rend. (8) 51, 44-52.
Srivastava, S. C. and Sinha, R. S.: 1972, Geodesic congruences in a Finsler space, Lincei Rend. (8) 52, 156-161.
Srivastava, S. C. and Sinha, R. S.: 1974, Projective transformation in recurrent and Ricci-recurrent Finsler spaces, Lincei Rend. (8) 57, 181-186.
Srivastava, T. N.: 1974, On a Finsler space. I, Tensor, N. S, 28, 13-18.
Srivastava, T. N. and Watanabe, S.: 1975, Some properties of indicatrices in a Finsler space, Canad. Math. Bull. 18, 715-721.
Stephenson, G.: 1953, Affine field structure of gravitation and electromagnetism, Nuovo Cimento (9) 10, 354-355.
Stokes, E. C.: 1938-1941, Applications of the covariant derivative of Cartan in the calculus of variations, Contrib. Calculus of Vari., Chicago, pp. 139-174.
Su, B.: 1949, Geodesic deviation in generalized metric spaces, Acad. Sinica Sci. Record 2, 220-226.
Su, B.: 1955, On the isomorphic transformations of minimal hypersurfaces in a Finsler space, Acta Math. Sinica 5, 471-488 (in Chinese).
Su, B.: 1958, Differential Geometry of Generalized Spaces, Science Publ., Peking, 162 p. (in Chinese).
Subramanian, S.: 1933, On Synge's paper, Bull. Acad. Sci. Allahabad 3, 61-64.
Sugawara, M.: 1940, A generalization of Poincaré-space, Proc. Imp. Acad. Tokyo 16, 373-377.
Suguri, T. and Hakayama, S.: 1955, Note on Riemannian spaces and contact transformations, Tensor, N. S. 5, 1-16.
Sulanke, R.: 1954/1955, Die eindeutige Bestimmtheit des von Hanno Rund eingeführten Zusammenhangs in Finsler-Räumen, Wiss. Z. Humboldt-Univ. Math.-Nat. R. 4, 229-233.
Sulanke, R.: 1955/1956, Die eindeutige Bestimmtheit des von Hanno Rund eingeführten Zusammenhangs in Finsler Räumen, Wiss. Z. Humboldt-Univ. Math.-Nat. R. 5, 269.
Sulanke, R.: 1957, Eine Ableitung des Cartanschen Zusammenhangs eines Finslerschen Raumes, Publ. Math. Debrecen 5, 197-203.

Sumimoto, T.: 1958, On the reduction of Minkowski space, J. Fac. Sci. Hokkaido Univ. I 14, 50-58.
Süss, W.: 1953, Über affine und Minkowskische Geometrie, Convegno Intern. Geom. Diff. Itali, pp. 55-63.
Süss, W.: 1954, Affine und Minkowskische Geometrie eines ebenen Variationsproblems, Arch. Math. 5, 441-446.
Suzuki, Y.: 1956, Finsler geometry in classical physics, College Arts. Sci. Chiba Univ. 2, 12-16.
Synge, J. L.: 1925, A generalisation of the Riemannian line-element, Trans. Amer. Math. Soc. 27, 61-67.
Szabó, Z. I.: 1977, Ein Finslerschen Raum ist gerade dann von skalarer Krümmung, wenn seine Weylsche Projektivkrümmung verschwindet, Acta Sci. Math. Szeged 39, 163-168.
Szabó, Z. I.: 1979, Relative Krümmungstheorie der Finslerschen Räume. I, Period. Math. Hungar. 10, 293-299.
Szabó, Z. I.: 1980a, Finslersche Projektivgeometrie. I. Eine globale Begründung der Finslerschen Projektivzusammenhänge, Acta Math. Acad. Sci. Hungar. 35, 79-96.
Szabó, Z. I.: 1980b, Finslersche Projektivgeometrie. II. Über Finslerische Projektivbündel mit Weylscher Projectivkrümmung Null, Acta Math. Acad. Sci. Hungar. 35, 97-108.
Szabó, Z. I.: 1980c, Über Zusammenhänge vom Finsler-Typ, Publ. Math. Debrecen 27, 77-88.
Szabó, Z. I.: 1981a, Positive definite Berwald space. (Structure theorems on Berwald spaces), Tensor, N. S. 35, 25-39.
Szabó, Z. I.: 1981b, Generalized spaces with many isometries, Geometrica Dedicata 11, 369-383.
Szabó, Z. I.: 1982, Structure theorems on positive definite Berwald spaces and some remark on a related problem, Proc. Nat. Sem. Finsler Spaces, Brasov, pp. 209-215.
Szenthe, J.: 1960, Über ein Problem von H. Busemann, Publ. Math. Debrecen 7, 408-413.
Szigeti, F.: 1970, Differentiable approximation on Banach manifolds, Richerche Mat., Napori 19, 171-178.
Szolcsányi, E.: 1975, Hyperflächen mit Riemannscher Massbestimmung im Finslerschen Raum, Publ. Math. Debrecen 25, 203-210.
Szolcsányi, E.: 1978, Hyperflächen mit Riemannscher Massbestimmung im Finslerschen Raum. II, Publ. Math. Debrecen 22, 133-150.
Tachibana: 1949, On normal coordinates of a Riemann space, whose holonomy group fixes a point, Tohoku Math. J. (2) 1, 26-30.
Tachibana: 1950, On Finsler spaces which admit a concurrent vector field, Tensor, N. S. 1, 1-5.
Takano, K.: 1950, Homogeneous contact transformations and a metric space, Rep. Univ. Electro-Commu. 2, 269-278 (in Japanese).
Takano, K.: 1952a, Homothetic transformations in Finsler spaces, Holonomy-gun Kenkyu 25, 10-13 (in Japanese).
Takano, K.: 1952b, Homothetic transformations in Finsler spaces, Rep. Univ. Electro-Commu. 4, 61-70.
Takano, K.: 1954, Contact transformations and generalized metric spaces, Tensor, N. S. 4, 51-66.
Takano, Y.: 1964, Spinor field in Finsler spaces, Progr. Theor. Phys. 32, 365-366.
Takano, Y.: 1967, Theory of fields in Finsler spaces. I. Geometrical foundation of non-local field theory, Soryushiron Kenkyu 36, 29-66 (in Japanese).
Takano, Y.: 1968, Theory of fields in Finsler spaces. I, Progr. Theor. Phys. 40, 1159-1180.
Takano, Y.: 1974a, Gravitational field in Finsler spaces, Lett. Nuovo Cimento (2) 10, 747-750.
Takano, Y.: 1974b, Variation principle in Finsler spaces, Lett. Nuovo Cimento (2) 11, 486-490.
Takano, Y.: 1975-1976, On the theory of fields in Finsler spaces, Proc. Intern. Symp. Relativity & Unified Field theory, Calcutta, pp. 17-26.
Takasu, T.: 1957, Generalized Riemannian geometry. I, Yokohama Math. J. 5, 115-169.
Takasu, T.: 1966, Non connection methods for some connection geometries based on canonical equations of Hamiltonian types of II-geodesic curves, Proc. Japan Acad. 42, 539-544.
Tamássy, L.: 1960/1961, Ein Problem der zweidimensionalen Minkowskischen Geometrie, Ann. Polon. Math. 9, 39-46.
Tamássy, L.: 1961, Ein Problem der zweidimensionalen Minkowskischen Geometrie, Ann. Polon. Math. 10, 175.
Tamássy, L.: 1973, Zur Metrisierbarkeit der affinzusammenhängenden Räume, Demonstratio Math. 6, 851-859.
Tamássy, L. and Kis, B.: 1984, Relations between Finsler and affine connections, Rend. Circ. Mat. Palermo, II. Suppl. 3, 329-337.
Tamássy, L. and Matsumoto, M.: 1979, Direct method to characterize conformally Minkowski Finsler spaces, Tensor, N. S. 33, 380-384.

Tamim, A.: 1978, Study of Special Types of Finsler Spaces: Thesis, Cairo, 45 p.

Tandai, K.: 1954, On areal spaces VI. On the characterization of metric areal spaces, Tensor, N. S. 3, 40-45.

Tarinà, M.: 1982, Finsler connections and associated algebras, Proc. Nat. Sem. Finsler Spaces, Brasov, pp. 221-224.

Tashiro, Y.: 1959, A theory of tranformation groups on generalized spaces and its applications to Finsler and Cartan spaces, J. Math. Soc. Japan 11, 42-71.

Taylor, J. H.: 1925a, A generalization of Levi-Civita's parallelism and the Frenet formulas, Trans. Amer. Math. Soc. 27, 246-264.

Taylor, J. H.: 1925b, Reduction of Euler's equations to a canonical form, Bull. Amer. Math. Soc. 31, 257-262.

Taylor, J. H.: 1926-1927, Parallelism and transversality in a sub-space of a general (Finsler) space, Ann. of Math. (2) 28, 620-628.

Tembharey, D. J. and Sachdeva, S. K.: 1971, A note of the two-point tensors in a generalized Finsler space, Math. Education A 5, 49-51.

Theodoresco, N.: 1938, Recherches sur les équations aux dérivées partielle, Linéaires d'ordre quelconque. Les solutions élémentaires, Ann. Sci. Univ. Jassy I 24, 263-321.

Theodoresco, N.: 1941, Sur les géodésiques de longueur nulle de certains éléments finslériens, Bull. Ecole Polytech. Bucarest 12, 9-16.

Theodoresco, N.: 1942, Géométrie finslérienne et propagation des ondes, Acad. Roum. Bull. Sci. 23, 138-144.

Theodoresco, N.: 1952, Introduction physico-mathématique à la théorie invariante de la propagation des ondes, Rev. Univ. 'C. I. Parhon' Politeh. Bucuresti, Nat. 1, 25-51.

Theodoresco, N.: 1962, A propos d'espace de Finsler, Acad. R. P. Romine Stud. Cerc. Mat. 13, 499-510 (in Roumanian).

Thompson, A. C.: 1975, An equiperimetric property of Minkowski circles, Bull. London Math. Soc. 7, 271-272.

Tomonaga, Y.: 1969, Jacobi, fields in a Finsler space, TRU Math., Tokyo 5, 37-42.

Tonooka, K.: 1956, On a geometry of three-dimensional space with an algebraic metric, Tensor, N. S. 6, 60-68.

Tonooka, K.: 1959, On three and four dimensional Finsler spaces with the fundamental forms $\sqrt[3]{a_{\alpha\beta\gamma} x'^{\alpha} x'^{\beta} x'^{\gamma}}$, Tensor, N. S. 9, 209-216.

Tonooka, K.: 1963, Theories of surface of three-dimensional space with an algebraic metric, Tensor, N. S. 14, 219-225.

Tonooka, K.: 1964-1965, Subspace theory of an n-dimensional space with an algebraic metric, J. Fac. Sci. Hokkaido Univ. I 18, 34-40.

Tonooka, K.: 1972, On a geometry of three-dimensional space based on the differential form of the fourth order, Tensor, N. S. 25, 148-154.

Toró, T.: 1969, On the spinorial equations in Finsler spaces, An. Univ. Timisoara, Sti. Fiz.-Chim. 7, 107-113 (in Roumanian).

Toró, T.: 1971, Sur l'équation spinorielle pour le neutrino dans les espaces non euclid-dens, An. Univ. Timisoara, Sti. Fiz.-Chim. 9, 131-138.

Tromba, A. J.: 1976, Almost-Riemannian structures on Banach manifolds: The Morse lemma and the Darboux theorem, Canad. J. Math. 28, 640-652.

Tucker, A. W.: 1934, On tensor invariance in the calculus of variations, Ann. of Math. 35, 341-350.

Tutaev, L. K.: 1965, On the differential geometry of curves and surfaces in a Minkowski space, Proc. 1st. Republ. Conf. Belorussian Math., pp. 290-307.

Tutaev, L. K.: 1969, The composed manifolds M_{11}, M_{12}, M_{13}, in Minkowski space, Proc. 2nd Republ. Conf. Belorussian Math., pp. 133-138.

Uhlenbeck, K.: 1970, Harmonic maps; a direct method in the calculus of variations, Bull. Amer. Math. Soc. 76, 1082-1087.

Uhlenbeck, K.: 1972a, Norse theory on Banach manifolds, J. Funct. Anal. 10, 430-445.

Uhlenbeck, K.: 1972b, Bounded sets and Finsler structures for manifolds of maps, J. Diff. Geom. 7, 585-595.

Underhill, A. L.: 1907, Invariants of the function F(x, y, x', y') under point and parameter transformations, connected with the calculus of variations: Thesis, Chicago.

Underhill, A. L.: 1908, Invariants of the function F(x, y, x', y') in the calculus of variations, Trans. Amer. Math. Soc. 9, 316-338.

Upadhyay, M. D. and Agnihotri, A. K.: 1971, K_{λ}-curves of order p in a Finsler space, Progr. Math. Allahabad 5, 62-75.

Upadhyay, M. D. and Agnihotri, A. K.: 1972a, Pseudoconjugate and pseudoasymptotic directions in a Finsler space, Ganita 23, 49-55.

Upadhyay, M. D. and Agnihotri, A. K.: 1972b, Pseudogeodesics and M-relative normal curvature in a Finsler space, An. Univ. Timisoara, Sti. Mat. 10, 103-110.

Upadhyay, M. D. and Sharma, N. K.: 1981a, On conformal nonsymmetric Finsler space, Demonstratio Math. 14, 201-207.

Upadhyay, M. D. and Sharma, N. K.: 1981b, On decomposed recurrent Finsler space, Demonstratio Math. 14, 1021-1029.
Urbonas, A. P.: 1966, Über Zusammenhängen des Raumes von Stützelementen, Liet. Mat. Rinkinys 6, 279-290 (in Russian).
Urbonas, A. P.: 1968, The differential invariants of a space of support elements, Kazan Gos. Univ. Uchen. Zap. 128, No. 3, 115-133 (in Russian).
Urbonas, A. P.: 1972, Les mouvements dans l'espaces des éléments linéaires, Liet. Mat. Rinkinys. 12, 225-230, 243 (in Russian).
Urbonas, A. P.: 1982, Maximally moving spaces of vector elements, Litovsk. Mat. Sb. 22, 202-208 (in Russian).
Vanstone, J. R.: 1962, A generalization of Finsler geometry, Canad. J. Math. 14, 87-112.
Varga, O.: 1936, Beiträge zur Theorie der Finslerschen Räume und der affinzusammenhängenden Räumen von Linienelementen, Lotos Prag 84, 1-4.
Varga, O.: 1941a, Zur Differentialgeometrie der Hyperflächen in Finslerschen Räumen, Deutsch. Math. 6, 192-212.
Varga, O.: 1941b, Bestimmung des invarianten Differentials in Finslerschen Räumen, Mat. Fiz. Lapok 48, 423-435 (in Hungarian).
Varga, O.: 1942-1943, Zur Herleitung des invarianten Differentials in Finslerschen Räume, Monatsh. Math. Phys. 50, 165-175.
Varga, O.: 1941-1943, Zur Begründung der Minkowskischen Geometrie, Acta Univ. Szeged Sci. Math. 10, 149-163.
Varga, O.: 1942, Aufbau der Finslerschen Geometrie mit Hilfe einer oskulierenden Minkowskischen Massbestimmung, Math. Naturwiss. Anz. Ungar. Akad. Wiss. 61, 14-22 (in Hungarian).
Varga, O.: 1946, Linienelementräume deren Zusammenhang durch eine beliebige Transformationsgruppe bestimmt ist, Acta Univ. Szeged Sci. Math. 11, 55-62.
Varga, O.: 1947, Über eine Klasse von Finslerschen Räumen, die die nichteuklidischen verallgemeinern, Comment. Math. Helv. 19, 367-380.
Varga, O.: 1949a, Über affinzusammenhängende Mannigfaltigkeiten von Linienelementen insbesondere deren Äquivalenz, Publ. Math. Debrecen 1, 7-17.
Varga, O.: 1949b, Affinzusammenhängende Mannigfaltigkeiten von Linienelementen, die ein Inhaltsmass besitzen, Nederl. Akad. Wetensch. Proc. A 52, 868-874 = Indag. Math. 11, 316-322.
Varga, O.: 1949c, Über das Krümmungsmass in Finslerschen Räumen, Publ. Math. Debrecen 1, 116-122.
Varga, O.: 1950a, Über den Zusammenhang der Krümmungsaffinoren in zwei eineindeutig aufeinander abgebildeten Finslerschen Räumen, Acta Sci. Math. Szeged 12, 132-135.
Varga, O.: 1950b, Normalkoordinaten in allgemeinen differentialgeometrischen Räumen und ihre Verwendung zur Bestimmung sämtlicher Differentialinvarianten, C. R. Premier Congr. Math. Hongrois, pp. 131-162.
Varga, O.: 1951, Eine geometrische Charakterisierung der Finslerschen Räume skalarer und konstanter Krümmung, Acta Math. Acad. Sci. Hungar. 2, 143-156.
Varga, O.: 1954a, Eine Charakterisierung der Finslerschen Räume mit absolutem Parallelismus der Linienelemente, Arch. Math. 5, 128-131.
Varga, O.: 1954b, Bedingungen für die Metrizierbarkeit von affinzusammenhängenden Linienelementmannigfaltigkeiten, Acta Math. Acad. Sci. Hungar. 5, 7-16.
Varga, O.: 1954c, Eine Charakterisierung der Finslerschen Räume mit absolutem Parallelismus der Linienelemente, Acta Univ. Debrecen 1, 105-108 (in Hungarian).
Varga, O.: 1955, Eine Charakterisierung der Finslerschen Räume mit absolutem Parallelismus der Linienelemente, Acta Univ. Debrecen 1, 16.
Varga, O.: 1956, Die Krümmung der Eichfläche des Minkowskischen Raumes und die geometrische Deutung des einen Krümmingstensors des Finslerschen Raumes, Abh. Math. Sem. Univ. Hamburg 20, 41-51.
Varga, O.: 1958, Hilbertsche Verallgemeinerte nicht-euklidische Geometrie und Zusammenhang derselben mit der Minkowskischen Geometrie, Intern. Congr. Math. Edinburgh, p. 111.
Varga, O.: 1960, Über die Zerlegbarkeit von Finslerschen Räumen, Acta Math. Acad. Sci. Hungar. 11, 197-203.
Varga, O.: 1961a, Bemerkung zur Winkelmetrik in Finslerschen Räumen, Ann. Univ. Sci. Budapest Sect. Math. 3-4, 379-382.
Varga, O.: 1961b, Über den inneren und induzierten Zusammenhang für Hyperflächen in Finslerschen Räumen, Publ. Math. Debrecen 8, 208-217.
Varga, O.: 1961c, Über eine Charakterisierung der Finslerschen Räume konstanter Krümmung, Monatsh. Math. 65, 277-286.
Varga, O.: 1962a, Zur Begründung der Hilbertschen Verallgemeinerung der nichteuklidischen Geometrie, Monatsh. Math. 66, 265-275.
Varga, O.: 1962b, Herleitung des Cartanschen Zusammenhangs in Finslerräumen mit Hilfe der Riemannschen Geometrie, Acta Phys. Chim. Debrecen 8, 121-124.
Varga, O.: 1962c, Eine einfache Herleitung der Cartanschen Übertragung der Finslergeometrie, Math. Notae 18, 185-196.

Varga, O.: 1963, Über Hyperflächen konstanter Normalkrümmung in Minkowskischen Räumen, Tensor, N. S. 13, 246-250.
Varga, O.: 1964, Hyperflächen mit Minkowskischer Massbestimmung in Finslerräumen, Publ. Math. Debrecen 11, 301-309.
Varga, O.: 1967, Die Methode des beweglichen n-Beines in der Finsler-Geometrie, Acta Math. Acad. Sci. Hungar. 18, 207-215.
Varga, O.: 1968, Hyperflächen konstanter Normalkrümmung in Finslerschen Räumen, Math. Nachr. 38, 47-52.
Varga, O.: 1969, Beziehung der ebenen verallgemeinerten nichteuklidischen Geometrie zu gewissen Flächen im pseudominkowskischen Raum, Aequationes Math. 3, 112-117.
Varga, O.: 1970, Zur Invarianz des Krümmungsmasses der Winkelmetrik in Finsler-Räumen bei Einbettungen, Math. Nachr. 43, 11-18.
Varga, T.: 1978, Über Berwaldsche Räume. I, Publ. Math. Debrecen 25, 213-223.
Varga, T.: 1979, Über Berwaldsche Räume, II, Publ. Math. Debrecen 26, 41-50.
Vasil'ev, A. M.: 1963, Invariant affine connections in a space of linear elements, Mat. Sb. 60 (102), 411-424 (in Russian).
Vasil'eva, M. V.: 1956, Geometric characteristics of certain invariants of Finsler geometry, Proc. 3rd All-Union Math. Congr. 2, 139.
Vasil'eva, M. V.: 1965a, Finslerian geometry in invariant presentation, Moskov. Gos. Ped. Inst. Uchen. Zap. 243, 38-54 (in Russian).
Vasil'eva, M. V.: 1965b, An invariant description of certain Finslerian geometries, Moskov. Gos. Ped. Inst. Uchen. Zap. 243, 55-68 (in Russian).
Vasil'eva, M. V.: 1967, The connections between the various Finsler geometries and the bundle of Finsler geometries, Moskov. Gos. Ped. Inst. Uchen. Zap. 271, 49-54 (in Russian).
Vasil'eva, M. V.: 1978, On the structure of n-dimensional Finsler spaces, Geom. of imbedded manifolds, Gos. Ped. Inst. Moscow, pp. 10-17 (in Russian).
Vasil'eva, M. V.: 1979, The holonomy group of an n-dimensional Finsler space, Geom. of imbedded manifolds, Gos. Ped. Inst. Moscow, pp. 14-21 (in Russian).
Vasil'eva, M. V.: 1980, Finsler spaces with degenerate holonomy group of Rund, Geom. of imbedded manifolds, Gos. Ped. Inst. Moscow, pp. 16-22 (in Russian).
Vasil'eva, M. V.: 1981, Some classification of Cartanian spaces, Proc. Cong. Honour 80th Anniversary Renato Calapso (Messina/Taormina, 1981), Veschi, Rome, pp. 312-319.
Vasil'eva, M. V.: 1982, Classification of Finsler spaces, in Differential Geometry, North-Holland, Amsterdam-New-York, pp. 769-782.
Velte, W.: 1953, Bemerkung zu einer Arbeit von H. Rund, Arch. Math. 4, 343-345.
Verma, M.: 1969, Union Gaussian curvature in Finsler spaces, Univ. Nac. Tucumán Rev. A 19, 147-150.
Verma, M.: 1970a, Hyper Darboux lines of subspaces of a Finsler space, Math. Nachr. 47, 293-298.
Verma, M.: 1970b, Generalised Gauss-Codazzi equations for Berwald's curvature tensors in subspaces of a Finsler space, Rev. Univ. Istanbul 35, 95-100.
Verma, M.: 1970c, Curvature tensors and Bianchi identities of the subspaces of a Finsler space, Proc. Math. Phys. Soc. A. R. E. 34, 87-91.
Verma, M.: 1971a, On the curvature properties of a hypersurface of a generalized Finsler space, Indian J. Pure Appl. Math. 2, 95-102.
Verma, M.: 1971b, Certain Investigations in the Differential Geometry of Finsler and Generalised Finsler Spaces: Thesis, Gorakhpur, 158 p.
Verma, M.: 1971c, The intrinsic and induced curvature theories of subspaces of a generalised Finsler space, Math. Nachr. 51, 35-41.
Verma, M.: 1971d, Union curvature tensors and union correspondence between two hypersurfaces of a Finsler space, Lincei Rend. (8) 51, 337-345.
Verma, M.: 1971e, Union Gaussian curvature in Finsler spaces. III, Univ. Nac. Tucumán Rev. A 21, 59-61.
Verma, M.: 1971f, Frenet formulae and some special curves in a Finsler space, Proc. Math. Phys. Soc. A. R. E. 35, 107-111.
Verma, M.: 1972, Generalised Laguerre functions of a hypersurface in a Finsler space, Tensor, N. S. 24, 169-172.
Verma, M.: 1977, Union Gaussian curvature in Finsler spaces, II, Acta Cienc. Indica 3, 279-280.
Verma, M.: 1980, Subspaces of a Finsler space admitting concurrent vector field, Indian J. Pure Appl. Math. 11, 988-993.
Verma, M.: 1981a, Laguerre function of a hypersurface in a Finsler space, Pure Appl. Math. 12, 807-813.
Verma, M.: 1981b, Generalised Gauss and Codazzi equations in a subspace of a Finsler (Minkowskian) Hermite space, Proc. Nat. Acad. Sci. India A 51, 277-291.
Verma, M. and Sinha, R. S.: 1971, Generalised Gauss-Codazzi equations for second of Cartan's curvature tensors for a subspace of a Finsler space, Tensor, N. S. 22, 321-325.

Vilhelm, V.: 1957, Kurven in Minkowskische Räumen, Casopis Pést. Mat. 82, 298-300 (in Czech).
Vilms, J.: 1967, Connections on tangent bundles, J. Diff. Geom. 1, 235-243.
Vilms, J.: 1968, Curvature of nonlinear connections, Proc. Amer. Math. Soc. 19, 1125-1129.
Vilms, J.: 1971, Nonlinear and direction connections, Proc. Amer. Math. Soc. 28, 567-572.
Vilms, J.: 1972, On the existence of Berwald connection, Tensor, N. S. 26, 118-120.
Vranceanu, G.: 1977, Espaces à connexion affine généralisée, Rev. Roum. Math. Pures Appl. 22, 1337-1342.
Wagner, V. V.: 1938a, A generalization of non-holonomic manifolds in Finslerian space, Uch. Zap. Saratow. Gos. Univ., Fiz.-Mat. 1(XIV), 67-96 (in Russian).
Wagner, V. V .: 1938b, Two-dimensional space with a cubic metric, Uch. Zap. Saratow. Gos. Univ. Fiz.-Mat. 1(XIV), 29-34 (in Russian).
Wagner, V. V.: 1938c, Über Berwaldsche Räume, Mat. Sbornik 3(2), 655-662 (in Russian).
Wagner, V. V.: 1943a, On generalizes Berwald spaces, Dokl. Akad. Nauk SSSR 39, 3-5 (in Russian).
Wagner, V. V.: 1943b, The inner geometry of non-linear non-holonomic manifolds, Mat. Sbornik 55 (13), 134-167 (in Russian).
Wagner, V. V.: 1943c, Two-dimensional Finsler space with finite continuous groups of holonomy, Dokl. Akad. Nauk SSSR 39, 99-102 (in Russian).
Wagner, V. V.: 1945a, Homological transformations of Finslerian metric, Dokl. Akad. Nauk SSSR 46, 263-265 (in Russian).
Wagner, V. V.: 1945b, Geometry of field of local curves in X_3 and the simplest case of Lagrange's problem in the calculus of variations, Dokl. Akad. Nauk SSSR 48, 229-232 (in Russian).
Wagner, V. V.: 1945c, Geometry of field of local central plane curves in X_3, Dokl. Akad. Nauk SSSR 48, 382-384 (in Russian).
Wagner, V. V.: 1947a, The geometrical theory of the simplest n-dimensional singular problem of the calculus of variations, Mat. Sb. 21 (63), 321-364 (in Russian).
Wagner, V. V.: 1947b, On the concept of the indicatrix in the theory of partial differential equations, Uspechi Mat. Nauk 18, 188-189 (in Russian).
Wagner, V. V.: 1947c, On the concept of the indicatrix in the theory of differential equations, Dokl. Akad. Nauk SSSR 57, 219-222 (in Russian).
Wagner, V. V.: 1948, Theory of field of local (n-2)-dimensional surfaces in X_n and its application to the problem of Lagrange in the calculus of variations, Ann. of Math. (2) 49, 141-188.
Wagner, V. V.: 1949a, On the embedding of a field of local surfaces in X_n in a constant field of surfaces in affine space, Dokl. Akad. Nauk SSSR 66, 785-788 (in Russian).
Wagner, V. V.: 1949b, The theory of a field of local hyperstrips in X_n and its application to the mechanics of a system with nonlinear anholonomic constraints, Dokl. Akad. Nauk SSSR 66, 1033-1036 (in Russian).
Wagner, V. V.: 1950a, The theory of composite manifolds, Trudy Sem. Vektor. Tenzor. Anal. 8, 11-72 (in Russian).
Wagner, V. V.: 1950b, The theory of a field of local hyperstrips, Trudy Sem. Vektor. Tenzor. Anal. 8, 197-272 (in Russian).
Wagner, V. V.: 1951, Two-dimensional Finsler space with finite continuous groups of holonomy, Holonomy-gun Kenkyu 22, 24-27 (in Japanese).
Wagner, V. V.: 1955, Differential-geometric methods in the calculus of variations, Uch. Zap. Kazansk. Gos. Univ. 115 (10), 4-7 (in Russian).
Wagner, V. V.: 1956a, Field theory of local surfaces, Proc. 3rd All-Union Math. Congr. 2, 57-60 (in Russian).
Wagner, V. V.: 1956b, Geometria del calcolo delle variazioni. II, Centro Intern. Mat. Estivo, Roma, 172 p.
Wagner, V. V.: 1959, Variational calcul as a theory of a field of cetral polycones, in Nauchnij Ezegodnik Saratowskogo Universiteta, Saratov University, pp. 27-34 (in Russian).
Walker, A. G.: 1934, The principle of least action in Milne's kinematical relativity, Proc. Roy. Soc. London A 147, 478-480.
Walker, A. G.: 1944, Completely symmetric spaces, J. London Math. Soc. 19, 219-226.
Wang, H.: 1947, On Finsler spaces with completely integrable equations of Killing, J. London Math. Soc. 22, 5-9.
Wang, Y.: 1958, Finsler spaces of constant curvature and totally extremal hypersurfaces, Sci. Record, N. S. 2, 211-214.
Warner, F. W.: 1965, The conjugate locus of a Riemannian manifold, Amer. J. Math. 87, 575-604.

Watanabe, S.: 1961, On special Kawaguchi spaces. III. Generalizations of affine spaces and Finsler spaces, Tensor, N. S. 11, 144-153.
Watanabe, S.: 1973, The indicatrices of a Finsler space, Tensor, N. S. 27, 135-137.
Watanabe, S. and Ikeda, S.: 1981, On some properties of Finsler spaces based on the indicatrix, Publ. Math. 28, 129-136.
Watanabe, S. and Ikeda, S.: 1980, On Finsler spaces satisfying the C-reducibility condition and the T-condition, Tensor, N. S. 34, 103-108.
Watanabe, S., Ikeda, S. and Ikeda, F.: 1983, On a metrical Finsler connection of a generalized Finsler metrik $g_{ij} = e^{2\sigma(x,y)}\gamma_{ij}(x)$; Tensor, N. S. 40, 97-102.
Wegener, J. M.: 1935, Untersuchungen der zwei- und dreidimensionalen Finslerschen Räume mit der Grundform $L = \sqrt[3]{a_{ikl}x'^i x'^k x'^l}$, Akad. Wetensch. Proc. 38, 949-955.
Wegener, J. M.: 1936a, Hyperflächen in Finslerschen Räumen als Transversalflächen einer Schar von Extremalen, Monatsh. Math. Phys. 44, 115-130.
Wegener, J. M.: 1936b, Untersuchungen über Finslersche Räume, Lotos Prag 84, 4-7.
Wei, H.: 1965, Veblen identities in Finsler spaces and in generalized Finsler spaces, Acta Sci. Nat. Univ. Amoiensis 12 (2), 23-31.
Wei, X.: 1981, Geodesic mappings between Finsler spaces, Acta Sci. Univ. Amoiensis 20, 146-154 (in Chinese).
Weinstein, A.: 1978, Periodic orbits for convex hamiltonian systems, Ann. of Math. (2) 108, 507-518.
Whitehead, J. H. C.: 1932, Convex regions in the geometry of paths, Quart. J. Math. Oxford 3, 33-42.
Whitehead, J. H. C.: 1933a, Convex regions in the geometry of paths, Quart. J. Math. Oxford 4, 226-227.
Whitehead, J. H. C.: 1933b, The Weierstrass E-function in differential metric geometry, Quart. J. Math. Oxford 4, 291-296.
Whitehead, J. H. C.: 1935, On the covering of a complete space by the geodesics through a point, Ann. of Math. (2) 36, 679-704.
Winternitz, A.: 1930, Über die affine Grundlage der Metrik eines Variationsproblems, S.-B. Preuss. Akad. Wiss. 26, 457-469.
Wirtinger, W.: 1923, Über allgemeine Massbestimmungen, in welchen die geodätischen Linien durch lineare Gleichungen darstallt werden, Monatsh. Math. Phys. 33, 1-14.
Witsenhausen, H. S.: 1972, On closed curves in Minkowski spaces, Proc. Amer. Math. Soc. 35, 240-241.
Wolf, D.: 1967, Imbedding a finite metric set in an n-dimensional Minkowski space, Nederl. Akad. Wetensch. Proc. A 70 = Indag. Math. 29, 136-140.
Wong, Y. and Mok, K.: 1976, Connections and M-tensors on the tangent bundle TM, in Topics in Differential Geometry, Acad. Press, New York, pp. 157-172.
Wrona, W.: 1938, Neus Beispiel einer Finslerschen Geometrie, Prace Mat. Fiz. 46, 281-290.
Wrona, W.: 1963a, On multi-isotropic Finsler spaces, Bull. Polon. Math. Astro. Phys. 11, 285-288.
Wrona, W.: 1963b, A necessary and sufficient condition for the Finsler space with distant parallelism of line elements to be multi-isotropic, Bull. Polon. Math. Astro. Phys. 11, 289-292.
Wrona, W.: 1963c, Generalized F. Schur theorem in Finsler spaces, Bull. Polon. Math. Astro. Phys. 11, 293-295.
Wrona, W.: 1968, On geodesics of a certain singular n-dimensional Finsler space, Zeszyty Nauk Politech. Warszawsk Mat. 11, 271-276.
Yamada, T.: 1980, On the indicatrized tensor $S_{ijkh|l}$ in semi-C-reducible Finsler spaces, Tensor, N. S. 34, 151-156.
Yamada, T. and Fukui, M.: 1981, On quasi-C-reducible Finsler spaces, Tensor, N. S. 35, 177-182.
Yamaguchi, S., Nemoto, H. and Kawabata, N.:1983, Extrinsic spheres in a Sasakian manifold, Tensor, N. S. 40, 184-188.
Yamandi, T., Kudo, T., Yamazaki, T. and Kawaguchi, M.: 1982, On the interpretation of some visual illusions by a geometrical model of Minkowski spaces, Tensor, N. S. 37, 257-262.
Yanagimoto, S.: 1980, A Minkowskian space with norm $\Sigma|x^i|$, Research Rep. Fukui Tech. College 14, 1-16.
Yano, K.: 1957, The Theory of Lie Derivatives and Its Applications, North-Holland, Amsterdam, 293 p.
Yano, K. and Davies, E. T.: 1954, On the connection in Finsler space as an induced connection, Rend. Circ. Mat. Palermo (2) 3, 409-417.

Yano, K. and Davies, E. T.: 1959, On some local properties of fibred spaces, Kōdai Math. Sem. Rep. 11, 158-177.
Yano, K. and Davies, E. T.: 1963, On the tangent bundles of Finsler and Riemannian manifolds, Rend. Circ. Mat. Palermo (2) 12, 211-228.
Yano, K. and Davies, E. T.: 1971, Metrics and connections in the tangent bundle, Kōdai Math. Sem. Rep. 23, 493-504.
Yano, K. and Davies, E. T.: 1975, Differential geometry on almost tangent manifolds, Ann. Mat. Pura Appl. (4) 103, 131-160.
Yano, K. and Motō, Y.: 1969, Homogeneous contact manifolds and almost Finsler manifolds, Kōdai Math. Sem. Rep. 21, 16-45.
Yano, K. and Okubo, T.: 1961, Fibred spaces and non-linear connections, Ann. Mat. Pura Appl. (4) 55, 203-243.
Yano, K. and Okubo, T.: 1970, On tangent bundles with Sasakian metrics of Finslerian and Riemannian manifolds, Ann. Mat. Pura Appl. (4) 87, 137-162.
Yano, S.: 1951, A generalization of the Frenet's formulae, J. Fac. Sci. Hokkaido Univ. I 12, 11-16.
Yasuda, H.: 1972a, On extended Lie systems, III (Finsler spaces), Tensor, N. S. 23, 115-130.
Yasuda, H.: 1972b, Finsler spaces as distributions on Riemannian manifolds, Hokkaido Math. J. 1, 280-297.
Yasuda, H.: 1976, On Finsler spaces with absolute parallelism of line-elements, J. Korean Math. Soc. 13, 179-188.
Yasuda, H.: 1979a, On the indicatrix bundle endowed with the K-connection over a Finsler space, Ann. Rep. Asahikawa Med. College 1, 117-124.
Yasuda, H.: 1979b, On the indicatrices of a Finsler space, Tensor, N. S. 33, 213-221.
Yasuda, H.: 1980a, On Landsberg spaces, Tensor, N. S. 34, 316-326.
Yasuda, H.: 1980b, On transformations of Finsler spaces, Tensor, N. S. 34, 316-326.
Yasuda, H.: 1981a, On Finsler geometry and analytical dynamics, Tensor, N. S. 35, 63-72.
Yasuda, H.: 1981b, On Lie derivatives on the indicatrix bundle over a Finsler space, Ann. Rep. Asahikawa Med. College 3, 1-9.
Yasuda, H.: 1981c, On transformations of Finsler spaces, II, Tensor, N. S. 35, 187-199.
Yasuda, H.: 1981d, On H(v)-mappings of the indicatrix bundle over a Finsler space, Tensor, N. S. 35, 319-327.
Yasuda, H.: 1982, On connections of a Finsler space, Ann. Rep. Asahikawa Med. College 4, 1-11.
Yasuda, H.: 1983, On TMA-connections of a Finsler space, Tensor, N. S. 40, 151-158.
Yasuda, H. and Fukui, M.: 1980, On the curvature tensors of the indicatrix bundle over a Finsler space, Ann. Rep. Asahikawa Med. College 2, 1-21.
Yasuda, H. and Shimada, H.: 1977, On Randers spaces of scalar curvature, Reports on Math. Phys. 11, 347-360.
Yasuda, H. and Yamanoi, T.: 1979, On curves in the indicatrix bundle over a Finsler space, Tensor, N. S. 33, 357-364.
Yoshida, M.: 1980, On an R3-like Finsler space and its special cases, Tensor, N. S. 34, 157-166.
Yoshida, M.: 1982, On Finsler spaces of Hp-scalar curvature, Tensor, N. S. 38, 205-210.
Youssef, N. L.: 1978, Distribution de Nullité du Tenseur de Courbure d'une Connexion: Thèse, Univ. Grenoble, 62 p.
Youssef, N. L.: 1980, Distribution de nullité du tenseur de courbure d'une connexion, C. R. Acad. Sci. Paris A 290, 653-656.
Youssef, N. L.: 1981, Etude de Certaines Connexions Linéaires sur le Fibré Tangent d'une Variété Finslérienne: Thèse, Grenoble-Caire, 79 p.
Youssef, N. L.: 1982, Quelques remarques sur le fibré tangent d'une variété finslérienne, An. Univ. Timisora, Sti. Mat. 20, 90-97.
Youssef, N. L.: 1984, Connexion Riemannienne sur le fibré tangent d'une variété finslérienne, Rev. Roum. Math. Pures Appl. 2, 195-208.
Zaguskin, V. L.: 1958, On a certain Finsler space and the motions in a Minkowski space, Nauchn. Dokl. Vyss. Skoly, Fiz.-Mat. Nauki No. 3, 50-52 (in Russian).
Zaguskin, V. L.: 1960a, Certain aspects of Finsler geometry, Uch. Zap. Yaroslavsk. Gos. Ped. Inst. 34, 83-110 (in Russian).
Zaguskin, V. L.: 1960b, On surfaces admitting affine transformations into themselves, J. Higher Educ. Colleges, Math., No. 2 (15), 96-103 (in Russian).
Zalgaller, V. A.: 1972, On k-dimensional directions which are singular for a convex body F in R^n, Zap. Nauchn. Sem. Leningrad, Mat. Inst. Steklov 27, 67-72 (in Russian).
Zaustinsky, E. M.: 1959, Spaces with Non-symmetric Distance, Memoirs of the American Math. Soc. No. 34, Providence, 91 p.
Zhang, M.: 1950a, Die mittlere Krümmung einer Fläche im dreidimensionalen Finslerschen Raum, Sci. Record 3, 35-39.

Zhang, M.: 1950b, Mean curvatures of a subspace in a Finsler space, Ann. Mat. Pura Appl. (4) 31, 297-302.

Zotikov, G. I.: 1965, The differential singular Finsler metric definable in X_n by a field of local singular hypersurfaces of singularity class n-m-1, Bashkir. Gos. Univ. Uchen. Zap. 20, Ser. Mat., No. 2, 32-45 (in Russian).

Biographies

Berwald, Ludwig (1883-1942).
Pinl, M: 1964, In memory of Ludwig Berwald, Scripta Math. 27, 193-203, Czechoslovakian: 1967, Casopis Pesto. Mat. 92, 229-238.
Davies, Evan Tom (1904-1973).
Clark, R. S.: 1974, Evan Tom Davies, Bull. London Math. Soc. 6, 370-376 = in: Topics in Differential Geometry, Acad. Press. N. Y. pp. 1-8.
Clark, R. S.: 1974, Professor E. T. Davies, Utilitas Math. 5, 5-13.
Kawaguchi, A.: 1974, Professor Evan Tom Davies, Tensor, N. S. 28, i-ii.
1974, Evan Tom Davies (1904-1973), Aequationes Math. 11, 121.
Bompiani, E.: 1976, Reminiscences of E. T. Davies, in Topics in Differential Geometry, Acad. Press, New York, pp. 9-14.
Finsler, Paul (1894-1970).
Van der Waerden, B. L.: 1970, Paul Finsler 1894-1970, Verhdl. Schweizer. Naturforsch. Ges., Wiss. 150, 285-286.
Gross, H.: 1970, Paul Finsler (1. 4. 1894-29, 4. 1970), Vjschr. Naturforsch. Ges. Zürich 115, 469-470.
Haimovici, Mendel (1906-1973).
Iesan, D. and Radu, A.: 1973, Mendel Haimovici, Gaz. Mat. Bucuresti A 78, 249-252 (in Roumanian).
1973, Academician Mendel Haimovici, An. Sti. Univ. Iasi I a 19, No. 2, i-vii.
Grindel, I. and Radu, A.: 1973, Mendel Haimovici (1906-1973), Fiz.-Mat. Spis. B''lgar. Akad. Nauk 16 (49), 225-226 (in Bulgarian).
Varga, Ottó (1909-1969).
Rapcsák, A.: 1970, Ottó Varga (1901-1969), Mat. Lapok 21, 19-30.
Tamássy, L.: 1970, Ottó Varga (1909-1969) in memoriam, Publ. Math. Debrecen 17, 19-26.

Index